Echinoderm Phylogeny and Evolutionary Biology

CURRENT GEOLOGICAL CONCEPTS

1. C. R. C. Paul and A. B. Smith (ed.) *Echinoderm phylogeny and evolutionary biology*

Echinoderm Phylogeny and Evolutionary Biology

Edited by

C. R. C. PAUL
Department of Earth Sciences
University of Liverpool

and

A. B. SMITH
Department of Palaeontology
British Museum (Natural History)

Published for the
LIVERPOOL GEOLOGICAL SOCIETY by
CLARENDON PRESS · OXFORD
1988

Oxford University Press, Walton Street, Oxford OX2 6DP
Oxford New York Toronto
Delhi Bombay Calcutta Madras Karachi
Petaling Jaya Singapore Hong Kong Tokyo
Nairobi Dar es Salaam Cape Town
Melbourne Auckland
and associated companies in
Berlin Ibadan

Oxford is a trade mark of Oxford University Press

Published in the United States
by Oxford University Press, New York

© *Liverpool Geological Society, 1988*

All rights reserved. No part of this publication may be reproduced, stored in a retrieval system, or transmitted, in any form or by any means, electronic, mechanical, photocopying, recording, or otherwise, without the prior permission of Oxford University Press

British Library Cataloguing in Publication Data
Echinoderm phylogeny and evolutionary biology.
1. Echinoderms. Evolution
I. Paul, C. R. C. (Christopher Ronald Charles)
II. Smith, Andrew B.   III. Liverpool
Geological Society   IV. Series
593.9'0438
ISBN 0-19-854491-X

Library of Congress Cataloging in Publication Data
Echinoderm phylogeny and evolutionary biology/edited by C. R. C. Paul and A. B. Smith.
p.   cm. —— (Current geological concepts; 1)
"A selection of presentations given at a symposium held in London in December 1986"—Pref.
Includes bibliographies and index.
1. Echinodermata–Evolution–Congresses.   2. Echinodermata–Congresses.   I. Paul, C. R. C. (Christopher R. C.)   II. Smith, Andrew B.   III. Series.
QL381.E3115   1988   593.9'0438—dc19   88-18689
ISBN 0-19-854491-X

Typeset by Cotswold Typesetting Ltd, Gloucester
Printed in Great Britain
at the Alden Press, Oxford

Preface

The contributions to this volume represent a selection of presentations given at a symposium held in London in December 1986. The idea for this meeting was prompted by the retirement of one of the most notable and influential workers in the field of fossil echinoderm studies, George Ubaghs, and received further impetus from the imminent retirement of two other major devotees of echinoderms, Porter Kier and Ailsa Clark. We felt that some publication to acknowledge our debt to these workers was needed, and hence the initial motivation for the conference. Having decided to hold a conference, the theme such a meeting should follow to attract papers posed little problem. We both felt that the time was ripe for a multidisciplinary assessment of echinoderm relationships. There seemed to be a revival of interest in this topic across many independent fields of research, all with the same ultimate goal of achieving a better understanding of the phylogenetic history of the group and so to unravel their evolutionary biology. This revival has come about through two important developments—the application of cladistic methodology in many fields and the major advances being made in molecular biology that enable large quantities of sequence data to be gathered rapidly. Each line of evidence has its strengths and weaknesses and, although each may provide a single most parsimonious solution, the only way to test the validity of such results is to compare them with results using different criteria or different data bases. Such multidisciplinary approaches are rarely achieved and it was the main aim of this symposium to bring together workers from diverse fields with an interest in the phylogeny and evolutionary biology of echinoderms just to see what, if any, consensus existed.

Even though this symposium resulted in no clear consensus about the major phylogenetic relationships of echinoderms, the various lines of evidence have been clearly laid out, and we now know where the problems lie and where future research effort needs to be directed. No matter what the true phylogeny of echinoderms turns out to be there must have been considerable convergent evolution in larval design (Strathmann, this volume) and so work needs to be done in re-assessing the traditional embryological data used in phylogenetic analysis, as has been done here by Smiley. Molecular biology is not without problems too. Holothurians, which are the most difficult and contentious group to place on morphological or palaeontological grounds, also appear to have an unusually fast molecular clock (Raff, this volume) which is difficult to interpret.

No matter what field of research work is undertaken, whether molecular biology, ecology, palaeontology, developmental biology, or comparative biochemistry, all are dependent upon a sound alpha-level taxonomy. No progress is possible in understanding the phylogeny or evolutionary biology of a group until the basic groundwork of a sound taxonomy has been laid. Thus, that this symposium is at all possible is a tribute to those systematists who have devoted their lives to the study of echinoderms. Here we would like to pay special tribute to the three workers already referred to in the opening paragraph.

George Ubaghs retired from Liège University in 1985. His publications span some 42 years and cover a wide range of echinoderm and carpoid groups. He initially worked on Devonian asteroids and ophiuroids, and Palaeozoic crinoids, but became interested in primitive echinoderms and is best known for his work on carpoids. His keen observation, meticulous attention to detail, and beautifully accurate line illustrations are an example to us all.

Porter Kier retired from his post at the Department of Paleobiology at the Smithsonian Institution, Washington in June 1985 and is sadly missed. He has a lively enthusiasm for echinoids that is both invigorating and contagious, and has provided stimulation and encouragement for many of us over the years. Few aspects of echinoid biology and palaeobiology have not at some time been investigated by Porter, from the Palaeozoic record to the distribution and behaviour of living Caribbean faunas.

Ailsa Clark retired from the Zoology Department of the British Museum (Natural History), London in May 1986. She has an unparalleled working knowledge of living echinoderms, and of starfish in particular, and has pursued their study with a tireless enthusiasm.

All those who struggle to understand echinoderms the better are indebted to the years of effort and devotion provided by these three workers, and to them we dedicate this book.

Finally, we should like to thank all who helped in the organization of this meeting and in the publication of this volume: the British Museum (Natural History) for providing many services free of charge; the Royal Society, British Council, and Liverpool Geological Society for funding towards the cost of speakers expenses; all of our referees who devoted their time to reviewing the typescripts presented; Roland Emson for help in organizing the accommodation, and David Lewis, the projectionist.

Liverpool   C.R.C.P
London   A.B.S.
1987

Contents

(Plates fall between pp 166 and 167)
List of contributors   ix

PART I   PHYLUM CHARACTERISTICS

1   How to characterize the Echinodermata—some implications of the sister-group relationship between echinoderms and chordates   R. P. S. JEFFERIES   3
2   The meaning of developmental asymmetry for echinoderm evolution: a new interpretation   NICHOLAS D. HOLLAND   13

PART II   CLASS RELATIONSHIPS

3   Molecular analysis of distant phylogenetic relationships in echinoderms   RUDOLF A. RAFF, KATHARINE G. FIELD, MICHAEL T. GHISELIN, DAVID J. LANE, GARY J. OLSEN, NORMAN R. PACE, ANNETTE L. PARKS, BRIAN A. PARR, and ELIZABETH C. RAFF   29
4   Collagen biochemistry and the phylogeny of echinoderms   TOSHIHARU MATSUMURA and MICHIO SHIGEI   43
5   Larvae, phylogeny, and von Baer's Law   RICHARD R. STRATHMANN   53
6   The phylogenetic relationships of holothurians: a cladistic analysis of the extant echinoderm classes   SCOTT SMILEY   69
7   Fossil evidence for the relationships of extant echinoderm classes and their times of divergence   ANDREW B. SMITH   85

PART III   MOLECULES AND RELATIONSHIPS

8   DNA evolution and echinoderm systematics   ROY J. BRITTEN   101
9   DNA–DNA hybridization, the fossil record, phylogenetic reconstruction, and the evolution of the clypeasteroid echinoids   CHARLES R. MARSHALL   107
10   Phylogenetic implications of genome rearrangement and sequence evolution in echinoderm mitochondrial DNA   HOWARD T. JACOBS, PETER BALFE, BERNARD L. COHEN, ANDREW FARQUHARSON, and LOREDANA COMITO   121
11   What molecular biology tells us about the genomic programme for development   ERIC H. DAVIDSON   139

PART IV ONTOGENY AND PHYLOGENY

12	Heterochrony and the evolution of echinoids KENNETH J. McNAMARA	149
13	Roles of allometry and ecology in echinoid evolution MICHAEL L. McKINNEY	165
14	A biomechanical approach to the ontogeny and phylogeny of echinoids JACOB DAFNI	175
15	Experimental embryology as a tool for studying the evolution of echinoderm life histories LARRY R. McEDWARD	189

PART V FOSSILS AND EVOLUTION

16	The phylogeny of the cystoids C. R. C. PAUL	199
17	The evolutionary palaeoecology of the Blastoidea JOHNNY A. WATERS	215
18	The early evolution of the Crinoidea STEPHEN K. DONOVAN	235
19	Ontogeny and phylogeny of disparid crinoids G. D. SEVASTOPULO and N. G. LANE	245
20	The evolution of feeding structures in Palaeozoic crinoids THOMAS W. BROADHEAD	255
21	The phylogeny of post-Palaeozoic crinoids M. J. SIMMS	269

PART VI EVOLUTIONARY BIOLOGY

22	The ultrastructure of tube foot epidermal cells and secretions: their relationship to the duo-glandular hypothesis and the phylogeny of the echinoderm classes J. DOUGLAS McKENZIE	287
23	Crystallographic axes of echinoid genital plates reflect larval form: some phylogenetic implications RICHARD B. EMLET	299
24	Mutable collagenous tissues and their significance for echinoderm palaeontology and phylogeny I. C. WILKIE and R. H. EMSON	311
25	Origins of the deep-sea holasteroid fauna B. DAVID	331
26	Feeding and respiratory strategies in Stylophora RONALD L. PARSLEY	347

Index 363

List of Contributors

Peter Balfe, Department of Genetics, University of Glasgow, Church Street, Glasgow G11 5JS, UK

Roy J. Britten, Division of Biology, California Institute of Technology, Pasadena, California 91125, and Carnegie Institution of Washington, USA

Thomas W. Broadhead, Department of Geological Sciences, University of Tennessee, Knoxville, Tennessee 37996-1410, USA

Bernard L. Cohen, Department of Genetics, University of Glasgow, Church Street, Glasgow G11 5JS, UK

Loredana Comito, Department of Genetics, University of Glasgow, Church Street, Glasgow G11 5JS, UK

Jacob Dafni, Interuniversity Institute of Eilat, PO Box 469, Eilat 88103, Israel

B. David, U.A. CNRS 157, Centre des Sciences de la Terre, 6 bd. Gabriel, F-21100 Dijon, France

Eric H. Davidson, Division of Biology, California Institute of Technology, Pasadena, California 91125, USA

Stephen K. Donovan, Department of Geology, University of the West Indies, Mona, Kingston 7, Jamaica

Richard B. Emlet, Department of Biology, University of California, Berkeley, California 94720, USA

R. H. Emson, Department of Biology, Kings College, Campden Hill Road, Kensington, London W8 7AH, UK

Andrew Farquharson, Department of Genetics, University of Glasgow, Church Street, Glasgow G11 5JS, UK

Katharine G. Field, Institute for Molecular and Cellular Biology, and Department of Biology, Indiana University, Bloomington, Indiana 47405, USA

Michael T. Ghiselin, Department of Invertebrate Zoology, California Academy of Sciences, Golden Gate Park, San Francisco, California 94118, USA

Nicholas D. Holland, Marine Biology Research Division, Scripps Institution of Oceanography, La Jolla, California 92093, USA

Howard T. Jacobs, Department of Genetics, University of Glasgow, Church Street, Glasgow G11 5JS, UK

R. P. S. Jefferies, Department of Palaeontology, British Museum (Natural History), Cromwell Road, London SW7 5BD, UK

David J. Lane, Institute for Molecular and Cellular Biology, and Department of Biology, Indiana University, Bloomington, Indiana 47405, USA.

N. G. Lane, Department of Geology, Indiana University, Bloomington, Indiana 47405, USA

Charles R. Marshall, Committee on Evolutionary Biology, c/o Department of Geophysical Sciences, University of Chicago, 5734 South Ellis Avenue, Chicago, Illinois 60637, USA

Toshiharu Matsumura, Meiji Institute of Health Science, 540 Naruda, Odawara, 250 Japan

Larry R. McEdward, Department of Zoology, University of Alberta, Edmonton, Alberta T6G 2E9, Canada

J. Douglas McKenzie, Department of Physiology, The Worsley Medical and Dental Building, The University, Leeds LS2 9NQ, UK

Michael L. McKinney, Department of Geological Sciences, University of Tennessee, Knoxville, Tennessee 37996-1410, USA

Kenneth J. McNamara, Western Australian Museum, Francis Street, Perth, Australia

Gary J. Olsen, Institute for Molecular and Cellular Biology, and Department of Biology, Indiana University, Bloomington, Indiana 47405, USA

Norman R. Pace, Institute for Molecular and Cellular Biology, and Department of Biology, Indiana University, Bloomington, Indiana 47405, USA

Annette L. Parks, Institute for Molecular and Cellular Biology, and Department of Biology, Indiana University, Bloomington, Indiana 47405, USA

Brian A. Parr, Institute for Molecular and Cellular Biology, and Department of Biology, Indiana University, Bloomington, Indiana 47405, USA

Ronald L. Parsley, Department of Geology, Tulane University, New Orleans, Louisiana 70118, USA

C. R. C. Paul, Department of Earth Sciences, Liverpool University, L69 3BX, UK

Elizabeth C. Raff, Institute for Molecular and

Cellular Biology, and Department of Biology, Indiana University, Bloomington, Indiana 47405, USA

Rudolf A. Raff, Institute for Molecular and Cellular Biology, and Department of Biology, Indiana University, Bloomington, Indiana 47405, USA

G. D. Sevastopulo, Department of Geology, Trinity College, Dublin, Ireland

Michio Shigei, Misaki Marine Biological Station, Faculty of Science, University of Tokyo, Misaki, Kanagawa, 238-02 Japan

M. J. Simms, Department of Earth Sciences, Liverpool University, L69 3BX, UK

Scott Smiley, Department of Biochemistry and Biophysics, University of California, San Francisco, California 94143, USA

Andrew B. Smith, Department of Palaeontology, British Museum (Natural History), Cromwell Road, London SW7 5BD, UK

Richard R. Strathmann, Friday Harbor Laboratories, 620 University Road, Friday Harbor, Washington 98250, USA

Johnny A. Waters, Department of Geology, West Georgia College, Carrollton, Georgia 30118, USA

I. C. Wilkie, Department of Biological Sciences, Glasgow College of Technology, Cowcaddens Road, Glasgow G4 0BA, UK

PART I
Phylum characteristics

How to characterize the Echinodermata—some implications of the sister-group relationship between echinoderms and chordates

R. P. S. JEFFERIES

Department of Palaeontology, British Museum (Natural History), Cromwell Road, London SW7 5BD, UK

In this paper I aim to consider some taxonomic implications of the fact that the echinoderms, among the extant fauna, are the sister group of the chordates. I call this a fact because I think it is very well established, although on the unfashionable basis of fossil evidence. I shall not argue for it here, but simply refer to my book (Jefferies 1986). To be more precise, I aim here to discover what features can truly be regarded as characteristic of the echinoderms as a group, i.e. which ones were evolved in the stem lineage of the Echinodermata and are therefore, by definition, autapomorphies of that group.

A TEXT-BOOK CHARACTERIZATION OF THE ECHINODERMATA

The echinoderms are widely, and correctly, regarded as a very well characterized group. I have consulted a number of highly respected texts (Cuénot 1948; Hyman 1955; Nichols 1975; Siewing 1984) and several features emerge as characteristic of echinoderms in the traditional view. These are as follows.

1. There is a water-vascular system, being a system of tubes of coelomic origin connected with podia (tube feet) and opening to the outside, through a hydropore or madreporite, via a stone canal.

2. There is a skeleton of calcite which takes the form of a three-dimensional network of stereom. This arises mesodermally and each plate is a single crystal of calcite.

3. As concerns ontogeny, the larva is funda-mentally trimerous in the manner of an adult hemichordate. This trimery, however, never shows itself in larval or adult echinoderms as three distinct body regions (the protosome, mesosome, and metasome of hemichordates). Rather, the evidence for fundamental trimery comes from the larval coeloms. For, with more or less clarity in different echinoderms, five different larval coeloms can be recognized, these being: the unpaired axocoel corresponding to the protocoel in the protosome of hemichordates; right and left hydrocoels corresponding to the right and left mesocoels in the mesosome of hemichordates; and right and left somatocoels corresponding to the right and left metacoels in the metasome of hemichordates. The axocoel of echinoderm larvae opens to the outside by a pore, the hydropore, and this, in the larva, is left and dorsal in position. The left hydrocoel comes to open in ontogeny by the so-called stone canal into the axocoel and thence, indirectly, via the hydropore, to the outside. At the end of larval life there is a striking metamorphosis in which:

(a) the right hydrocoel is lost;
(b) the left and right somatocoels come to lie, not at left and right, but one above the other;
(c) the mouth shifts from the mid-ventral line of the larva to a new position corresponding to the centre of the left larval side (usually the mouth disappears and later a new mouth appears in the new position);
(d) the left hydrocoel becomes the water-vascular system of the adult with a circum-oesophageal ring and radial water vessels, primitively and usually five radial water vessels. (Several other organ systems adopt a similar lay-out, induced to do so by the water-vascular system itself.)

Echinoderm phylogeny and evolutionary biology (ed. C. R. C. Paul and A. B. Smith). Clarendon Press, Oxford, 1988.

4 *R. P. S. Jefferies*

As regards the change in orientation at metamorphosis, the right somatocoel comes to lie below the left somatocoel in crinoids, which probably represent the primitive condition in this respect, although in all other extant echinoderms (the Eleutherozoa) the right somatocoel adopts a position above the left somatocoel—this eleutherozoan condition almost certainly results from a true turning-over (inversion) of the whole animal in phylogeny.

4. There is striking radial symmetry, primitively and usually based on the number five. As already mentioned, this shows itself in the fact that several organ systems are arranged as a circumoesophageal ring with five radial extensions.

Other features sometimes cited as characteristic of echinoderms by texts for students are gross symplesiomorphies. An example is the fact that echinoderms live in the sea (see, for example, Nichols 1975, p. 3) and thus retain a condition primitive for life itself. I shall ignore such features in what follows. They merely show that the text books have not yet caught up with Hennig.

Which of the features listed above are truly characteristic of the Echinodermata?

SOME METHODOLOGICAL REMARKS

The use of cladistic methods with fossils involves some concepts which are still not widely understood and which I shall therefore summarize (Fig. 1.1). Given two still extant monophyletic groups, 1 and 2, which are sister groups of each other, there are two obvious ways of delimiting group 2, for instance, if fossil forms are taken into account. The narrower delimitation regards the group as consisting of the latest common stem species (y in Fig. 1.1) of all the extant members of group 2 and all descendants of that species, whether living or extinct. I have called this narrower delimitation the crown group (Jefferies 1979) and it corresponds almost exactly to Hennig's ∗ group (1969, 1981, p. 25). The wider delimitation of group 2 is not so easy to define. It regards group 2 as comprising all those descendants of the latest common stem species of [1 + 2] (x in Fig. 1.1) which are not members of group 1, and this wider delimitation can be called the total group 2 (*Gesammtgruppe* of Hennig). (This definition of the total group is not entirely satisfactory because it is partly self-referential, but Fig. 1.1

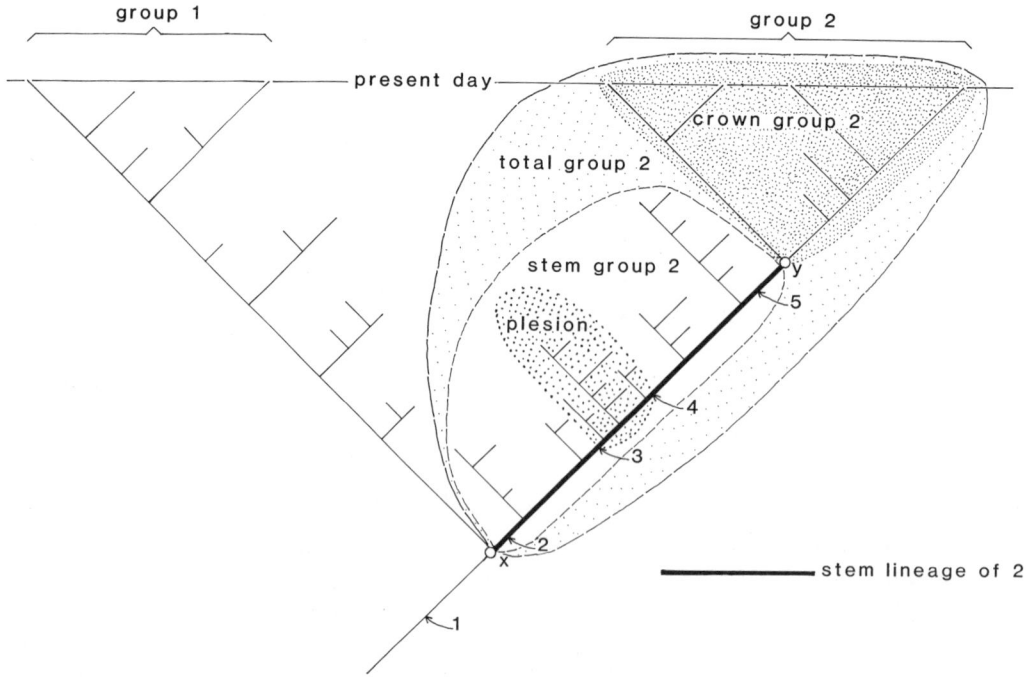

Fig. 1.1 The definitions of total group, crown group, stem group, stem lineage, and plesion. Groups 1 and 2 are sister groups of each other in the extant fauna.

should make things clear.) If the crown group of 2 is subtracted from the total group 2, then an extinct paraphyletic group remains which is the stem group of 2 as I define it. Within the stem group, it is important to distinguish the stem lineage of 2—the direct sequence of ancestors and descendants which led from the latest common stem species (x) of [1+2] to the latest common stem species (y) of the Recent species of 2. In my definition, the stem lineage will exclude both these common stem species. Features which evolved within this stem lineage and which were retained in y (the first member of the crown group of 2) are known as the autapomorphies of group 2. The importance of the stem-lineage concept has recently been emphasized in a brilliant book by Ax (1984a, 1987). The autapomorphies of the Echinodermata are what I seek to establish in this paper.

The autapomorphies of a monophyletic group ought, for clarity's sake, to be distinguished from its synapomorphies. The autapomorphies of such a group are those features which evolved in its stem lineage. The synapomorphies of the group, on the other hand, are those advanced features which it shares with its extant sister group. They are the autapomorphies of the higher group comprised of these two sister groups. Thus, in Fig. 1.1, the autapomorphies of group 2 evolved within the stem lineage of group 2 and existed in the latest common ancestor (y) of the extant members of group 2. The synapomorphies of 1 with 2, on the other hand, are autapomorphies of the group [1+2], evolved within *its* stem lineage and existed in the latest common ancestor (x) of [1+2]. Thus, feature 1 in Fig. 1.1 is an autapomorphy of the group [1+2], but a synapomorphy of group 1 and group 2. The distinction between autapomorphy and synapomorphy, made by Hennig (1966) and stressed by Ax (1984a, 1987), has been widely ignored in the past, by me among others. Sometimes the two words have been used interchangeably (e.g. Jefferies 1986, p. 9), which now seems to me deplorable.

Also for clarity, I stress that a feature which is an autapomorphy of a group need not be present in all members of the group, for it may secondarily have been lost. It may even have disappeared from all extant members of the group. What matters in defining a feature as an autapomorphy of a group is that it arose in the stem lineage of the group and persisted in the stem lineage until the end of that lineage, so as to be still present in the latest common stem species of the extant members of the group.

The above definitions of crown group, stem group, and stem lineage differ slightly, as I now realize and contrary to Jefferies (1986, p. 10), from those of Hennig (1969) and Ax (1984a). For Hennig's ∗ group differs from the crown group by excluding the latest common stem species of the living forms (y of group 2); it thus comprises only the descendants of this latest common stem species, whether these are living or extinct. Correspondingly, this latest common stem species is included in the stem group as Hennig defined it and is also included in the stem lineage as defined by Ax. These differences in definition make very little difference in practice since the latest common stem species of the living members of a group could never be recognized as such if found fossilized.

In a group containing only one extant member species, as for instance the Hominidae with the extant species *Homo sapiens*, the stem group can conveniently be taken as comprising all those individuals which can be studied by palaeontological methods only. Still living members of a group and, for most purposes, members preserved in museums (perhaps in formalin or as dried skins or pinned insects), would be the only ones which do not belong to the stem group under this definition.

The particular strength of the stem-group concept is that every fossil must belong to one stem group only. Identifying this stem group may be difficult, but will always be worth attempting. One of the most important tasks now confronting palaeontologists is to place every known fossil in its correct stem group.

The plesion concept was introduced by Patterson and Rosen (1977) as a way of subdividing a stem group. In my view, a plesion is best defined as: 'All those members of a stem group which, so far as can be discerned, are equally closely related to the crown group'. It will include two constituents, as shown in the plesion illustrated in Fig. 1.1. These constituents are:

(1) a portion of the stem lineage of a crown group—this portion begins when an evolutionary novelty (feature 3 in Fig. 1.1) arises in the stem lineage by mutation and ends at the next recognizable evolutionary novelty (feature 4 in Fig. 1.1) in the stem lineage;
(2) all side branches from this portion of the stem lineage.

If so understood, a plesion is in principle paraphyletic, since it includes some members—the members of the stem lineage—which are ancestral to non-members of the plesion. A plesion is not overtly paraphyletic, however. For if, within a formerly accepted plesion, some members can be shown to be more closely related to the crown group than others are, by demonstrating a previously unnoticed evolutionary novelty in the included part of the stem lineage, then the former plesion will divide into two plesions, one more closely related to the crown group than the other.

I have proposed the term 'more crownward' to signify that one plesion, or one fossil, is more closely related to the crown group than another (Jefferies 1986, p. 13). A plesion therefore comprises all those members of a stem group which, so far as can be discerned, are equally crownward.

A final methodological remark concerns the names of categorial ranks (class, phylum, etc.). These cannot be assigned objectively, cause endless and useless quarrelling, and are often, though wrongly, assumed to be meaningful. For these reasons they are best abandoned (Ax 1984a, 1987). In discussing the phylogenetic system of the echinoderms and their relatives, I shall sometimes mention the categorial rank conventionally assigned to a group, but do so only for the sake of completeness.

In summary, fossils should be assigned to the stem groups of Recent groups; within those stem groups they should be assigned to as many different plesions as can be recognized; and to delimit plesions we must reconstruct the sequence of evolutionary novelties which arose in the stem lineage of the relevant extant group. Also, for clarity of argument, it is necessary to stress that the terms 'autapomorphy' and 'synapomorphy', with respect to any one group, do not mean the same thing. Moreover, categorial rank names are meaningless.

THE PHYLOGENETIC SYSTEM FOR
ECHINODERMS WITHIN THE BILATERIA

The echinoderms and chordates (Fig. 1.2), being sister groups, together comprise a group of higher rank which I have named the Dexiothetica (Jefferies 1979). The extant sister group of the Dexiothetica is unknown—it may be the Hemichordata, the Hemichordata *plus* Chaetognatha, or some group within the Hemichordata. All that can usefully be said about this at present is that the Dexiothetica, Hemichordata, and Chaetognatha together comprise a group of still higher rank, conventionally seen as a superphylum, and named the Deuterostomia (Grobben 1908). The Deuterostomia have deuterostomy as their only known autapomorphy, i.e. the blastopore gives rise in ontogeny to the anus while the mouth is a secondary perforation. Deuterostomy is almost certainly more advanced than the alternative condition (protostomy) for the latter exists in the Coelenterata and Plathelminthomorpha and these Metazoa primitively lack an anus (see discussion in Ax 1984a, p. 274, 1984b) and so cannot be descended from ancestors with deuterostomy. The origin of deuterostomy is indicated by (2) in Fig. 1.2.

The Deuterostomia form part of a larger group, of higher rank, the Radialia. The name-giving autapomorphy of this group is radial cleavage of the egg. This feature is almost certainly more advanced than the alternative state of spiral cleavage, for the latter occurs in those Bilateria which primitively lack an anus (the Plathelminthomorpha), as well as in many of the Bilateria which have an anus (Eubilateria) (Ax 1984a). This strongly suggests that spiral cleavage is primitive for the Eubilateria and, therefore, that the Radialia, within the Eubilateria, have lost it. Another probable autapomorphy of the Radialia is the trimerous condition, i.e. a body divided into the protosome, the mesosome (with tentacles), and the metasome, and containing, respectively, a protocoel, a pair of mesocoels, and a pair of metacoels. Trimery occurs in clearly recognizable form in the so-called lophophorate phyla (Brachiopoda, Phoronida, and Bryozoa), which are protostomatous, and also in the Hemichordata, which are deuterostomatous. In modified form, as already mentioned, it is found also in the Echinodermata. It is therefore likely that trimery is primitive for the Deuterostomia. The origins of radial cleavage and trimery are indicated by (1) in Fig. 1.2.

As to the three 'lophophorate phyla', I know of no synapomorphy which connects them together as a monophyletic group. Perhaps they together comprise such a group or perhaps some of them are more closely related to the Deuterostomia than others are.

THE STEM LINEAGES IN WHICH THE TEXTBOOK FEATURES OF ECHINODERMS AROSE

The problem which this paper aims to solve can therefore be re-phrased as follows: which of the

Characteristics of echinoderms 7

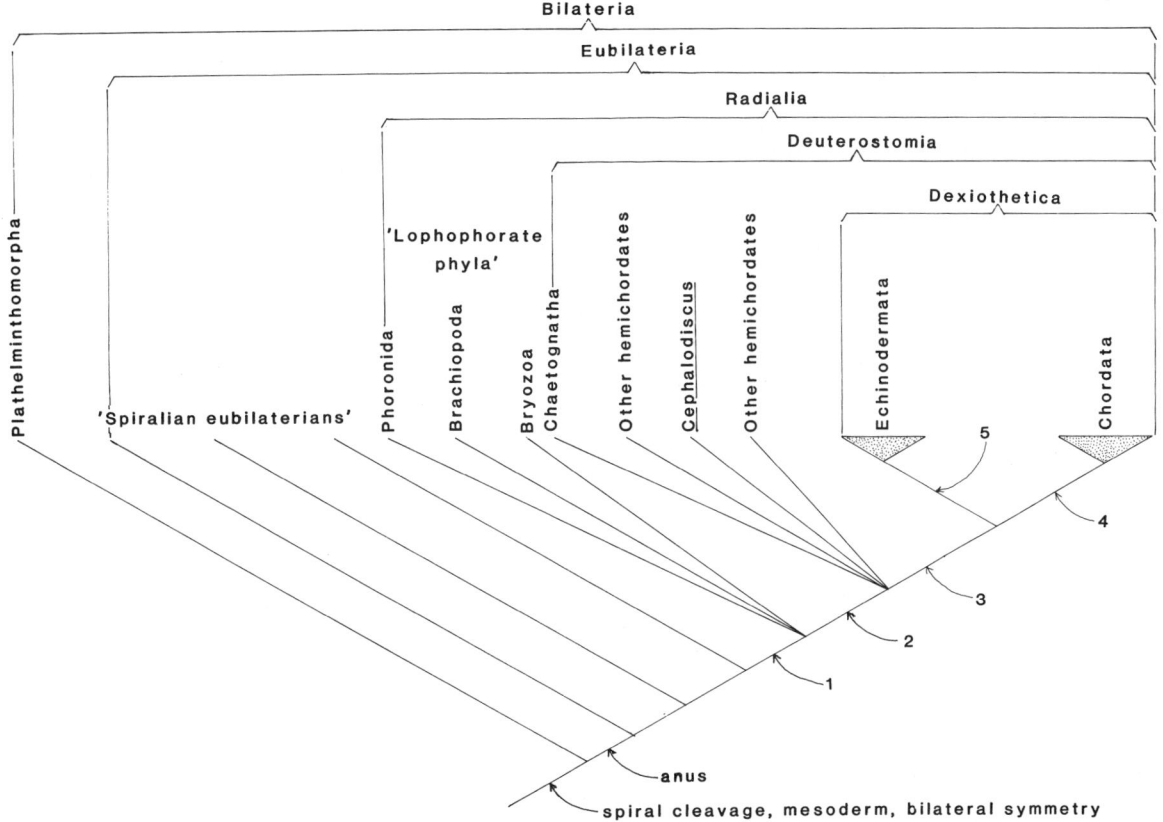

Fig. 1.2 The phylogenetic position of Echinodermata within the Bilateria. (1) = Autapomorphies of the Radialia, i.e. radial cleavage of the egg and trimery (with the origin of the water-vascular system = left mesocoel); (2) = autapomorphy of the Deuterostomia, i.e. deuterostomy; (3) = autapomorphies of the Dexiothetica, i.e. dexiothetism, the calcitic skeleton, the stone canal, absence by loss of the left and right mesocoel pores; (4) = autapomorphies of the Chordata e.g. locomotory tail, notochord; (5) = autapomorphies of the Echinodermata i.e. torsion, pentameral symmetry, loss of all remaining gill slits. The numbers point at the phylogenetic segment in which the autapomorphy arose. Position within the segment does not signify.

text-book features of the echinoderms are genuine autapomorphies of that group, evolved in the stem lineage of the Echinodermata? And which, on the other hand, evolved in other stem lineages, such as those of the Dexiothetica, the Deuterostomia or even the Radialia? In discussing this problem I shall, so far as possible, treat the various text-book features in the sequence listed above.

The water-vascular system is not, strictly speaking, an echinoderm autapomorphy. In making such a statement I am not merely spinning a paradox, for the water-vascular system is generally and rightly accepted as homologous with the left mesocoel (the coelom of the left mesosome) in hemichordates and in the lophophorate phyla.

The water-vascular system can, therefore, logically be said to have originated in the stem lineage of the Radialia and to be an autapomorphy of the Radialia, i.e. it arose at (1) of Fig. 1.2.

The water-vascular system of echinoderms has some features which, in the extant fauna, are known in echinoderms alone. Thus, it develops in all non-pathological living echinoderms from the left mesocoel (left hydrocoel) only, while the right mesocoel aborts. Also, unlike any extant hemichordates, the left mesocoel of echinoderms does not open to the outside directly by a mesocoel pore, but debouches into the protocoel (axocoel) by the stone canal. It is this protocoel which opens externally by way of the left protocoel pore

(=hydropore) so that the connection of the water-vascular system to the surrounding sea water is, in fact, indirect. *Prima facie* it might seem likely that these asymmetrical echinoderm peculiarities are true echinoderm autapomorphies, evolved within the stem lineage of the Echinodermata.

The same might be thought of another peculiarity of extant echinoderms: the fact that the right and left metacoels (right and left somatocoels) lie in the post-larva below and above the gut, instead of right and left of it as in echinoderm larvae and the adults of hemichordates. Like the stone canal and the water-vascular system derived solely from the larval left hydrocoel, this in the extant fauna is known in echinoderms only.

The fossils, however, give a different answer. I have argued, in Jefferies (1986) and elsewhere (e.g. most recently in Jefferies *et al.* 1987), that the Cornuta (Middle Cambrian to latest Ordovician) are stem-group chordates and that the least crownward cornute known is the Middle Cambrian genus *Ceratocystis*. This form has a head and a tail, like all other cornutes, and the openings in the head can be identified on the basis of various arguments (Jefferies 1986, chap. 7). It is instructive, in this connection, to compare the openings of *Ceratocystis* seen in dorsal aspect with those of the pterobranch hemichordate *Cephalodiscus* seen from the left side (Fig. 1.3). For, in moving clockwise around the sketch of *Cephalodiscus* we pass: a pair of gill slits, the mouth, a pair of protocoel pores ('hydropore' in Fig. 1.3), a pair of mesocoel pores, a pair of gonopores, and the anus. While in passing clockwise round the sketch of *Ceratocystis* we pass: gill slits, the mouth, the hydropore (presumably equivalent to the left protocoel pore), a single gonopore, and the anus. The two diagrams seem to be broadly equivalent topologically except that *Ceratocystis*, like echinoderms, shows no mesocoel pore (or pores), and is also lacking any equivalent of the right gill slit of *Cephalodiscus*, of the right gonopore or of the right protocoel pore—its hydropore is presumably homologous with the hydropore of echinoderms and, therefore, with the left axocoel pore of *Cephalodiscus*. This broad topological agreement suggests that the dorsal surface of *Ceratocystis* corresponds to the left surface of *Cephalodiscus*. This, in turn, suggests that, in the ancestry of *Ceratocystis*, a form like *Cephalodiscus* took to lying down on its originally right side, in the habit which I have called dexiothetism (Jefferies 1979). A

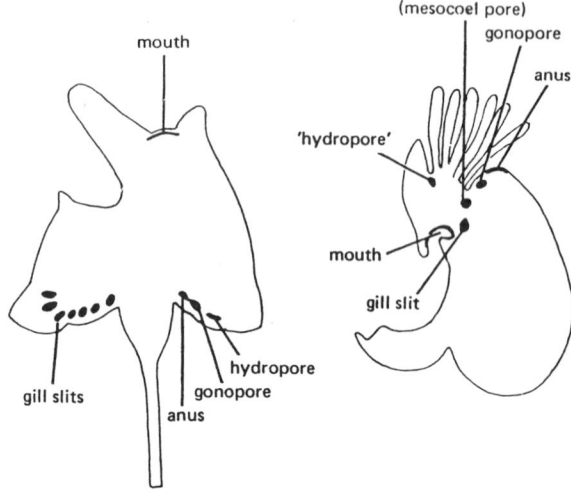

Fig. 1.3 A diagram to illustrate the broad topological equivalence between the head openings of the cornute *Ceratocystis* (left), in dorsal aspect, and the pterobranch hemichordate *Cephalodiscus* (right) in left aspect. (From Jefferies 1986, p. 221.)

dexiothetic phase (Fig. 1.4) is also likely in the ancestry of extant echinoderms which would explain their various embryological asymmetries, i.e. the facts that the left metacoel (somatocoel) is above the right metacoel in crinoids, that the water-vascular system develops from the left larval side only in all echinoderms, and that only the left protocoel (axocoel) pore (not the right one) gives rise to the hydropore in all non-pathological echinoderms. It is reasonable to assume that this dexiothetic phase was one and the same in echinoderms and in chordates and, if so, the various already mentioned embryological asymmetries of echinoderms are not echinoderm autamorphies, but autapomorphies of the group Dexiothetica which the extant chordates have lost. The only likely traces of them in extant chordates are a number of asymmetries, such as the fact that larval amphioxus has left gill slits only, as summarized in Jefferies (1986, p. 79ff.). Dexiothetism originated at (3) in text Fig. 1.2.

The same arguments suggest one echinoderm autapomorphy which has never been proposed in the text books—the absence of gill slits. For, among the hemichordates, gill slits exist in *Cephalodiscus*, *Atubaria*, and the enteropneusts and also occur, as presumed homologues with these, in the chordates. Extant echinoderms, however, and

Characteristics of echinoderms 9

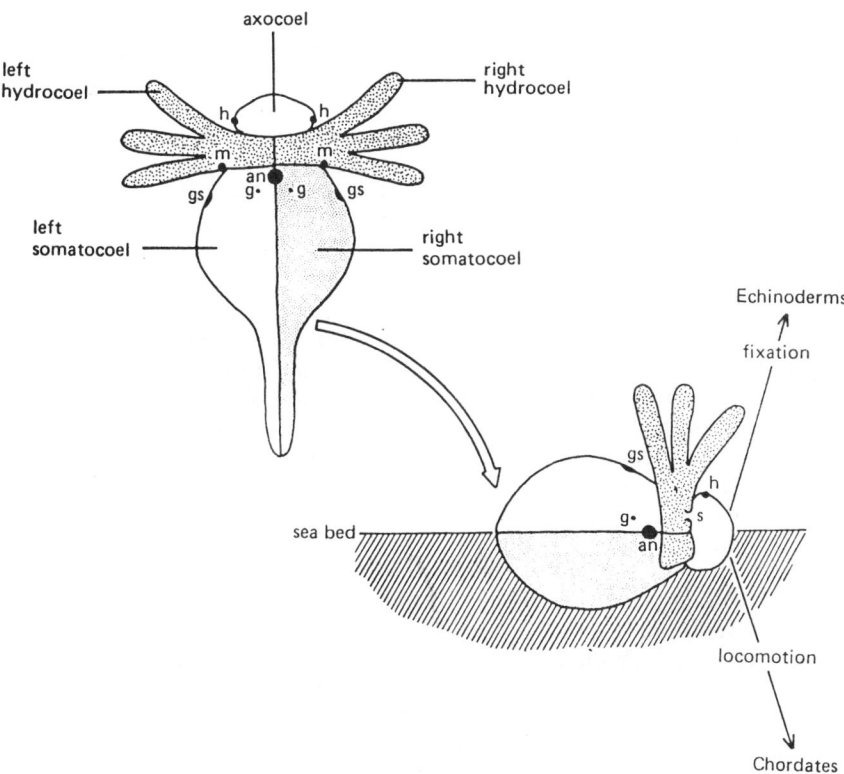

Fig. 1.4 Dexiothetism. In the stem lineage of the Dexiothetica, a *Cephalodiscus*-like ancestor (left of figure, shown in dorsal aspect) came to lie, right-side-downwards, on the sea floor (right of figure) and lost its stalk. As a result: (1) the right somatocoel (right metacoel) came to be below the left somatocoel; (2) the right hydrocoel, being forced downwards into the mud, became useless and disappeared; (3) the openings of the original right side, having come to face ventrally, into the mud, closed up. Abbreviations: an = anus; g = gonopore; gs = gill slit; h = hydropore; m = mesocoel pore. At about the same time the originally left mesocoel pore closed up and the stone canal (s) arose to connect the left hydrocoel with the axocoel. (From Jefferies 1986, p. 52.)

even the stem-group echinoderms called helicoplacoids, have no gill slits (Paul and Smith 1984). Given that echinoderms are more closely related to the chordates than the hemichordates are, then gill slits were lost in the stem lineage of echinoderms, less crownward than the helicoplacoids. The absence by loss of all remaining gill slits would therefore be a genuine echinoderm autapomorphy, the origin of which is indicated by (5) in Fig. 1.2.

The calcitic skeleton of extant echinoderms, in which each plate is built of stereom mesh and is crystallographically a single crystal, is parallel to dexiothetism in its systematic distribution. It exists in stem-group and crown-group echinoderms (including extant ones), but is also found in stem-group chordates (cornutes) and primitive crown-group chordates (mitrates) though absent in all non-mitrate crown-group chordates including all extant ones. Despite its systematic distribution in the extant fauna, it is therefore not in truth an autapomorphy of the Echinodermata, but evolved in the stem lineage of the Dexiothetica, and is an autapomorphy of the Dexiothetica which the extant chordates have lost. As such, its origin is shown by (3) in Fig. 1.2. Indeed, the calcitic skeleton was lost three times independently in the chordates, in the stem lineages of the acraniates, the tunicates, and the vertebrates—this is shown by the fact that all known mitrates can be assigned to the stem groups of the extant chordate subphyla, being either stem-group acraniates, stem-group tunicates, or stem-group vertebrates (Jefferies 1986, chapters 8 and 9).

Thus, none of the text-book echinoderm features so far considered is an echinoderm autapomorphy in fact, and one such autapomorphy is not a text-book echinoderm feature.

The only text-book echinoderm features still to be discussed are: radial, primitively pentameral, symmetry; and the situation of the definitive mouth in a position corresponding to the centre of the larval left side. As already mentioned, radial symmetry shows itself in the fact that various organ-systems comprise a circumoesophageal ring surrounding the mouth, with radial extensions coming off this ring. These extensions are primitively five in number, or ten where the structures are paired in each ray as in the case of the perihaemal vessels and Lange's nerves, for example. The fact that the definitive mouth of echinoderms corresponds in position to the centre of the larval left side, rather than to the ventral mid-line of the larva, is linked with radial symmetry, but should probably be seen as resulting from a separate evolutionary event called torsion. This would have shifted the mouth from the larval ventral position to the definitive central one. Torsion would primitively cause the mouth to open upwards, as it still does in crinoids. In the stem lineage of the Eleutherozoa, however, the subsequent evolutionary event of inversion (total inversion of the whole animal) caused the mouth of Eleutherozoa to open downwards (Paul and Smith 1984; Smith 1984).

Radial symmetry and torsion are almost certainly true autapomorphies of the Echinodermata, their origins being shown by (5) in Fig. 1.2. On the one hand, the stem-group chordates of the Cornuta and the primitive crown-group chordates of the Mitrata lack any radial symmetry and have the mouth situated anteriorly and level with the sea bed, in the pre-torted position, and not in the centre of the dorsal surface. On the other hand, the likely evolutionary origin of torsion and radial symmetry can be reconstructed in the stem group of the Echinodermata.

Thus, Paul and Smith (1984) have shown that the Lower Cambrian helicoplacoids, which were stem-group echinoderms, had a triradiate water-vascular system while the mouth was situated at the mid-length of the animal and one end of the animal was inserted in the mud of the sea bottom. This means that, if a helicoplacoid in its life-position could be viewed from above, the mouth would be situated peripherally as in stem-group chordates, not centrally as in crinoids. It is logical to suppose that this peripheral position of the mouth is primitive for Echinodermata as a group since it corresponds to that of all chordates and all hemichordates. Also relevant is the Lower Cambrian fossil *Camptostroma*, which Paul and Smith (1984) regard as either a crownward member of the echinoderm stem group, or else as a primitive member of the echinoderm crown group. This animal had a pentameral water-vascular system while the mouth was central and faced upward. In the stem lineage of the Echinodermata, therefore, crownward of the helicoplacoids, but less crownward than *Camptostroma*, the triradial water-vascular system had become quinqueradial and the mouth had undergone torsion so as to open upwards. Both these changes were probably the result of fixation of the animal in one place. Such fixation may mean cementation to a hard substrate or merely the inability to locomote on a soft one. Such a static mode of life favours radial symmetry because the organism is thereby able to profit equally from all sources of food, oxygen, etc., irrespective of the direction from which they come. It is interesting that Bather (1900) had already supposed that the quinqueradiality of crown-group echinoderms was preceded by a triradial stage. He thought this because many early fossil echinoderms have the water-vascular system arranged in a $2+1+2$ pattern. His view is confirmed by Bell's discovery (1976) that some edrioasteroids (stem-group eleutherozoans) had a triradiate stage in their ontogeny.

A point of morphological terminology must now be dealt with. The dorsal surface of a cornute such as *Ceratocystis* is homologous with the dorsal surface of a fish or a mammal. This is implied by all the arguments which show that cornutes and mitrates are chordates. However, my discussion in this paper also implies that the dorsal surface of a cornute is homologous with the upper surface of such a primitive echinoderm as *Camptostroma* or of a crinoid. This upper surface is conventionally labelled ventral in many echinoderm texts, as for example in Cuénot (1948), and Paul and Smith (1984), a convention based on the fact that in Eleutherozoa the corresponding surface is downward. There can be no doubt that the chordate usage has priority (Latin *dorsum*=back, *venter*=belly) and is colloquial to far more people than the echinoderm usage. It therefore seems to me that the use of 'dorsal' and 'ventral' in echinoderm morphology should be abandoned. Such a course has already been advocated by Hyman (1955) on the grounds that nobody then knew how dorsal and ventral corresponded between echinoderms and vertebrates. Her argument is strengthened by knowing that echinoderm dorsal is, in fact, equivalent to chordate ventral. I have already pointed this out in Jefferies *et al.* (1987).

Torsion and pentameral symmetry thus emerge as probable autapomorphies of the Echinodermata, arising at point (5) in Fig. 1.2. However, I must mention one doubt. In all non-pathological extant echinoderms, the water-vascular system, as already said, consists of a complete circum-oesophageal ring with radial extensions. It is parsimonious, and may be correct, to suppose that the circum-oesophageal ring was likewise complete in the latest common ancestor of extant echinoderms. If so, then it would have been a true autapomorphy of the echinoderms since it would have arisen in the echinoderm stem lineage. On the other hand, the circum-oesophageal ring develops in ontogeny from a U-shaped *anlage* which forms a closed circle by growing around the oesophagus with fusion of the two free ends of the U. To my mind, it would not, therefore, be surprising if the transition of this central part of the water-vascular system from a U to a ring had happened more than once in the echinoderm crown group, implying that the latest common ancestor of extant echinoderms had a U. It will not be easy to find relevant evidence, but perhaps the stem-group eleutherozoans should be examined with this doubt in mind.

CONCLUSIONS—THE TRUE AUTAPOMORPHIES OF ECHINODERMATA AND RELATED GROUPS

The features which emerge as true autapomorphies of the Echinodermata, originating at (5) in Fig. 1.2, are therefore as follows:

1. Torsion, with the mouth opening upward.
2. Pentameral symmetry, with five radial water vessels extending outwards from a central ring-shaped (or possibly primitively U-shaped) circum-oesophageal water vessel.
3. Absence, by loss, of any gill slits.

Features which are autapomorphies of the Chordata, arising at (4) in Fig. 1.2, are not the particular subjects of this paper, but include the locomotory tail, notochord, segmental muscle blocks, dorsal nerve chord, the brain, the trigeminal ganglia, endostyle, and filter-feeding pharynx, right gill slits, right and left atria and atrial openings, and several others as discussed in Jefferies (1986, Chapt. 9). The right gill slits of chordates are not homologous with the right gill slits of Hemichordata, but represent new formations—this is shown by the fact that the known stem-group chordates (cornutes) have left gill slits only.

Features which are autapomorphies of the group Dexiothetica, arising at (3) in Fig. 1.2, are as follows:

1. The calcitic skeleton of stereom mesh.
2. Dexiothetism; this was an evolutionary event in which a *Cephalodiscus*-like ancestor fell over on its right side. This resulted in a complex of features which include: the right somatocoel being below the gut and the left somatocoel above it; the absence, by loss, of the right hydrocoel; the absence, by loss, of the originally right (now ventral) gill slit, right gonopore, right gonad, and right hydropore.
3. The stone canal, being a connection between the left hydrocoel and the axocoel.
4. The loss of the left and right mesocoel pores.

Autapomorphies 1–3 of the group Dexiothetica have left no trace in extant chordates and therefore mimic echinoderm autapomorphies in their extant systematic distribution.

Dexiothetic autapomorphies 3 and 4 may have been the results of the evolutionary change of orientation called dexiothetism, but this is uncertain.

The only recognizable autapomorphies of the Deuterostomata are the blastopore giving rise to the anus and the secondary perforation of the mouth. These features together constitute deuterostomy. They originated at (2) in Fig. 1.2.

There are two autapomorphies of the Radialia, originating at (1) in Fig. 1.2.

1. Trimery, with three body regions and the corresponding five coeloms (protocoel, left and right mesocoel, left and right metacoel). One of the autapomorphies of the Radialia, paradoxically, is therefore the origin of the water-vascular system under the guise of the left mesocoel. (It would be possible to retain the water vascular system as an echinoderm autapomorphy by defining it as a system of coelomic tubes derived from the left mesocoel, opening to the outside by a stone canal

and arranged on a radiate plan. Such a stratagem, however, seems more legalistic than useful.)

2. Radial cleavage of the egg.

The radialian autapomorphy of the water-vascular system has been lost in all extant members of the Chordata.

METHODOLOGICAL CONCLUSIONS

The above discussion shows that fossils are sometimes all-important in phylogenetic research. This is particularly the case when the groups in question are high-ranking and the phylogenetic splits which produced them are therefore very old. A phylogenetic segment which lasted 10 million years in the Cambrian will usually be more difficult to detect, by comparing still living organisms, than one of the same duration in the Miocene. The main reason is that features which arose in a very old phylogenetic segment are likely to have been lost in some of the modern descendants of that segment; the older the segment, the more likely the loss. A striking example is the calcite skeleton, a dexiothetic autapomorphy which has been lost in all modern chordates. This loss has, in fact, happened three times and results in the skeleton being regarded, by almost all zoologists, as characteristic of the Echinodermata. This is perfectly correct for Recent animals, but the obvious phylogenetic conclusion is wrong.

I agree with Patterson (1981) that phylogenetic analysis has to begin with the Recent organisms but I feel certain, unlike him, that fossils can sometimes be crucial in phylogenetic reconstruction. In other words, I positively disagree with Løvtrup when he asserts (1977, p. 21), as a principle, that: 'The discovery of a new fossil has no impact on classification'. Fossils are certainly difficult to use in phylogenetic research, but given correct methodology, they can sometimes solve problems which cannot be settled otherwise.

REFERENCES

Ax, P. 1984a. *Das Phylogenetische System*. Fischer, Stuttgart.

Ax, P. 1984b. The position of the Gnathostomulida and Plathelminthes in the phylogenetic system of the Bilateria. In *The origins and relationships of lower invertebrates* (ed. S. Conway Morris, J. D. George, K. Gibson, and H. M. Platt), pp. 168–80. Oxford University Press, Oxford.

Ax, P. 1987. *The phylogenetic system*. Wiley, Chichester.

Bather, F. A. 1900. Chapters 8–12 in *A treatise on zoology Part III, Echinoderma* (ed. E. R. Lankester), pp. 1–216. A. and C. Black, London.

Bell, B. M. 1976. Phylogenetic implications of ontogenetic development in the class Edrioasteroidea (Echinodermata). *Journal of Paleontology* **50**, 1001–9.

Cuénot, L. 1948. Anatomie, ethologie et systematique des echinodermes. In *Traité de Zoologie. Tome XI. Echinodermes, Stomocordés, Protocordés* (ed. P.-P. Grassé), pp. 1–275. Masson, Paris.

Grobben, K. 1908. Die systematische Einteilung des Tierreiches. *Verhandlunger der zoologischen Gesallschaft zu Wien* **58**, 491–511.

Hennig, W. 1966. *Phylogenetic systematics*. University of Illinois Press, Urbana.

Hennig, W. 1969. *Die Stammesgeschichte der Insekten*. Kamer, Frankfurt-am-Main.

Hennig, W. 1981. *Insect phylogeny* (trans. A. C. Pont). Wiley, Chichester.

Hyman, L. 1955. *The invertebrates: Vol. IV Echinodermata*. McGraw-Hill, New York.

Jefferies, R. P. S. 1979. The origin of chordates—a methodological essay. In *The origin of major invertebrate groups*, Systematics Association Special Volume 12 (ed. M. R. House), pp. 443–77. Academic Press, New York.

Jefferies, R. P. S. 1986. *The ancestry of the vertebrates*. British Museum, Natural History, London.

Jefferies, R. P. S., Lewis, M., and Donovan, S. K. 1987. *Protocystites menevensis*—a stem-group chordate (Cornuta) from the Middle Cambrian of South Wales. *Palaeontology*, **30**, 429–84, pls 54–60.

Løvtrup, S. 1977. *The phylogeny of the Vertebrata*. Wiley, Chichester.

Nichols, D. 1975. *The uniqueness of the echinoderms*, Oxford Biology Readers, No. 53. Oxford University Press, London.

Patterson, C. 1981. Significance of fossils in determining evolutionary relationships. *Annual Review of Ecology and Systematics* **12**, 195–223.

Patterson, C. and Rosen, D. E. 1977. Review of the ichthyodectiform and other Mesozoic teleost fishes and the theory and practice of classifying fossils. *Bulletin of the American Museum of Natural History* **158**, 85–172.

Paul, C. R. C. and Smith, A. B. 1984. The early radiation and phylogeny of echinoderms. *Biological Reviews* **59**, 443–81.

Siewing, R. 1984. Echinodermata. *Lehrbuch der Zoologie. Band 2. Systematik* (ed. H. Wormbach and R. Siewing), pp. 292–338. Fischer, Stuttgart.

Smith, A. B. 1984. Classification of the Echinodermata. *Palaeontology* **27**, 431–59.

2

The meaning of developmental asymmetry for echinoderm evolution: a new interpretation

NICHOLAS D. HOLLAND

Marine Biology Research Division, Scripps Institution of Oceanography, La Jolla, California 92093, USA

INTRODUCTION

During development of extant echinoderms, the primary bilateral symmetry of the larva changes to asymmetry due to atrophy of structures along the right side of the body and, subsequently, the already complicated anatomy becomes overlain by radial symmetry. This succession of symmetries during ontogeny has often been used as evidence in attempts to trace the origin and early evolution of the phylum Echinodermata. For a century, evolutionary scenarios that have taken the ontogenetic evidence into account—as opposed to those that have not (e.g. Nichols 1967; Gutmann 1981)—have constituted the main stream of thought on echinoderm evolution. The ontogeny-based scenarios, although long popular with textbook writers, have been criticized soundly (Hyman 1955; Clark 1964). The present paper is written from the viewpoint that such scenarios are indeed unsatisfactory in some respects, but they contain much worth saving and should be improved by modification. Therefore, my purposes are to review the subject, to identify flaws in previous arguments, and to make appropriate corrections. Finally, I examine the utility of my corrections by following their consequences forward in echinoderm evolution and, to a lesser extent, in chordate evolution. My arguments are primarily zoological, with special attention to feeding biology, but they are also consistent, in their broad outline, with current ideas of echinoderm palaeontology (Philip 1979; Paul and Smith 1984; Smith 1984).

Echinoderm phylogeny and evolutionary biology (ed. C. R. C. Paul and A. B. Smith). Clarendon Press, Oxford, 1988.

REVIEW

Metschnikoff (1869) was the first to point out that an asymmetrical stage intervenes between the bilateral stage and the radial stage in echinoderm development, although he did not speculate on what this sequence of symmetries might mean for echinoderm phylogeny. Such speculations were first made by Semon (1888, 1889), and were continued and elaborated upon by his successors. The more important references on this subject are listed in Table 2.1.

Most of the authors in Table 2.1 derived echinoderms from a bilaterally symmetrical deuterostome that was either a dipleurula or a pterobranch hemichordate. These authors were concerned with evolution in post-larval animals, yet much of the evidence was obtained from developing larvae of extant echinoderms. This paradox arises because, during evolution, some post-larval characters were transferred into the larval stages by a type of heterochronic development that Jägersten (1972) has termed adultation.

The dipleurula (Fig. 2.1A) was imagined to resemble an early larva of an echinoderm, but with a more perfect bilateral symmetry. There was an uncoiled gut running from an anteroventral mouth to a posteroventral anus, and the coeloms, from posterior to anterior, were a pair of somatocoels, a pair of hydrocoels, and a pair of axocoels (this last coelom was single according to some authors). Each axocoel was connected to the exterior via a narrow hydropore duct that opened near the dorsal midline and to the ipsilateral hydrocoel by a narrow stone canal. Depending on the reference (Table 2.1), the dipleurula was often presumed to have either tentacles associated with the hydrocoels or epidermal tracts of cilia. In

TABLE 2.1 References assuming that the bilateral, asymmetrical, and radial stages in echinoderm ontogeny are evidence for a corresponding sequence of symmetries during echinoderm phylogeny. Throughout the table, the terms dorsal, ventral, right, left, anterior, and posterior correspond to regions of the bilaterally symmetrical stage*, and the terms upper, under, and lateral designate body surfaces according to their orientation to the substratum

	Bilaterally symmetrical ancestor			
Reference	Body plan	Association with substratum	Asymmetrical stage in evolution	Radially symmetrical stage in evolution
Semon (1888)	Dipleurula with ciliated tracts	By dorsal surface; firm attachment	Ignored	Pentactula with tentacle ring around mouth on upper (ventral) surface
Semon (1889)	Dipleurula with ciliated tracts	By right surface; firm attachment	Ignored	As above, except mouth and tentacles migrated from lateral (ventral) surface to upper (left) surface
Bütschli (1892)	Dipleurula with right and left tentacle groups	By right surface; right tentacles adhere; later firm attachment	Structures apposed to substratum tended to atrophy	Mouth migrated from lateral (ventral) surface to upper (left) surface and was encircled by left tentacle group
Lang (1894)	Dipleurula	By right anteroventral region; firm attachment	Asymmetry resulted from stalk formation on right side	Mouth migrated from approximate lateral (ventral) surface to upper (left) surface; became encircled by left hydrocoel, which sprouted tentacles
Bury (1896)	Ignored	Unspecified, but evidently no firm attachment	Ignored	Unattached pentactula with mouth on under (ventral) surface; mouth surrounded by ring of tentacles (phylogenetic origin not explained)
MacBride (1896)	Dipleurula with right and left tentacle groups	By anterior end; firm attachment	Stage mentioned, but its cause† ignored	Text not informative; figures show mouth migrated from one lateral (ventral) surface to another lateral (left) surface and became encircled by left tentacle group

Bather (1900)	Dipleurula with general ciliation	By right anteroventral region; firm attachment	Attachment caused visceral rearrangement that, in turn, caused atrophy of right side	Mouth migrated posteriorly along left-ventral surface and became encircled by the left hydrocoel, which then produced the radially arranged water vascular system
Masterman (1902)	Dipleurula (not so named in this reference) with ciliated tracts	By right anteroventral region; attachment temporary at first, later firm	Structures apposed to substratum tended to atrophy	Mouth migrated from ventral surface to upper (approximately left) side and became encircled by left hydrocoel, which then sprouted tentacles
Heider (1912)	Pterobranch-like modified *Rhabdopleura*	Not explicit, but claimed to agree with Bather (1900)	Stage mentioned, but its causes ignored	Starting conditions vague, but mouth moved temporarily to left surface and became encircled by left hydrocoel, which sprouted additional tentacles
Grobben (1923)	Pterobranch-like modified *Cephalodiscus*	Anteroventral; firm attachment	Mouth migrated from under (ventral) side to lateral (left) surface and pulled right tentacles under body where they atrophied	The mouth, in migrating to the left side approached the left hydrocoel and pushed it into a ring; the left group of tentacles thus became orientated radially
Jefferies (1979, 1981)	Pterobranch much like *Cephalodiscus*	By right surface; not attached	Structures apposed to substratum tended to atrophy	Not dealt with, since author was mainly concerned with the origin of carpoids, which lack radial symmetry

*In the present paper, the terms dorsal and ventral will never be used as synonyms for aboral and oral, respectively; the terms anterior and posterior will never be used as for orienting irregular echinoids (see Hyman 1955, p. 417); and the terms anterior, posterior, right, and left will never be used as in Bather's system for describing fossils (see Ubaghs 1967, pp. 8–9) or as used for describing the carpoid theca (Jefferies 1981).

†Elsewhere, MacBride (1914) wrote, 'It can only be described as an idiosyncrasy of echinoderms that bilateral symmetry is unstable, and that, therefore, radial symmetry was arrived at by the overgrowth of the organs of the left side and a partial suppression of those on the right side'.

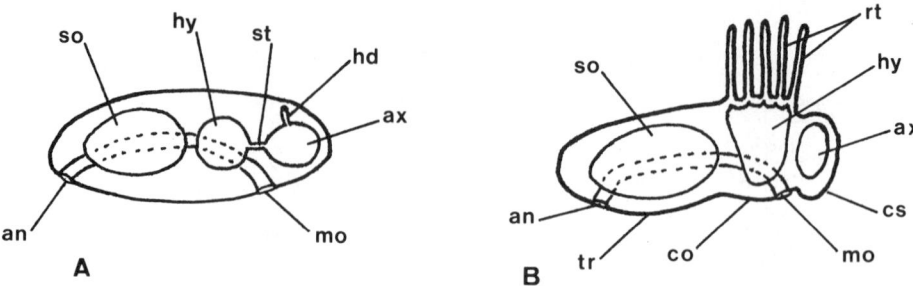

Fig. 2.1 (A) Ancestral dipleurula (after Bather 1900) viewed from right side; (B) ancestral hemichordate (after Grobben 1923) viewed from right side (coelomic ducts not shown); an, anus; ax, axocoel; co, collar; cs, cephalic shield; hd, hydropore duct; hy, hydrocoel; mo, mouth; rt, right tentacle group; so, somatocoel; st, stone canal; tr, trunk.

comparison to dipleurulae, the proposed pterobranch ancestors of echinoderms leave less to the imagination. Heider (1912) and Grobben (1923) conceived of these creatures as pterobranchs modified in having a posteroventral anus and in having neither a peduncle nor gill slits (Fig. 2.1B), and Jefferies (1979, 1981) proposed a pterobranch ancestor that closely resembled a modern *Cephalodiscus* in having a peduncle, gill slits, and an anterodorsal anus. In all these pterobranchs, the regions of the body (excluding the peduncle if present), from posterior to anterior, were the trunk, the collar, and the cephalic shield. The ventral mouth opened at the junction of the collar and the cephalic shield. Tentacles projected from the right and left dorsolateral regions of the collar. The coeloms were a pair of somatocoels in the trunk, a pair of hydrocoels in the collar, and a paired or unpaired (depending on the reference) axocoel in the cephalic shield. The axocoels and hydrocoels opened to the exterior via narrow ducts that opened dorsolaterally.

The authors in Table 2.1 subscribed to the common, although not universal, view that ontogeny may recapitulate phylogeny; they generally agreed that the sequence of symmetries observed during ontogeny is evidence for a corresponding sequence of symmetries that appeared during the early evolution of echinoderms. In the present review, it will be convenient to refer to phylogenetic stages as bilaterally symmetrical, asymmetrical, or radially symmetrical until I develop my arguments; thereafter, I will replace these vague terms with more precise ones.

Many of the authors in Table 2.1 were concerned with possible causes for the evolutionary transitions (1) from bilateral symmetry to asymmetry and (2) from asymmetry to radial symmetry. The authors who speculated on the causes of the asymmetrical stage proposed that it resulted when the right surface of the body became associated with the substratum. Depending on the author, the strength of this association varied from a sessile, but unattached state to a permanently attached state. Bather (1900) thought that attachment at the right anteroventral end caused visceral rearrangement, which in turn, somehow resulted in the atrophy of structures of the right side. Most other authors in Table 2.1 assumed that the atrophy of the right side resulted directly from the impairment of the functions of structures on this side soon after they became turned toward the substratum. Unfortunately, none of these authors unequivocally identified the functions in question. Masterman (1902) came closest to discussing this important subject; on p. 406, he wrote of the asymmetrical stage that, 'Its diet was unquestionably microscopic floating organisms . . .', and he clearly implied that lying on the right side impaired the function of the suspension feeding structures turned toward the substratum. It is at least fair to suggest that most of the other authors in Table 2.1 believed that the bilaterally symmetrical stage, whether a dipleurula or a pterobranch, was a suspension feeder and this feeding mode was retained during the transition to the asymmetrical stage.

For the radially symmetrical stage in evolution, the ultimate cause is generally agreed to have been the transition to a sessile or attached habit. Many of the authors in Table 2.1 made the reasonable assumption that an animal remaining in one place and having food equally available from all sides would be likely to develop radial symmetry. To these ecological reasons, some authors added a mechanical one: namely, that the migrating anterior part of the gut pushed the left hydrocoel into a crescent that later closed as a ring and became the centre of the radially orientated, water-vascular system.

PROBLEMS WITH PREVIOUS EVOLUTIONARY SCENARIOS BASED ON ECHINODERM DEVELOPMENT

The references in Table 2.1 suffer from at least one of two chief problems. The first is that most of the scenarios began with the permanent attachment of the bilaterally symmetrical ancestor, which is not in agreement with current palaeontological evidence that the earliest echinoderms were not permanently attached. The second, and more serious, problem arises at the transition between the bilaterally symmetrical ancestor and the asymmetrical stage. Whether the authors in Table 2.1 started with a dipleurula or a pterobranch ancestor, this bilaterally symmetrical creature was evidently a suspension feeder gathering food with ciliated tracts or tentacles located in whole or in part on the right and left sides of the body. Therefore, as a consequence of turning one side of the body toward the substratum, the animal put a substantial portion of its suspension feeding apparatus out of commission. The inescapable conclusion is that the asymmetrical stage in evolution, as conceived by the references in Table 2.1, is highly improbable because it began with a sudden and pronounced decrease in fitness.

A MODIFIED EVOLUTIONARY SCENARIO FOR THE ORIGIN OF THE ASYMMETRICAL STAGE

I will begin my modified evolutionary scenario with a bilaterally symmetrical ancestor that includes what I think are the best features of the hemichordate ancestors of Grobben (1923) and Jefferies (1979, 1981). As depicted in Fig. 2.2, the body was divided into the cephalic shield, the collar and the trunk, but there was no peduncle. A cluster of tentacles arose from the right and left dorsolateral regions of the collar, and the gut ran from an anteroventral mouth to a posterior anus. The pair of pharyngeal gill slits probably functioned for respiration and for evacuation of water

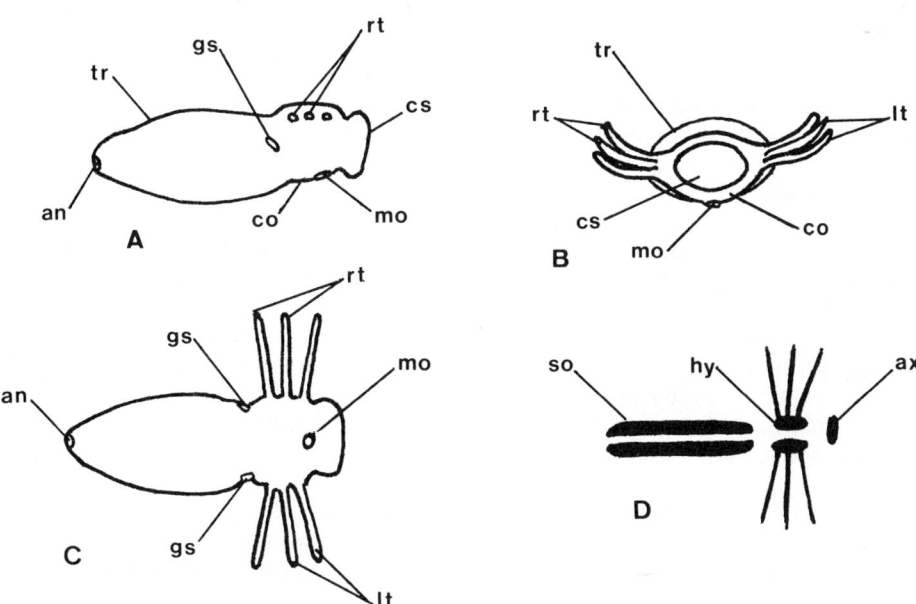

Fig. 2.2 Bilaterally symmetrical stage (stem deuterostome). (A) viewed from right side with anterior toward right; (B) viewed from anterior end; (C) viewed from ventral side; (D) coeloms viewed from ventral side; an, anus; ax, axocoel; co, collar; cs, cephalic shield; gs, gill slit; hy, hydrocoel; lt, left, tentacle group (only tentacle bases shown in A); mo, mouth; rt, right tentacle group; so, somatocoel; tr, trunk.

entering the gut with the food. The creature was an epibenthic suspension feeder that crawled sluggishly on the substratum.

From this starting point, I will alter the traditional evolutionary scenarios in two major ways. First, I will propose that suspension feeding did not carry over from the bilaterally symmetrical stage into the asymmetrical stage. Instead, the suspension feeding, bilaterally symmetrical ancestor became a *deposit feeder* during the transition to the asymmetrical stage. The type of deposit feeding I have in mind for the start of the asymmetrical stage is the picking up of particles from the substratum by tentacles or tube feet of an epibenthic animal. Good examples of this are known for many ophiuroids (Warner 1982) and some echinoids (De Ridder and Lawrence 1982; Telford *et al.* 1985).

Transitions from suspension feeding to deposit feeding are not rare events. In the opinion of Jørgensen (1966, p. xiii), suspension feeders have commonly evolved into deposit feeders and vice versa on an evolutionary time scale. Moreover, studies of extant echinoderms have shown that, on time scales of minutes or hours, a given individual may alternate between suspension feeding and deposit feeding, evidently depending on the current speed and flux of suspended matter (Roberts and Bryce 1982; Jangoux 1982; Warner 1982). Importantly, such work demonstrates that, in some species at least, the same food gathering structures can either remove suspended particles from the surrounding sea water or pick up particles from the substratum.

The transition from suspension feeding to deposit feeding at the start of the asymmetrical stage can reasonably be interpreted as an adaptation to take advantage of a benthic food supply that had become more abundant and/or more stable than the planktonic food supply. Levinton (1974) and Valentine (1986) have suggested that, over geologic time, planktonic food supplies have been somewhat less stable than benthic (especially bacterial) food supplies. Therefore, suspension feeders have tended to become extinct when their planktonic food supply has failed, while deposit feeders have tended to survive. Many deposit feeding species probably had suspension feeding larvae in their life histories, but the larvae could have survived extended interruptions in the supply of phytoplankton and zooplankton by suspension feeding on bacterioplankton in the manner described by Rivkin *et al.* (1986).

In my scenario, the bilaterally symmetrical stage originally used the tentacles on both sides of the body for suspension feeding. Then, probably during an extended period of decreased planktonic food supply, the animals began to augment their usual diet of suspended food with matter the tentacles picked up from the substratum by deposit feeding. At about this point in evolution, a posterior appendage (the stele or tail) developed dorsal to the anus and no doubt functioned for locomotion. This appendage is evidently not homologous to the hemichordate peduncle, which is located ventral to the anus.

After the transition to deposit feeding, the feeding efficiency was greatly improved when the animals rolled over through 90° to position one group of tentacles beneath the body where they could more readily pick up particles from the substratum (Fig. 2.3A). These tentacles, having become the chief food collecting organs, would not have atrophied. In contrast, the tentacles of the other side, now pointing upward, would have captured comparatively little food and thus atrophied (Fig. 2.3B) along with nearby coelomic structures that constituted their hydraulic systems (Fig. 2.3D, arrow). The atrophy of the upper tentacles probably did not significantly diminish the total respiratory surface because body size was small and gill slits were present. Following the conversion to deposit feeding, it is likely that the gill slits retained a respiratory function, but became unimportant for the evacuation of water taken into the gut with the food.

My first modification—that the bilaterally symmetrical stage in phylogeny became a deposit feeder with well developed tentacles beneath and atrophied tentacles above—when considered with the predominance of the left side during ontogeny, requires a second major modification in the traditional scenario of echinoderm evolution. This modification is that the bilaterally symmetrical ancestor turned not its right, but its *left* side toward the substratum during the transition to the asymmetrical stage (Fig. 2.3A–D).

The modified evolutionary scenario, in comparison to the references in Table 2.1, is in better agreement with echinoderm ontogeny. In extant echinoderms, the right side of the larva never becomes turned toward the substratum at the end

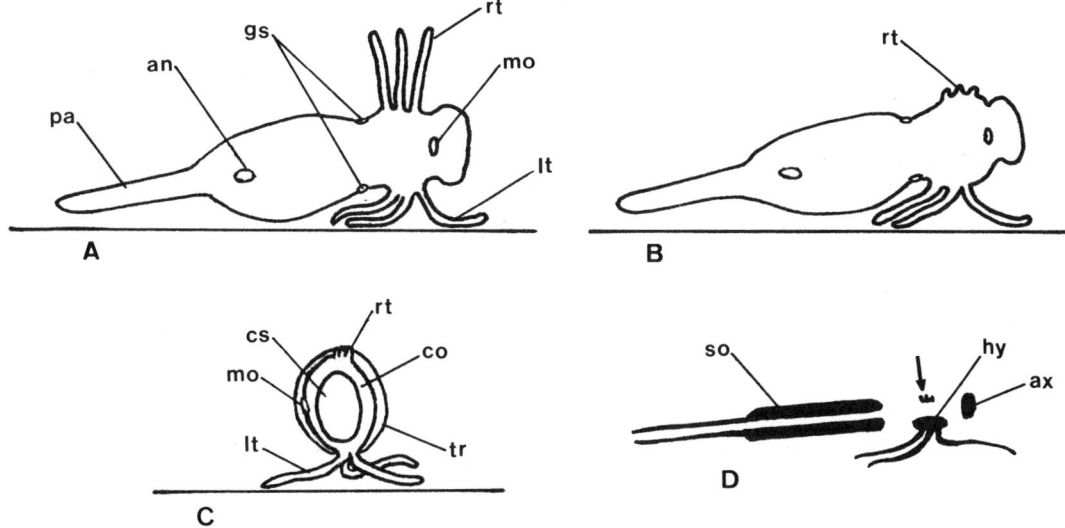

Fig. 2.3 (A) Bilaterally symmetrical stage after development of posterior appendage dorsal to anus and after transition to deposit feeder lying on its left side; viewed from lateral (ventral) side; terms of orientation explained in heading of Table 2.1; (B) early asymmetrical stage in lateral (ventral) view after atrophy of right tentacle group; (C) early asymmetrical stage viewed from anterior end; (D) coeloms of early asymmetrical stage shown in lateral (ventral) view; each somatocoel sends an extension into the posterior appendage, and the arrow indicates atrophied right hydrocoel; an, anus; ax, axocoel; co, collar; cs, cephalic shield; gs, gill slits; hy, left hydrocoel; lt, left tentacle; mo, mouth; rt, right tentacle; so, somatocoel; pa, posterior appendage; tr, trunk.

of metamorphosis. Instead, in two and possibly three of the five classes, the left side becomes apposed to the substratum.

EVOLUTION STARTING FROM A DEPOSIT FEEDING, ASYMMETRICAL STAGE LYING ON ITS LEFT SIDE: ECHINODERM EVOLUTION AND A SHORT DIGRESSION ON THE ORIGIN OF CHORDATES

My two modifications in the evolutionary scenario for the origin of echinoderms lead to no insurmountable difficulties when followed forward in geological time. A deposit feeding, asymmetrical creature lying on its left side makes a plausible ancestor for all the echinoderms and probably also for the chordates, as the present section of this paper will demonstrate. For what follows, the reader should understand the awkward, but necessary terms of orientation for the asymmetrical stage, which are explained in the heading of Table 2.1.

The asymmetrical stage almost certainly did not become permanently attached, since, as an epibenthic deposit feeder, it needed to move along the substratum to locate new food after depleting the supply within reach. The motive force for these movements was probably supplied in part by the tentacles, which thus added locomotion to their original feeding function, and in part by the body wall musculature. Specialized structures for attachment, if present, would have been for temporary adhesion—for example, muscular suckers or duo-gland organs (*sensu* Hermans 1983)—and would probably have been located on the cephalic shield and at the tip of the posterior appendage. The animal could have used such attachment structures in co-ordination with its body wall muscles to move after the fashion of a geometrid caterpillar (inchworm).

Early in the asymmetrical stage, the deposit feeding tentacles must have delivered food particles (1) directly to the mouth on the lateral (ventral) surface or (2) to the ciliated food groove of the oral lamella connecting the tentacular bases to the mouth. In either case, one of the earliest events of the asymmetrical stage was evidently a migration of the mouth from the lateral (ventral) surface to the under (left) surface to permit a more direct transfer of food from the tentacles to the mouth (Fig. 2.4A–B). Bury (1896) thought it likely that, at this stage in evolution, an epidermal invagination (the vestibule) developed to house the retracted tentacles; however, Bury admitted that the vestibule might not have appeared until much later and, for simplicity, I will not consider it further.

Having reached this point in evolution, one has at hand plausible common ancestors of the phylum Echinodermata and the phylum Chordata. I will combine these two phyla into a group of higher rank called Laeothetica (laeos = left; thetikos = suitable for lying down), and I will call these common ancestors the stem laeothetes (Fig. 2.4A–B). With the introduction of this terminology, I can conveniently present a phylogenetic tree of the deuterostomes (Fig. 2.5). The origin of the deuterostomes is currently highly controversial (e.g. Nielsen 1985; Bergstrom 1986) and will not be considered here. The tree in Fig. 2.5 begins with the stem deuterostomes (corresponding to the bilaterally symmetrical stage), which split into two sister groups: the hemichordates and the laeothetes. I will not discuss hemichordate evolution here, except to point out that the peduncle, which is a posterior appendage located ventral to the anus, appears to be an important apomorphy of the group.

A detailed discussion of chordate evolution (reviewed by Jollie 1973, 1982; Bone 1979, 1981) is also beyond the scope of the present study. However, I will digress briefly to show that a satisfactory transition can be made between stem laeothetes and stem chordates as conceived by MacBride (1910) and by Selys Longchamps (1936). The stem chordate of Selys Longchamps was somewhat like a tadpole larva of a tunicate lying on its left side, and with the cephalenteron (as seen from the anterior end) twisted 90° counter-clockwise relative to the tail. The stem chordate of MacBride (1910) was somewhat like an amphioxus larva lying on its left side, and with the mouth twisted to the under (left) surface 'to improve its opportunities of feeding ...' (p. 310). To both authors, the stem chordate was a pharyngeal suspension feeder and this new feeding mode was related to the reduction or loss of the tentacles. The stem chordates evidently lacked atrial cavities, and neither MacBride (1910) nor Selys Longchamps (1936) was explicit as to whether notochords or dorsal nerve cords were present. On the basis of nucleotide sequences from 18s rRNA, it is likely that the stem chordates split into (1) urochordates and (2) cephalochordates plus vertebrates (Ghiselin et al. 1986). The stem chordates, which can plausibly be derived from a stem laeothete, cannot have evolved in any reasonable way from a dipleurula or pterobranch that had adopted the habit of lying on the right side (the dexiothete condition of Jefferies 1979, 1981). I will close this short digression on the origin of the chordates by calling attention to an important insight of Gislén (1930); namely, that the marked asymmetry of the ancestral chordates has tended to evolve toward bilateral symmetry during the subsequent history of the phylum.

Stem echinoderms arose from stem laeothetes by the evolution of the calcitic stereom skeleton and by the loss of the gill slits (Fig. 2.6A). It is possible that the gill slits disappeared because their role in respiration had diminished: perhaps life with a skeleton was less active, requiring less oxygen, or perhaps the tentacles had enlarged and increased the respiratory surface of the animal enough to make gill respiration superfluous. The mouth still opened on the under (left) surface close to the left hydrocoel (hereafter referred to simply as the hydrocoel), which as yet showed no signs of

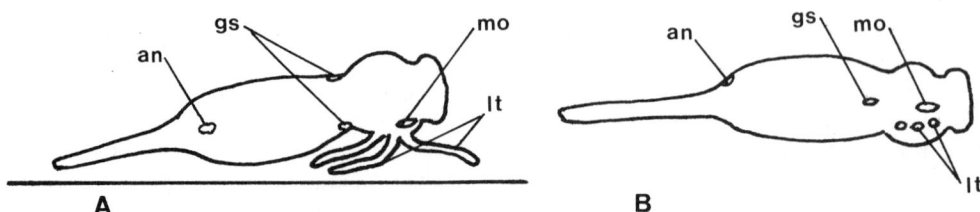

Fig. 2.4 Asymmetrical stage (stem laeothete). (A) lateral (ventral) view after migration of mouth to under (left) surface; (B) left side view (as if one were looking up from the substratum); an, anus; gs, gill slits; lt, left tentacles (only tentacular bases are shown in B); mo, mouth.

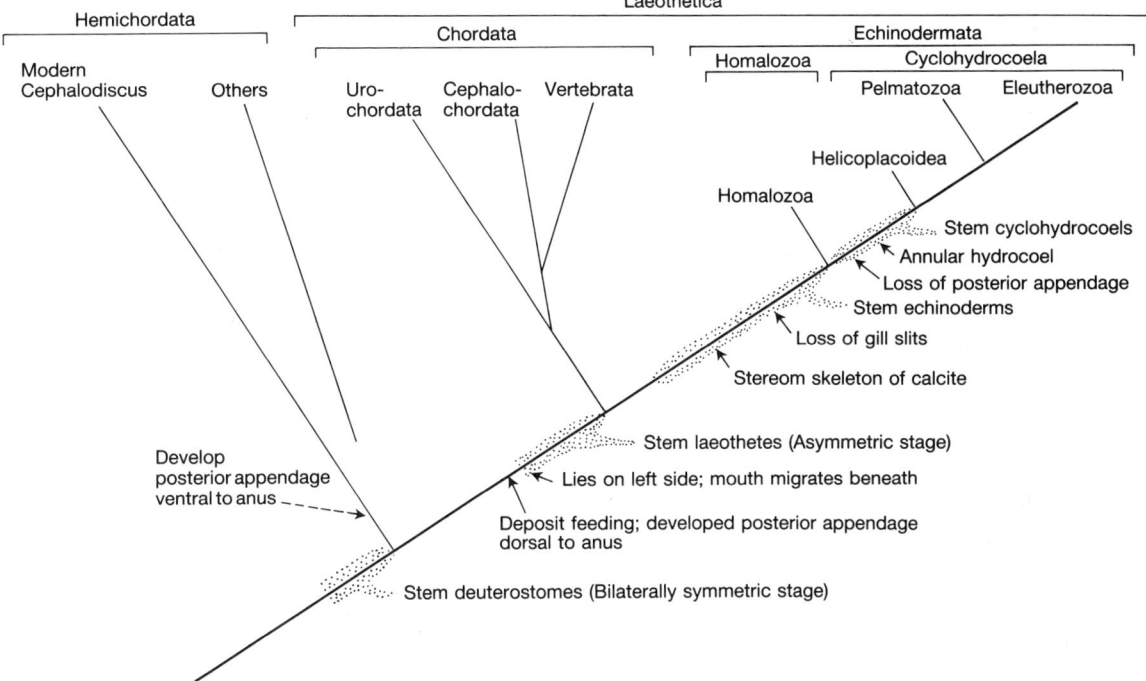

Fig. 2.5 Reconstructed phylogenetic tree of the deuterostomes.

becoming radially symmetrical. The arrangement of coeloms in stem echinoderms was unchanged from that of the stem laeothetes.

At this juncture, stem echinoderms evolved on the one hand toward homalozoans (which I consider to be echinoderms) and on the other hand toward the radially symmetrical echinoderms. In the homalozoan line (Fig. 2.6B–D), the important initial event was the migration of the mouth and hydrocoel along the under (left) surface to the anterior pole of the body with the concomitant shrinkage or disappearance of the oral shield. The posterior appendage was retained as the stele of the solutan, cinctan, and stylophoran carpoids, but it was lost in ctenocystoids. In solutan carpoids, the non-radial hydrocoel and its tentacles probably extended linearly along the length of the brachiole. In contrast, the hydrocoel of other homalozoans tended to atrophy and often disappeared entirely. In solutan carpoids, the anus, when demonstrable, opened laterally (ventrally), whereas in cinctan carpoids and many other homalozoans, the anus appears to have migrated anteriorly to open near the mouth, although other arrangements for the gut have also been proposed (reviewed by Philip 1979). The location of the coeloms would probably have been as diagrammed in Fig. 2.6D. I will not assign functions to all the holes in the homalozoan theca, but will point out that enough functions have been proposed for each hole (see the menu in Philip 1979) to permit considerable freedom in reconstructing the soft parts of a workable animal. Homalozoan locomotion was effected not by the tentacles, but by the muscular stele. The diverse ideas on the modes of feeding and locomotion that have been proposed for homalozoans have been reviewed by Philip (1979).

The stem echinoderms, in addition to giving rise to the homalozoans, evolved into the radially symmetrical echinoderms, which comprise the major branch of the phylum. These echinoderms (which, for convenience, I will call the stem cyclohydrocoels) were characterized by (1) the disappearance of the posterior appendage, (2) a ring-shaped, circum-oesophageal hydrocoel from which the ambulacra radiated, and (3) biserial flooring plates forming an integral part of the body wall. Whereas, earlier in evolution, the stem laeothetes and stem echinoderms were deposit

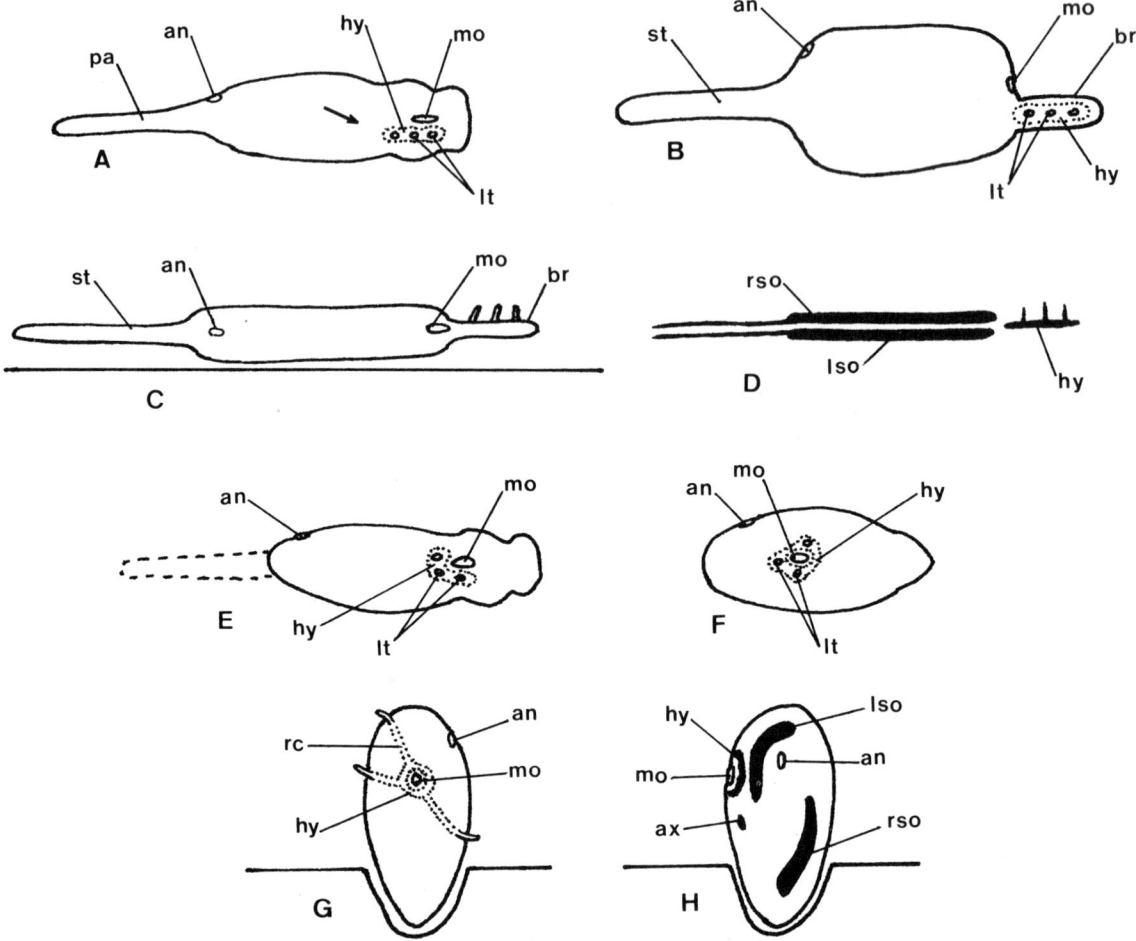

Fig. 2.6 (A) Stem echinoderm viewed from under (left) surface; arrow indicates loss of gill slits; (B) solutan carpoid from under (left) surface; (C) solutan carpoid from lateral (ventral) side; (D) coeloms of solutan carpoid from lateral (ventral) side; (E) stem echinoderm during transition to stem cyclohydrocoel as seen from under (left) surface; disappearance of posterior appendage indicated by dashed lines; posterior migration of mouth pushes hydrocoel into crescent that later becomes annular; anus migrates from posterodorsal to posteroventral position; (F), as in 6E, but with transition to stem cyclohydrocoel completed; (G) stem cyclohydrocoel from left side; (H) coeloms of stem cyclohydrocoel from ventral side; an, anus; ax, axocoel; br, brachiole; hy, left hydrocoel; lso, left somatocoel; lt, left tentacles (only tentacular bases show in A, B, E, and F); mo, mouth; pa, posterior appendage; rc, radial canal; rso, right somatocoel; st, stele.

feeders, it is probable that the stem cyclohydrocoels had reverted to suspension feeding. This reversion could well have followed a transitional period during which a given animal could alternate between both modes of feeding, as in some extant sand dollars.

In becoming suspension feeders, the stem cyclohydrocoels probably assumed a relatively immobile way of life, although current paleontological evidence strongly indicates that they were not permanently attached to the substratum. At most, these animals may have had organs of temporary attachment. The stem cyclohydrocoels appear to have lived erected on their anterior end, which was inserted into a depression in the substratum, rather the way an egg sits in an egg-cup. The posterior appendage, being no longer important for locomotion and being an unbalancing influence, would have disappeared (Fig. 2.6E, F). Anteriorly, the cephalic shield was much reduced, but may have continued to serve as an organ of temporary attachment as in pterobranchs (Lester 1985, 1986). Temporary adhesion may have played a part in the excavation of a new depression in the substratum after an animal had been upset by physical or biological disturbances.

During the transition from deposit feeding to suspension feeding, the mouth and the hydrocoel with its tentacles (which correspond to primary azygous tube feet) would have migrated posteriorly along the left side of the body to a position roughly two-thirds of the way between the anterior and posterior poles of the animal. During this migration, the hydrocoel came to encircle the oesophagus, evidently because the latter pushed against the former (Fig. 2.6E, F). The migration of the hydrocoel and the anterior part of the gut was part of an overall visceral torsion (in a clockwise direction when the animal is viewed from its ventral surface). Figure 2.6H shows the effects of this visceral torsion, which has carried the coeloms of the left side posteriorly and those of the right side anteriorly. After visceral torsion, no extension of the axocoel remained in the anterior region of the body (this conclusion is based on recent work on stalked crinoids, reviewed by Grimmer et al. 1985).

Three of the water vascular connectives leading to the tentacles would have lengthened into radial canals (Fig. 2.6G). Each radial canal was presumably overlain by uncalcified, ciliated epidermis acting as a food groove and was underlain by biserial flooring plates that developed by calcification of the mesentery, separating the radial canals from the left somatocoel. It is convenient to refer to each complex of radial canal, food groove, and flooring plates as an ambulacrum. The food collecting efficiency of each ambulacrum was probably increased by the sprouting of new tentacles as lateral branches from each radial canal (although, for simplicity, no lateral tentacles have been included in the figures of the present paper).

During the evolution of the stem cyclohydrocoels, the first major dichotomy separated the helicoplacoids from all the other groups, which are collectively termed the pentaradiate echinoderms by Paul and Smith (1984) or the crown echinoderms by Smith (1984). Helicoplacoids are best defined by the derived character state of spiral plating. (Whether the anus is truly absent, and thus a second derived character, is open to question.) In other respects, helicoplacoids share the primitive character states of the stem cyclohydrocoels, including three-part radial symmetry. The anterior end may have been provided with an organ for temporary adhesion that played a part in excavating a depression in the substratum. The mouth was located laterally (probably on the left side, although this orientation is much obscured by the spiral plating). The three ambulacra, which were used for suspension feeding in the opinion of Paul and Smith (1984), radiated from the mouth and followed the spiral contours of the skeleton. It is not known to what extent the superficial spirality was imposed on the internal soft parts; if these parts were relatively unaffected, the coeloms would have been arranged approximately as in Fig. 2.6H.

The pattern of evolution of the pentaradiate echinoderms is shown in the useful cladograms of Paul and Smith (1984, p. 421) and of Smith (1984, p. 456). The pentaradiate echinoderms are defined by the derived character states of five-part radial symmetry, and the differentiation of the skeleton into oral and aboral surfaces. In the Lower Cambrian, all pentaradiate echinoderms were suspension feeders, and their visceral torsion had gone farther than in helicoplacoids; as a result, the coeloms of the left side and the gut openings had rotated all the way to the posterior end of the animal, while the right somatocoel came to occupy much of the anterior end. The posterior end (also called the oral side) faced away from the substratum, and the anterior end (also called the aboral side) faced toward the substratum.

The pentaradiate echinoderms split into two distinct groups (the attached pelmatozoans and the unattached eleutherozoans) between which *Camptostroma* occupied an intermediate position. There is no evidence that *Camptostroma* and the eleutherozoans were permanently attached. At most, they had an organ for temporary attachment at the anterior (aboral) pole, which may have played a role in righting behaviour if, by mischance, the animal were overturned. The early eleutherozoans were suspension feeders which ultimately gave rise to echinoderms that included deposit feeders, macrophagous feeders, and suspension feeders.

In the pelmatozoans, attachment at the anterior (aboral) end became permanent. It seems likely that the organ of permanent attachment did not arise *de novo*, but originated from an organ of temporary attachment that could be traced back in time to the cephalic shield of the pterobranch ancestor. Permanent attachment tended to be accompanied by a narrowing and lengthening of the anterior region of the body into a stalk. The early pelmatozoans were suspension feeders, and this remained the chief mode of feeding throughout their subsequent evolution, although some reversions to deposit feeding did occur.

For my purposes, it is not necessary to follow echinoderm evolution beyond the Lower Cambrian radiation. There is nothing in more recent echinoderm evolution to affect my argument that a deposit feeding, asymmetrical creature lying on its left side makes a plausible ancestor for echinoderms and probably also for chordates. Moreover, in my evolutionary scenario, unlike most of those in Table 2.1, radial symmetry first appeared in echinoderms which had evolved from a series of ancestors that were never permanently attached to the substratum; this is consistent with current paleontological ideas about the origin and early evolution of echinoderms.

ACKNOWLEDGEMENTS

I am indebted to C. P. Galt, L. Z. Holland, and J. M. Lawrence for their constructive criticisms.

REFERENCES

Bather, F. A. 1900. The Echinoderma. In *A Treatise on Invertebrate Zoology*, Part III (ed. E. R. Lankester), pp. 1–344. Adam and Charles Black, London.

Bergstrom, J. 1986. Metazoan evolution—a new model. *Zoologica Skripta* **15**, 189–200.

Bone, Q. 1979. *The origin of chordates*, Carolina Biology Readers, 2nd edn. Burlington, North Carolina.

Bone, Q. 1981. The neotenic origin of chordates. *Atti dei Convegni Lincei* **49**, 465–86.

Bury, H. 1896. The metamorphosis of echinoderms. *Quarterly Journal of Microscopical Science* **38**, 45–135, pl. III–IX.

Bütschli, O. 1892. Versuch der Ableitung des Echinoderms aus einer bilateralen Urform. *Zeitschrift für Wissenschaftliche Zoologie* **53**, (Suppl.), 136–60, pl. IX.

Clark, R. B. 1964. *Dynamics in metazoan evolution. The origin of the coelom and segments*. Clarendon Press, Oxford.

De Ridder, C. and Lawrence, J. M. 1982. Food and feeding mechanisms: Echinoidea. In *Echinoderm Nutrition* (ed. M. Jangoux and J. M. Lawrence), pp. 57–115. Balkema, Rotterdam.

Ghiselin, M. T., Field, K. G., Olsen, G. J., Lane, D. J., Raff, R. A., Raff, E. C., and Pace, N. R. 1986. A phylogenetic tree of the chordate subphyla based on 18s ribosomal RNA sequences. *American Zoologist* **26**, 484 (abstract).

Gislén, T. 1930. Affinities between Echinodermata, Enteropneusta, and Chordonia. *Zoologiska Bidrag från Uppsala* **12**, 199–304.

Grimmer, J. C., Holland, N. D., and Hayami, I. 1985. Fine structure of the stalk of an isocrinid sea lily (*Metacrinus rotundus*). *Zoomorphology* **105**, 39–50.

Grobben, K. 1923. Theoretische Erörterungen betreffend die phylogenetische Ableitung der Echinodermen. *Sitzungsberichte der Akademie der Wissenschaften in Wien* **132**, 263–90.

Gutmann, W. F. 1981. Relationships between invertebrate phyla based on functional-mechanical analysis of the hydrostatic skeleton. *American Zoologist* **21**, 63–81.

Heider, K. 1912. Uber Organverlagerungen bei der Echinodermen-Metamorphose. *Verhandlungen Deutschen Zoologischen Gesellschaft* **22**, 239–51.

Hermans, C. O. 1983. The duo-gland adhesive system. *Oceanography and Marine Biology Annual Reviews* **21**, 283–339.

Hyman, L. H. 1955. *The invertebrates, Vol. IV, Echinodermata*. McGraw-Hill, New York.

Jägersten, G. 1972. *Evolution of the metazoan life cycle*. Academic Press, London.

Jangoux, M. 1982. Food and feeding mechanisms: Asteroidea. In *Echinoderm nutrition* (ed. M. Jangoux and J. M. Lawrence), pp. 117–59. Balkema, Rotterdam.

Jefferies, R. P. S. 1979. The origin of chordates—a methodological essay. In *The origin of major invertebrate groups* (ed. M. R. House), pp. 443–77. Academic Press, London.

Jefferies, R. P. S. 1981. Fossil evidence on the origin of the chordates and echinoderms. *Atti dei Convegni Lincei*, **49**, 487–561.

Jollie, M. 1973. The origin of the chordates. *Acta zoologica, Stockholm* **54**, 81–100.

Jollie, M. 1982. What are the 'Calcichordata'? and the larger question of the origin of the chordates. *Zoological Journal of the Linnean Society* **75**, 167–88.

Jørgensen, C. B. 1966. *Biology of suspension feeding*. Pergamon Press, Oxford.

Lang, A. 1894. *Lehrbuch der vergleichenden Anatomie der wirbellosen Thiere*. Gustav Fischer, Jena.

Lester, S. M. 1985. *Cephalodiscus* sp. (Hemichordata: Pterobranchia): observations on the functional morphology, behavior and occurrence in shallow water around Bermuda. *Marine Biology* **85**, 263–8.

Lester, S. M. 1986. The reproductive biology and development of *Rhabdopleura normani* (Hemichor-

# PART II
Class relationships

3

Molecular analysis of distant phylogenetic relationships in echinoderms

RUDOLF A. RAFF, KATHARINE G. FIELD, MICHAEL T. GHISELIN*,
DAVID J. LANE, GARY J. OLSEN, NORMAN R. PACE, ANNETTE L. PARKS,
BRIAN A. PARR, and ELIZABETH C. RAFF

*Institute for Molecular and Cellular Biology, and
Department of Biology, Indiana University, Bloomington,
Indiana 47405, USA*

*Department of Invertebrate Zoology,
California Academy of Sciences, Golden Gate Park,
San Francisco, California 94118, USA

RIBOSOMAL RNA AND ECHINODERM PHYLOGENY

Phylogenetic relationships among distantly related animal groups have long puzzled systematists. The echinoderms have been no exception and present two major unsolved phylogenetic problems: (1) the relationship of the echinoderms to other phyla (i.e. are deuterostomes a natural group and echinoderms specifically related to chordates?); and (2) the evolutionary relationships among living echinoderm classes.

Distant phylogenetic relationships have been difficult to resolve because adult body plans of distant groups are so distinct. Darwin, Fritz Müller, and Haeckel all recognized that embryos could provide information on evolutionary relationships not available from adult anatomy. It is not obvious that an adult barnacle is a crustacean; a larval barnacle can be nothing else. Haeckel's proclamation that ontogeny recapitulates phylogeny supplied the impetus for a major research programme in which detailed descriptions of the embryology of animal phyla provided the data for construction of phylogenetic trees. Haeckel's programme fell out of favour by the end of the 19th century, as embryologists began to explore mechanistic problems in development (Gould 1977; Raff and Kaufman 1983). Nevertheless, the Haekelian approach has continued to be regarded as a powerful and often the only tool available to link very distant groups. Molecular sequence data offer a powerful alternative basis for phylogenetic reconstructions, and an independent check on classic morphological and embryological criteria. Molecular analysis is particularly useful in groups like echinoderms, where application of classic criteria has yielded inconsistent results.

It is generally accepted that phylogenies should be based upon shared, derived characters, especially complex features that have a low probability of convergence. However, convergence does occur occasionally, and there are frequent parallelisms among distinct, but related evolutionary lineages (Gosliner and Ghiselin 1984). Further problems are caused by reversions and losses of features, and the difficulties of determining the polarity of such changes. There is little doubt that such events have occurred in the evolution of echinoderms, because different morphological features yield contradictory phylogenies. This is illustrated by Fig. 3.1, which presents some possible trees for the five major living echinoderm classes. Depending upon the character trait chosen, different echinoderm phylogenies are supported. For instance, if overall body plan were used, as shown in Fig. 3.1, tree 1 would result. Holothurians and echinoids have their tube feet arranged in rows on the body surface, whereas asteroids and ophiuroids have tube feet on arms which suggests independent

Echinoderm phylogeny and evolutionary biology (ed. C. R. C. Paul and A. B. Smith). Clarendon Press, Oxford, 1988.

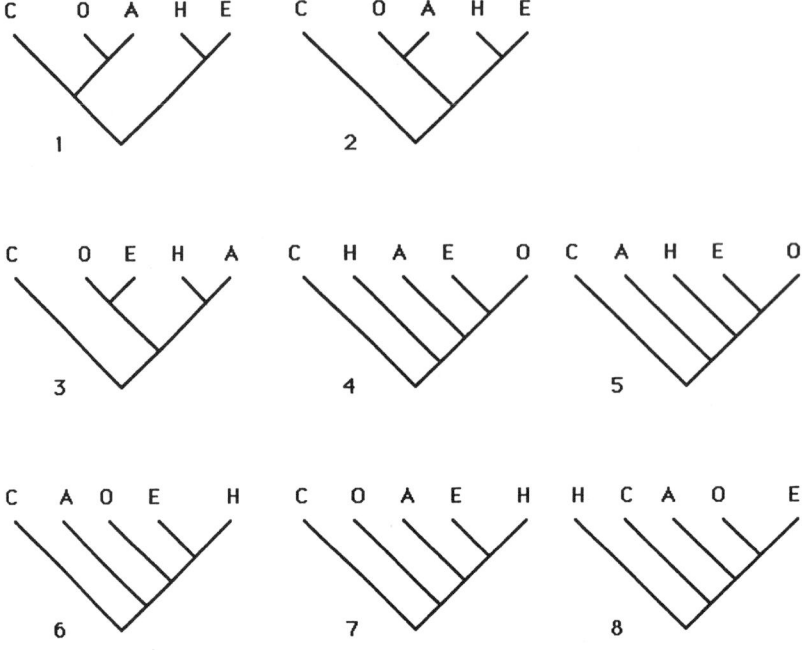

Fig. 3.1 Possible phylogenetic trees for living echinoderm classes. The classes are asteroids (A), crinoids (C), echinoids (E), holothurians (H), and ophiuroids (O). These trees are qualitative branching diagrams, not intended to show evolutionary distance or time.

derivation from the crinoids (Fell 1963). Tree 8, on the other hand, is consistent with the proposal that the single gonad of the holothurians is primitive, whereas the five gonads present in other classes are an advanced trait. A more generally held view is shown in tree 2, which separates the stalked echinoderms (Pelmatozoa) from the free moving echinoderms (Eleutherozoa). Tree 2 combines stelleroid forms into one group of eleutherozoans and globular forms into a second. Accepting the division of echinoderms into pelmatozoans and eleutherozoans, there are still 15 possible dichotomous trees for the four eleutherozoan classes. For example, tree 3 is consistent with the presence of delta 7 steroids, batyl alcohol, and saponins in asteroids and holothurians (Bolker 1967; Goad et al. 1972), whereas trees 3, 4, 5, and 8 are consistent with the possession of a pluteus larva by echinoids and ophiuroids.

There is a second problem (see Smiley, this volume), that if characters are to be useful for phylogenetic purposes they must be not only homologous, but they must also be shared by two or more classes. The antiquity of the separation of echinoderm classes compounds the problem of deciding if traits are indeed homologous, particularly when developmental features are used.

We are using molecular sequence data derived from the 18S rRNA of the small ribosomal subunit to estimate phylogenetic relationships among major animal phyla and classes (Field et al., 1988). The 18S rRNA molecule is ideally suited for studies of distantly related organisms because it is rich in information and the sequencing methodology permits the rapid accumulation of a very large database (see Lane et al. 1985; Olsen et al. 1986; Pace et al. 1985, 1986; Raff et al. 1987; Olsen 1987; Field et al. 1988).

Pace et al. (1985) have presented the major theoretical and practical reasons for the use of 18S rRNA sequence data for phylogenetic inference.

1. These molecules are conservative in overall structure and constitute a single gene family, so that problems of establishing homology among paralogous genes are avoided.
2. The conservative nature of rRNA structure extends to the nucleotide sequence level. Some regions of the molecule are highly conserved among distantly related species (see below).

3. Enough data are provided from 18S rRNA sequences for statistically significant comparisons. Since each of the approximately 1000 sequenced nucleotides constitutes a character state, the evolutionary information content of an rRNA sequence is very high. Sequences for 5S RNA are available for a large number of animal groups, but these RNAs are very short (about 120 nucleotides) and very conserved, resulting in limited phylogenetic resolving power (Olsen et al. 1986). See Ohama et al. (1983) for an illustration of the limits of 5S RNA sequence data in resolving echinoderm relationships.
4. The rRNA genes seem to be free of artifacts of lateral gene transfer between phylogenetically distant organisms (Stackebrandt and Woese 1981). This and the absence of paralogous genes mean that 18S rRNA sequences accurately reflect phylogenetic relationships among the *organisms* from which the rRNAs were prepared.
5. The rRNAs are present in large amount in all organisms and are easily isolated. Since rRNA can be sequenced directly, data can be obtained rapidly without recourse to cloning.
6. Direct rapid sequencing of 18S rRNA is possible because, as noted above, there are universally conserved regions in the molecule. Deoxyoligonucleotides have been synthesized that are complementary to the conserved regions. These serve as primers for sequencing by a modification of the Sanger et al. (1977) 'dideoxy' method, in which reverse transcriptase is used to copy directly 18S rRNA (Lane et al. 1985). Since 18S rRNA comprises about 25 per cent of the mass of cellular RNA, no additional purification is required and sequence reactions are done with total cellular RNA.

To construct phylogenetic trees from 18S rRNA sequence data, it is necessary to compare homologous nucleotides (Olsen 1987). This is done by a process of sequence alignment to find the closest correspondence of sequences to be compared. The strongly conserved nature of the 18S molecule aids considerably. Where necessary, alignment gaps may be inserted. Furthermore, because the secondary structure of 18S rRNA is well known, it is possible to fit less well conserved primary sequences into their appropriate secondary structural region and thus align them (Gutell et al. 1985).

The computation of our phylogenetic trees is based on estimates of the total evolutionary distance between pairs of contemporary sequences (Fitch and Margoliash 1967). An 'evolutionary distance tree' provides the branching order that best matches the estimated evolutionary distances in all pairwise sequence comparisons (Olsen 1987). Since, by definition, all evolutionary distances are additive, if all pairwise evolutionary distances between contemporary sequences are precisely known, it is possible to construct a unique tree. Evolutionary distances are defined as the average number of fixed point mutations per sequence position. However, the actual number of nucleotide substitutions will be greater than the number observed because multiple substitutions occur at some positions. To correct for these events, we have used the equation of Jukes and Cantor (1969) to estimate evolutionary distance from observed data. Among additive trees to the evolutionary distance estimates, we define the 'best tree' as that tree with the minimum weighted mean square error (Olsen 1987).

Our confidence in extending this methodology to phylogenetic relationships among echinoderm classes is based upon phylogenetic determinations we have made with other groups (Pace et al. 1986; Field et al. 1988). The evolutionary distance trees constructed on the basis of 18S rRNA sequence data result in coherent hierarchical relationships that often correspond to those established on other grounds. For instance, kingdom, phylum, class, and order level relationships are clearly indicated by 18S rRNA trees. Thus, eukaryotes stand clearly distinct from archaebacteria and eubacteria. The major divisions within each of these kingdoms are coherent. Among the animals, classes group within phyla, and orders within their classes.

A second indication of the general validity of 18S rRNA trees for phylogenetic studies of distant animal groups is implicit in our analysis of chordate 18S rRNAs. Vertebrates are unique among major groups in that their fossil record is sufficiently complete that an accurate reconstruction of the phylogenetic relationships of most classes has been possible (Gregory 1951). Furthermore, the times of divergence of vertebrate classes in the Palaeozoic are reasonably well known (Thomson 1977). The origins of at least some of the chordate classes discussed above were sequential and widely separated in time. Thus, jaw-bearing fishes diverged from the ancestral jawless 'fishes' about 500×10^6 years ago, bony fishes and tetrapods split about 350×10^6 years ago, amphibians from reptiles about 320×10^6 years ago, and the mammalian lineage from other tetrapods about 300×10^6 years ago. The relationships of non-vertebrate chordates are not known from fossil evidence, but their embryology and anatomical homologies provide strong evidence of phylogenetic positions. Thus, the major branching order of vertebrate phylogeny is available from morphological data and the fossil record. The tree constructed from available chordate 18S rRNA sequence data (an ascidian, amphioxus, an amphibian, and a mammal) unambiguously corresponds to the known phylogenetic relationships of these animals (Field *et al.* 1988). This is especially significant when evaluating the reliability of 18S rRNA trees for major groups for which the fossil record does not document a phylogenetic history. This is certainly the case for the origin of echinoderm classes, which occurred in a radiation event early in the Palaeozoic that is only beginning to be understood (Paul 1977; Paul and Smith 1984).

AN 18S rRNA TREE FOR ECHINODERMS

We have determined the partial 18S rRNA sequences of members of the five available living classes of echinoderms (crinoids, ophiuroids, asteroids, holothurians, and echinoids). We unfortunately have no data for the newly discovered living echinoderm class Concentricycloidea (Baker *et al.* 1986). The tree is shown in Fig. 3.2. In this kind of tree, evolutionary distance is proportional only to the horizontal component of the branches. The tree is rooted to a distant outgroup. Several features of the tree are significant. First, the 18S rRNA-derived phylogeny confirms the monophyly of echinoderms. Secondly, the tree confirms the separation of crinoids from the eleutherozoans. Thirdly, the tree is hierarchical. The classes all fall into the echinoderm clade, and the several representatives of distinct orders of sea urchins all fall onto the echinoid branch (Fig. 3.3).

Of the eleutherozoan groups, the holothurian sequences (represented by two orders in the tree) exhibit a higher rate of nucleotide substitution than sequences from the other classes. Despite that, a unique tree including the holothurian sequences can be inferred (see discussion of problems of inferring trees from sequence data, below).

CONTRASTS BETWEEN ECHINODERM PHYLOGENIES

Three of the trees in Fig. 3.1 have been proposed in recent studies of phylogenetic relationships among living echinoderm classes. Tree 7 corresponds to our tree based upon 18S rRNA sequencing. Tree 6 corresponds to Smith's (1984) most parsimonious tree based on cladistic analysis of morphological and embryological characters. Tree 8 is consistent with Smiley's (this volume) cladistic analysis of morphological and developmental characters. The three studies have reached very different conclusions. Given the long-standing disagreements about echinoderm genealogy, this is perhaps not too surprising. These disagreements derive in large measure from the fact that *no* tree can be

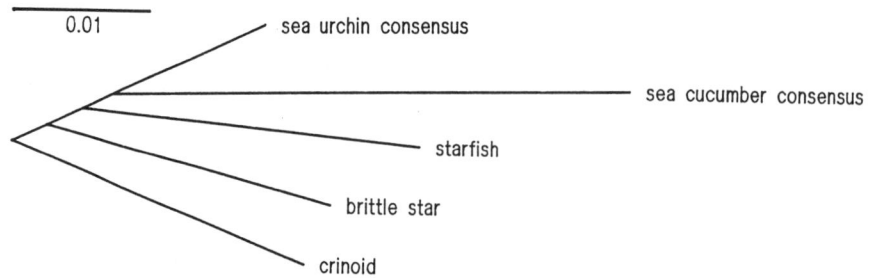

Fig. 3.2 Phylogenetic tree for living echinoderm classes derived from 18S rRNA sequence data. Organisms used were crinoid (*Lamprometra palmata*), asteroid (the starfish, *Asterias forbesi*), ophiuroid (the brittlestar, *Ophiocoma wendti*), holothurians (species of two sea cucumber orders, *Thyone briareus* and *Leptosynapta inhaerens*), and echinoids (species of four sea urchin orders, as in Fig. 3.3). The scale bar corresponds to an average of 0.01 fixed point mutations per sequence position, or about 10 mutational events per 1000 positions in the sequence.

Molecular analysis of phyletic relationships 33

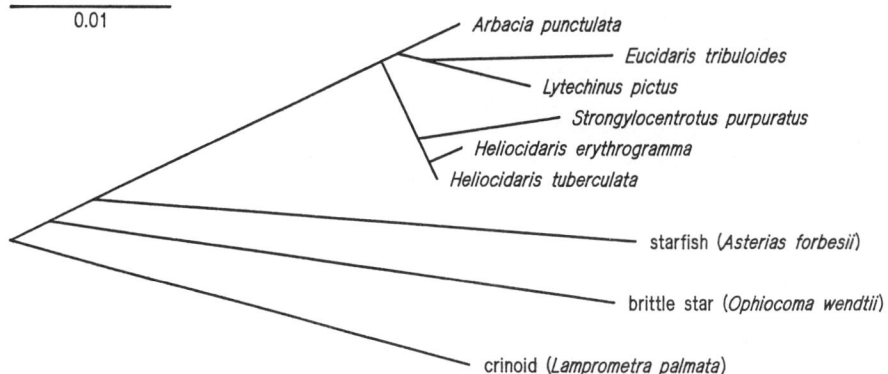

Fig. 3.3 Ribosomal RNA phylogenetic tree for echinoderms showing inferred relationships for echinoid sequences. It is important to note that the inferred branching of *Eucidaris* (Cidaroida) from *Lytechinus* (Temnopleuroida) is not statistically significant because the branch segment represents too few base positions. Thus, the 18S rRNA data cannot resolve branching orders in the Triassic radiation of echinoids (see Fig. 3.4). *Strongylocentrotus* and *Heliocidaris* are in the same order (Echinoida). Scale bar as in Fig. 3.2.

reconciled with all of the available evidence; extensive parallel or convergent evolution must have occurred in the history of the phylum. The lack of consensus about echinoderm phylogeny suggests that the relative strengths and weaknesses of the different methodologies should be evaluated.

We are particularly disturbed that two cladistic analyses of comparative morphology and development should yield such discordant phylogenies. Although the molecular approach has definite limitations (see below), we feel that traditional methods may be subject to even greater biases. Our first three concerns involve the difficulty of proving true homology and the problems raised by recurrent parallel/convergent evolution.

Problems of determining homology in embryonic traits

Features of embryonic or larval development are not static characters, but are elements of complex dynamic processes. Unless these processes are well understood, comparisons of supposedly homologous structures are apt to be misleading (Raff 1987). For example, Smith (1984) uses presence of a larval mouth. As Smiley (this volume) correctly points out, this is not valid because the group lacking a mouth, the crinoids, have direct developing larvae which are not homologous to the feeding planktotrophic larvae typical of other classes. Smiley (this volume) notes that the failure to distinguish between direct and indirect modes of development has created other problems for Smith (1984) and Fell (1963).

The pluteus larva illustrates how parallel or convergent evolution complicates the determination of homology. Both Smith (1984) and Smiley (this volume) use the presence/absence of a calcareous larval skeleton as a character. This requires that the echinopluteus and ophiopluteus larvae of sea urchins and brittle stars are homologous and thus comparable. However, there is evidence that these larvae are a good example of parallel evolution. The arms of plutei are probably homologous to the non-calcified arms of the auricularia larva, which probably represents the primitive condition in echinoderms (Ohshima 1911, discussed by Kume and Dan 1968; Raff *et al.* 1987). The novel feature of deposition of a calcareous skeleton is presaged by the ability of some auricularia larvae to deposit small calcitic spicules in their arms (Kume and Dan 1968).

The independent origin of the ophiopluteus and echinopluteus is indicated by the significant differences in their course of development. The echinopluteus skeletal arms are generated by spicule secretion initiated at several sites on each side of the archenteron, whereas the ophiopluteus has only one site of initiation on each side, and grows all of its skeletal arms by branching (Mortensen 1921; Hyman 1955). Moreover, the patterns of arm growth differ. In ophioplutei, the postero-lateral arms are the first to form, and are the dominant arms. By contrast, in the echinopluteus the postoral and anterolateral arms form first. The posterolateral arms form late, if at all (Balfour 1885; Mortensen 1921; Hyman 1955; Kume and Dan 1968). The strong similarity in overall form may be dictated by underlying developmental constraints (Raff 1987), as well as by the mechanical constraints imposed by the pluteus feeding mode (Strathmann 1978). It would be interesting to know the genetic mechanisms underlying this instance of parallel evolution, but whatever they may be, the presence of a calcareous larval skeleton is not a useful character for determination of phylogenetic relationships among echinoderm classes.

Inadequate range of variation sampled

Character states within a class are usually determined from a limited sample of species. This raises the possibility that a broader survey of the class would uncover a wider range of variation and upset previous notions of character distribution. For example, the larval vestibule is absent in asteroids, but was thought to be present in all other echinoderms. However, it has become clear that embryos of the primitive sea urchin order Cidaroida also lack a vestibule (R. B. Emlet, personal communication 1986). Thus, absence of a vestibule cannot be used as a character that distinguishes classes. Instead multiple losses (or acquisitions) must have occurred.

Employment of the absence of a feature as a character

Although the presence or absence of a feature is often used as a distinguishing characteristic, independent loss of features in different lineages is a distinct possibility, and contributes to the problems of determining homology and polarity.

Uncertainties in establishing polarity

The assignment of polarities to character states is a key step in the construction of cladograms. In some cases, the fossil record or other grounds provide solid support for the polarities assigned by Smith (1984) and Smiley (this volume), but other decisions appear arbitrary. For example, Smiley (this volume) uses an outgroup (hemichordates) or parsimony to determine polarities. However, one cannot be sure that hemichordates are indeed an appropriate sister group, and the parsimony criteria may yield arbitrary assignments. Furthermore, parsimony may not always be a reliable guide in groups displaying many instances of parallelism or convergence. Both authors discuss this issue, but the problems may be difficult to overcome in practice.

The polarity of some traits is not only difficult to establish, but also may be uninformative once a choice is made. Consider Smiley's (this volume) traits 30 and 32.

	Primitive state	*Derived state*
30.	Batyl alcohol present H, C, A	Batyl alcohol not present O, E
32.	Tube feet uncalcified C, A, O	Tube feet calcified H, E

If the polarity of character 30 is correct, the character is uninformative because it represents the absence of a feature. There is no way to tell if batyl alcohol was lost by the common ancestor of ophiuroids and echinoids (synapomorphy) or by the two groups independently (convergence). If the polarity were reversed the character would be more informative, although it still would not be possible to tell if the absence of this character in two classes is a shared primitive trait or a convergent loss by one or the other group. Character 32 presents similar problems: if the proposed polarity is correct, one still has to judge whether calcified tube feet evolved twice. Since all echinoderms possess skeleton-secreting mesenchyme cells (Hyman 1955), the possibility cannot be dismissed out of hand.

Inconsistencies in the choice of characters

Smith (1984) and Smiley (this volume) employ a total of 19 independent embryological characters, but only 14 and 12 are used, respectively; only seven are used by both studies (Table 3.1). The sets of adult characters likewise differ between the two studies. This does not mean that the studies have employed invalid traits, but it raises questions about the criteria for inclusion of specific characters. Why are some characters used and other, equally informative ones, neglected? For example, the development of pedicellaria is not included in either study, although it is a complex character and present in only two classes (asteroids and echinoids). The selective use of characters is a problem because it may bias the final outcome, especially when the number of characters is small.

The problems noted above seem to plague studies of comparative morphology and embryology. There are some other difficulties in the

TABLE 3.1 Embryological characters used by Smith and by Smiley for cladistic analyses of relationships of echinoderm classes

No.	Trait*	Smith† (43 total traits)	Smiley (33 total traits)
1	Right larval coeloms suppressed	+ (2, 16)	+ (9)
2	Larvae with processes	+ (4)	
3	Larval vestibule	+ (11)	
4	Larva attaches by pre-oral lobe	+ (18)	+ (13)
5	Site of gonad origin	+ (20)	+ (21)
6	Site of endomesoderm formation	+ (22)	
7	Direction radial water vessel grows	+ (23)	
8	Adult axis symmetry vs larval	+ (31)	+ (4, 5)
9	Calcareous larval skeleton	+ (30)	+ (14)
10	Larval mouth present	+ (26)	
11	Origin larval anus	+ (40)	
12	Open/closed vestibule	+ (43)	
13	Larva bilaterally symmetrical		+ (3)
14	Mode of ambulacral development		+ (7)
15	Persistent adult rudiment in larva		+ (12)
16	Larval hydropore persists or not		+ (23)
17	Site hyponeural coelom formation	+ (15)	+ (29)
18	Time peripheral and perianal coeloms form	+ (32, 33)	+ (31)
19	Paired v. single coelomic pouches		+ (8)
Total number of embryological traits‡		14 (16)	12 (13)
Fraction of total traits§		0.33 (0.37)	0.36 (0.39)
Number embryological traits shared¶		7 (10)	

*From character tabulations of Smith (1984) and Smiley (this volume).
†A + indicates that the trait is used. Numbers in parentheses give character numbers listed by Smith and Smiley.
‡Numbers in parentheses indicate totals when duplicated features were tabulated as independent in the original papers.
§Embryological features as a fraction of all traits.
¶Traits listed in common by the two papers.

characters listed by Smith (1984) and Smiley (this volume). Some are unsuitable for cladistic analysis because they are common to all echinoderms (symplesiomorphies, e.g. bilateral larval symmetry), unique to a single group (autapomorphies, e.g. the absence of a left axocoel in holothurians), or not truly independent of other characters in the analyses (see Table 3.1).

If we combine embryological characters that are not independent, and eliminate features that are clearly non-homologous, plesiomorphic, or autapomorphic, the number of potentially reliable traits is reduced. We delete nine of the embryological characters used by Smith (1984) and seven of those listed by Smiley (this volume). This would leave eight of Smith's embryological characters, and ten of Smiley's. Interestingly, Smith's remaining embryological characters are most consistent with a close relationship between echinoids, ophiuroids, and holothurians, whereas Smiley's characters are most consistent with holothurians standing distinct from other echinoderms. Smiley (this volume), in fact, carefully discussed why his characters differ from Smith (1984). Most significantly, he introduced new information derived from his own recent analysis of holothurian metamorphosis. These data tend to separate holothurians and echinoids. However, just how sensitive the inference of a tree is to choice of which embryonic characters are chosen is shown by the still different trees arrived at by Strathmann (this volume).

The 'best' tree should be the most parsimonious (Kluge 1984). However, in a group with so many parallel and convergent evolutionary features, and a relatively limited number of characters available for analysis, is parsimony really a reliable guide? This problem is explicitly considered by Smith (1984) and Smiley (this volume), but it is very difficult to avoid in practice.

We do not deny the importance of morphological or embryological features in evolution. In fact, the evolution of these features poses some of the most exciting problems in biology (Gould 1977;

Raff and Kaufman 1983). If we are to understand such changes, it is vital to know the phylogenetic relationships of the group being studied. To avoid circularities, phylogenetic inferences should be based on characters other than those being studied. Furthermore, in cases where parallel and convergent evolution are rife, as in echinoderms, there is clearly a need for methods which transcend morphology. We hope that use of 18S rRNA sequencing may be one such method. The advantages of this approach were presented earlier in this chapter. The potential difficulties must now be evaluated.

Establishment of a phylogenetic tree by sequence data potentially suffers from four effects:

(1) The rate of phylogenetic splitting relative to the rate of nucleotide substitution in the clock molecule;
(2) imperfections in the actual sequence data;
(3) problems in identification of sequence homologies;
(4) differences in rate of sequence evolution between lineages.

The seriousness of these potential difficulties will vary depending upon the particular phylogeny, and the method used for analysis of data.

Effect of rapid phyletic splitting

Although molecular phylogenies transcend organismal morphology, the resolution of branching order in molecular trees is sensitive to the rate of splitting of lineages versus the rate of accumulation of base substitutions. Molecular trees are based upon the number of nucleotide sequences fixed since the two organisms being compared diverged from a common ancestor. To distinguish branch points (and hence branching orders) it is necessary to examine sequences which change rapidly enough to accumulate a significant number of nucleotide substitutions in the interval between the branch points. On the other hand, in order to observe these changes through comparisons of contemporary sequences, it is also necessary that the substitutions which distinguish the branch points not be obscured by subsequent changes at the same nucleotide positions. The balance between these effects sets the ultimate limit on our ability to resolve radiations in the distant past. Thus, if the radiation was rapid, the derived tree will not capture the topological relationship of the actual tree (Fig. 3.4). Such an effect has been

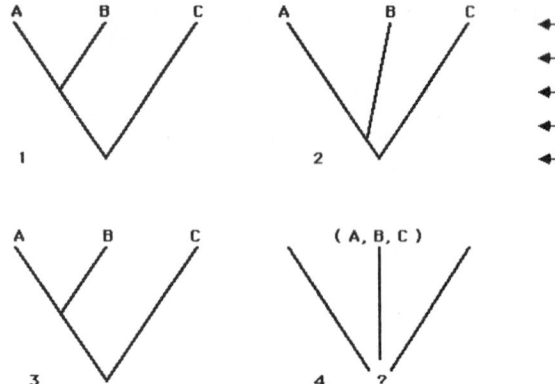

Fig. 3.4 Effects of rate of phyletic branching on inference of phylogeny from sequence data. Two model phylogenies are shown (trees 1 and 2). In phylogeny 1, the time between branching events is slow relative to fixations of base changes in 18S rRNA (arrows), and the phylogeny derived from sequence data (tree 3) accurately reflects organismal branching. In phylogeny 2, the time between branching events is rapid relative to base changes in 18S rRNA, and the inferred tree (tree 4) cannot resolve a branching order.

noted by Goodman *et al.* (1982) in their examination of the molecular phylogeny of mammalian orders, and reflects the apparently rapid radiation of mammals following the extinction of the dinosaurs. This is a potential difficulty with 18S rRNA trees, and can only be evaluated empirically for any tree.

Imperfections in sequence data

Molecular sequence data are not perfect. Data may be absent (1–2 per cent of the nucleotides in an 18S rRNA sequence cannot be unambiguously identified) or erroneous (less than 1 per cent of the nucleotide identifications suffer from anomalies in DNA fragment mobility, gel interpretation, or data recording). If these errors are not the result of a systematic (group-specific) bias, then the missing data have little effect on branching order or branch lengths, whereas sequence errors tend to increase the inferred evolutionary distances between groups.

Identification of homologous traits

For valid comparisons of 18S rRNA sequences, accurate identification of homologous nucleotides (i.e. sequence alignment) is required. In the absence of sequence length variations resulting from insertions and deletions, the correspondence of residues in sequences from different species should be clear. When inferred sequence length differences are positioned to maximize the apparent sequence similarities (in terms of some combination of primary and secondary structure), there is a systematic tendency to overestimate sequence similarity in the regions of sequence length variation. To maximize the reliability of sequence alignment, regions of 18S rRNA which display the greatest variability in structure and/or sequence length were excluded from the analysis. We estimate that a very small fraction of the data analysed is improperly aligned, and these errors should be random, rather than favouring any specific phylogeny.

Effect of differences in the rate of sequence evolution between lineages

In practice, none of the current methods of phylogenetic tree inference have been demonstrated to be free of systematic errors when confronted with a combination of lineage-to-lineage variations in rates of sequence change and substantial amounts of sequence divergence. Most methods tend to underestimate the number of fixed mutations that are responsible for an observed difference between contemporary sequences, and this underestimation increases with the amount of sequence divergence. We have chosen to use a distance method of tree inference, in part, because of the relative sensitivity of parsimony methods to differences in lineage-to-lineage clock rate (Felsenstein 1978).

In direct pairwise comparisons of echinoderm sequences, the distances between holothurians and any other class are greater than for any other pairwise comparisons. Superficially, this might be taken to indicate that holothurians separated from other echinoderms at the earliest bifurcation of the phylum, and would accord with the view of Smiley (this volume). However, when outgroup comparisons are made, the holothurian sequences are substantially less similar to any outgroup sequence than are any of the other echinoderm sequences. This observation indicates a greater average rate of sequence change (a 'faster clock') in the holothurians than in other echinoderm lineages. This phenomenon is true for sequences of two different orders of holothurians. An accurate analysis of echinoderm phylogeny must be able to cope with this large difference in observed rates of sequence change.

To work around this dilemma, we have broken the problem to be solved into two parts:

(1) examination of the relationships within the echinoderms by inferring a phylogenetic tree that includes the fast clock holothurian sequences, but which minimizes the total amount of sequence divergence by excluding the non-echinoderms;
(2) examining the relationships of the echinoderms to the rest of the Metazoa by inferring a phylogenetic tree which includes non-echinoderm eukaryotes, but which minimizes variations in clock speed by excluding the holothurian sequences.

The common elements of these two phylogenies (the relationship of the crinoid, brittle star, starfish, and sea urchins) were found to be congruent. Furthermore, because the connection to the other eukaryotes (the point representing the most recent common ancestor of these echinoderms) and the connection to the holothurians fell on different branches of this four group core tree, there is a unique composite tree which is compatible with both of these analyses.

The results of these analyses must be regarded as our best interpretation of the available rRNA data. There remains an intrinsic statistical uncertainty in the order of branching in the vicinity of some of the short tree branch segments. This effect is evident in the placement of the cidaroid *Eucidaris* shown in Fig. 3.3. This tree would, if the branchings among orders were taken at face value, contradict the more usual relationships among sea urchin orders. However, the branching is best regarded as ambiguous, and reflects the uncertainty of short branch lengths as well as the rapid splitting of sea urchin orders (Fig. 3.3). Thus, the correct reading of Fig. 3.3 is that the four orders represented (Cidaroida, *Eucidaris*; Arbacioida, *Arbacia*; Temnopleuroida, *Lytechinus*; Echinoida, *Strongylocentrotus*, and *Heliocidaris*) radiate from a single point. We cannot assign a branching order from these data. Also, because of the large amount of sequence change in the holothurian lineages, the statistical error in their placement is significant. In particular, placing holothurian divergence before the separation of sea urchins from starfish cannot be excluded (although our preferred branching order is supported by both of the two holothurian sequences). The placement of the root of the echinoderm tree, such that the crinoid lineage is the first to separate in echinoderm evolution, is probably the next least certain conclusion. However, it is supported by the relative independence of the observed branching order from the outside reference group chosen (be it a plant, fungus, or another metazoan phylum).

ECHINODERM DEVELOPMENTAL PATTERNS AND PHYLOGENY

We began this chapter by proposing that the use of embryological traits for phylogenetic analysis should be re-evaluated in the light of an independent methodology. It is clear that phylogenetic inferences drawn from larval traits of echinoderms are far from unequivocal. We have further investigated the relationship between development and phylogenetic position in echinoderms by examining direct developing echinoids.

Development via a planktotrophic larva is primitive in echinoderms as in many other phyla, but evolution of non-feeding direct developing larvae has occurred in many groups including all echinoderm classes. This shift provides a remarkable opportunity to study the processes that underlie radical evolutionary changes in development (Raff 1987). Some of the patterns of development observed in echinoids are presented in Fig. 3.5. Direct development has evolved independently in several lineages of echinoids. In some cases, very closely related species have strikingly different patterns of development (Raff 1987; Parks *et al.* 1988). In the case of the endemic Australian genus *Heliocidaris*, *H. erythrogramma* is a direct developer, developing from a large egg without a pluteus (Fig. 3.5), whereas *H. tuberculata* develops in typical fashion from a small egg via a typical feeding pluteus. The major differences between typical and direct modes of development arise primarily from heterochronies in the appearance of adult features. Essentially, larval features are suppressed and adult features begin to appear at gastrulation or soon after (Raff 1987). However, modifications are not limited to these heterochronies. There are also significant modifications in oogenesis, cleavage pattern, and blastulation. The change from an unequal fourth cleavage to an equal fourth cleavage, and thus the production of no micromeres is a very drastic change for euechinoids (Williams and Anderson 1975; Raff 1987), and suggests that other cherished traits of early development that have been given phylogenetic value may be likewise variable.

We have determined over 900 bases of 18S rRNA sequence from both species of *Heliocidaris*. As seen in Fig. 3.5, the sequences of the two species are extremely close, and very similar to those of other sea urchins. Two conclusions can be drawn. First, *H. tuberculata* and *H. erythrogramma* are truly very closely related, despite their radically different developmental programs. Secondly, there appears to be no correlation between radical changes in development and evolutionary rate as measured by the 18S rRNA sequence clock. This suggests that the fast 18S rRNA evolutionary rates of holothurians are not related in any obvious way to evolution of their developmental patterns or morphological features. This is consistent with analogous cases in other organisms (King and Wilson 1975).

Another point of relationship raised at the beginning of this chapter was the question of the relationship of echinoderms to other phyla. The two especially important questions are those of the naturalness of the deuterostomes and the relationship of echinoderms to chordates *per se*. We have determined the 18S rRNA sequences of representatives from most major animal phyla and classes. These were used to infer deep relationships among phyla by the methods discussed above for echinoderms. A full presentation and discussion of these phylogenetic inferences are presented by Field *et al.* (1988). A major result of that work is that the eucoelomate phyla represent a coherent radiation event, the branching order of which cannot be resolved by our data. Our results (see Field *et al.* 1988) support the concept of protostomes (annelids, molluscs, brachiopods, and some minor phyla) as an evolutionarily closely allied group. However, our data neither confirm nor controvert the concept of deuterostomes as a clade.

We can, however, comment on the specific affinity of echinoderms to chordates. Jefferies (1975, 1979) has interpreted several peculiarities of internal and external structures of one class of lower Palaeozoic fossil echinoderms, the enigmatic stylophorans, as representing a group of animals transitional between echinoderms and chordates. Thus, the stylophorans become the calcichordates, and are interpreted as having a complex nervous system and notochord, as well as several echinoderm features such as a skeleton

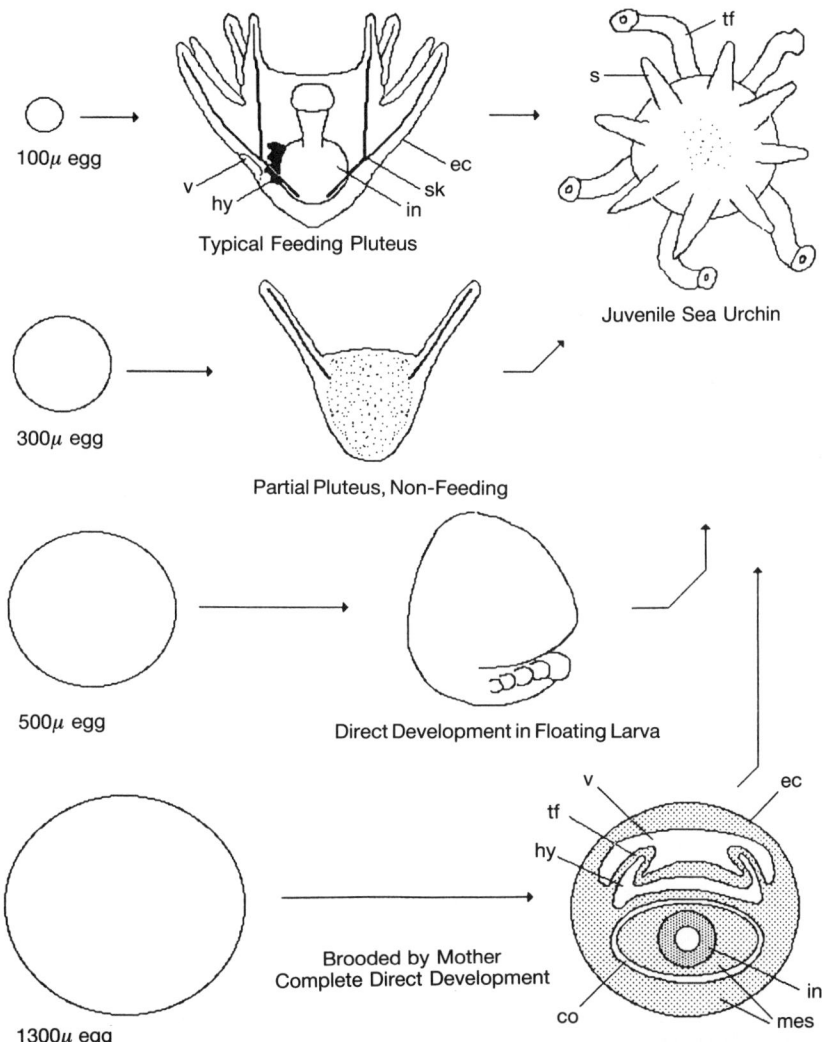

Fig. 3.5 Variation in modes of echinoid development. A typically developing species is diagrammed at the top. The mature eight-armed pluteus contains a developing echinus rudiment: ec, ectoderm; hy, hydrocoel; in, intestine; sk, skeleton; v, vestibule. Structures of the juvenile sea urchin include tube feet (tf) and spines (s). The second diagram is of *Peronella japonica*, which produces a partial two-armed pluteus that lacks a gut. The third diagram shows *Heliocidaris erythrogramma*, which produces a non-feeding larva lacking pluteus structures. The bottom diagram shows *Abatus cordatus*, which undergoes direct development in a maternal brood chamber. Structures as in the typical rudiment, with mesenchyme (mes) and coelom (co) as indicated in the section through the *Abatus* embryo. *P. japonica* modified from Okazaki and Dan (1954); *H. erythrogramma* modified from Williams and Anderson (1975); *A. cordatus* modified from Schatt (1985).

consisting of plates composed of single crystals of calcite and a water vascular system. These forms are seen as giving rise to acraniates, ascidians, and vertebrates in the early Palaeozoic. Our 18S rRNA data fail to support any ancestor-descendant relationship between these phyla (Field *et al.* 1988). The ascidians represent the deepest split from the chordate stem, followed by the acraniates. The ascidian branch is almost as deep in the tree as the divergences between the major groups of phyla, and clearly deeper than the divergence points among living echinoderm classes, which are certainly lower Palaeozoic in timing.

CONCLUSIONS

The determination of phylogenetic relationships among the living classes of echinoderms by cladistic analyses of morphological and embryological criteria has been plagued by the prevalence of parallel and convergent features between the classes, and by subjective difficulties in the choice and interpretation of characters. The object of our study has been to infer a phylogenetic tree for the echinoderm classes by use of molecular sequence data, which are both information-rich (about 1000 base positions per organism), and which transcend morphology. The development of reliable techniques for collection and analysis of molecular sequence data should allow the establishment of solid trees for groups refractory to other methods. This will allow us to test many phylogenetic hypotheses as to both order and relative timing of branching of lineages.

Finally, the complex convergences and parallelisms revealed by morphological and embryological studies of echinoderms have proved difficult to interpret in inferring phylogenetic relationships. However, this complexity reflects the complexity of evolution in the phylum, and is thus of considerable interest aside from phylogenetic perspectives. If a reliable phylogeny can be constructed on independent grounds, the evolution of morphological and embryonic traits can be studied effectively within the context of a known evolutionary history. This may be the most significant contribution of molecular phylogenetics.

ACKNOWLEDGEMENTS

We thank S. Giovannoni and R. Emlet for helpful discussions, and S. Smiley for making his table of characters available to us before publication. This study was supported by National Science Foundation Grant BSR 85-16582.

REFERENCES

Baker, A. N., Rowe, F. W. E., and Clark, H. E. S. 1986. A new class of Echinodermata from New Zealand. *Nature, London* **321**, 862–4.

Balfour, F. M. 1885. *A treatise on comparative embryology*. Macmillan, London.

Bolker, H. I. 1967. Phylogenetic relationships of echinoderms: biochemical evidence. *Nature, London* **213**, 904–5.

Fell, H. B. 1963. The evolution of echinoderms. *Annual Report of the Smithsonian Institute* **1962**, 457–90.

Felsenstein, J. 1978. Cases in which parsimony or compatibility methods will be positively misleading. *Systematic Zoology* **27**, 401–10.

Field, K. G., Ghislen, M. T., Lane, D. J., Olsen, G. J., Pace, N. R., Raff, E. C., and Raff, R. A. 1988. Molecular phylogeny of the animal kingdom. *Science, New York* **239**, 748–53.

Fitch, W. M. and Margoliash, E. 1967. Construction of phylogenetic trees: A method based on mutation distances as estimated from cytochrome C sequences is of general applicability. *Science, New York* **155**, 279–84.

Goad, L. J., Rubenstein, I., and Smith, A. G. 1972. The sterols of echinoderms. *Proceedings of the Royal Society of London* **B180**, 223–46.

Goodman, M., Weiss, M. L., and Czelusniak, J. 1982. Molecular evolution above the species level: branching pattern, rates, and mechanisms. *Systematic Zoology* **31**, 376–99.

Gould, S. J. 1977. *Ontogeny and phylogeny*. Belknap Press/Harvard University Press, Cambridge, Mass.

Gosliner, T. M. and Ghiselin, M. T. 1984. Parallel evolution in opisthobranch gastropods and its implications for phylogenetic methodology. *Systematic Zoology* **33**, 255–74.

Gregory, W. K. 1951. *Evolution emerging*. Macmillan, New York.

Gutell, R. R., Weiser, B., Woese, C. R., and Noller, H. F. 1985. Comparative anatomy of 16-S-like ribosomal RNA. *Progress in Nucleic Acid Research and Molecular Biology* **32**, 155–216.

Hyman, L. H. 1955. *The invertebrates: Echinodermata*, Vol. IV. McGraw-Hill, New York.

Jefferies, R. P. S. 1975. Fossil evidence concerning the origin of the chordates. *Symposium of the Zoological Society of London* **36**, 253–318.

Jefferies, R. P. S. 1979. Methodology and the origin of chordates. In *The origin of major invertebrate groups* (ed. M. R. House), pp. 443–77. Academic Press, New York.

Jukes, T. H. and Cantor, C. R. 1969. Evolution of protein molecules. In *Mammalian protein metabolism* (ed. H. N. Munro), pp. 21–132. Academic Press, New York.

King, M.-C. and Wilson, A. C. 1975. Evolution at two levels in humans and chimpanzees. *Science, New York* **188**, 107–16.

Kluge, A. G. 1984. The relevance of parsimony to phylogenetic inference. In *Cladistics: perspectives on the reconstruction of evolutionary history* (ed. T. Duncan and T. F. Stuessy), pp. 24–38. Columbia University Press, New York.

Kume, M. and Dan, K. 1968. *Invertebrate embryology*. Nolit Publishing House, Belgrade.

Lane, D. J., Pace, B., Olsen, G. J., Stahl, D. A., Sogin, M. L. and Pace, N. R. 1985. Rapid determination of 16S ribosomal RNA sequences for phylogenetic analyses. *Proceedings of the National Academy of Sciences, USA* **82**, 6955–9.

Mortensen, T. 1921. *Studies of the development and larval forms of echinoderms*. G. E. C. Gao, Copenhagen.

Ohama, T., Hori, H., and Osawa, S. 1983. The nucleotide sequence of 5S RNAs from a sea cucumber, a starfish, and a sea urchin. *Nucleic Acids Research* **11**, 5181–4.

Ohshima, H. 1911. Larva of Echinoderma. *Zoological Magazine* **23**, 377–94. (In Japanese.)

Okazaki, K. and Dan, K. 1954. The metamorphosis of partial larvae of *Peronella japonica* Mortensen, a sand dollar. *Biological Bulletin of the Marine Biological Laboratory, Woods Hole*, **106**, 83–99.

Olsen, G. J. 1988. Phylogenetic analysis using ribosomal RNA. *Methods in Enzymology* (in press).

Olsen, G. J., Lane, D. J., Giovannoni, S. J., Pace, N. R., and Stahl, D. A. 1986. Microbial ecology: A ribosomal RNA approach. *Annual Review of Microbiology* **40**, 337–65.

Pace, N. R., Stahl, D. A., Lane, D. J., and Olsen, G. J. 1985. Analyzing natural microbial populations by rRNA sequences. *American Society of Microbiology News* **51**, 4–12.

Pace, N. R., Olsen, G. J., and Woese, C. R. 1986. Ribosomal RNA phylogeny and the primary lines of evolutionary descent. *Cell* **45**, 325–6.

Parks, A. L., Parr, B. A., Chin, J.-E., Leaf, D. S., and Raff, R. A. 1988. Molecular analysis of heterochronic changes in the evolution of direct developing sea urchins. *Journal of Evolutionary Biology* **1**, 27–44.

Paul, C. R. C. 1977. Evolution of primitive echinoderms. In *Patterns of evolution as illustrated by the fossil record* (ed. A. Hallam), pp. 123–58. Elsevier, Amsterdam.

Paul, C. R. C. and Smith, A. B. 1984. The early radiation and phylogeny of echinoderms. *Biological Reviews* **59**, 443–81.

Raff, R. A. 1987. Constraint, flexibility, and phylogenetic history in the evolution of direct development in sea urchins. *Developmental Biology* **119**, 6–19.

Raff, R. A. and Kaufman, T. C. 1983. *Embryos, genes, and evolution*. Macmillan, New York.

Raff, R. A., Anstrom, J. A., Chin, J. E., Field, K. G., Ghislen, M. T., Lane, D. J., Olsen, G. J., Pace, N. R., Parks, A. L., and Raff, E. C. 1987. Molecular and developmental correlates of macroevolution. In *Development as an evolutionary process* (ed. R. A. Raff and E. C. Raff), pp. 109–38. Alan R. Liss, New York.

Sanger, F., Nicklen, S., and Coulson, A. R. 1977. DNA sequencing with chain-terminating inhibitors. *Proceedings of the National Academy of Sciences, USA* **74**, 5463–7.

Schatt, P. 1985. L'edification de la face oral au cours du development direct de *Abatus cordatus*, oursin incubant subantarctique. In *Echinodermata. Proceedings of the Fifth International Echinoderm Conference, Galway* (ed. B. F. Keegan and D. S. O'Connor). A. A. Balkema, Rotterdam.

Smith, A. B. 1984. Classification of the Echinodermata. *Palaeontology* **27**, 431–59.

Stackebrandt, E. and Woese, C. R. 1981. The evolution of prokaryotes. In *Molecular and cellular aspects of microbial evolution* (ed. M. J. Carlisle, J. R. Collins, and B. E. B. Moseley), pp. 1–31. Cambridge University Press, Cambridge.

Strathmann, R. 1978. The evolution and loss of feeding larval stages of marine invertebrates. *Evolution* **32**, 894–906.

Thomson, K. S. 1977. The pattern of diversification among fishes. In *Patterns of evolution as illustrated by the fossil record* (ed. A. Hallam), pp. 377–404. Elsevier, Amsterdam.

Williams, D. H. C. and Anderson, D. T. 1975. The reproductive system, embryonic development, larval development and metamorphosis of the sea urchin *Heliocidaris erythrogramma* (Val.) (Echinoidea: Echinometridae). *Australian Journal of Zoology* **23**, 371–403.

4

Collagen biochemistry and the phylogeny of echinoderms

TOSHIHARU MATSUMURA and MICHIO SHIGEI*

Meiji Institute of Health Science, 540 Naruda, Odawara, 250 Japan
*Misaki Marine Biological Station, Faculty of Science, University of Tokyo, Misaki, Kanagawa, 238-02 Japan

FROM COLLAGEN BIOCHEMISTRY TO EVOLUTIONARY BIOLOGY

Data on the primary amino acid sequence of evolutionarily related proteins indicate that the substitution of amino acids during evolution is frequent among chemically similar amino acids. This led Kimura to propose the theory of neutral mutation: that non-Darwinian fixation of random mutation in a protein is much larger than fixation due to Darwinian pressure. The above findings, together with evidence from many families of proteins that the rate of substitution is approximately constant per site per year, have provided us with a tool to relate protein chemistry to evolutionary biology (see Kimura 1968; Doolittle 1979 for extensive reviews).

Although primary sequence data are desirable, amino acid composition data have also been informative for evolutionary biology. This is because the difference in the amino acid composition between two closely-related proteins, and the numbers of mutations accumulated during evolution in the two proteins from their ancestor protein, are related, when the number of mutation is not too large (Metzger 1968; Harris and Teller 1973).

Previously, we reported that collagens from echinoderm body wall tissues belong to the same group of collagens, i.e. interstitial collagens, as mammalian skin type I collagens (Matsumura 1972). We also reported that in terms of collagen amino acid composition sea cucumbers and starfish are closely related, as are sea urchins and brittle stars, whereas feather stars are distantly related to all four other classes (Matsumura *et al.* 1979).

In this paper, we briefly review collagen chemistry in relation to evolutionary biology, then extend these previous studies to a comparison within order-level groups of echinoderms. A general agreement will be shown between relatedness in collagen amino acid composition and the currently accepted echinoderm phylogeny, confirming that the application of collagen chemistry to evolutionary biology is feasible. On the other hand, factors limiting the current approach will also become apparent. For further introduction to echinoderm collagens and their phylogenetic aspects, readers should refer to Bacetti (1985), Bailey (1985), and Mathews (1985).

PROTEINS OF COLLAGEN FAMILY

Although collagen has been defined classically by its special X-ray diffraction pattern, biochemical and biophysical studies have visualized its molecular conformation in detail (Bornstein and Traub 1979). The X-ray diffraction pattern reflects a triple helical structure called the collagen helix. The collagen helix can be explained in the way that three parallel peptide chains, in which individual chains are bound to each other by hydrogen bonds perpendicular to the chain axis, are folded along the chain axis to form a cylinder, being sealed by perpendicular hydrogen bonds, and then twisted to form a triple helix. All existing molecules with this type of helix as their major structural component are called collagens (Fig. 4.1).

The interstitial collagens, which are the major

Echinoderm phylogeny and evolutionary biology (ed. C. R. C. Paul and A. B. Smith). Clarendon Press, Oxford, 1988.

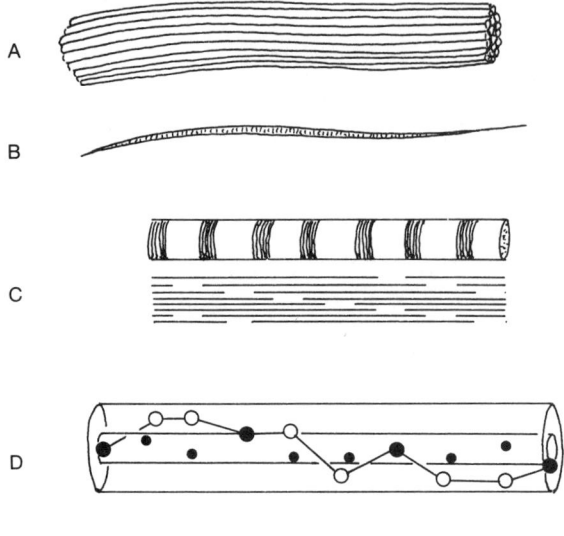

Fig. 4.1 Structural components of interstitial collagens. From top to bottom. (A) A part of a collagen fibre as a bundle of collagen fibrils. (B) A collagen fibril with its dimensions *c.* 100–200 μm long and 0.1 μm thick. (C) A part of collagen fibril as quarter staggered paracrystal of collagen molecules. A fibril shows 640 Å cross striation pattern in the electron microscope. (D) A collagen molecule as a triple helix of three component peptide chains. It is 3000 Å long and 14 Å thick, with molecular weight 300 000. In this scheme, only one chain of the three components chains is shown. The filled circles on the chain show glycine residues, which are facing inside the molecule. Open circles are facing outside the molecule. Small filled circles show glycine residues on the other two chains which are not shown here. (E) Part of a collagen peptide chain with glycine in every third position and abundant imino acids. X and Y are certain amino acid residues.

components of body connective tissue protein in a variety of animals including sponges and echinoderms, as well as vertebrates, share even more similarities to each other: the collagen helix forms an entire molecule except for small entities at its two terminal regions. A collagen molecule is thus a thin rod with its dimension described in Fig. 4.1. In the primary sequence of a collagen peptide, glycine occupies every third position of the amino acid residues throughout the triple helical part of the molecule. The occurrence of glycine in every third position is essential for the collagen helix. In other words, if this position were occupied by an amino acid residue with a side chain, the collagen helix could no longer be maintained. The interstitial collagen molecules associate side-by-side in a quarter-staggered fashion to form a paracrystal of molecules, i.e. a collagen fibril, with a particular cross-striation pattern, as seen in a stained preparation under an electron microscope. The striation pattern, which reflects the distribution of polar amino acids on collagen chains, is maintained over a wide range of animals including echinoderms.

WHAT IS SPECIAL ABOUT COLLAGEN MOLECULES?

In the comparative biochemistry of proteins, collagen has special advantages that are not shared with many other proteins. Firstly, the variation in amino acid composition is remarkably large (see Doolittle 1979 for amino acid variation in other families of proteins). A substitution of more than 20% of total amino acid is noted among echinoderm collagens (Matsumura *et al.* 1979). This high variation is probably related to the fact that two-thirds of the total amino acids, i.e. those other than glycine, are all facing outward from the molecule, being free from the restriction of a structural role. Also, it is probably related to the fact that the repeat of the simple rod-like molecules in a collagen fibril permits a large amount of freedom of amino acid substitution along its vertical axis with little change of the structural and chemical properties of the fibril. Secondly, amino acid substitution in collagen has been known to occur mostly between two amino acids with similar chemical nature. This has been diagrammatically shown by plotting the ratios of polar amino acid content, non-polar amino acid content and hydroxylic amino acid content in a triangular graph: the ratios for interstitial collagens fall within a small area of the graph for a wide variety of animals from sponges to human beings, including echinoderms (Matsumura 1972; Murray *et al.* 1982; Franc 1985). The mode of amino acid substitution has been described further from protein sequence data (Bornstein and Traub 1979). The above two lines of evidence support the idea that the substitution of collagen amino acids has in fact occurred during evolution, as predicted by the neutral mutation theory, without large changes in the chemical nature of the molecule.

There are, however, several possible factors

which may complicate the evolutionary study of collagen. Collagen is not a single protein, but forms a family of proteins existing in a single body. Particularly in mammals and other highly evolved animals, more than ten types of collagen are known (Miller and Gay 1982; Ricard-Blum and Herbage 1985). Information at present is scarce on the types of echinoderm collagens. More than one type of collagen is known to be present (Bailey 1985). The second complication comes from the fact that the hypothesis of constant rate of mutation may never be applied evenly to every kind of component amino acid. The relationship known between the average environmental temperature for a species, and the imino acid contents of its collagen suggests the presence of Darwinian pressure on the imino acid content (Josse and Harrington 1964). Thus, we should be careful in comparing collagens from animal species in respect of the collagen types and the temperature environments. These complicating features will be discussed further in a later section.

PREPARING ECHINODERM COLLAGENS

Because some echinoderm tissues are very rich in collagen, washed tissue fragments themselves and hot water extract of body tissues, i.e. gelatin, can be used for the determination of collagen amino acid composition (Piez and Gross, 1959; Travis *et al.* 1969). Collagen can be solubilized and purified after the limited application of certain proteinases to the tissue, if it loses a small part of the terminal portion of the molecule (Matsumura 1980; Bailey 1985). A chemical method for disaggregating body tissues, and for isolating and purifying collagen in the form of insoluble fibrils has been developed by us, as described in detail previously (Matsumura *et al.* 1973; Matsumura 1974). In the present study, we used the amino acid composition data for collagen fibrils that had been obtained by this tissue disaggregation method and published in part previously (Matsumura 1973; Matsumura *et al.* 1979). Briefly, specimens of 27 echinoderm species were obtained, mostly from Sagami Bay near Misaki, except for purchased *Stichopus japonicus* from Nanao Bay, Ishikawa. Body wall tissues of starfish and sea cucumbers, tegmina from feather stars, and discs from brittle stars, were sliced into pieces, which were then subjected to a procedure of tissue disaggregation and collagen fibril isolation. Cuvierian organs, male gonad tissues, and intestines were also obtained, but not used in this study except for irregular sea urchins from which body wall tissues were difficult to isolate. The tissue fragments were mixed in a disaggregating solution containing 0.5 M NaCl, 0.05 M ethylenediaminetetraacetic acid, 0.2 M β-mercaptoethanol and 0.1 M tris-HCl buffer (pH 8.0). The mixture, rich in collagen fibrils, was then subjected to a purification sequence consisting of filtration, centrifugation and resuspension.

A purified fibril preparation was heat-hydrolysed in 6N HCl, and the hydrolysate analysed for its amino-acid composition using a conventional double column chromatographic method. The contents for threonine, serine, and glutamic acid were corrected for the hydrolytic losses. A value in a set of amino acid composition data is presented as an average from two or more determinations, except for those from single determinations for *Holothuria monacaria* and *Toxopneustes pileolus*. The content of glycine was used as a purity criterion for the collagen preparations. Those preparations with more than 300 glycine residues per 1000 total amino-acids were used in this study, except for the one for *Comanthus japonica* which contained 299.4 (Table 4.1 for species and tissues used in this study, and Tables 4.2 and 4.3 for the amino acid composition of collagens). In addition to the above amino acid composition data, some data published in the literature were used and are referred to in Tables 4.1 and 4.2. Echinoderm species were classified in Table 4.1 by reference to Mortensen 1928–1952; Hyman 1955; Durham and Melville 1957; Moore 1966; Irimura 1982; Smith 1984b and Shigei 1986.

AMINO ACID DATA SETS

A value given by the following equation was calculated from two amino acid composition data sets, $\langle AAij \rangle$ and $\langle AAik \rangle$, and will be referred to as $DSjk$, i.e. the difference sum defined for the two collagen preparations j and k:

$$DSjk = \sum_i |AAij - AAik|.$$

Here, a value $AAij$ is the number of residues per 1000 total residues for the i kind of amino acid in j kind of collagen preparation, and a value $AAik$ is that for the i kind of amino acid in k kind of collagen. In this calculation, the values for glutamine and glutamic acid contents were combined,

Table 4.1 Source animals and tissues

Class	Order	Species	Tissue
Crinoidea	Articulata	*Comanthus japonica*	tegmen
Holothuroidea	Aspidochirotida	*Holothuria monacaria*	body wall
		Stichopus japonicus	body wall
	Dendrochirotida	*Thyone* sp.	body wall*
Asteroidea	Paxillosida	*Astropecten scoparius*	body wall
	Valvatida	*Certonardoa semiregularis*	body wall
	Forcipulatida	*Coscinasterias acutispina*	body wall
		Asterias amurensis	body wall
Ophiuroidea	Myophiurida	*Ophioplocus japonicus*	disc
Echinoidea	Diadematoida	*Diadema setosum*	peristome
	Temnopleuroida	*Temnopleurus toreumaticus*	peristome
		Toxopneustes pileolus	peristome
	Echinoida	*Pseudocentrotus depressus*	peristome
		Hemicentrotus pulcherrimus	peristome
		Anthocidaris crassispina	peristome
	Clypeasteroida	*Peronella japonica*	intestine

*Gelatin preparation by Piez and Gross (1959).
Order Camarodonta consists of temnopleurids and echinids.

as were those for asparagine and aspartic acid, because the analytical method gave these combined values only. Also, the lysine and hydroxylysine contents were combined, as were the proline and hydroxyproline contents, the genetic codes for both combinations being respectively the same. A *DS* value is a direct indicator of the similarity of the two sets of amino acid data.

For the analyses of amino acid data, we will introduce the following three basic assumptions on the bases of the neutral mutation theory: in Assumption A, amino acid substitutions during evolution are due to non-Darwinian neutral mutation; in Assumption B, the rate of amino acid substitution is the same for every kind of amino acid; in Assumption C, a hypothetical set of amino acid composition data, which is generated by averaging two sets of amino acid composition data, respectively, for two collagens, is closer to the amino acid composition of the ancestor collagen than those of the two collagens. Here, the ancestor collagen is the collagen of the ancestor species from which the two recent collagens have evolved.

GENERAL FEATURES OF AMINO ACID COMPOSITION

In the 16 data sets of amino acid composition shown in Tables 4.2 and 4.3, a variation appeared in the imino acid content, i.e. the sum of the proline and hydroxyproline contents, depending on the class-level group of echinoderms. The imino acid content ranged from the highest for echinoids, between 173.3 and 211.2 per 1000 residues, to the lowest of 135 per 1000 for the crinoid.

For the Japanese echinoderms within a class-level group shown in Table 4.1, their collagen imino acid contents fell in a distinctly small range. This is in accordance with the reported correlation between the imino acid composition and the body temperature of animals (Josse and Harrington 1964). However, for a reported collagen we noted a different imino acid content, i.e. 162.6 for a cold-water echinoid *Strongylocentrotus droebachiensis* (Travis et al. 1967) as compared to 173.3 for the average of the five camarodont echinoids from Japan. These results suggest that the content of imino acids, i.e. proline and hydroxyproline, has changed at a high rate during evolution under Darwinian pressure. The contents of lysine and hydroxylysine also did not appear to drift in a non-Darwinian fashion: the contents of lysine and hydroylysine differed from class to class, with the highest hydroxylysine content ranging from 110 to 140 for holothuroids, down to 40 to 71 for asteroids. Since lysine and hydroxylysine, as well as proline and hydroxyproline, are coded by the same codons, respectively, these differences cannot

Table 4.2 Amino acid composition of collagen preparations*

Species	Hyp	Asx	Thr	Ser	Glx	Pro	Gly	Ala	Cys	Val	Met	Ile	Leu	Tyr	Phe	Hyl	Lys	His	Arg
Peronella japonica	100.3	67.1	33.7	55.7	93.7	110.9	318.0	71.3	0.0	20.3	0.3	12.0	29.5	6.1	6.7	7.1	7.0	2.7	57.9
Camarodont species average	73.2	58.8	35.4	66.6	99.0	100.1	316.3	86.1	0.8	19.5	13.3	11.7	28.0	7.1	6.2	6.2	9.2	4.6	57.9
Diadema setosum	83.3	60.7	34.7	49.7	91.7	104.6	306.2	103.1	1.5	23.4	10.9	10.5	31.8	9.1	5.5	7.9	6.8	3.0	56.0
Ophioplocus japonicus	75.4	50.2	33.0	49.3	104.1	108.7	336.3	96.0	2.4	18.8	3.0	12.3	20.2	6.7	5.8	6.2	7.9	1.8	62.5
Asterias amurensis	59.0	60.5	25.3	71.7	85.9	90.7	329.8	116.8	4.0	22.3	10.4	16.8	17.9	8.5	3.6	7.1	16.0	2.8	50.9
Coscinasterias acutispina	69.2	59.5	30.5	58.1	84.0	99.2	318.2	122.4	3.3	23.7	7.7	17.2	18.6	9.1	5.0	4.0	17.2	2.9	50.6
Certonardoa semiregularis	53.6	54.5	35.4	48.6	87.6	99.8	326.5	131.7	2.6	27.5	7.1	11.2	16.0	11.2	3.4	6.2	16.7	2.9	57.9
Astropecten scoparius	66.0	62.1	34.6	48.1	90.4	90.4	300.0	124.6	2.8	25.4	7.9	14.0	23.9	11.6	8.2	6.4	20.6	4.5	59.1
Thyone sp.	60.0	62.0	35.0	43.0	110.0	109.0	306.1	113.0	2.5	30.0	2.2	13.0	22.0	7.9	8.9	11.0	7.5	2.8	54.0
Stichopus japonicus	65.5	71.8	37.6	48.5	102.8	94.4	324.2	104.1	3.4	22.1	3.0	16.6	18.6	8.5	7.3	14.0	5.9	2.1	49.6
Holothuria monacaria	77.7	69.2	39.9	44.4	97.7	97.0	325.2	118.6	0.0	20.2	0.0	7.7	18.1	5.4	7.3	12.0	4.8	2.3	52.3
Comanthus japonica	62.1	50.1	27.7	53.5	118.3	72.9	299.4	113.3	1.9	24.5	15.2	15.6	27.2	13.7	8.5	6.5	24.4	5.8	59.8

*Residues per 1000 total amino acid residues.

Table 4.3 Amino acid composition of collagen preparations from camarodont species*

	Temnopleurus toreumaticus	Toxopneustes pileolus	Psuedocentrotus depressus	Hemicentrotus pulcherrimus	Anthocidaris crassispina	Average
Hyp	75.1	66.1	67.5	73.0	84.2	73.2
Asx	64.3	55.7	57.1	59.8	56.9	58.8
Thr	32.3	37.4	35.6	35.4	36.5	35.4
Ser	64.6	62.5	71.5	71.7	63.1	66.6
Glx	97.0	101.0	99.4	102.2	95.3	99.0
Pro	98.7	102.5	98.3	96.0	105.3	100.1
Gly	311.3	321.7	327.6	305.8	315.3	316.3
Ala	91.4	88.6	81.0	81.0	88.8	86.1
Cys	0.0	0.0	0.0	2.5	1.6	0.8
Val	19.2	21.1	16.2	20.9	20.1	19.5
Met	11.9	13.6	18.0	17.2	5.8	13.3
Ile	12.1	12.0	11.7	11.1	11.5	11.7
Leu	29.0	27.1	28.0	30.1	26.2	28.0
Tyr	7.3	6.0	7.4	7.8	7.2	7.1
Phe	6.5	5.2	6.3	7.3	5.5	6.2
Hyl	5.1	7.2	6.0	6.4	6.4	6.2
Lys	9.7	9.5	9.0	8.4	9.1	9.2
His	4.3	6.4	4.3	3.9	4.1	4.6
Arg	60.4	56.7	55.5	60.0	57.2	57.9

*Residues per 1000 total amino acid residues.

be related to nucleotide substitution on DNA, but may be related to the change of chemical nature of collagen molecules. These results show that the substitution of amino acids in echinoderm collagens cannot be regarded as being due solely to neutral mutation, and that there are substitutions due to Darwinian pressure during evolution of class-level groups, and even within order-level groups. In order to remove the complication due to imino acid drift, we excluded data from *Strongylocentrotus droebachiensis*, and used data from those temperate Japanese species, and from *Thyone* sp. (Piez and Gross 1959) which have a similar imino acid composition.

RELATEDNESS IN TERMS OF AMINO ACID COMPOSITION

A matrix of DS values is shown in Table 4.4 for the 12 amino acid data sets in Table 4.2. In this table, the DS values for the collagen of the five camarodonts are those for the single data set representing the five collagens. A separate DS matrix is shown in Table 4.5 for the seven data sets, respectively, for the seven echinoid collagens. A general tendency shown in these tables is that any set of two species in an order-level group gives a considerably smaller DS value: in holothuroids, the DS value between the two aspidochirotids was small in comparison to a DS value between an aspidochirotid and a dendrochirotid. Also in asteroids, the DS value between the two forcipulatids was small in comparison to the DS value between a forcipulatid and either a valvatid or a paxillosid. However, it also happened that a DS value between two species respectively from two different order-level groups was as small as a DS value between two species within an order-level group. In echinoids, this was apparent for the temnopleurids and echinids (Table 4.5): DS values were distinctly small for all combinations of species studied within these two order-level groups. These results suggest that these two orders are very close.

Table 4.4 A DS matrix of echinodermal collagens

Species	1	2	3	4	5	6	7	8	9	10	11	12	
1 Comanthus japonica	0												crinoids
2 Holothuria monacaria	204	0											holothuroids
3 Stichopus japonicus	168	71	0										
4 Thyone sp.	133	88	87	0									
5 Astropecten scoparius	108	127	113	88	0								asteroids
6 Certonardoa semiregularis	159	114	108	123	74	0							
7 Coscinasterias acutispina	162	101	94	102	88	82	0						
8 Asterias amurensis	161	135	110	150	120	91	68	0					
9 Ophioplocus japonicus	189	110	105	121	164	140	141	165	0				ophiuroids
10 Diadema setosum	164	117	119	95	103	136	110	154	104	0			
11 Camarodont species average*	171	122	124	123	142	154	117	143	104	88	0		echinoids
12 Peronella japonica	224	141	157	166	189	203	167	212	128	108	102	0	

*DS values are for the amino acid data set shown in Table 4.1 for the camarodont species average.

Table 4.5 A DS matrix of echinoid collagens

Species	1	2	3	4	5	6	7	
1 Diadema setosum	0							diadematoids
2 Temnopleurus toreumaticus	77	0						temnopleurids
3 Toxopneustes pileolus	101	54	0					
4 Pseudocentrotus depressus	125	70	48	00				
5 Hemicentrotus pulcherrimus	96	55	62	48	0			
6 Anthocidaris crassispina	68	55	53	81	81	0		echinids
7 Peronella japonica	108	100	118	128	125	81	0	clypeasteroids

It should be noted here that enzyme immunochemistry shows a close relationship between temnopleurids and echinids. (Matsuoka 1986).

As to the relatedness of class-level groups, the following features were noted (Table 4.4):

(1) the relationship between ophiuroids and asteroids is more distant than that between ophiuroids and echinoids or that between ophiuroids and holothuroids;
(2) the relationship between crinoids and any of the other four classes is more distant than that between any two of the other four groups;
(3) the relationship between asteroids and echinoids is more distant than that between echinoids and ophiuroids, or that between echinoids and holothuroids, and that between asteroids and holothuroids.

When we took assumption A and assumption B as described above to find relatedness in DS values on the bases of neutral mutation theory, the above results were in agreement with the two different phylogenetic trees proposed respectively by Smith and Paul (Paul and Smith 1984; Smith 1984a, Fig. 4.2a), and by us Matsumura et al. 1979 Fig. 4.2c).

Two further features are apparent from the DS matrices:

(1) echinoids and ophiuroids are closer to each other than to any other echinoderms;
(2) asteroids and holothuroids are closer to each other than to any other echinoderms.

These findings led us to introduce cluster analysis of amino acid compositions. Here, assumption C as described above was introduced, and a hypothetical amino acid composition for a combination of two closely related groups, holothuroids and asteroids, for example, was generated by averaging all sets of amino acid data in the two groups. For the cluster analysis, two different methods were used. In the first method, we averaged all amino acid data sets within a class first, and obtained a data set which we call a class-representative collagen, e.g. holothuroid-representative collagen, and then averaged two data sets respectively for two closely related classes to generate a data set for a combined-class representative collagen, e.g. holothuroid - asteroid - representative collagen (Matsumura et al. 1979). In the second method, the clustering process was also introduced within a class-level group. Thus, the two amino acid data sets closest to each other within a class, e.g. *Asterias amurensis* and *Coscinasterias acutispina* in asteroids, were combined first, and then the combined set of data and other sets of data were compared for further clustering in the same manner, until all data sets in a class-level group were combined to generate a class-representative data set. For either of the methods, after the generation of a class-representative collagen for each of the five classes, the clustering process was continued until a single data set representing all the five echinoderm groups was generated. The DS values obtained by the first methods were shown previously (Matsumura et al. 1979). The values obtained by the second method were similar to those obtained by the first method, and are not shown here. The above two methods gave the same phenogram shown in Fig. 4.2c. This collagen phenogram, and Smith and Paul's phylogenetic tree are similar except that the holothuroids share a common ancestor with echinoids in Smith and

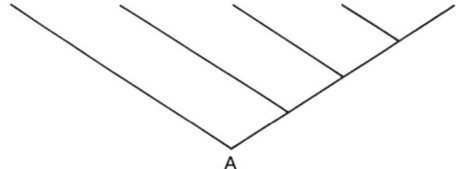

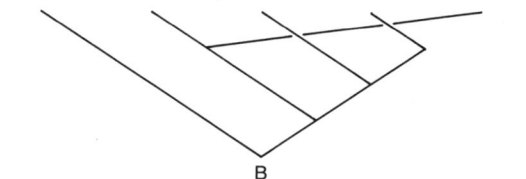

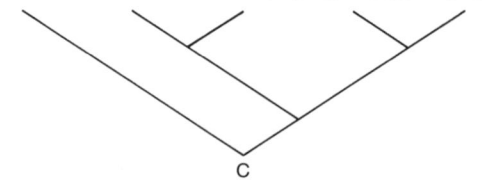

Fig. 4.2 Phylogenetic trees for the five extant classes of echinoderms. A is from Smith (1984a) and Paul and Smith (1984). C is from Matsumura et al. (1979). B is the same as C except that the five classes are arranged in the same order as in A.

Paul's phylogenetic tree, while they share it with asteroids in the collagen phenogram (Fig. 4.2b).

RELATIONSHIP WITH OTHER PHYLOGENETIC STUDIES

Embryology suggests that there is a close relationship between ophiuroids and echinoids, since they share the same type of larva (pluteus) throughout their larval life, and their ambulacral grooves are closed in the course of development. Asteroids and holothuroids also appear to be closely related, since they pass through similar early larval stages (the bipinnaria and auricularia). The development of the doliolaria, which appears in the early larval stage of crinoids and in the late larval stage of holothuroids, may suggest some relationship between crinoids and holothuroids (Hyman 1955). Taking into account data on embryogenesis, metamorphosis, and comparative anatomy, Smiley (this volume) considers holothuroids not to be closely related to echinoids, but as a sister group of crinoids.

In cladistic studies of living and fossil groups, Smith (1984a) and Paul and Smith (1984) recognized a number of similarities between holothuroids and echinoids and regarded these two as closest to each other among the five living groups. In relation to the phylogenetic trees derived from morphological and embryological approaches, we should note the limitation of the collagen phylogenetic tree of Fig. 4.2c: apparently, this diagram can be regarded as only one possible phylogenetic tree based on collagen biochemistry: DS values between holothuroids and echinoids are in fact not very much larger than those between holothuroids and asteroids (Table 4.4), and therefore, the tree diagram of Fig. 4.2a cannot be readily dismissed from the collagen study. Holothuroids, as well as the other three classes are too distant from crinoids in terms of DS values. In general, the more distant the phylogenetic relationship the more inaccurate the collagen amino acid comparison. This is in contrast to the embryological approach in which early stages of development have often been successfully related to early stages of evolution, and thus, again a collagen study cannot be used to reject proposed phylogenetic trees based on embryology.

It is obvious that protein and nucleic acid sequence study can overcome some difficulties, particularly in comparing closely related species for which the difference in amino acid composition is too small to be detected accurately. On the other hand, for the comparison of distantly related species, collagen might not be an appropriate target, because its high rate of substitution may make it difficult to determine this rate, and to trace back the sequence of ancestral forms. Therefore, a sequence study of proteins with different rates of substitution, such as collagen and histone, may be informative in a further study.

In the present study, we used collagen fibril preparations from body tissues, except for intestine collagen from *Peronella japonica*. The similarity among them in the ratios of polar amino acid content, hydrophobic amino acid content, and hydroxylic amino acid content (0.451, 0.148, and 0.401, respectively, for *Peronella japonica*, and 0.446, 0.154, and 0.400, respectively, for *Anthocidaris crassispina* for example, as calculated by the method described previously by Matsumura (1972), at least rejects the possibility that these preparations contain a large amount of collagens like mammalian type IV. A more detailed study is needed to scrutinize collagen preparations for their types, and to reject any possible comparison of different types of collagen.

CONCLUSION

There is little doubt that collagen amino acid composition contains substantial information on the phylogeny of echinoderms. A drift in amino acid composition among echinoderm classes, as well as among order-level groups, can largely be explained on the basis of neutral mutation theory, and thus provides us with information on the relatedness of class-level and order-level groups. The phylogenetic tree of echinoderms proposed on the basis of collagen study differs in part from Smith and Paul's phylogenetic tree, and that of Smiley. Critical analyses, however, showed that collagen study does not contain decisive information to exclude either of these two models. During the course of analysis, we noted several factors which limited the application of the neutral mutation theory to echinoderm collagens. To further this line of evolutionary study, sequence analyses of collagens, in combination with other proteins with low rates of amino acid substitution, would be promising.

ACKNOWLEDGEMENTS

Our thanks are due to Dr Masami Hasegawa for his discussion on molecular evolution.

REFERENCES

Baccetti, B. 1985. Collagen and animal phylogeny. In *Biology of invertebrate and lower vertebrate collagens* (ed. A. Bairati and R. Garrone), pp. 29–47. Plenum Press, New York.

Bailey, A. J. 1985. The collagen of the Echinodermata. In *Biology of invertebrate and lower vertebrate collagens* (ed. A. Bairati and R. Garrone), pp. 369–88. Plenum Press.

Bornstein, P. and Traub, W. 1979. The chemistry and biology of collagen. In *The proteins* (ed. H. Neurath and R. L. Hill), pp. 411–632, New York.

Doolittle, R. F. 1979. Protein Evolution. In *The proteins* (ed. H. Neurath and R. L. Hill), pp. 1–118. Academic Press, New York.

Durham, J. W. and Melville, R. V. 1957. A classification of echinoids. *Journal of Paleontology* **31**, 242–72.

Franc, S. 1985. Collagen of coelenterates. In *Biology of invertebrate and lower vertebrate collagens* (ed. A. Bairati and R. Garrone), pp. 197–210. Plenum Press, New York.

Harris, C. E. and Teller, D. C. 1973. Estimation of primary sequence homology from amino acid composition of evolutionary related proteins. *Journal of Theoretical Biology* **38**, 347–62.

Hyman, L. H. 1955. Echinodermata. The invertebrates. McGraw-Hill, New York.

Irimura, S. 1982. *The brittle stars of Sagami Bay*. Maruzen, Tokyo.

Josse, J. and Harrington, W. F. 1964. On the arrangement of the hydrogen bonds in the structure of collagen. *Journal of Molecular Biology* **9**, 269–87.

Kimura, M. 1968. Evolutionary rate at the molecular level. *Nature, London* **217**, 624–6.

Mathews, M. B. 1985. Evolution and collagen. In *Biology of invertebrate and lower vertebrate collagens* (ed. A. Bairati and R. Garrone), pp. 545–60. Plenum Press, New York.

Matsumura, T. 1972. Relationship between amino-acid composition and differentiation of collagen. *International Journal of Biochemistry* **3**, 265–74.

Matsumura, T. 1973. Shape, size and amino acid composition of collagen fibril of the starfish *Asterias amurensis*. *Comparative Biochemistry and Physiology* **44B**, 1197–205.

Matsumura, T. 1974. Collagen fibrils of the sea cucumber, *Stichopus japonicus*: Purification and morphological study. *Connective Tissue Research* **2**, 117–25.

Matsumura, T. 1980. Hydrodynamic properties of collagen fibril and aging. In *Aging phenomena* (ed. K. Oota, T. Makinodan, M. Iriki, and L. S. Baker), pp. 47–54. Plenum Press, New York.

Matsumura, T., Shinmei, M., and Nagai, Y. 1973. Disaggregation of connective tissue: Preparation of fibrous components from sea cucumber body wall and calf skin. *Journal of Biochemistry, Tokyo* **73**, 155–62.

Matsumura, T., Hasegawa, M., and Shigei, M. 1979. Collagen biochemistry and phylogeny of echinoderms. *Comparative Biochemistry and Physiology* **62B**, 101–5.

Matsuoka, N. 1986. Further immunological study on the phylogenetic relationship among sea-urchins of the order Echinoida. *Comparative Biochemistry and Physiology* **84B**, 465–8.

Metzger, H., Shapiro, M. B., Mosimann, J. E., and Vinton, J. E. 1968. Assessment of compositional relatedness between proteins. *Nature, London* **219**, 1166–8.

Miller, E. J. and Gay, S. 1982. Collagen: An overview. *Methods in Enzymology* **82**, 3–32.

Moore, R. C. 1966. Treatise on invertebrate paleontology, Echinodermata, part U. Geological Society of America and University of Kansas Press, Lawrence, Kansas.

Mortensen, Th. 1928–1952. *Monograph of the Echinoidea*, Vols 1–5. C. A. Reitzel, Copenhagen.

Murrey, L. W., Waite, J. H., Tanzer, M. L., and Hauschka, P. V. 1982. Preparation and characterization of invertebrate collagens. *Methods in Enzymology* **82**, 65–96.

Paul, C. R. C. and Smith, A. B. 1984. The early radiation and phylogeny of echinoderms. *Biological Reviews* **59**, 443–81.

Piez, K. A. and Gross, J. 1959. The amino acid composition and morphology of some invertebrate and vertebrate collagens. *Biochemica Biophysica Acta* **34**, 24–39.

Ricard-Blum, S. and Herbage, D. 1985. The different types of collagen present in cartilaginous tissues. In *Biology of invertebrate and lower vertebrate collagens* (ed. A. Bairati and R. Garrone), pp. 53–64. Plenum Press, New York.

Shigei, M. 1986. *The sea urchins of Sagami Bay*. Maruzen, Tokyo.

Smith, A. B. 1984a. Classification of the Echinodermata. *Palaeontology* **27**, 431–59.

Smith, A. B. 1984b. *Echinoid palaeobiology*. George Allen and Unwin, London.

Travis, F. S., Francois, C. J., Bonar, L. C., and Glimcher, M. J. 1967. Comparative studies on the organic matrices of invertebrate mineralized tissues. *Journal of Ultrastructure Research* **18**, 519–50.

5

Larvae, phylogeny, and von Baer's Law

RICHARD R. STRATHMANN

*Friday Harbor Laboratories, 620 University Road,
Friday Harbor, WA 98250, USA*

INTRODUCTION

Speculation on the relationship of animal classes and phyla are often based on larval forms. This paper examines the implications of larval forms for the phylogeny of echinoderm classes. An isolated treatment of larvae is possible because many of the puzzles posed by echinoderm larvae can be analysed without reference to specific adult features. An isolated treatment of larvae is useful because traits of larval echinoderms have often been misunderstood. The conclusion of the first section of this paper is that incongruities among larval characters indicate that striking convergences in development have occurred within the phylum and that parsimonious phylogenetic trees for classes based on larval morphology are sensitive to small changes in the larval character set and, therefore, do not provide strong indications of homology and convergence.

The limited usefulness of embryonic and larval traits for the inferences of relationships among echinoderm classes arises, in part, from exceptions to von Baer's first law. I therefore examined evolutionary explanations of von Baer's first law and the reasons exceptions are so numerous. Von Baer's first law (as cited by Russell 1916) states that the more general characters of the large group of animals to which the animal belongs appear in development earlier than the special characters. A more general version of von Baer's law is that the developmental history of the individual is the history of the growing individuality in every respect. Exceptions to von Baer's first law tell us that constraints on evolution of traits appearing early in development are weak because many early traits are terminal, developmental processes are in part redundant, and development includes regula-

Echinoderm phylogeny and evolutionary biology (ed. C. R. C. Paul and A. B. Smith). Clarendon Press, Oxford, 1988.

tory mechanisms for recovery from early perturbations.

LARVAL CHARACTERS AND PHYLOGENY OF ECHINODERM CLASSES

Inferences about phylogeny involve hypotheses about the direction of change (polarity) and hypotheses about similarity from common ancestry (homology) as opposed to similarity from convergence. A common procedure is first to find traits that are at least possible homologues (Rieppel 1980) and that are shared by more than one and less than all of the taxa in question and then to apply criteria of parsimony or compatibility (Felsenstein 1982, 1983) to distinguish homology from convergence and thereby arrive at a phylogenetic tree. Unless one wants to weight characters, it is desirable to have independent characters. This section examines problems of homology, independence, direction, and taxonomic distribution of larval characters for echinoderms.

Mode of development and homology

Before comparing larval characters, one must determine which larval forms are comparable. Larvae are known from five distinct groups of living echinoderms, the holothuroids, echinoids, ophiuroids, asteroids, and crinoids. Both feeding and non-feeding larvae are known from the first four classes, but feeding larvae are unknown from crinoids. Are similarities among feeding larvae likely to be homologues, and to what extent are non-feeding larvae comparable to each other or to feeding larvae?

Are feeding larvae comparable?

Comparisons within and among echinoderm classes suggest that non-feeding larvae are derived from feeding larvae (Jägersten 1972; Strathmann 1974; Hendler 1982). The hypothesis that feeding larvae of echinoderms evolved more than once within the phylum involves a fantastic number of convergences, given the similarities among all the feeding larvae, the complete absence of the larval feeding mechanism in most lecithotrophic larval forms, and the absence of the larval feeding mechanism in other stages of the life history.

Comparison with the tornaria larva of enteropneusts suggests that many features of the feeding echinoderm larvae originated before the separation of echinoderms and enteropneust hemichordates. Apparent homologies shared among feeding larvae of echinoderms and enteropneusts include:

(1) a circumoral ciliated band of simple cilia that separates the larval surface into a downstream aboral field of inflated appearance and an upstream adoral field that includes the mouth, is depressed in most portions, and extends to both preoral and postoral portions of the body;
(2) an anus formed from the blastopore and mouth formed at a new site;
(3) a coelom formed by outpocketing from an invaginated archenteron with early formation of a single hydropore between the coelom and the dorsal surface;

(MacBride 1914; Stiasny Wijnhoff and Stiasny 1927; Hyman 1955, 1959; Jägersten 1972; Strathmann 1971; Strathmann and Bonar 1976).

Apparent homologues shared by feeding larvae of four classes of echinoderms (but absent in the tornaria larva of enteropneusts) include an approximately triangular adoral band of cilia and a recurved gut with ventral anus. Many similarities among the feeding larvae of echinoderms could be derived from the feeding larva of the common ancestor of echinoderms and enteropneusts. Comparisons among feeding larvae should, therefore, be admitted for tests of hypotheses on relationships among classes.

Can crinoid larvae be compared to larvae of other classes?

Comparisons involving characters of non-feeding larvae involve several potential pitfalls. Many non-feeding larvae are not directly comparable among classes because they are obviously derived from feeding forms. These non-feeding larvae possess vestiges of structures employed in feeding, such as the circumoral ciliated band or skeletal rods that support extensions of this band in plutei. Reduction of these structures is highly variable in larvae that do not require particulate food. In echinoid larvae that do not require food for development through metamorphosis, a larval skeleton may be entirely lacking (Williams and Anderson 1975), present to a variable extent (Mortensen 1921), or (in a facultative feeder) complete (Emlet 1986). Similarly, reduction of arms of ophioplutei varies within the ophiuroid family Amphiuridae (Fenaux 1963; Hendler 1977, 1978). Non-feeding larvae that are obviously derived from feeding larvae are not directly comparable among classes. In non-feeding larvae traits such as skeletal rods or larval arms are not useful for testing hypotheses about phylogeny of higher taxa. The number of coelomic compartments and their formation is also often modified in non-feeding echinoderm larvae.

A few traits are peculiar to non-feeding larvae. Can these be compared? Some traits, such as the buoyancy of many planktonic non-feeding embryos of asteroids and holothuroids or the late stage at hatching of many brooded embryos, are clearly convergent and not useful for inferences on relations among classes. However, at least one trait of lecithotrophic larvae might possibly be retained from a common ancestor. This is the transverse ciliated bands of the non-feeding doliolaria stage. These bands are strikingly similar in the doliolaria stages of crinoids, ophiuroids, and holothuroids. Convergent similarity (Jägersten 1972; Strathmann 1974) and homology (Grave 1903) have both been suggested. An argument for convergence is that in most descriptions of ophiuroid metamorphosis there is no trace of a doliolaria stage and that in the doliolaria of *Ophionereis annulata* there are posterior skeletal rods that appear to be vestiges of a pluteus skeleton (Hendler 1982). These observations suggest that the doliolaria of ophiuroids is an alternative to a pluteus stage, not subsequent to it, and that the doliolaria form evolved from the pluteus. Under this view traits of the doliolaria of ophiuroids are not homologous with traits of other doliolaria larvae. This view is plausible, because transverse bands are a common larval adaptation for swimming. Series of transverse bands occur in ciliated larvae quite unrelated to echinoderms, such as protobranch bivalves, chaetopterid polychaetes, and gymnosomatous pteropods (Jägersten 1972).

On the other hand, the ancestors of ophiuroids, crinoids, and holothuroids might have had a feeding larval stage followed by a doliolaria stage, and then the transverse bands of the doliolaria could be homologous (Ivanova-Kasas 1973). An early stage in the formation of the transverse bands in the crinoid *Florometra* can be interpreted as a remnant of an ancestral convoluted band, though this is not the only interpretation (Lacalli and West 1986). On the other hand Mladenov (1985) has shown that in *Ophiocoma pumila* the metamorphosing ophiopluteus transforms into a swimming stage resembling the doliolaria larvae in the genera *Ophiolepis*, *Ophioderma*, and *Ophionereis*. This opens the possibility that the form of metamorphosis in *O. pumila* is the ancestral form for ophiuroids and that another common form of ophiuroid metamorphosis, in which the posterolateral arms are retained till a late stage, is secondary. I doubt that transformations of feeding larvae of *O. pumila* and holothuroids to a doliolaria are derived from a common ancestor with this trait, but I cannot reject the hypothesis. Those ophiuroids with a doliolaria but no pluteus might have retained a doliolaria stage from an ancestral metamorphosis as in *O. pumila*, and the ophiuroids with a pluteus might have lost the subsequent doliolaria stage.

Comparisons among doliolaria stages of different classes must therefore be considered in a search for possible homologues that could relate classes (synapomorphies). Functional analysis of the doliolaria does not resolve the question of similarity by convergence or ancestry. Formation of a doliolaria at metamorphosis could be an adaptation for swimming speed in larvae competent to settle, because the doliolaria of holothuroids swims faster than the auricularia (H.-t. Lee, personal communication; R. Strathmann, unpublished observations). Some echinoplutei also develop transverse bands posteriorly and increase speed as they become competent to settle, and in another taxon, the cirripedes, a slow swimming nauplius metamorphoses into a faster swimming cyprid that is competent to settle. It is possible that transverse bands are the most simple and effective way for competent echinoderm larvae to be fast swimmers. Thus, it is undetermined whether doliolaria stages in different echinoderm classes are convergent or derived from a common ancestor.

One of the larval traits of asteroids is commonly retained through the evolutionary transition from feeding to non-feeding larval development, and in this case a comparison between feeding and non-feeding forms is reasonable. Most non-feeding larvae of forcipulate and spinulosid asteroids retain an adhesive disc and brachiolar arms that attach the larva to the substratum. This raises the possibility that the adhesive pit of crinoids and adhesive disc of asteroids are homologous, but homology of these structures is by no means obvious. The ciliation and form of the asteroid disc and crinoid pit look different in scanning and transmission electron micrographs (Barker 1978; Mladenov and Chia 1983; Chia *et al.* 1986). Also, in the asteroids the adhesive disc forms when the axohydrocoel approaches the epidermis; in the crinoids Bury (1888) found that the anterior coelom sometimes approaches the anterior end of the larva, but Bury's sections through the adhesive pit do not show a coelomic epithelium or cavity in the vicinity. Instead there is much nearby mesenchyme (Bury 1888; MacBride 1914). In the asteroids the adhesive disc is accompanied by brachiolar arms and papillae that are absent in crinoids. Jägersten (1972) considered brachiolar arms to have arisen separately from the adhesive disc, but thought it likely on functional grounds that an adhesive disc was a precondition for evolution of brachiolar arms in echinoderms. The similarity of brachiolar arms to podia (Jägersten 1972; Smiley, personal communication) raises the possibility that the brachiolar arms are induced in the same way as podia when the large larval coelom unique to asteroids contacts the epidermis. The brachiolar arms are indeed podia-like in development, coelomic lining, and function. Homology of the adhesive disc of asteroids and the adhesive pit of crinoids is doubtful, but cannot be rejected definitely on present evidence.

Absence of feeding crinoid larvae severely limits comparisons between larval crinoids and larvae of the other classes.

Remarkable convergences must be accepted for echinoderms of four classes

The feeding larvae of ophiuroids, echinoids, holothuroids, and asteroids are grouped differently depending on which of two very conspicuous larval traits are considered. Larval form and skeleton suggest one set of relationships; site of formation of the adult mouth suggests another; and the classes are grouped yet a third way if similarity of general adult shape is the criterion

(Strathmann 1971). If skeletal arm rods and the pluteus form evolved once, and a larval form like the auricularia and bipinnaria evolved once, then the classes are related as in the unrooted tree in Fig. 5.1A. The rooted phylogenetic tree of Smiley (this volume) is consistent with this tree. If development of the adult mouth at a new site on the lower left side of the larva evolved once, and development of the adult mouth at or near the site of the larval mouth evolved once, then the classes are related as in the unrooted tree in Fig. 5.1B. If the adult body forms (globose with ends of meridional ambulacra near each other versus star shaped with ends of ambulacra ending at arm tips, far from each other) evolved once each, then the classes are related as in the unrooted tree in Fig. 5.1C. The rooted trees of Smith (1984) and Raff et al. (this volume) are consistent with the tree in Fig. 5.1C. Each tree implies at least one remarkable convergence in larval development.

None of the characters appears to be especially labile. The fossil record suggests that all these classes diverged in the Palaeozoic and some as early as the Ordovician, and existing forms indicate that all of these characters have been very conservative within classes. Nevertheless, any phylogenetic scheme for echinoderms must imply some remarkable convergences.

A character set and class phylogenetic tree from larvae

Does existing information on larval forms indicate which characters are homologous and which convergent? One way to resolve incongruities is to add more characters and accept whichever scheme involves the fewest changes of character state or is compatible with the greatest number of characters. The assumptions behind these methods and problems of statistical evaluation are reviewed by Felsenstein (1982, 1983). I shall follow conventional methods of tabulating characters and deriving a parsimonious trees, and shall then examine the implications of the trees and their sensitivity to minor changes in interpretation of characters.

Table 5.1 lists larval characters. The traits of the outgroup (enteropneust larvae) are scored '0' with '1' the alternative, and '?' designates traits that are either unknown or not applicable. Traits known to vary among species with feeding larvae within a class have been scored B. If larvae were known from more echinoderm species, more traits might be seen to vary within classes. The tornaria larva of enteropneusts is included as an outgroup, which provides one method of rooting a phylogenetic tree (Watrous and Wheeler 1981; Brooks and Wiley 1985; de Queiroz 1985). The selection of traits was explained in part above. Some further explanation of traits in the table is included here, and some that are omitted are discussed in the following section.

Smith (1984) considered early ingression of mesenchyme a trait shared by echinoids, asteroids, and holothuroids. Though ingression of mesenchyme is earlier in gastrulation of holothuroids than in asteroids, descriptions in the literature indicate that ingression of mesenchyme in holothuroids generally occurs after invagination begins (as in Runnström 1927), later than in euechinoids or ophiuroids (MacBride 1914). Grouping holothuroids with euechinoids and ophiuroids for this trait is, therefore, somewhat arbitrary. Variation within classes complicates matters. Ingression of

A LARVAL FORM AND SKELETON

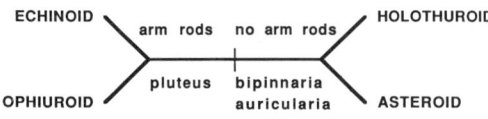

B SITE OF ADULT MOUTH

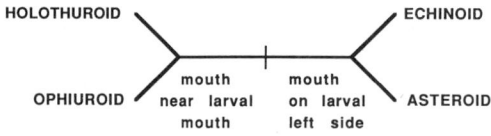

C ADULT FORM

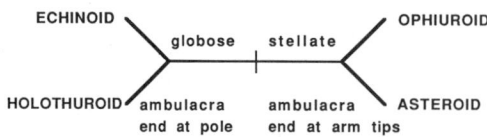

Fig. 5.1 Unrooted trees, each based on a single trait, for four classes known to have feeding larvae.

TABLE 5.1 Distribution of larval characters among classes: 0, trait is absent; 1, trait is present; B, trait present or absent in different groups within the class; ?, Character state unknown or not applicable to that class.

1. Early ingression of mesenchyme; presence of any larval ossicle or spicule.
2. Coelom first formed as pair of pouches from archenteron (as opposed to a single dorsal pouch from the archenteron).
3. Skeletal rods that support arms with loops of the ciliated band; anterior oral dilator muscles attaching to anterolateral rods.
4. Paroral ciliated bands (from preoral to postoral transverse bands on each side of mouth).
5. Anterior adhesive disc or pit for attachment.
6. Doliolaria stage with transverse bands of cilia.
7. Adult mouth forms on lower left side of larva; larval mouth retained during formation of the adult rudiment.

Class	Character						
	1	2	3	4	5	6	7
Enteropneusta	0	0	0	0	0	0	0
Crinoidea	?	?	?	?	1	1	?
Asteroidea	0	1	0	0	B	0	1
Holothuroidea	1	0	0	B	0	1	0
Ophiuroidea	1	1	1	1	0	B	0
Echinoidea	1	1	1	1	0	0	1

mesenchyme occurs after the beginning of invagination of the archenteron in a cidaroid echinoid (Schroeder 1981), before invagination in a molpadiid holothuroid that lacks a feeding larva (McEuen 1986), and perhaps at the beginning of invagination in some holothuroids with feeding larvae (Hörstadius 1939). Earlier ingression of mesenchyme may be connected to early formation of the skeleton of echinoids and ophiuroids (MacBride 1914; Dan 1960) and thus may be partly dependent on skeletal traits that are separately listed in Table 5.1. I have lumped early ingression of mesenchyme with another trait whose distribution can be variously interpreted. The skeletal ossicle of the auricularia could be counted as a posterior skeleton whose occurrence is independent of arms, though it could also be argued that this counts the skeleton of ophiuroids and echinoids twice. Some recognition of these similarities among holothuroids, echinoids, and ophiuroids seems appropriate, and they are listed as a single trait here.

Early formation of a single dorsal coelom with hydropore is a trait suggested by Smiley (this volume); homology can certainly be questioned for this trait, but it still seems possible.

Anterior dilator muscles are listed with skeletal rods. These muscles widen the oral cavity (Strathmann 1971), and are known only from ophioplutei. Because their function and evolution depends on skeletal rods, they cannot be listed separately as an independent trait.

I did not include similarities in overall form among early stages of bipinnaria, auricularia, and tornaria larvae because this trait is not independent of the presence or absence of skeletal rods.

Paroral bands are a pair of ciliated ridges to each side of the mouth (Strathmann 1971). Their occurrence is variable in holothuroids, poorly known for most groups, and quite possibly the result of convergence.

Some asteroids (Luidiidae, Astropectenidae, Pterasteridae) lack an adhesive disc and brachiolar arms (Kaufman 1968; Komatsu 1975, 1982; Oguro et al. 1976). MacBride (1923) thought that the adhesive disc and brachiolar arms were lost in the asteroid lineage leading to *Luidia* and *Astropecten*; Mortensen (1923) thought that the common ancestor of all asteroids lacked these structures. Because origin and loss of the trait in asteroids is uncertain, I scored the adhesive disc 'B'.

Most events occurring late in the transformation from larva to juvenile were excluded as nonlarval, but site of formation of the adult mouth was included because formation of the adult mouth affects duration of larval feeding during metamorphosis and is, therefore, an important part of larval form and function (Strathmann 1971, 1974).

A doliolaria stage follows a feeding stage only in

those classes in which the adult mouth forms at the site of the larval mouth. When the adult mouth forms at a new site on the lower left, larval feeding structures are retained while the adult rudiment develops. That is why an echinopluteus in which epaulettes form transverse bands is not called a doliolaria. This taxonomic distribution of traits raises the possibility that a doliolaria stage is incompatible with formation of the adult mouth on the lower left side of a feeding larva. Table 5.1 may give double weight to a single trait.

The phylogenetic implications of this character set were explored by simple inspection of Table 5.1 and also with the aid of the programs PENNY, MIX, and CLIQUE in a software package called PHYLIP (written by J. Felsenstein and distributed by G. D. F. Wilson). PENNY searched for trees most parsimonious by Wagner parsimony. Wagner parsimony includes the assumptions that changes of traits in either direction are equally likely, that changes in all the listed characters are extremely improbable over the time spans involved, and that two changes in a long segment of a tree are far less probable than one change in a short segment. CLIQUE uses a compatibility criterion and selects trees congruent for the largest set of characters. The compatibility method in CLIQUE assumes that each character might be one that evolves rapidly or has been misinterpreted, and that the probability of two changes in a low rate character in much less than the probability that it is a high rate character (Felsenstein, software documentation). MIX determined the number of changes in traits when phylogenetic trees of other authors were applied to the traits in Table 5.1.

Wagner parsimony produced two equally parsimonious trees of 12 steps (Fig. 5.2A,B). I suspect neither tree will appeal to echinoderm specialists, though tree A is close to Smiley's tree (this volume). The trees are not decisively parsimonious, however, even for the larval character set. Minor changes in assumptions change the trees.

One source of instability is that the only potential synapomorphies for larvae of crinoids and other classes are the anterior attachment pit or disc and the doliolaria stage. The crinoids and holothuroids formed a monophyletic group when the doliolaria stage was scored B (both present and absent for holothuroids; Fig. 5.2A,B), but when the doliolaria stage was scored 1 for ophiuroids, on the assumption that it was ancestral for that

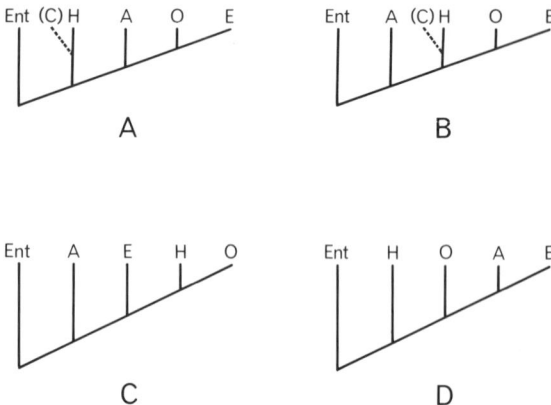

Fig. 5.2 Phylogenetic trees based on larval traits. Ent, enteropneusts; C, crinoids; H, holothuroids; A, asteroids; O, ophiuroids; E, echinoids. Trees A and B have the same number of changes in traits for traits in Table 5.1, with or without crinoids. Trees A, B, and C have the same number of changes in traits without crinoids and with the ophiuroid doliolaria scored 1. Trees A, B, and D have the same number of changes in traits without crinoids and without the paroral band. There would be seven changes in traits if there were no reversals or convergence.

group, trees proliferated to ten equally parsimonious 12-step trees, in which the crinoids formed monophyletic groups with holothuroids, ophiuroids, and echinoids, and occupied diverse other positions. Because the crinoids could not be placed reliably on the basis of larval traits, trees were also constructed without them.

When crinoids were omitted from consideration, the adhesive disc became an autapomorphy for asteroids with no implications for their relation to other classes. This reduced the character set to six. The two most parsimonious trees were simply 10-step trees like Fig. 5.2A,B, but without the crinoids. This tree was also sensitive to small changes in assumptions about traits, however. If a dololiaria was assumed to be ancestral for ophiuroids (scored 1), an additional tree (Fig. 5.2C) became equally parsimonious for a total of three 10-step trees. If, instead, the paroral band (a doubtful character) was omitted, an additional tree (Fig. 5.2D) became equally parsimonious for a total of three 8-step trees. In all these trees only two traits changed only once.

To obtain trees with the largest cliques of traits, I omitted crinoids and assumed all traits previously scored B were 1. The largest cliques were three traits, with the two equally most compatible trees being 5.2B (but without crinoids) and 5.2C. Omitting the paroral band resulted in five trees with the largest clique being two traits. These included all the relationships between holothuroids, ophiuroids, asteroids, and echinoids in Fig. 5.2, plus one with two monophyletic groups (holothuroids with ophiuroids and asteroids with echinoids).

Thus, the trees based on Wagner parsimony or on compatibility were quite sensitive to minor changes in the traits included and in the way they were scored. This analysis has brought us back to our beginning. The incongruities of larval traits among echinoderm classes are striking and cannot be resolved on the basis of known larval traits.

Comparison with other trees

Two relationships never appeared in the trees from larval traits in Table 5.1. One was a monophyletic group including holothuroids and echinoids, but excluding both asteroids and ophiuroids. This group appears in trees of Smith (1984), Raff *et al.* (this volume), and Fell and Pawson (1966). The other was a monophyletic group including asteroids and ophiuroids, but excluding both holothuroids and echinoids. This group appears in the tree of Fell and Pawson, but not in the trees of Smith or Raff *et al.* Neither group appears in Smiley's tree (this volume). Smith's tree emphasizes adult traits, while the tree of Raff *et al.* is based on molecular data.

The recently suggested trees for echinoderm classes require more changes in the larval traits in Table 5.1 than do the most parsimonious trees from this larval character set (Fig. 5.3). A change can be either an initial change or later reversal of a character state because some trees admit several possible ways that traits could have evolved with the same total number of changes. For example, in Smith's (1984) tree the skeletal rods could have either evolved twice or evolved once, and been lost once, but in either case there are two changes. The tree of Raff *et al.* (this volume) implies the most changes, with all the larval traits changing at least twice. Smiley's tree implies the fewest changes. The trees in Fig. 5.3 imply more changes for the doliolaria form than for the other larval traits. Smith's tree is consistent with the view that the pluteus form evolved once and that the auricularia is derived from it. Smiley's tree is consistent with the view that the pluteus evolved once and the

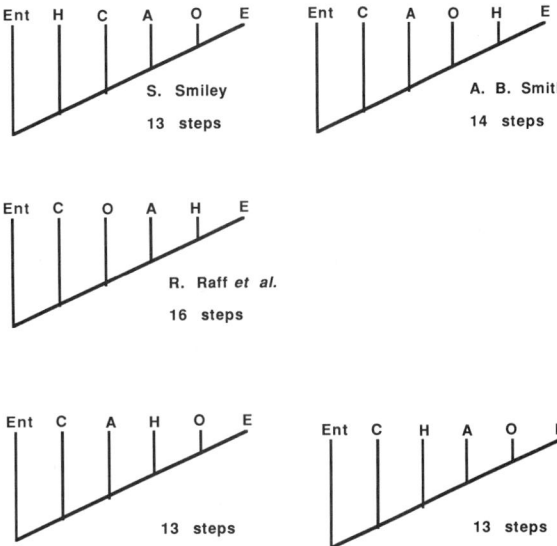

Fig. 5.3 Taxa as in Fig. 5.2. Steps are the number of changes in traits for traits in Table 5.1. There would be seven changes in traits if there were no reversals or convergence.

auricularia is primitively simple. Both seem equally reasonable scenarios for larval evolution. The tree of Raff *et al.* implies that plutei evolved twice or that both the bipinnaria and auricularia are derived from a pluteus, a sequence not as parsimonious as the other trees, but possible. All of the trees require two changes in site of formation of the adult mouth. The trees based solely on larval traits (Fig. 5.2A,B) are slightly more parsimonious for evolution of larval traits than trees based on morphology in general (upper two trees in Fig. 5.3) or than the larval tree with the crinoids attached as a sister group (lower two trees in Fig. 5.3).

The larval traits are so incongruent in themselves that they can be reconciled with several trees based on other character sets. What I do not yet know, is whether trees based on adult morphology or molecular data are so robust that they can resolve questions of homology and homoplasy for the larval traits. I know of no analysis for sensitivity of trees based on adult morphology or molecular data, but the variety of opinions among echinoderm specialists suggests that trees based on adult morphology could also be sensitive to small changes in assumptions about traits and only slightly more parsimonious than alternative trees.

Is the set of useful larval characters really small?

Further comparative studies could expand the set of potential synapomorphies. Some might be found, for example, in the nervous system and sensory structures (Burke 1978, 1983a,b; Burke et al. 1986; Nakajima 1986). Holland (1979, 1981) has compared extraembryonic membranes and coats, and concluded that ophiuroids and echinoids are most similar in initial formation of the hyaline layer. This relation is also implied by the traits in Table 5.1, and inclusion of this trait (scored as 1 for ophiuroids, B for echinoids, and O for the rest) resulted in the trees in Fig. 5.2A,B (without the crinoids). Possible homologies among extraembryonic structures are difficult to evaluate because few species have been sampled with electron microscopy, results depend in part on techniques (Crawford and Abed 1986), and development of the hyaline layer may differ between cidaroids and euechinoids (Schroeder 1981, 1986). Judgements on homology of many structures of known taxonomic distribution could be improved by combining experimental and comparative methods in studies of development. Nevertheless, larval characters presently known from echinoderms have stronger implications for relations within classes (Mortensen 1921, 1938) and among phyla (Hyman 1955, 1959) than for relationships between classes. There may be a genuine shortage of potential synapomorphies that relate one subset of classes while excluding another.

Many larval characters are peculiar to a class (autapomorphies). Examples from feeding larvae of asteroids include anterior median separation and rejoining of the ciliated band at an early stage, fusion of left and right axohydrocoels anterior to the mouth, narrow intestine, and dorsal flexor muscles. Traits peculiar to feeding larvae of echinoids include posterior dilator muscles between mouth and skeletal rods, and more than two separately initiated skeletal rods. Larval traits peculiar to single classes clearly distinguish the classes, but do not help resolve relationships between the classes.

Many other larval traits vary within classes, and some have implications for relationships within classes. These include anterior attachment organs of the asteroid brachiolaria (absent in the Luidiidae and Astropectinidae, and arranged differently in forcipulates and spinulosids) (Mortensen 1931, 1937, 1938; Barker 1978) and reorganization of direction of ciliary beat on the paired arms of late stage larvae of forcipulate asteroids (Gemmill 1914; Strathmann 1971). In echinoids fenestration of postoral and posterodorsal arm rods, the varying sites of formation and branches of skeletal rods, rearrangement of the ciliary band to form locomotory epaulettes or lobes, and early formation of pedicellariae have been useful in distinguishing relationships within the class (Mortensen 1921, 1937), but have no bearing on relationships between classes.

Have other investigators found better larval characters for inferring relationships among classes? None that I have seen published. Smith (1984) used a dozen embryonic and larval characters from the literature, and these also collapse to few independent characters useful in inferring relationships among classes. Smith's trait No. 2 (suppression of larval coeloms on the right side) occurs in all five extant classes and, therefore, does not resolve relationships among them. Trait Nos. 4, 26, and 40 (presence of larval processes, presence of larval mouth, closure of blastopore, respectively) confound comparisons among classes with comparisons of feeding and non-feeding larval forms. If trait No. 11 (no larval vestibule) were accepted, it would now be scored as both present and absent in echinoids, because R. Emlet (personal communication) has found no vestibular invagination in a species of the cidaroid *Eucidaris*. I excluded this trait because the homology of the euechinoid amniotic invagination, and the alterations of the larval oral cavity of holothuroids and ophiuroids at metamorphosis seems extremely doubtful, despite Hyman's (1955) suggestion. I excluded trait No. 43 (vestibule sealed from exterior or remaining open) because Smith considered an open vestibule an autapomorphy of ophiuroids, and the trait therefore does not resolve relationships between classes. I used Smith's trait No. 22 (reinterpreted here as time of ingression of mesenchyme relative to time of invagination of the archenteron). Trait Nos. 29 and 30 (elongate processes, processes supported by calcite rods, respectively) are not independent if they are taken as exclusive traits of ophioplutei and echinoplutei, because in these larvae arm extension depends on the skeletal rods. If some asteroids are included, then variation among asteroid larvae becomes a problem, because some have arms as long as many plutei, but others have arms almost as short as the processes of auriculariae. These two traits therefore collapsed to one in my table. I used trait No. 31 (adult mouth at a new site on the lower left side of the larvae), but corrected its distribution among classes. Smith's trait No. 16 (presence or absence of a vestige of the right hydrocoel) was excluded because its state in the crinoids could be influenced by lack of larval feeding, and it would, therefore, have been a trait restricted to holothuroids in my list. It should also be noted that the post-metamorphic fate of the right axohydrocoel varies among classes that have it (Olsen 1942; Hyman 1955). Table 5.1 includes a modification of Smith's trait No. 18 (presence or absence of attachment by preoral lobe). Thus, Smith's set also reduces to few functionally and developmentally independent synapomorphies.

The larval traits listed by Smiley (this volume) also reduce to few that are both independent of each other and relate subsets of five echinoderm classes. I used his suggestion on formation of the first coelomic cavity. I did not use the presence or absence of a right anterior coelom; because I could not determine the comparable trait in crinoids, absence of the right anterior coelom became unique to holothuroids (an autapomorphy) and, therefore, of no use in resolving relationships among classes. I did not use presence or absence of the left axocoel because Smiley scored it is as an autapomorphy of holothuroids. I used site of formation of the adult mouth instead of presence of a persistent adult rudiment on the larva, because a persistent rudiment results from formation of the adult mouth at a new site in echinoids and asteroids. In ophiuroids in which the posterolateral arms are retained as a locomotory device, one has the impression of a persistent adult rudiment, but retention of posterolateral arms seems different from maintaining a full larval body that contains the rudiment of a juvenile. Also retention of posterolateral arms till late in metamorphosis is a trait that varies within the class Ophiuroidea. The effect of my interpretation was to bring ophiuroids closer to holothuroids. I used presence of an anterior attachment organ, but scored asteroids as having both character states and scored it as absent in the outgroup because I did not include the pterobranchs. A hydrostatic (or gelatinous) larval skeleton is the same as absence of skeletal rods in my character set. Smiley's traits are distributed somewhat differently from my own list but also reduce to few independent synapomorphies. Our differences in interpretation of traits are a further indication of uncertain implications of echinoderm larval traits for relations among classes.

A few problems of polarity

Outgroups are often used to root phylogenetic trees (Brooks and Wiley 1985; de Queiroz 1985). The tornaria is a reasonable choice for this purpose, but it is unfortunate that many of the traits that the tornaria larva shares with echinoderm larvae can be construed as the absence of a structure and therefore imply less about the directions of transitions among echinoderm classes than would complex structures shared with a subset of echinoderm classes. This is the case for the anterior attachment organ, formation of the adult mouth on the lower left, skeletal rods, the doliolaria stage, and the paroral bands. Only the timing of ingression of mesenchyme, and the formation and position of the first coelomic cavity appear to be alternative character states in the enteropneusts, not merely absence of a structure. (Smiley has improved this situation slightly by using a composite of enteropneust and pterobranch traits for the outgroup. In his scheme the pterobranchs provided an anterior attachment organ as a possible homologue.) This limits the tornaria larva as a device for rooting a phylogenetic tree for echinoderm classes.

The ontogenetic criterion for polarity is an extension of von Baer's law that the more general characters of the large group of animals to which the animal belongs appear in development earlier than the special characters (Patterson 1983; de Queiroz 1985; Kluge and Strauss 1985). The traits in Table 5.1 do not form a pattern of successive divergences in order of development, but rather a pattern of successive or simultaneous incongruencies. Later ingression of mesenchyme separates asteroids from three other classes, then coelom formation separates holothuroids from three other classes, and so forth. There is no pattern of successive divergences upon which to base a phylogenetic tree.

For each trait one can ask whether transitions are equally likely in either direction. It is easy to find an evolutionary bias in the transition from a feeding larva to a nonfeeding larva because of the loss of complex structures (Strathmann 1978a,b) but much more difficult to find a bias in the transition from one kind of feeding larva to another. Skeletons are an example. The pluteus skeleton is conservative within classes, only lost when the capacity to feed is lost. There are striking similarities in detail, such as the anterior dilator muscles between mouth and anterolateral arms, and differences in detail, such as different numbers of cells in these muscles in echinoplutei and ophioplutei (Strathmann 1971). The larval skeleton is so complex in both ophiuroids and echinoids that hypotheses on functional intermediate stages in its evolutionary origin seem strained, yet the formation and branching of the skeleton is so different between classes that homology can be questioned. Within the Echinoidea, the parsimonious interpretation is that unfenestrated postoral and posterodorsal rods are secondarily simplified; thus, skeletal evolution is not always in the direction of greater complexity. The posterior ossicle in the holothuroid larva could be a vestige of a pluteus skeleton, or perhaps it evolved as a counterweight that helps orientate the larva in the water and is functionally similar to the precursor of the pluteus skeleton. At present, evolution from simple to complex and from complex to simple seem equally likely for the traits in Table 5.1, but further studies of development and function might indicate biases in evolutionary change.

EARLY DIVERGENCE AND LATER CONVERGENCE IN ONTOGENY

According to von Baer's first law, the more general characters in a group appear early and the more specific characters appear later. For species within a taxon embryos should resemble each other more than adults, and anatomical differences should increase during development. This is not the case for many anatomical features of echinoderms, nor for many other taxa, yet von Baer's law has been well regarded by the same commentators (Russell 1916; Garstang 1922; de Beer 1940; Waddington 1956; Gould 1977; Raff and Kaufman 1983) who consider Haeckel's biogenetic law (that ontogeny recapitulates phylogeny) to be of limited application. Most of these commentators have noted exceptions to von Baer's law, but nevertheless, considered it to have greater generality than Haeckel's biogenetic law. This opinion appears to be based on an interpretation of Haeckel's law that emphasizes addition of new characters at the end of development and a transfer of formerly adult characters to earlier stages so that development represents a sequence of forms similar to the evolutionary sequence of adult ancestors. In contrast, the common evolutionary interpretation of von Baer's law is that earlier developmental stages are more similar than later stages, but embryos may simply resemble ancestral embryos. A developmental sequence need not resemble an evolutionary sequence of ancestral adult forms. Most commentators' good opinion of von Baer's law is not based on sampling organisms but rather on an assessment of the evolutionary processes that might produce the predicted pattern in development of form (see also Løvtrup 1978; Gould 1979; Patterson 1983). Von Baer's law would be more useful if we knew where we should expect it to apply and where not. A first step is to examine possible evolutionary causes for the predicted progressive divergence in form during development. This may give some insight into conditions under which this law applies.

Hypothesis 1: a change in an early step in morphogenesis affects later steps

The most frequently suggested evolutionary basis for von Baer's law is that a change in one step in development affects subsequent steps (Darwin 1872; Goldschmidt 1940; de Beer 1951; Haldane 1932; Waddington 1956; Simpson 1953; Rensch 1959; Stebbins 1974; Riedl 1978; Maynard Smith 1983, and many others). Most of those who suggested this evolutionary basis also qualified their statement by noting exceptions in which a series of steps terminates early in development and an evolutionary change in an early stage is, therefore, not propagated to later stages.

It is widely recognized that a change in early development will not propagate beyond the end of the sequence of steps; however, it is easy to overlook the large number of structures that appear early in development, but are terminal in that they are not transformed into later structures. These include most parts of sperm and polar bodies (Sander 1983), extraembryonic membranes and their precursors in the ovum, nutrient supplies for prefeeding stages of development, and most of the larval body. Parts of the larval body that are usually resorbed or cast away at metamorphosis include structures for feeding and locomotion when these differ from those in the post-metamorphic juvenile. In echinoderms, this is most of the body of a feeding larva. In echinoderm species without feeding larvae most of the structures of feeding larvae are lost, with no effect on the adult form. Giard (1904) remarked on this developmental divergence in larvae followed by a convergence in form of adult.

Structures that are terminal in larval or embryonic stages provide the exceptions to von Baer's law that are most often cited, but these exceptions are even more numerous than is commonly recognized because development is usually viewed retrospectively. In the retrospective view, one picks a structure and then sees how it develops. In this case development appears to be a sequence of steps leading to an end. For many biologists the end is the adult, sometimes called the 'definitive form'. If, instead, one samples structures at a given stage, one finds many structures and processes that are stage specific, and whose function is restricted to that stage. The retrospective view of development contributes to faith in von Baer's law through a sampling bias. If terminal structures were the only exceptions to von Baer's law, one could apply a single criterion to determine where the law applies and where not, but structures terminating in early stages are not the only ones that vary independently of later structures.

Exceptions to von Baer's law also occur among structures that are directly involved in the development of later structures. These should be structures of high burden in Riedl's (1978) sense, but they are not. In vertebrates the 'pharyngula' stage is more conservative than the preceding gastrula (Ballard 1976, 1981; Elinson 1987). In insects, the germ band stage is more uniform than preceding stages (Sander 1983). Waddington (1956) noted that 'it is at an intermediate period, early but not right at the beginning, that embryos are most alike'. Most of the earlier embryonic structures are not terminal. One structure is transformed into another, and one might expect later steps to depend on earlier steps; yet the embryos diverge in form and then converge at a later stage.

Sander (1983) suggests that evolution of variation in early embryos may result from redundancy in developmental processes. With redundancy, even though a mutation changes one pathway, an alternate pathway exists to assure the same later result. Another possibility is that regulative processes tend to bring animals to the same stage and form despite evolutionary changes in earlier stages. Experiments on embryos of echinoids demonstrate processes that return embryos to normal development after severe perturbations in early development. Some experimental perturbations of embryos can be taken as models of mutations that might affect processes in early development.

Cortical granule breakdown at fertilization of euchinoids provides a protective fertilization membrane that excludes additional sperm and a hyaline layer that holds cells in position. When these extraembryonic membranes are removed or when cortical granule breakdown is blocked by high pressure (Chase 1967), the cells of the embryo form loose aggregates quite different in form from normal embryos; but other developmental processes at the cell's surfaces continue (Schroeder 1986), and the cells can reaggregate and form a normal embryo at the blastula stage. From there morphogenesis appears normal. Chase (1967) obtained 16 normal plutei out of 38 monospermic embryos with cortical granules intact. The hyaline layer may increase reliability in early development, but at least some embryos can develop without it. Any arrangement that keeps cells touching and prevents polyspermy would appear adequate for a high percentage of normal development. A hyaline layer is reduced or absent in cidaroid echinoids (Schroeder 1986) and in several classes of echinoderms (Holland 1981). In asteroids, the cells of the early embryo appear to hold more loosely, often assuming a tetrahedral arrangement at the four cell stage. The echinoid hyaline layer is an unessential structure that nevertheless affects the form of the echinoid embryo during development.

Dissociation of echinoid embryos at the mesenchyme blastula stage or later is also followed by reaggregation (Giudice 1973), and under suitable conditions more than half the reaggregated embryos continue development to nearly normal plutei (Spiegel and Spiegel 1980, and personal communication). Gastrulae of asteroids can also be dissociated into small groups of cells which reaggregate to form bipinnariae (Dan-Sohkawa et al. 1986; Yamanaka et al. 1986), although Dan-Sohkawa et al. showed that bipinnariae of normal form came from smaller than normal reaggregates and that reaggregates of normal size tended to form larvae with more than one anus. The normal morphogenetic movements of gastrulation can be experimentally circumvented to a remarkable extent. Experiments such as these suggest that extensive evolutionary changes in early development may occur without an effect on later stages.

A non-essential structure or process is often dismissed as unimportant in experimental embryology, but this conclusion may be premature. Some features of development increase reliability even though some embryos develop without them. The percentage of normal larvae is greater in normal development than from disaggregated embryos. An extreme possibility is that all processes in some developmental events are redundant when taken a few at a time. In such a case, any subset of a group of processes could be removed and the remainder would provide a return to normal development, while elimination of the whole group would be lethal. In that case, any one trait would be redundant, but some subset would be essential. Selection for developmental reliability or speed might maintain a degree of redundancy that provides latitude for evolutionary change in early stages.

Regulation can also return an embryo to a conserved later form after an earlier perturbation. Comparative and experimental studies combine to demonstrate that a change in egg size changes the initial form of a pluteus, but that this early divergence in form is followed by a regulatory convergence at a later stage. Species of *Strongylocentrotus* vary in size of eggs, but not in size at metamorphosis (McEdward 1986). Sinervo and McEdward (1984) experimentally reduced the egg size of *S. droebachiensis* by separation of blastomeres at the two-and four-cell stage. The early plutei from the quarter embryo resembled the normal pluteus of *S. purpuratus*, whose egg is about one-sixth of the volume of that of *S. droebachiensis*. As quarter and half embryos of *S. droebachiensis*, and normal embryos of *S. purpuratus* developed, they converged on the size and form of the normal *S. droebachiensis* pluteus until all plutei were of similar size and form at about the six-armed stage. An acute experimental change in egg size therefore mimics an evolutionary change in egg size. The two species converge on the same larval form during development, and regulation following separation of blastomeres follows the same pattern. This is a clear exception to von Baer's law, and it occurs within a continuous developmental sequence. This is not a case of variation in terminal structures or a case of redundant pathways. Both experimental and evolutionary changes in egg size produced early deviation in form followed by developmental convergence to the same form.

Hypothesis 2: earlier stages live in less demanding or diversified habitats

Often earlier stages are protected within an egg membrane or the mother's body. Requirements for defence, feeding, or locomotion are reduced. One might expect a simpler and less divergent form for such protected stages, but divergence in early development does not appear to be correlated with degree of parental protection. In forcipulate and spinulosid asteroids, externally brooded larvae usually resemble non-feeding planktonic larvae even though they are protected by the mother and hatch at a later stage. Brooded ophiuroid larvae (Mortensen 1921; Fell 1946) can differ greatly in form between species. A correlation between divergence in form and change in habitat during development is not apparent.

Hypothesis 3: progressive divergence is correlated with increasing complexity during development

Holmes (1944) suggested that divergence is more likely to be recognized in later stages because they are more complex. This is similar to the argument of Schopf et al. (1975) that degree of perceived taxonomic change in lineages is related to their general morphological complexity. Schopf et al. used the number of descriptive terms per taxon to measure morphological complexity. In a large group there could be more terms for adults than larvae or embryos because adults have diverged more, but a count of terms for larva and adult of a single species avoids this bias and does suggest a difference in perceived complexity. Mortensen (1928, 1937) used 15 terms for the skeleton of the eight-armed pluteus and 29 terms for a description of the adult of the cidarid echinoid *Eucidaris metularia*. This example could be multiplied for other species and for soft parts with the same qualitative result. For echinoderms and many other taxa, the morphological complexity recognized in the adults is greater than the morphological complexity recognized in the embryos and larvae. An increase in perceived morphological complexity during development could give an impression of greater divergence between adults of different species than between larvae or embryos.

The perceived increase in morphological complexity during development probably represents a real increase in morphological complexity that does provide greater potential for divergence of later stages. One reason for increased complexity during development is increase in size (Bonner 1965). Growth and its consequences may provide more routes for evolutionary divergence in later stages of development.

Number of parts often increases as an organism grows, and the increasing number of parts may become more than sufficient to carry out the original function. Some of the parts added as a consequence of growth may, therefore, be free to take on other forms and functions. Larger stages can have more specialized parts simply because they have more parts. An adult echinoid with many skeletal plates and spines may have more ways to diverge in subsets of these skeletal elements than a larva with fewer skeletal rods.

Parts often increase in size as organisms grow. Components of these parts must continue to function on a small scale while the entire part functions on a large scale. This usually results in increased complexity of parts with increasing size of parts. The skeleton in adult echinoderms remains complex on the small scale of the stereom (Smith 1980) while possessing additional features on a larger scale. The difference in complexity reflected in Mortensen's terms for the larval and the adult skeleton underestimates the actual increase in complexity because he used a high magnification in viewing the larva than in viewing the adult.

Larger stages have more scope for divergence both within and between parts. This basis for von Baer's law provides the potential for progressive divergence during development, but does not require it as a necessary consequence of developmental processes.

Conclusion on the evolutionary basis of von Baer's law

The proposed evolutionary causes of progressive divergence during development point to tendencies only, not strong developmental constraints. Exceptions to von Baer's law are not limited to terminal structures. Divergence in form followed by convergence in form is common within embryonic and larval development.

CONCLUSIONS

1. Absence of feeding larvae in crinoids restricts the use of larval traits in relating crinoids to other classes.
2. Phylogenetic trees for the four classes with feeding larvae necessarily imply striking convergences.
3. Phylogenetic trees based on larval traits of the four classes with feeding larvae are sensitive to small changes in assumptions about traits.
4. The taxonomic distribution of larval traits includes violations of von Baer's first law, that animals progressively diverge in form during development.
5. Suggested evolutionary causes for von Baer's first law are weak because of terminal structures, redundancy, and regulatory processes in development. Divergence followed by convergence in development is common, and cases where von Baer's law is a predictable guide cannot yet be predicted.

ACKNOWLEDGEMENTS

NSF grant OCE8606850, the Friday Harbor Laboratories, and the Department of Zoology of the University of Washington supported this study. I was helped by conversations with R. Emlet, J. Felsenstein, R. Huey, R. P. S. Jefferies, K. Kardong, L. McEdward, F. Schram, T. Schroeder, S. Smiley and C. Staude, and visiting scientists at the Friday Harbor Laboratories.

REFERENCES

Ballard, W. W. 1976. Problems of gastrulation: real and verbal. *Bioscience* **26**, 36–9.

Ballard, W. W. 1981. Morphogenetic movements and fate maps of vertebrates. *American Zoologist* **21**, 391–9.

Barker, M. F. 1978. Structure of the organs of attachment of brachiolaria larvae of *Stichaster australis* (Verrill) and *Coscinasterias calamara* (Gray) (Echinodermata: Asteroidea). *Journal of Experimental Marine Biology and Ecology* **33**, 1–36.

Beer, C., de 1940. *Embryos and ancestors.* Clarendon, Oxford.

Bonner, J. T. 1965. *Size and cycle.* Princeton University Press, Princeton, N.J.

Brooks, D. R. and Wiley, E. O. 1985. Theories and methods in different approaches to phylogenetic systematics. *Cladistics* **1**, 1–11.

Burke, R. D. 1978. The structure of the nervous system of the pluteus larva of *Strongylocentrotus purpuratus*. *Cell and Tissue Research* **191**, 233–47.

Burke, R. D. 1983a. Development of the larval nervous system of the sand dollar *Dendraster excentricus*. *Zoomorphologie* **98**, 209–25.

Burke, R. D. 1983b. The structure of the larval nervous system of *Pisaster ochraceus* (Echinodermata: Asteroidea). *Journal of Morphology* **178**, 23–35.

Burke, R. D., Brand, D. G. and Bisgrove, B. W. 1986.

Structure of the nervous system of the auricularia larva of *Parastichopus californicus*. *Biological Bulletin of the Marine Biology Laboratory, Woods Hole*, **170**, 450–60.

Bury, H. 1888. The early stages in the development of *Antedon rosacea*. *Philosophical Transactions of the Royal Society of London, Series B* **179**, 257–301, pl. 43–7.

Chase, D. G. 1967. Inhibition of the cortical reaction with high hydrostatic pressure and its effects on the fertilization and early development of sea urchin eggs. PhD Thesis, University of Washington.

Chia, F. S., Burke, R. D., Koss, R., Mladenov, P. V., and Rumrill, S. S. 1986. Fine structure of the doliolaria larva of the feather star *Florometra serratissima* (Echinodermata: Crinoidea), with special emphasis on the nervous system. *Journal of Morphology* **189**, 99–120.

Crawford, B. and Abed, M. 1986. Ultrastructural aspects of the surface coatings of eggs and larvae of the starfish, *Pisaster ochraceus*, revealed by alcian blue. *Journal of Morphology* **187**, 23–37.

Dan, K. 1960. Cyto-embryology of echinoderms and amphibia. *International Reviews in Cytology* **9**, 321–67.

Dan-Sohkawa, M., Yamanaka, H., and Watanabe, K. 1986. Reconstruction of bipinnaria larvae from dissociated embryonic cells of the starfish, *Asterina pectinifera*. *Journal of Embryology and Experimental Morphology* **94**, 47–60.

Darwin, C. 1872. *The origin of species*, 6th edn, reprinted 1958. Mentor, New York.

Elinson, R. P. 1987. Change in developmental patterns: embryos of amphibians with large eggs. In *Development as an evolutionary process* (ed. R. A. Raff and E. C. Raff). Alan R. Liss, New York.

Emlet, R. B. 1986. Facultative planktotrophy in the tropical echinoid *Clypeaster rosaceus* (Linnaeus) and a comparison with obligate planktotrophy in *Clypeaster subdepressus* (Gray) (Clypeasteroida: Echinoidea). *Journal of Experimental Marine Biology and Ecology* **95**, 183–202.

Fell, H. B. 1946. The embryology of the viviparous ophiuroid *Amphipholis squamata* (delle Chiajei). *Transactions of the Royal Society of New Zealand* **75**, 419–64.

Fell, H. B. and Pawson, D. L. 1966. General biology of echinoderms. In *Physiology of Echinodermata* (ed. R. A. Boolotian), pp. 1–48. Interscience, New York.

Felsenstein, J. 1982. Numerical methods for inferring evolutionary trees. *Quarterly Reviews in Biology* **57**, 379–404.

Felsenstein, J. 1983. Parsimony in systematics: biological and statistical issues. *Annual Review in Ecology and Systematics* **14**, 313–33.

Fenaux, L. 1963. Note préliminaire sur le développement larvaire de *Amphiura chiajei* (Forbes). *Vie et Milieu* **14**, 91–6.

Garstang, W. 1922. The theory of recapitulation: a critical restatement of the biogenetic law. *Journal of the Linnean Society, Zoology* **35**, 81–101.

Gemmill, J. F. 1916. The development and certain points in the adult structure of the starfish *Asterias rubens* L. *Philosophical Transactions of the Royal Society of London, Series B*, **205**, 213–94.

Giard, A. 1904. La poecilogonie. *Sixième Congres international de Zoologie, Compte Rendue des Séances* **6**, 617–46.

Giudice, G. 1973. *Developmental biology of the sea urchin embryo*. Academic Press, New York.

Goldschmidt, R. 1940. *The material basis of evolution*. Yale University Press, New Haven.

Gould, S. J. 1977. *Ontogeny and phylogeny*. Harvard University Press, Cambridge.

Gould, S. J. 1979. On the importance of heterochrony for evolutionary biology. *Systematic Zoology* **28**, 224–6.

Grave, C. 1903. On the occurrence among echinoderms of larvae with cilia arranged in transverse rings, with a suggestion as to their significance. *Biological Bulletin of the Marine Biology Laboratory, Woods Hole*, **5**, 169–86.

Haldane, J. B. S. 1932. The time of action of genes and its bearing on some evolutionary problems. *American Naturalist* **66**, 5–24.

Hendler, G. 1977. Development of *Amphioplus abditus* (Verrill) (Echinodermata: Ophiuroidea). I. Larval Biology. *Biological Bulletin of the Marine Biology Laboratory, Woods Hole*, **152**, 51–63.

Hendler, G. 1978. Development of *Amphioplus abditus* (Verrill) (Echinodermata: Ophiuroidea). II. Description and discussion of ophiuroid skeletal ontogeny and homologies. *Biological Bulletin of the Marine Biology Laboratory, Woods Hole*, **154**, 79–95.

Hendler, G. 1982. An echinoderm vitellaria with a bilateral larval skeleton: evidence for the evolution of ophiuroid vitellariae from ophioplutei. *Biological Bulletin of the Marine Biology Laboratory, Woods Hole*, **163**, 431–7.

Holland, N. D. 1979. Electron microscope study of the cortical reaction of an ophiuroid echinoderm. *Tissue and Cell* **11**, 445–55.

Holland, N. D. 1981. Electron microscope study of development in a sea cucumber, *Stichopus tremulus* (Holothuroidea) from an unfertilized egg through hatched blastula. *Acta Zoologica* **62**, 89–111.

Holmes, S. J. 1944. Recapitulation and its supposed causes. *Quarterly Reviews in Biology* **19**, 319–31.

Hörstadius, S. 1939. Über die Larve von *Holothuria poli* Delle Chiaje. *Arkiv für Zoologie* **31A(14)**, 1–15.

Hyman, L. H. 1955. *The invertebrates. Vol. IV. Echinodermata*. McGraw-Hill, New York.

Hyman, L. H. 1959. *The invertebrates. Vol. V. Smaller coelomate groups.* McGraw-Hill, New York.

Ivanova-Kasas, O. M. 1973. The nature of barrel-shaped larvae of echinoderms. *Zoologicheskii Zhurnal* **52**, 883–90.

Jägersten, G. 1972. *Evolution of the metazoan life cycle.* Academic Press, New York.

Kaufman, Z. S. 1968. Postembryonic period of development of some White Sea starfishes. *Doklady Akademii Nauk Soyuza Sovetskikh Sotsialisticheskikh Respublik* **181**, 1009–12.

Kluge, A. G. and Strauss, R. E. 1985. Ontogeny and systematics. *Annual Review in Ecology and Systematics* **16**, 247–68.

Komatsu, M. 1975. On the development of the sea-star, *Astropecten latespinosus* Meissner. *Biological Bulletin of the Marine Biology Laboratory, Woods Hole* **148**, 49–59.

Komatsu, M. 1982. Development of the sea-star *Ctenopleura fisheri. Marine Biology* **66**, 199–205.

Lacalli, T. C. and West, J. E. 1986. Ciliary band formation in the doliolaria larva of *Florometra. Journal of Embryology and Experimental Morphology* **96**, 303–23.

Løvtrup, S. 1978. On von Baerian and Haeckelian recapitulation. *Systematic Zoology* **27**, 348–52.

MacBride, E. W. 1914. *Text-book of embryology. Vol. 1. Invertebrata.* Macmillan, London.

MacBride, E. W. 1923. Echinoderm larvae and their bearing on classification. *Nature, London* **111**, 47, 323–4.

Maynard Smith, J. 1983. Evolution and development. In *Development and evolution* (ed. B. C. Goodwin, N. Holder, and C. C. Wylie), pp. 33–45. Cambridge University Press, Cambridge.

McEdward, L. R. 1986. Comparative morphometrics of echinoderm larvae. II. Larval size, shape, growth, and the scaling of feeding and metabolism in echinoplutei. *Journal of Experimental Marine Biology and Ecology* **96**, 267–86.

McEuen, F. S. 1986. The reproductive biology and development of twelve species of holothuroids from the San Juan Islands, Washington. PhD Thesis, University of Alberta.

Mladenov, P. V. 1985. Development and metamorphosis of the brittle star *Ophiocomina pumila*: evolutionary and ecological implications. *Biological Bulletin of the Marine Biology Laboratory, Woods Hole* **168**, 285–95.

Mladenov, P. V. and Chia, F. S. 1983. Development, settling behaviour, metamorphosis and pentacrinoid feeding and growth of the feather star *Florometra serratissima. Marine Biology* **73**, 309–23.

Mortensen, T. 1921. *Studies on the development and larval forms of echinoderms.* G. E. C. Gad, Copenhagen.

Mortensen, T. 1923. Echinoderm larvae and their bearing on classification. *Nature, London*, **111**, 322–3.

Mortensen, T. 1928. *Monograph of the Echinoidea. Vol. 1. Cidaroidea.* C. A. Reitzel Publishing, Copenhagen.

Mortensen, T. 1931. Contributions to the study of the development and larval forms of echinoderms. 1–11. *Kongelige Danske Videnskabernes Selskabs Skrifter* **4(1)**, 1–39.

Mortensen, T. 1938. Contributions to the study of the development and larval forms of echinoderms. IV. *Kongelige Danske Videnskabernes Selskabs Skrifter* **7(3)**, 1–59.

Mortensen, T. 1938. Contributions to the study of the development and larval forms of echinoderms. IV. *Kongelige Danske Videnskabernes Selskabs Skrifter* **7(3)**, 1–59.

Nakajima, Y. 1986. Presence of a ciliary patch in preoral epithelium of sea urchin plutei. *Development Growth and Differentiation* **28**, 243–9.

Oguro, C., Komatsu, M., and Kano, Y. T. 1976. Development and metamorphosis of the sea-star, *Astropecten scoparius* Valenciennes. *Biological Bulletin of the Marine Biology Laboratory, Woods Hole*, **151**, 560–73.

Olsen, H. 1942. The development of the brittle-star *Ophiopholis aculeata* (O. Fr. Müller), with a short report on the outer hyaline layer. *Bergens Museum Årbok Afhandlingar og Arsberetning* **6**, 1–107.

Patterson, C. 1983. How does phylogeny differ from ontogeny? In *Development and evolution* (ed. B. C. Goodwin, N. Holder, and C. C. Wylie), pp. 1–31. Cambridge University Press, Cambridge.

Queiroz, K., de. 1985. The ontogenetic method for determining character polarity and its relevance to phylogenetic systematics. *Systematic Zoology* **34**, 280–90.

Raff, R. A. and Kaufman, T. C. 1983. *Embryos, genes, and evolution.* Macmillan, New York.

Rensch, B. 1959. *Evolution above the species level.* John Wiley and Sons, New York.

Riedl, R. 1978. *Order in living organisms* (translated by R. P. S. Jefferies). John Wiley and Sons, New York.

Rieppel, O. 1980. Homology a deductive concept? *Zeitschrift für Zoologische Systematik und Evolutionsforschung* **18**, 315–9.

Runström, S. 1927. Über die Entwicklung von *Leptosynapta inhaerens. Bergens Museum Årbok Afhandlingar og Arsberetning* **(1)**, 1–80.

Russell, E. S. 1916. *Form and function. A contribution to the history of animal morphology*, reprinted 1982. University of Chicago Press, Chicago.

Sander, K. 1983. The evolution of patterning mechanisms: gleanings from insect embryogenesis and spermatogenesis. In *Development and evolution* (ed. B. C. Goodwin, N. Holder, and C. C. Wylie), pp. 137–59. Cambridge University Press, Cambridge.

Schopf, T. J. M., Raup, D. M., Gould, S. J., and Simberloff, D. S. 1975. Genomic versus morphologic rates of evolution: influence of morphologic complexity. *Paleobiology* **1**, 63–70.

Schroeder, T. E. 1981. Development of a "primitive" sea urchin (*Eucidaris tribuloides*): irregularities in the hyaline layer, micromeres, and primary mesenchyme. *Biological Bulletin of the Marine Biology Laboratory, Woods Hole* **161**, 141–51.

Schroeder, T. E. 1986. The egg cortex in early development of sea urchins and starfish. In *Developmental biology. Vol.* 2 (ed. L. W. Browder), pp. 59–100. Plenum, New York.

Simpson, G. 1953. *The major features of evolution.* Colombia University Press, New York.

Sinervo, B. R. and McEdward, L. R. 1984. The effect of experimentally reduced egg size on form, function, and rate of development of planktotrophic larval echinoids. *American Zoologist* **24**, 131A.

Smith, A. B. 1980. Stereom microstructure of the echinoid test. *Special Papers in Palaeontology* **25**, 1–81.

Smith, A. B. 1984. Classification of the Echinodermata. *Paleontology* **27**, 431–59.

Spiegel, E. and Spiegel, M. 1980. The internal clock of reaggregating embryonic sea urchin cells. *Journal of Experimental Zoology* **213**, 271–81.

Stebbins, G. L. 1974. *Flowering plants: evolution above the species level.* Harvard University Press, Cambridge, Mass.

Stiasny-Wijnhoff, G. and Stiasny, G. 1927. Die Tornarien. *Ergebnisse und Fortschritte der Zoologie* **7**, 41–204.

Strathmann, R. R. 1971. The feeding behavior of planktotrophic echinoderm larvae: mechanisms, regulation, and rates of suspension-feeding. *Journal of Experimental Marine Biology and Ecology* **6**, 109–60.

Strathmann, R. R. 1974. Introduction to function and adaptation in echinoderm larvae. *Thalassia Jugoslavica* **10**, 321–39.

Strathmann, R. R. 1978a. The evolution and loss of feeding larval stages of marine invertebrates. *Evolution* **32**, 894–906.

Strathmann, R. R. 1978b. Progressive vacating of adaptive types during the Phanerozoic. *Evolution* **32**, 907–14.

Strathmann, R. R. and Bonar, D. 1976. Ciliary feeding of tornaria larvae of *Ptychodera flava* (Hemichordata: Enteropneusta). *Marine Biology* **34**, 317–24.

Waddington, C. H. 1956. *Principles of embryology.* Allen and Unwin, London.

Watrous, L. E. and Wheeler, Q. D. 1981. The out-group comparison method of character analysis. *Systematic Zoology* **30**, 1–11.

Williams, D. H. C. and Anderson, D. T. 1975. The reproductive system, embryonic development, larval development and metamorphosis of the sea urchin *Heliocidaris erythrogramma* (Val.) (Echinoidea: Echinometridae). *Australian Journal of Zoology* **23**, 371–403.

Yamanaka, H., Tanaka-Ohmura, Y., and Dan-Sohkawa, M. 1986. What do dissociated embryonic cells of the starfish, *Asterina pectinifer 1*, do to reconstruct bipinnaria larvae? *Journal of Embryology and Experimental Morphology* **94**, 61–71.

6

The phylogenetic relationships of holothurians: a cladistic analysis of the extant echinoderm classes

SCOTT SMILEY

Department of Biochemistry and Biophysics, University of California, San Francisco, California 94143, USA

INTRODUCTION

The phylogenetic relationship of holothurians to the other extant echinoderm classes has been debated for nearly a century. Bell (1891) excluded the holothurians from the Eleutherozoa, and Bather (1900), while making a strong case for the inclusion of asteroids and ophiuroids in the Stellarozoa, considered holothurians quite distinct from other living or fossil classes. Study of echinoderm embryology, and particularly metamorphosis, led Bury (1895) to note that, while holothurians shared several characters with echinoids, he could not resolve their placement within the phylum until more information was available. However, MacBride (1906, 1914), another echinoderm embryologist, proposed that holothurians and echinoids should be allied in the Echinozoa.

More recently, Fell (1963a,b) grouped the extant non-crinoid classes into two subphyla; the Asterozoa consisting of the asteroids and ophiuroids, and the Echinozoa, consisting of holothurians and echinoids. Fell's arguments primarily concerned uncovering the close relationship between the asterozoans, but he used similarities between the echinozoans as ancillary corroboration for his thesis. Four main lines of evidence are used to support the hypothesis that holothurians and echinoids share a unique common ancestor.

(1) They both possess a calcified structure surrounding the anterior gut. This is the Aristotle's lantern of echinoids, and the calcareous ring of holothurians.

(2) Both have enclosed ambulacra, and the process of enclosure was reported to be the same (Hyman 1955).
(3) Both are said to possess a vestibule, an epidermal invagination within which the larval hydrocoel is said to organize the pentaradial symmetry of the adult during metamorphosis.
(4) Adults of both share meridional growth gradients, contrasted with radial growth gradients in asterozoans (Fell 1963a,b).

My recent investigation into the process of metamorphosis of the holothurian *Stichopus californicus* has shown that the ambulacra of this holothurian form in a distinctly different manner from that reported for ophiuroids and echinoids (Smiley 1986). Furthermore, since holothurians have neither axial nor visceral torsion during metamorphosis, the symmetry patterns underlying adult growth are quite distinct from all other extant echinoderms, whether these grow meridionally or radially (Smiley 1986). Finally, the structure called a vestibule in holothurians is nothing more than the larval oral cavity. This invaginates as the stomodaeum during late gastrulation and is homologous with the stomodaeal invagination that becomes the larval oral cavity in all the extant classes. It is not homologous with the echinoid vestibule (Smiley 1986). With many of the points which traditionally supported the concept of the Echinozoa now in question, it is timely to retest the hypothesis of ancestral relationship between holothurians and echinoids. A cladistic analysis of the relationships between the extant classes was recently published (Smith 1984a). However, lacking this new information on holothurian metamorphosis, and failing to make a clear distinction between direct and indirect developers,

Echinoderm phylogeny and evolutionary biology (ed. C. R. C. Paul and A. B. Smith), Clarendon Press, Oxford, 1988.

Smith inadvertantly misrepresented echinoderm embryological relationships.

In this paper, I will test the hypothesis that holothurians and echinoids share a unique common ancestor, a hypothesis implicit in the idea of the subphylum Echinozoa. I will present a suite of characters amassed from Smith's (1984a) paper, an exhaustive literature search, and my own recent investigation of holothurian metamorphosis (Smiley 1986). I include detailed arguments for the choice of each character, and the determination of each character state. I analyse these characters using recently developed computer methods (Felsenstein 1983, 1985). My results are not consistent with the hypothesis that holothurians share a unique ancestor in common with echinoids. They do support the interpretation that holothurians diverged from the ancestral echinoderm stock before the remaining extant classes.

MATERIALS AND METHODS

The major treatises of invertebrate zoology (Ludwig 1889–1892; Bather 1900; MacBride 1906; Cuénot 1948; Dawydoff 1948; Hyman 1955, 1959; Beklemischev 1969) are often too brief or are inaccurate on points of comparative embryology. Consequently, I have relied on primary sources in building the character suite. Since my expertise is in comparative embryology and anatomy, I restricted the character suite to extant classes.

Principles used in choosing the characters

In choosing embryological and metamorphic characters, I concentrated on indirectly developing species because their patterns of development are less abbreviated than those found in directly developing species (Hyman 1955; Strathmann 1976). More importantly, major alterations in the patterns of direct development, such as the absence of larval gut structures and of catastrophic morphogenic metamorphosis, make comparisons between larvae very difficult. I used characters from direct developers only when I was able to ascertain that the character was relatively invariant regardless of the mode of development (e.g. character 8). This choice complicated my analysis of the crinoids, because all known extant forms are direct developers.

It is difficult to identify an adequate number of characters for class level phylogenies, because, to be useful, the character must be shared by several classes. The antiquity of the separation of the echinoderm classes is reflected in the limited number of characters used in this analysis and in Smith's (1984a). Convergences pose only modest problems for species level phylogeny, because the time elapsed since individual species diverged within a genus is usually limited, and only occasionally do convergences on the derived character state crop up. The situation is significantly different with class level phylogenies. Numerous instances of character state convergence have occurred in the more than 400 million years since the divergence of the extant echinoderm classes. Because there are relatively few unambiguous useful characters, convergences cannot be outweighed by other trends in the data, and their effect becomes magnified. When we assert that a character state is derived and shared by two or more groups, we assert that the condition is homologous in these groups. I rejected characters where I found substantial evidence to support the contention that shared derived character states are the result of convergence not homology. I view homology as the most restrictive comparison between similar structures in related groups, and my definition of homology includes similarity in morphology, similarity in position, and similarity in ontogeny.

In the main, I avoided weighting characters. However, repeating characters is weighting of a sort, and I described several characters that all impinge on the process of metamorphic torsion. I did this because I interpret this torsion as part of an important chapter in the evolution of the phylum. I did not use the weighting option available in the PHYLIP programs for any character.

I chose hemichordates as the outgroup for determining the ancestral condition of a number of the characters. The similarity between the tornaria larva of enteropneusts and the larvae of indirectly developing echinoderms has long been noted (Bather 1900). The phylum Hemichordata consists of two classes; the enteropneusts and the pterobranchs, which are considered to be more ancestral (Dawydoff 1948; Hyman 1959; Beklemischev 1969). I had some difficulties assessing the character states in the outgroup because pterobranchs are wholly direct developers (Hyman 1959), although enteropneusts have a number of indirectly developing species. Therefore, I used looser restrictions in interpreting the characters and their states in the outgroup than I did in the echinoderm classes. Where information from the outgroup was useful, I used it. However, where establishing homology between the condition in the outgroup and the echinoderms was not possible, I used other criteria to determine character state polarization or I did not determine it, and used Wagner parsimony in the analysis.

Computer analysis

Computer analysis was done with the programs available in the PHYLIP package (Felsenstein 1985). Detailed explanation of the assumptions and logic used in these programs is available in the PHYLIP documentation. The programs in PHYLIP that I used allow one either to choose a defined ancestral state option (Camin–Sokal parsimony), or to make no choice (Wagner parsimony). Camin–Sokal parsimony restricts the direction of changes in the character state primarily toward the derived condition, whereas Wagner makes no such restriction (Felsenstein 1983, 1985). When I was unable to determine the ancestral state, for any reason, I used the Wagner option for that character. I ran all the programs without the outgroup data and using Wagner parsimony criteria alone, as a control to test the consequences of my character state determinations. This resulted in no different trees.

RESULTS

General

Each computer analysis generated the single tree depicted in Fig. 6.1. The characters and their state in each of the extant echinoderm classes and the outgroup are listed in Table 6.1, and the computer input data are listed in Table 6.2. An iterated version of the MIX program, run both with and without the outgroup; and using either mixed Camin–Sokal and Wagner parsimony criteria or Wagner criteria alone for each character, produced Tree 1 (Table 6.3). MIX builds trees by additions and rearrangements. An identical tree was obtained with the program PENNY, which uses a branch and bound method to determine the shortest tree, under the same sets of parsimony criteria. METRO also produces the same tree, under the same conditions. The METRO program uses a simulated annealing algorithm and is better, but slower than MIX. BOOTSTRAP (BOOTM)

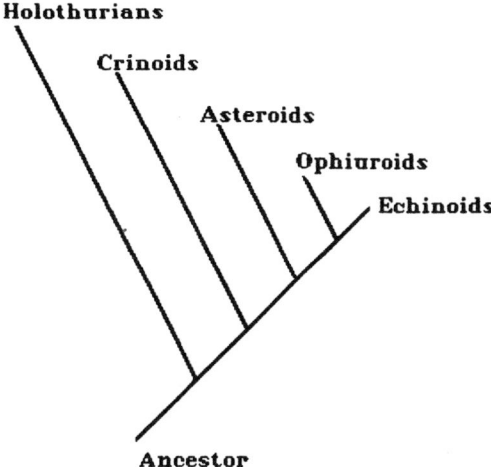

Fig. 6.1 Tree 1: the most parsimonious phylogenetic tree.

iterated 50 times gave the pair *ophiuroid and echinoid* 50 times, the trio *asteroid, ophiuroid, and echinoid* 50 times, and the quartet *crinoid, asteroid, ophiuroid, and echinoid* 48 times. Using BOOTM, with Wagner parsimony criteria exclusively, and without the hemichordate data, the pair *ophiuroid and echinoid* appeared 50 times out of 50, as did the trio *asteroid, ophiuroid, and echinoid* (Table 6.3). The results from the BOOTM program can be converted to probabilities (Felsenstein 1985), but I have not done so.

Description of the characters

Characters 1–3 define the differences between echinoderms and the hemichordates and result in no synapomorphies. Characters 4–14 concern embryological and metamorphic processes and structures; these characters are discussed fully in Smiley (1986). Characters 15–33 primarily involve points of comparative anatomy.

Characters 4, 5, and 12: torsion and rudiment

The absence of metamorphic torsion in holothurians is reflected in characters 4, 5, and 12. By torsion, I mean that the intersection of the anterio-posterior and dorso-ventral axes of the larva is 90° opposed to the axis of pentaradial symmetry of the adult (Smiley 1986). Character 4 specifically refers to axial torsion; character 5 to visceral torsion. The distinction between these is evident in Bury (1895) and Smiley (1986). Although the crinoids,

TABLE 6.1 Characters and character states for cladistic analysis

Primitive character state	Derived character states
1. Skeleton absent in adult: Hm	1a Adult calcareous stereomic skeleton present: Ho, C, A, O, E
2. Ambulacra absent: Hm	2a Ambulacra present: Ho, C, A, O, E
3. Adult symmetry bilateral: Hm	3a Adult symmetry pentaradial: C, A, O, E
	3b Adult symmetry pentaradial and bilateral: Ho
4. Adult axis of symmetry congruent with axis of larval bilateral symmetry: Hm, Ho	4a Adult axis of symmetry perpendicular to axis of larval bilateral symmetry: C, A, O, E
5. Plane of fusion of somatocoels parallel with adult axis of symmetry: Hm, Ho	5a Plane of fusion of somatocoels transverse to adult axis of symmetry: C, A, O, E
6. Primary podia circumoral: Hm, Ho	6a Primary podia terminal: A, O, E
	6b Primary podia circumoral and terminal: C
7. Primary podia derived from water vascular ring canal: Hm, Ho	7a Primary podia derived from radial canals: A, O, E
	7b Primary podia derived from ring and radial canals: C
8. One dorsal enterocoelic pouch formed at gastrulation: Hm, Ho, (C)	8a Paired lateral enterocoelic pouches formed: A, O, E
9. Right axocel present: Hm, A, O, E	9a Right axocoel absent: Ho, C
10. Left axocoel present: Hm, C, A, O, E	10a Left axocoel absent: Ho
11. Ambulacra unenclosed: C, A	11a Ambulacrum internalized by epineural overfolding: O, E
	11b Ambulacra develop internally: Ho
12. No persistent adult rudiment grows on larva: Hm, Ho, C	12a Persistent adult rudiment grows on larva: A, (O), E
13. Larval axohydrocoelic attachment organ present: Hm, Ho, (C), A	13a Larval axohydrocoelic attachment organ absent: O, E
14. Larval skeleton hydrostatic: Hm, Ho, (C), A	14a Larval skeleton calcareous and hydrostatic: O, E
15. Adult shape cylindrical: Hm, Ho	15a Adult shape stellate or globostellate to discoidal: C, A, O, E
16. Growth subterminal: Hm, Ho, A, O, E	16a Growth terminal: C
17. Tiedemann's bodies absent: Ho, C	17a. Tiedemann's bodies present: A, O, E
18. Polian Vesicles present: Ho, A, O	18a. Polian vesicles absent: C, E
19. Axial complex present: C, A, O, E	19a. Axial complex absent: Ho
20. Gonad single or bilaterally paired and dorsal: Hm, Ho	20a Gonad pentameric and not dorsal: C, A, O, E
21. No genital rachis forms: Ho	21a Genital rachis forms: C, A, O, E
22. No outer sac surrounds gonad: Ho, C	22a. Outer sac surrounds gonad: A, O, E
23. Larval hydropore persists in adult: Hm, A, O, E	23a Larval hydropore does not persist in adult: Ho, C
24. Madreporite internal: Ho, C	24a. Madreporite external: A, O, E
25. Perivisceral coelomic pores persist in adult: Ho, C	25a. Perivisceral coelomic pores absent in adult: A, O, E
26. Apical skeletal system absent: Hm, Ho, C	26a. Apical skeletal system present: A, O, E
27. No movable spines: Ho, C	27a Movable spines or pedicellariae present: A, O, E
28. Oralmost ossicles form semi-flexible frame: Ho, C, A	28a. Oralmost ossicles form muscular jaw: O, E
29. Hyponeural coelom formed from left and right somatocoels: Ho	29a. Hyponeural coelom formed from left somatocoel alone: A, O, E
30. Batyl alcohol present: Ho, C, A	30a. Batyl alcohol absent: O, E
31. Peripharyngeal and perianal coeloms not formed during juvenile growth: C, A, O	31a. Peripharyngeal and perianal coeloms formed during juvenile growth: Ho, E
32. Tube feet uncalcified: C, A, O	32a. Tube feet calcified: Ho, E
33. Anterior gut not surrounded by calcareous structure: C, A, O	33a. Anterior gut surrounded by calcareous structure: Ho, E

Code: Hm, Hemichordates. Ho, Holothurians. C, Crinoids. A, Asteroids. O, Ophiuroids. E, Echinoids. ( ) indicates data unavailable or incomplete.

TABLE 6.2 Data as entered into the programs

Character	1 2 3 4 5	6 7 8 9 0	1 2 3 4 5	6 7 8 9 0	1 2 3 4 5	6 7 8 9 0	1 2 3
Hemichordates	0 0 0 0 0	0 0 0 0 0	? 0 0 0 0	0 ? ? ? 0	? ? 0 ? ?	0 ? ? ? ?	? ? ?
Holothurians	1 1 B 0 0	0 0 0 1 1	B 0 0 0 0	0 0 0 1 0	0 0 1 0 0	0 0 0 0 0	1 1 1
Crinoids	1 1 1 1 1	B B 0 1 0	0 ? ? ? 1	1 0 1 0 1	1 0 1 0 0	0 0 0 ? 0	0 0 0
Asteroids	1 1 1 1 1	1 1 1 0 0	0 1 0 0 1	0 1 0 0 1	1 1 0 1 1	1 1 0 1 0	0 0 0
Ophiuroids	1 1 1 1 1	1 1 1 0 0	1 1 1 1 1	0 1 0 0 1	1 1 0 1 1	1 1 1 1 1	0 0 0
Echinoids	1 1 1 1 1	1 1 1 0 0	1 1 1 1 1	0 1 1 0 1	1 1 0 1 1	1 1 1 1 1	1 1 1
Parsimony	C C C C C	C C C C C	C W W C W	W W W W C	C C W C W	C W W W W	W W C

Camin–Sokal (C) or Wagner (W) parsimony criteria. ? Information unknown or incomplete. B fits both character states or character state b.

TABLE 6.3 Results from each of the computer analyses

Program	Outgroup mixed parsimony		No outgroup Wagner parsimony	
ITERMIX	Tree 1		Tree 1	
PENNY	Tree 1		Tree 1	
METRO	Tree 1		Tree 1	
BOOTM	O-E	50/50	O-E	50/50
	A-O-E	50/50	A-O-E	50/50
	C-A-O-E	48/50	C-A-O-E	****

Code: C, crinoids; A, asteroids; O, ophiuroids; E, Echinoids; ****, did not appear.

asteroids, ophiuroids, and echinoids share the derived condition for both these characters, there are some differences between the classes in how these torsions are achieved. This is most apparent in the metamorphosis of ophiuroids (Bury 1895; Fedotov 1924; Hyman 1955). Character 12 refers to the persistent adult rudiment. Presence of a rudiment indicates that development has been compressed (Smiley 1986), because these are adult tissues on the larval body. The presence of a persistent rudiment in ophiuroids is a matter of interpretation. While I have seen no photomicrographic evidence for such a structure on any ophiuroid larva, I adhered to the interpretations of Hendler (1982) and Mladenov (1985) on this point. It is important to distinguish between a rudiment, an *anlage* composed of strictly adult tissues growing on the larval body, and the direct development of the adult itself. These have been confused recently (Raff 1987). The polarization of characters 4 and 5 was determined by the outgroup. With character 12, I used Wagner parsimony. These characters substitute for Smith's characters 2, 6, 11, 31, and 43.

Characters 6 and 7: primary podia

These characters compare the origin and position of the primary podia in pterobranchs and echinoderms. The relationship between primary podia and buccal podia in the extant classes is discussed in Smiley (1986). Semon (1888), noted that the holothurian buccal podia were primary, and were formed directly from the water vascular ring. With this, Semon argued a phylogeny for the echinoderms based on his interpretation that holothurian buccal podia were homologous with the ambulacral radii of the other classes. This hypothesis was not supported by Bury (1895) or MacBride (1914), who maintained that the ambulacral radii of all echinoderm classes were homologous. I concur with these later assessments and, like MacBride, I distinguish between the primary buccal podia of holothurians and the primary podia of the other classes, which although homologous by structure, by an altered ontogeny, and a modified function, are not homologous by position.

I consider that feeding was an ancestral function of echinoderm tube feet, and argue that primary circumoral podia are primitive. This is supported by the presence of primary feeding podia in the pterobranchs, and the outgroup state polarized this character. The B character state described the condition in crinoids, where in addition to the terminal azygous primary podia formed at the tips of the radii, evaginations of the circumoesophageal water ring give rise to small epidermal protrusions located around the oral opening. These circumoral 'podia' degenerate soon after settlement and metamorphosis (MacBride 1914; Hyman 1955). Smith (1984a) does not use a character involving the primary podia in his analysis.

Characters 8 and 9: enterocoel and right axocoel

Character 8 refers to a difference in the pattern of formation of the enterocoelomic vesicle during gastrulation. In holothurians, all complete descriptions of gastrulation and coelom formation indicate that the enterocoel (the first coelomic pouch) develops as a single dorsal structure. This is invariant whether the animal is a direct or indirect developer (Inaba 1930; Ludwig 1898; Newth 1916; Ohshima 1921; Rustad 1940; Selenka 1876), with the single exception of *Molpadia intermedia* (McEuen and Chia 1985). In the asteroids, ophiuroids, and echinoids, two bilaterally opposed enterocoels form at gastrulation in all indirect and nearly all direct developers (Hyman 1955). Because the condition found in direct development is almost always identical to that found in indirect development within each class, I extended this character to include the crinoids (Hyman 1955) and the hemichordates (Dawydoff 1948; Hyman 1959), which also form a single dorsal enterocoel at gastrulation. The right axocoel, described in character 9 may be the same as Smith's (1984a) character 16 in a different wording.

Character 10: left axocoel

For the purposes of this analysis, I have assumed that the state of this character in holothurians represents a derived condition. I did this for convenience only, and have argued at length elsewhere (Smiley 1986) why I believe that the holothurian coelomic condition is not derived. This point was not addressed by Smith, although his character 35 may pertain to it. Polarization is through the outgroup.

Character 11: internal ambulacra

Character 11 describes the mode by which ambulacra are enclosed in holothurians, ophiuroids, and echinoids. Recently, I reported (Smiley 1986) that the process of formation of the epineural sinus in the holothurian *Stichopus californicus* was by a mechanism distinct from the folding over of epineural flaps characteristic of echinoids, and hypothesized by Runnström and Runnström (1919), and Runnström (1927) to occur in holothurians. Holothurian ambulacra form well within the body and never have a direct connection with the epidermis during their formation. This new information makes it difficult, if not impossible, to derive the mechanism of formation of the internal ambulacra of holothurians from that reported for echinoids and ophiuroids. However, I find it relatively easy to derive the condition found in echinoids and ophiuroids from that found in asteroids. Smith uses Hyman's misquote of Runnström's (1927) hypothesis as evidence for his character 19. Since hemichordates do not have ambulacra, I used the character state in crinoids to polarize this character.

Character 13: axohydrocoelic attachment organ

I interpret the proboscis of enteropneusts (Morgan 1891) and the cephalic shield of pterobranchs (Harmer 1905; Lester 1985) as axohydrocoelic attachment organs in character 13. Furthermore, I consider the buccal podia of holothurians as homologous attachment structures. Settling postmetamorphic *Stichopus californicus* (Smiley 1986), and a number of directly developing holothurian species, attach to the substratum with their buccal podia. I have argued (Smiley 1986) that since the coelomic linings of holothurian buccal podia arise from an undivided axohydrocoel, and since they have no obvious homologues among the other extant classes; the separate axocoel of the crinoids, asteroids, ophiuroids, and echinoids had its origin in the anterior evaginations of the undivided axohydrocoel of a holothurian-like ancestor (Smiley 1986). I also ascribe a homologous function to the brachiolar podia of asteroids in that they are homologous attachment podia derived from an anterior evagination of the axohydrocoel. Due to their tissue of origin, their primary development, azygous state, and the sensory functions they bear, holothurian buccal podia are also homologous with the terminal primary podia of the asteroids, ophiuroids, and echinoids. I did not determine the ancestral state of this character and used Wagner criteria in the analysis. A different interpretation of this point is used by Smith in his character 18.

Character 14: larval skeleton

No direct evidence confirms the hypothesis that hydrostatic force determines the shape of hemichordate and echinoderm larvae. It is a reasonable hypothesis in the absence of other skeletal elements, and recent work by Ruppert and Balser (1986) may support it. Most auricularia larvae contain calcareous elements, but these do not have a skeletal function (Hyman 1955). Furthermore, the criterion of homology by position fails to support the contention that the calcareous elements of auriculariae and plutei are homologous. This point is dealt with in characters 4, 29, and 30 by Smith (1984a). The polarization of this character is through the condition in the outgroup.

Character 15: adult shape

I use this character to distinguish between the globostellate shape of crinoids, asteroids, ophiuroids, and echinoids on one hand; and the cylindrical shape of hemichordates and holothurians on the other. Traditionally, the term growth gradient has been applied to descriptions of the development of the adult shape in echinoderms. I have not included a growth gradient character because I interpret growth of all post-metamorphic echinoderms to be determined by the inductive influence of the water vascular system. Since growth is directed by inductive influence, and since the adult mouth of holothurians is not homologous, either on the basis of position or ontogeny, with that of echinoids (Smiley 1986); invocation of elusive developmental gradients seems vague and unnecessary to describe the phenomenon. Finally, since the inductive influence driving adult growth patterns is found in all echinoderms, the character leads to no synapomorphies.

I consider the globose shape of echinoids distinct from the cylindrical shape of holothurians. I contend that the globose shape of echinoids can be achieved through slight modifications in the shape of asteroids (Smiley 1986); the lack of torsion precludes a similar interpretation in holothurians. The derivation of the globose echinoid from the stellate asteroid form is supported by a series of asteroid species whose shapes nearly grade into that of regular echinoids (M. Downey, personal communication); and the arguments of Bookstein et al. (1985), concerning the derivation of complex shapes from one another. I assigned Wagner parsimony criteria to the analysis of this character. This idea is expressed differently in Smith's characters 12 and 28.

Character 16: location of growth zone

This character addresses the question of growth pattern, but results in no synapomorphies. It may be of value when applied to extinct groups. This is Smith's character 8 in a different wording. I used Wagner parsimony.

Character 17: Tiedemann's bodies

It is likely that the absence of both an axial complex and Tiedemann's bodies in holothurians means that the splenic function usually attributed to these organs is probably not localized in these animals. Perhaps this function is mediated by the action of free coelomocytes or by the perivisceral coelomic pores recently described (Shinn 1985a,b; Smiley 1986). I used Wagner parsimony for this character. This is Smith's character 42.

Character 18: Polian vesicle

Hyman (1955) reports that small spongy bodies, that may be homologues of the Polian vesicles, are found on the water vascular ring in echinoids. If analysis substantiates Hyman's hypothesis, it would help resolve the phylogenetic problems imposed by both crinoids and echinoids sharing the loss of this organ. This is Smith's character 41. I used Wagner parsimony in evaluating this character.

Character 19: axial complex

The echinoderm axial complex consists of both a coelomic and a connective tissue component (Hyman 1955). Although Erber (1983b) has recently shown that the connective tissue component of the axial complex is present in adults of several holothurian species, an axocoelic component is absent (Smiley 1986). Because a fully formed axial complex does not occur in holothurians, and because the connective tissue component is not formed until some time after metamorphosis, I separated the holothurians from the other classes. In crinoids, the coelomic component of the axial complex does not arise from the axocoel (Hyman 1955; N. D. Holland, personal communication), contrary to the claims of Erber (1983b). Whether or not the axial complex of crinoids is exactly homologous with that of asteroids, ophiuroids, and echinoids, is a matter of some dispute (Hyman 1955), and further investigation on this point would be helpful. Available evidence indicates that the holothurian condition is distinct from any of these. Smith describes these points in his character 35. I used Wagner parsimony criteria.

Characters 20, 21, and 22: the gonad

These characters address the structure and histology of the gonad. Character 20 concerns the single holothurian gonad, which is considered primitive by virtually all echinoderm comparative anatomists (Smiley and Cloney 1985). I address the *genital rachis* in character 21. There are no reports of a genital rachis in the hemichordates (Dawydoff 1948; Hyman 1959). Théel (1901) noted that a *germinal cord* extends within the dorsal mesentery along the gonoduct in adult *Mesothuria intestinalis*, but distinguished this from the genital rachis which is present in the other classes only as a developmental stage. There is neither a germinal cord nor a genital rachis in *Stichopus californicus*, either in the adult or in the juvenile (Smiley 1984). In holothurians, the most likely homologue of the genital rachis is the entire germinal epithelium, which extends throughout the gamete-bearing tubules (Smiley and Cloney 1985). For character 22, information on the presence and absence of the outer sac surrounding the gonad in echinoderm classes can be found in Davis (1971), Atwood (1973), and Smiley and Cloney (1985). The gonads of neither pterobranchs (Lester 1988) nor crinoids possess an outer sac. I established the polarization of these three characters on the basis of the condition in holothurians, and I used Camin–Sokal parsimony criteria for all three. The single gonad of holothurians is found in Smith's character 20.

Characters 23, 24, and 25: hydropore, madreporite, and coelomic pores

These characters address the connection of the water vascular system to the sea. In character 23, I make clear the distinction between the larval hydropore and pores which run through the adult madreporite. In the asteroids, ophiuroids, and echinoids, the larval hydropore persists in the adult (Hyman 1955). Erber (1983a) reported that larval hydropores persist in some adult holothurians of the orders Elasipoda and Molpadida, but these two orders are thought derived by most holothurian taxonomists (Ludwig 1889–1892; Ekman 1925; Hyman 1955), and represent only a minority of holothurian species. In the dendrochirotes and aspidochirotes, the two orders thought to represent the most basic holothurian condition, larval hydropores rarely persist (Smiley 1986). In character 24, I considered the internal madreporite ancestral because only ophiocistioids, of the extinct classes, have external madreporites. The rediscovery of perivisceral coelomic pores in holothurians by Shinn (1985a,b), and the demonstration of their surprisingly early appearance (Smiley 1986), gain phylogenetic significance when compared to the condition in crinoids and extinct forms. This conclusion is supported by the fact that the stone canal opens into the perivisceral coelom in both holothurians and crinoids. I did not polarize characters 23 and 25. Parts of these points are addressed in Smith's character 5.

Characters 26, 27, and 28: adult skeleton and pedicellariae

I consider the apical skeletal system (character 26) in the asteroids, ophiuroids, and echinoids homologous, following Hyman (1955). This homology is discounted if one adheres to Fell's (1963a,b) concept of growth gradients, which for reasons previously explained, I consider erroneous. Character 26 is polarized on the basis of information from the fossil record (Fell and Pawson 1966). Characters 27 and 28 are assessed using Wagner criteria. Characters 27 and 28 are used by Smith (his characters 10 and 36, respectively).

Character 29: hyponeural coelom

This character reflects changes in the ontogeny of the hyponeural coelom (MacBride 1914) probably caused by metamorphic torsion. The loss of the hyponeural coelom in crinoids may be part of the general reduction in coelomic space within this class. This corresponds to Smith's character 15. I used Wagner parsimony criteria.

Character 30: batyl alcohol

I drew this information from Bolker (1967), and use it where Smith does not. Other biochemical data are more difficult to interpret. For example, the data on amino acid composition of echinoderm collagens (Matsumura et al. 1979) are not easily interpretable in terms of ancestral and derived character states. This information contradicts the hypothesis that holothurians and echinoids are closely related, but neither Smith nor I use it. Other recent molecular information, such as on the holothurian H1 histone (Azorin et al. 1985), and on the sequence of echinoderm repetitive DNAs (Poltaraus 1981; Poltaraus et al. 1980, 1981), also contradicts the hypothesis that echinoids and holothurians are closely related. However, these data are difficult to interpret in these terms, and I did not use them. This character follows Wagner parsimony criteria.

Characters 31, 32, and 33: putative homologies with echinoids

These three characters are from Smith (1984a), among others (Hyman 1955). Because perianal coeloms occur in all groups which have peripharyngeal coeloms, and both are incompletely separated from the perivisceral coelom in the juvenile (Hyman 1955), I have worded these into one character where Smith uses two (32 and 33). When coeloms are formed early in embryogenesis, few question their phylogenetic significance; but there is little compelling information available on the significance of secondarily derived coelomic spaces (Hyman 1940). However, I cannot falsify the hypothesis that they are homologous in echinoids and holothurians. I also combined information from Smith's characters 24 and 25 into character 32 describing calcifications in tube feet. Character 33 (a different wording of Smith's character 37) addresses the question of whether the Aristotle's lantern of echinoids is homologous with the calcareous ring of the holothurian aquapharyngeal bulb. This question is vexing, and although I interpret this as an example of convergence, there is not enough detailed functional morphological or developmental information on this point to contradict the hypothesis that they are homologous. I assigned Camin–Sokal parsimony criteria to this character, but Wagner criteria to characters 31 and 32.

DISCUSSION

The subphylum Echinozoa

The results presented in this paper do not support the hypothesis that holothurians and echinoids share a unique common ancestor. None of the trees generated by the PHYLIP programs allied the holothurians and the echinoids. Because of these results, and those presented in Smiley (1986), I maintain that the taxon uniting these classes, the Echinozoa, should be abandoned.

My results are in direct conflict with those of Smith (1984a) with respect to the relationship between holothurians and echinoids. One explanation for the difference may be that I have relied heavily on new information; most of which was drawn from my recent study of holothurian metamorphosis and which clarified a number of points traditionally used to ally holothurians and echinoids (Smiley 1986). A second reason for the difference, probably includes the clear distinction I have made between direct and indirect development, where Smith did not. The results from my analysis agree well with Smith (1984a) on most points other than joining holothurians and echinoids in the Echinozoa.

The subphylum Eleutherozoa

Hyman (1955) and Smith (1984a,b) have suggested that the concept of the Stellarozoa or Asterozoa is misleading. I believe that if we abandon the Echinozoa, we must also abandon the Stellarozoa. Once echinoids are removed from an association with the holothurians, their relationship to the asteroids and ophiuroids can be examined more closely. Support for this interpretation comes from the BOOTM results where the pair *ophiuroid and echinoid* occurred 50 times in 50 trials, and the trio *asteroid, ophiuroid, and echinoid* also occurred 50 times out of 50. Smith (1984a,b) similarly suggested that it is not necessary to exclude echinoids from a taxon including the asteroids and ophiuroids. Emlet (1988) reports that not all indirectly developing echinoids form a vestibule, indicating that echinoid and asteroid embryologies may not be as distinct as previously thought (Bury 1895; MacBride 1914). The presence of an external madreporite coupled with the absence of perivisceral coelomic pores in the asteroids, ophiuroids, echinoids, and ophiocistioids, supports the tentative hypothesis that these classes share a unique ancestor.

Paul and Smith (1984), and Smith (1985) have argued that two subphyla, the Pelmatozoa and the Eleutherozoa comprise all the classes of echinoderms, with possibly the exception of the helicoplacoids. The results presented here and in Smiley (1986) are in accord with this view that the asteroids, ophiuroids, and echinoids all belong to the subphylum Eleutherozoa. These same results suggest that the holothurians should be removed from this subphylum as was originally suggested by Bell (1891).

Are holothurians eleutherozoans?

It is useful to list the characteristics which distinguish holothurians and eleutherozoans considering the information presented in this paper and in the analysis presented in Smiley (1986). If holothurians are derived from an eleutherozoan ancestor, then the basic pattern of eleutherozoan development has been substantially altered. In holothurians the mode of coelom formation is distinct. It is shared by the hemichordates and possibly the crinoids, but by none of the Eleutherozoa. Holothurians lack metamorphic torsion, and, hence, the basic axial relationships between larvae and adults are distinct from those found in the Eleutherozoa and the crinoids. The process of metamorphosis is simplified in holothurians compared to what is found in the eleutherozoans. Holothurians show no trace of a rudiment on the larval left side. The mesenteries formed by the fusion of the left and right somatocoels in holothurians are coplanar with the plane of bilateral and pentaradial symmetry, not perpendicular to it as is found in the eleutherozoans. Holothurians have no axocoel, and no fully formed axial complex. Those functions subsumed by the axocoel alone in the eleutherozoans are subsumed exclusively by the hydrocoel in holothurians. The gonads of holothurians are single, and dorsal, not aboral; and they are not surrounded by an outer sac, a condition shared only by the crinoids. Holothurian primary podia arise from the circumoesophageal water ring and are located around the mouth, not at the distal ends of the radii; the case in the crinoids and the eleutherozoans. Vestigial circumoral podia are also found in crinoids, but do not occur in the eleutherozoans. The ambulacra of holothurians form internally and the epineural sinus is formed by cavitation of a superficial tissue layer. Holothurians bear perivisceral coelomic pores and have an internal madreporite formed within a syncytium, showing a greater similarity to the condition in crinoids than the eleutherozoans.

In addition to these alterations, common eleutherozoan structures such as pedicellariae or movable spines, an apical skeletal system, pentaradial gonads surrounded by an outer sac, and terminal primary podia have left no vestigial trace. During development, there is not a hint or ancestral reminiscence of characteristically eleutherozoan processes such as metamorphic torsion, incipient rudiment formation or epidermal alteration of the lower left side of the larva, incipient external ambulacral formation, incipient formation of an external madreporite, or formation and subsequent regression of a separate axocoel. Those characteristically eleutherozoan structures that could be argued to be present in vestigial form in holothurians are: the persistent hydropore in several adult holothurians, the inward movement of the madreporic vesicle, and the feint to the left of the opening of the oral cavity during the second phase of metamorphosis (Smiley 1986). Although not an eleutherozoan property, the rapid growth of the mid-ventral radius during the development of all holothurians appears to be an ancestral reminiscence, whose significance can, as yet, only be guessed. The reported holothurian vestibule is more appropriately considered homologous with the oral cavities of other echinoderm larvae than the secondary epidermal invagination which forms on the left side of many echinoid larvae.

Many of the structures that are present in holothurians bear primitive characteristics, and must be interpreted as spontaneous atavisms. These include the primitive mode of enterocoel formation, the single gonad unsurrounded by an outer sac, circumoral primary podia, the internal end to the stone canal, and the perivisceral coelomic pores.

This combination of both failures to demonstrate eleutherozoan affinities in development and in adult anatomy, coupled with the requirement for spontaneous atavisms, I believe, makes the hypothesis that holothurians descended from eleutherozoans untenable. It could be argued that the distinctive holothurian metamorphosis alone explains a large number of these differences, and that the holothurians can be derived from eleutherozoans given the differences in metamorphosis. However, this explanation is too limited; because the metamorphic differences are almost certainly too great to be caused by a simple mutation, and because other events which occur even earlier in development, such as the formation of the enterocoel, are also primitive. The phylogenetic problems posed by the similarity of the holothurian calcareous ring and the Aristotle's lantern of echinoids remain considerable. Only detailed developmental and functional morphological investigations of these structures will help determine if they actually are homologous, or if their similarity is superficial and they only represent analogous calcifications surrounding the oesophagus. As far as I am aware, this has not yet been done. In summary, whatever advantages are gained by concluding that holothurians are eleutherozoans are outweighed by the simplicity offered if we conclude that holothurians are uniquely primitive among the extant echinoderms.

The relationship between the holothurians and the crinoids is more difficult to assess. The fact that holothurians and crinoids share perivisceral coelomic pores, internal ends to their stone canals, no outer sac surrounding the gonads, and early circumoral primary podia, may only indicate that they arose from a similar stock, not that they share a unique ancestor. The importance of the perivisceral coelomic pores in holothurians and their relationship to those found in crinoids needs to be evaluated more completely. Crinoids are torted as adults, like asteroids, ophiuroids, and echinoids, while holothurians are not. Our understanding of important events in crinoid embryology and metamorphosis is currently poor because, there are no known indirectly developing species, there is little information on the embryology of the class aside from the comatulids, and available information is often contradictory (Hyman 1955).

The phylogenetic position of holothurians

Figure 6.2 places holothurians within the classification of the echinoderms according to the results reported here. To determine the true affinities of

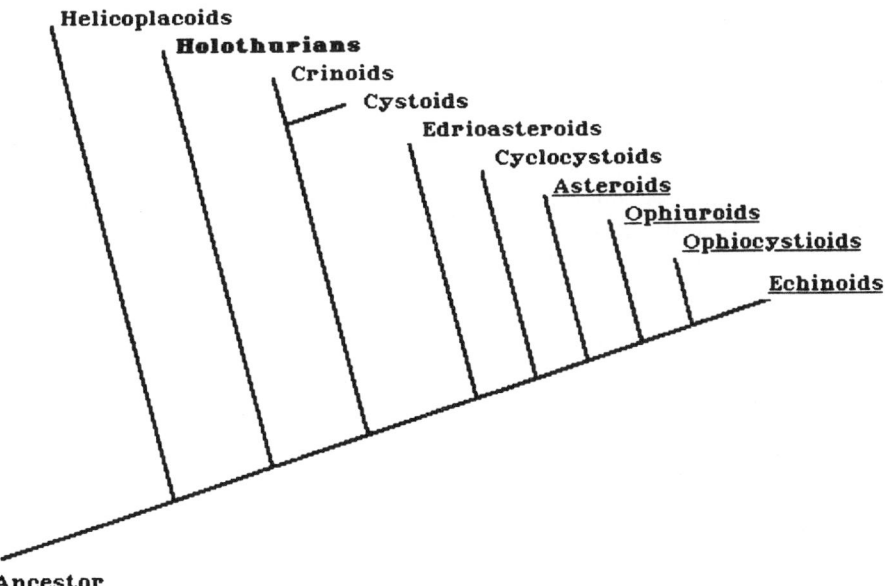

Fig. 6.2 Proposed phylogenetic position of the holothurians. Schematic depiction of the relationship of holothurians to the other echinoderm classes, drawn largely from Paul and Smith (1984). Those eleutherozoans possessing an external madreporite are underlined.

holothurians, we must search for other echinoderms which share several of the following characteristics. They would be untorted, and perhaps bilaterally symmetrical as adults, with perivisceral coelomic pores, and an internal madreporite, or internal ends to the stone canal(s). The gonad would be single and dorsal, and the gonoduct would be located anteriorly in interambulacrum CD. They would have a preoral lobe, sensory buccal podia, and a midventral mouth, but no movable spines.

The class that most closely fits these requirements is the helicoplacoids (Fig. 6.2). Unfortunately, a number of points concerning the anatomy of helicoplacoids are still a matter of dispute. Paul and Smith (1984) assert that the mouth is found at the focus of the three ambulacra, but Durham (1967) has argued that the mouth is apical. The anus, gonopore, external madreporite, and the hydropore have not been found in helicoplacoids. The helical patterning of the skeleton suggested to Durham (1967) that it was flexible, that skeletal growth was terminal, and that it occurred at the posterior end. Paul and Smith (1984) argue that growth was subterminal from the ends of each of the three ambulacra. If growth of the animal was spiral, as indicated by the pattern of the skeletal plates, then the connections between the viscera and the epidermis must have grown in compensation. This spiral growth may be important in determining the affinities of holothurians and helicoplacoids.

If Paul and Smith (1984) are correct in their placement of the mouth in helicoplacoids, then that portion of the body anterior to the mouth may represent an adult preoral lobe. This could be homologous with the preoral lobe of larval holothurians. In the embryology of all indirectly developing holothurians, and in those directly developing holothurians that have a larval mouth, the mouth migrates to a terminal anterior position during metamorphosis or settlement. This may represent an ancestral reminiscence of the usurpation of the helicoplacoid preoral lobe. It is clear, however, that whether development is direct or indirect, in all holothurian larvae, the preoral lobe is a prominent feature which degrades after metamorphosis, indicating it is in regression (Selenka 1876; Semon 1888; Bury 1895; Newth 1916; Ohshima 1921; Inaba 1930; Rustad 1940; McEuen and Chia 1985; Smiley 1986).

Another universal feature of holothurian embryology is the marked difference in the rate of growth of the mid-ventral (A) radius compared to the other four. This is true for both directly and indirectly developing species (Ohshima 1921; Smiley 1986). Embryological evidence from holothurians indicates that the mid-ventral holothurian radius (A) would correspond to the unpaired ambulacrum of helicoplacoids. If holothurians are related to helicoplacoids, this information may prove useful in interpreting both the identity of the pore at the blunt apex and the orientation of helicoplacoids in life (Paul and Smith 1984).

The differences in skeletal formation in helicoplacoids and modern holothurians might well reflect the difficulties of organizing helical skeletal growth. Recent investigations into the formation of holothurian ossicles (Stricker 1985, 1986) showed that there are many similarities in the mode of growth of anchor plate ossicles from adult holothurians and both juvenile echinoids (Emlet 1985) and asteroids (Komatsu 1982). Perhaps this reflects a persistent juvenile state to the skeletal organization in holothurians; a state that was evolutionarily preferable to the condition found in adult helicoplacoids (see Gould 1977 for a discussion of this mechanism). If the hypothesis that holothurians and helicoplacoids share a unique ancestor is supported by further investigations, Paul and Smith's (1984) interpretation of the helicoplacoid mouth would be correct.

Implications for palaeontology

The recent rediscovery that holothurians, in addition to crinoids, have perivisceral coelomic pores (Shinn 1985a,b), internal ends to their stone canals, and no persistent hydropores in the adult may have importance when considering extinct classes. Although palaeontologists have concluded that the gonopore and hydropore were fused in many of the extinct classes, this assumption may not be necessary in light of these observations. If these animals had internal ends to their stone canals and no persistent adult hydropore, then the opening assumed to be a fused gonopore-hydropore may be simply the gonopore. The hydropore and gonopore are located in the same interambulacrum in holothurians. Alternatively, if these extinct forms bore their gonads on projections outside the theca as is the case in crinoids and possibly cyclocystoids (Smith and Paul 1982), then gonopores may have been unnecessary, but the condition in the extant classes make this a less likely alternative. A reassessment of the identity of these fused gonopore-hydropores, in light of this new information, would be helpful.

One palaeontological point where embryological clarification is in order concerns recent speculations on the coelomic organization of a number of extinct echinoderm classes. Haugh and Bell (1980a,b) have suggested that a respiratory coelom was present in a number of diverse extinct echinoderms, and proposed elaborate speculations on its coelomic antecedents. Since coelomic pores connect the perivisceral coelom to the sea in both holothurians and crinoids, these extant animals have a true respiratory coelom which probably functions similarly to that described by Haugh and Bell (1980a,b). The perivisceral coelom of holothurians, crinoids and all the other extant classes is homologous with that found in the hemichordates (Hyman 1955, 1959). Therefore, the coelomic antecedent of the respiratory coelom in holothurians and crinoids is the perivisceral coelom. Haugh and Bell (1980a,b) proposed an extensive classification of the echinoderms on the basis of putative coelomic primordia which exist only in the hypothetical ancestral dipleurula larva, and for which there is absolutely no evidence. I believe that there is little justification for the assumption that a perfectly symmetrical, bilateral, tripartite coelomic configuration existed in any larval echinoderm (Smiley 1986). Just as the symmetry of adult echinoderms has been perfected over evolutionary time, it is more reasonable to conclude that larval coelomic symmetry has been perfected as well. If this view is supported by further investigations, then it would be possible to explain the left side dominance of echinoderm larval coeloms as a consequence of alterations in larval shape requiring dorso-ventral flattening for feeding and locomotory efficiency; where the choice of left side over right was biased by: spiral growth of the water vascular system in an ancestor such as the helicoplacoids, reflected today in the growth of the hydrocoel around the oesophagus from the left, as in the holothurians, or through the ingression of adult tissues in a rudiment, whose axial properties are a consequence of metamorphic torsion, as is the case in the crinoids and Eleutherozoans. My arguments concerning the relationship between the single anterior coelom in holothurians and the several anterior coeloms in the Eleutherozoa are compatible with this view (Smiley 1986). In this light, the elaborate speculations concerning the origin of the respiratory coelom in extinct echinoderms, and the classification system drawn from these, seem unnecessary.

CONCLUSIONS

The evidence presented here and in Smiley (1986) supports the hypothesis that holothurians are ancient beasts, and that they are quite distinct from the Eleutherozoa (*sensu* Bell 1891). The most important characters concern coelomogenesis and the coelomic configuration, the distinct axial symmetries, the circumoral position of the primary, sensory, azygous, buccal podia, and the presence of perivisceral coelomic pores (Smiley 1986). Also of importance is the discovery that the ambulacra of holothurians form internally, they do not actually close, and that the epineural sinus forms using a mechanism distinct from that reported for echinoids (Smiley 1986). Finally, although it is possible to derive the two anterior coeloms (axocoel and hydrocoel) of crinoids, asteroids, ophiuroids, and echinoids from the undivided axohydrocoel in holothurians, it is difficult to explain how this structure could be lost in holothurians without catastrophic results (Smiley 1986). These characters set holothurians apart from all the remaining extant echinoderm classes.

The application of data drawn from detailed studies of development to evolutionary questions has long been out of vogue. It has been my goal to demonstrate that our definition of homology (the explanation of structural similarity by common descent) includes an ontogenetic component that cannot be replaced by other information. Today, we understand that each cell in a body contains all information necessary to create any structure in that body. Furthermore, we are now aware that cells themselves are capable of great movements. If we assert that homology is the most restrictive comparison between similar structures in different animals, then it seems to me that we must apply developmental information to our assessment of homology. This is made much more complicated by recent advances in developmental genetics (Scott and O'Farrell 1986), which show that structural genes contain regulatory regions controlled by separate regulatory genes. Given the enormous amount of time since the divergence of the extant echinoderm classes, shuffling of regulatory genes and regulatory regions of structural genes, surely could have created similar morphological structures from identical structural genes without functional intermediates. Therefore, it is important to test if two structures, similar in shape and position, share ontogenic pathways; to determine if developmental evidence supports the hypothesis of common descent. If they do not, then a strict interpretation would argue that they are not homologous. However, if such alterations arose from differences in developmental mechanisms which are themselves desirable from one another, such as gastrulation by invagination and gastrulation by epiboly, then homology could be supported.

ACKNOWLEDGEMENTS

I would particularly like to thank K. P. Irons, for the many productive discussions and all her thoughtful criticisms that went into this paper. I also thank Dr J. Felsenstein, Genetics Department, University of Washington, Seattle, for the gift of the PHYLIP programs. Much of this work was done at the Friday Harbor Labs, and I thank the faculty, staff, visiting graduate students and scientists for their help and encouragement. I wish to thank Andrew Smith for inviting me to this symposium and for very helpful criticism and stimulating comments which improved this paper substantially. M. Downey, J. Durham, R. Emlet, M. Hille, N. Holland, J. Lawrence, W. Moody, G. Paulay, D. Pawson, L. Riddiford, E. Ruppert, R. Strathmann, and J. Truman all offered challenging comments and editorial advice. Support for this work came from a Developmental Biology Traineeship from NIH (HD-00266).

REFERENCES

Atwood, D. G. 1973. Ultrastructure of the gonad wall of the sea cucumber *Leptosynapta clarki*. *Zeitschrift für Zellforschung und mikroscopische Anatomie* **14**, 319–30.

Azorin, F., Rocha, E., Cornudella, L. and Subirana, J. A. 1985. Anomalous nuclease digestion of *Holothuria* sperm in chromatin containing histone H1 variants. *European Journal of Biochemistry* **148**, 529–32.

Bather, F. A. 1900. The Echinodermata. In *A treatise of zoology* III (ed. E. R. Lankester), pp. 1–344. Adam & Charles Black, London.

Beklemischev, W. N. 1969. *Principles of comparative anatomy of invertebrates*. University of Chicago Press, Chicago.

Bell, F. J. 1891. On the arrangement and interrelations of the classes of Echinodermata. *Annals and Magazine of Natural History, Series 6* **8**, 206–15.

Bolker, H. I. 1967. Phylogenetic relationships of echinoderms; biochemical evidence. *Nature, London* **213**, 904–5.

Bookstein, F. L., Chernov, B., Elder, R. L., Humphries, J. M., Smith, G. R., and Strauss, R. E. 1985. *Morphometrics in evolutionary biology*. Special Publications of the Academy of Science and Philosophy No. 15.

Bury, H. 1985. The metamorphosis of Echinoderms. *Quarterly Journal of Microscopical Science* **38**, 45–135.

Cuénot, L. 1948. Anatomie, éthologie, et systématique des échinoderms. In *Traité de Zoologie* **11** (ed. P. G. Grassé), pp. 3–275. Masson, Paris.

Davis, H. S. 1971. The gonad wall of the Echinodermata: a comparative study based on electron microscopy. MSc Thesis. University of California, San Diego.

Dawydoff, C. 1948. Classe des Pterobranchs. In *Traité de Zoologie* **11** (ed. P. G. Grassé), pp. 454–89. Masson, Paris.

Durham, J. W. 1967. Notes on the Helicoplacoidea and early echinoderms. *Journal of Paleontology* **41**, 97–102.

Ekman, S. 1925. Systematische—Phylogenetischen Studien über Elasipoden und Aspidochiroten. *Zoologische Jahrbucher Abteilung für Anatomie und Ontogenie der Tiere* **47**, 429–540.

Emlet, R. B. 1985. Crystal axes in recent and fossil adult echinoids indicate trophic mode in larval development. *Science, New York* **230**, 937–40.

Emlet, R. B. 1988. Larval form and metamorphosis of a "primitive" sea urchin *Eucidaris thourasi* (Echinodermata: Echinoidea: Cidaroidea), with implications for development and phylogenetic studies. *Biological Bulletin* **174**, 4–19.

Erber, V. W. 1983a. Der Steinkanal der Holothurien: Ein Morphologische Studie zum Problem der Protocoelampulle, *Sonderheft aus Zeitschrift für zoologische Systematik und Evolutionsforschung* **21**, 217–34.

Erber, V. W. 1983b. Zum Nachweis des Axialkomplexes bei Holothurien. *Zoologica Scripta* **12**, 305–13.

Fedotov, D. M. 1924. Einige beobachtung über die Biologie und Metamorphose von *Gorgonocephalus*. *Zoologische Anzeiger* **61**, 303–11.

Fell, H. B. 1963a. The evolution of Echinoderms. *Reports of the Smithsonian Institution* **1962**, 457–90.

Fell, H. B. 1963b. Phylogeny of sea stars. *Philosophical Transactions of the Royal Society of London, Series B.* **246**, 381–435.

Fell, H. B. and Pawson, D. L. 1966. General biology of echinoderms. Ch. 1. In *Physiology of Echinodermata* (ed. R. A. Boolootian), pp. 1–48. Interscience, New York.

Felsenstein, J. 1983. Parsimony in systematics: Biological and statistical issues. *Annual Review in Ecology and Systematics* **14**, 313–25.

Felsentein, J. 1985. Confidence levels on phylogenies: an approach using a bootstrap. *Evolution* **39**, 783–91.

Gould, S. J. 1977. *Ontogeny and phylogeny*. Belknap Press, Cambridge, Mass.

Harmer, S. F. 1905. The pterobranchs of the Siboga Expedition. *Siboga Expeditie.* **26**, 1–132.

Haugh, B. W. and Bell, B. M. 1980a. Fossilized viscera in primitive echinoderms. *Science, New York* **209**, 653–7.

Haugh, B. W. and Bell, B. M. 1980b. Classification Schemes. In *Echinoderms: notes for a short course.* University of Tennessee Department of Geological Sciences Studies in Geology **3** (ed. T. W. Broadhead and J. A. Waters), pp. 94–105.

Hendler, G. 1982. An echinoderm vitellaria with a bilateral larval skeleton: evidence for the evolution of ophiuroid vitellariae from ophioplutei. *Biological Bulletin of the Marine Biology Laboratory, Woods Hole* **163**, 431–7.

Hyman, L. H. 1940. *Protozoa through Ctenophora. Vol. I The invertebrates*. McGraw-Hill, New York.

Hyman, L. H. 1955. *The echinoderms. Vol. IV The invertebrates*. McGraw-Hill, New York.

Hyman, L. H. 1959. *Smaller coelomate groups. Vol. V The invertebrates*. McGraw-Hill, New York.

Inaba, D. 1930. Notes on the development of a holothurian, *Caudina chilensis. Science Reports of Tohôku University Series 4* **5**, 215–47.

Komatsu, M. 1982. Development of the sea star *Ctenopleura fisheri. Marine Biology* **66**, 199–205.

Lester, S. M. 1985. *Cephalodiscus* sp.; observations on functional morphology, behavior and occurrence in shallow water around Bermuda. *Marine Biology* **85**, 263–8.

Lester, S. M. 1988. Ultrastructure of the adult gonads and development and structure of the larva of *Rhabdopleura normani* (Hemichordata: Pterobranchia). *Acta Zoologica* (in press).

Ludwig, H. 1889–1892. Bronn's *Klassen und Ordnungen des Thier Reiches*. Band 2, Buch 1. *Die Seewaltzen*. C. F. Winter'sche Verlagshandlung, Leipzig.

Ludwig, H. 1898. Bruptflege und Entwicklung von *Phyllophorus urna. Zoologische Anzeiger* **21**, 95–9.

MacBride, E. W. 1906. The Echinodermata. In *Cambridge natural history* (ed. S. F. Harmer and A. E. Shipley), pp. 423–623. Macmillan, London.

MacBride, E. W. 1914. *The invertebrates. Textbook of embryology Vol. I*. Macmillan Co., London.

Matsumura, T., Hasegawa, M. and Shigei, M. 1979. Collagen biochemistry and the phylogeny of Echinoderms. *Comparative Biochemistry and Physiology* **62B**, 101–5.

McEuen, F. S. and Chia, F-S. 1985. Larval development of a molpadid holothuroid, *Molpadia intermedia. Canadian Journal of Zoology* **63**, 2553–9.

Mladenov, P. V. 1985. Development and metamorphosis of the brittlestar *Ophiocoma pumila*: evolutionary and ecological implications. *Biological Bulletin of the Marine Biology Laboratory, Woods Hole* **168**, 285–95.

Morgan, T. H. 1891. The growth and metamorphosis of tornaria. *Journal of Morphology* **5**, 407–50.

Newth, H. G. 1916. The early development of *Cucumaria*: a preliminary account. *Proceedings of the Zoological Society of London*. **2**, 631–41.

Ohshima, H., 1921. On the development of *Cucumaria echinata. Quarterly Journal of Microscopical Science* **65**, 173–246.

Paul, C. R. C. and Smith, A. B. 1984. The early radiation and phylogeny of echinoderms. *Biological Reviews* **59**, 443–81.

Poltaraus, A. B. 1981. The estimation of relative connections between 9 species of echinoderms by molecular hybridization of their DNA. *Zhurnal Obshchei Biologii* **42**, 55–9.

Poltaraus, A. B., Petrov, N. B. and Antonov, A. S. 1980. Divergence of repetitive sequences in Echinodermata. 1. Comparison of sequences with a high degree of intra-genomic divergence. *Molecular Biology (Moscow)* **14**, 661–74.

Poltaraus, A. B., Petrov, N. B. and Antonov, A. S. 1981. Divergences of reiterated DNA sequences in echinoderms. 2. Comparison of sequences with a low degree of intra-genomic divergence. *Molecular Biology (Moscow)* **14**, 824–32.

Raff, R. A. 1987. Constraint, flexibility, and phylogenetic history in the evolution of direct development in sea urchins. *Developmental Biology* **119**, 6–19.

Runnström, J. and Runnström, S. 1919. Über die Entwicklung von *Cucumaria frondosa* und *Psolus phantapus. Bergens Museum Årbok Afhandlingar og Arsberetning* **1918–1919** (5), 1–99.

Runnström, S. 1927. Über die Entwicklung von *Leptosynapta inhaerens. Bergens Museum Årbok Afhandlingar og Arsberetning* **1927** (1), 1–80.

Ruppert, E. E. and Blaser, E. J. 1986. Nephridia in the larvae of hemichordates and echinoderms. *Biological Bulletin of the Marine Biology Laboratory, Woods Hole* **171**, 188–96.

Rustad, D. 1940. The early development of *Stichopus tremulus. Bergens Museum Årbok Afhandlingar og Arsberetning* **1938** (3), 1–23.

Scott, M. P. and O'Farrell, P. H. 1986. Spatial programming of gene expression in early *Drosophila* embryogenesis. *Annual Review of Cell Biology* **2**, 49–80.

Selenka, E. 1876. Zur Entwicklung der Holothurien (*Holothuria tubulosa* und *Cucumaria doliolum*). *Zeitschrift für wissenschaftlichen Zoologie* **27**, 155–187.

Semon, R. 1888. Die Entwicklung der *Synapta digitata*

und ihre Bedeutung für die Phylogenie der Echinodermen. *Jenaische Zeitschrift für Naturwissenschaft herausgegeben von der medizinische naturwissenschaftlichen Gesellschaft zu Jena* **22**, 175–309.

Shinn, G. L. 1985a Reproduction of *Anoplodium hymanae*, a turbellarian flatworm inhabiting the coelom of sea cucumbers, production of egg capsules, and escape of the infective stages *without* evisceration of the host. *Biological Bulletin of the Marine Biology Laboratory, Woods Hole* **169**, 182–98.

Shinn, G. L. 1985b. Ultrastructure of the transrectal coelomoducts of a sea cucumber. *American Zoologist* **25**, 114a.

Smiley, S. 1984. A description and analysis of the structure and dynamics of the ovary, of ovulation, and of oocyte maturation in the sea cucumber *Stichopus californicus*. MSc Thesis, University of Washington, Seattle.

Smiley, S. 1986. Metamorphosis in *Stichopus californicus* and its phylogenetic implications. *Biological Bulletin of the Marine Biology Laboratory, Woods Hole* **171**, 611–31.

Smiley, S. and Cloney, R. A. 1985. Ovulation and the fine structure of the *Stichopus californicus* fecund ovarian tubules. *Biological Bulletin of the Marine Biology Laboratory, Woods Hole* **169**, 342–63.

Smiley, S., McEuen, F. S., and Chaffee, C. 1988. Holothurian reproductive biology. In *A treatise of invertebrate reproduction* (ed. A. C. Geise and J. S. Pearse) (in press).

Smith, A. B. 1984a. Classification of the Echinodermata. *Palaeontology* **27**, 431–59.

Smith, A. B. 1984b *Echinoid palaeobiology*. George Allen & Unwin, London.

Smith, A. B. 1986. Cambrian eleutherozoan echinoderms and the early diversification of the edrioasteroids. *Palaeontology* **28**, 715–56.

Smith, A. B. and Paul, C. R. C. 1982. A revision of the class Cyclocystoidea. *Philosophical Transactions of the Royal Society of London, Series B.* **296**, 577–684.

Strathmann, R. R. 1976. Introduction to function and adaptation in echinoderm larvae. *Thalassia Jugoslavica* **10**, 321–39.

Stricker, S. A. 1985. Ultrastructure and formation of the calcareous ossicles in the body wall of the sea cucumber *Leptosynapta clarki*. *Zoomorphology* **105**, 209–222.

Stricker, S. A. 1986. The fine structure and development of calcified skeletal elements in the body wall of holothurian echinoderms. *Journal of Morphology* **188**, 273–88.

Théel, H., 1901. A case of hermaphroditism in Holothuroids. *Bihang till Kongliga Svenska Vetenskaps-Akademiens Handlingar* **27**, Af. 4, No. 6.

7

Fossil evidence for the relationships of extant echinoderm classes and their times of divergence

ANDREW B. SMITH

Department of Palaeontology, British Museum (Natural History), Cromwell Road, London SW7 5BD, UK

INTRODUCTION

The fossil record provides the only unambiguous record of life in the past and can, when analysed correctly, throw light on the relationships between echinoderm classes. Furthermore, it provides the most direct method of estimating divergence times between sister taxa. In this review, I will therefore try to summarize fossil evidence concerning the relationships of the five echinoderm classes and then to use this to bracket their times of divergence as accurately as possible. This scheme can then be used for comparison with independently derived phylogenies.

Fossils in phylogenetic analysis

Fossils cannot and do not provide simple answers to questions of phylogeny because phylogenetic relationships have to be determined on an interpretation of the character suite that each taxon displays. It is through an analysis of morphological similarity that ideas of relationship develop, and the definition and analysis of a suite of characters in fossil taxa are subject to as many problems of interpretation as are ontogenetic, comparative morphological, and molecular studies. Phylogeny certainly cannot simply be read directly from the stratigraphical record, though, intuitively, stratigraphical data can provide short cuts to discovering the most parsimonious solution.

The first important point to be made is that, in order to be able to use fossil evidence in determining questions of relationships amongst extant taxa on more than just a trial and error basis, the character distribution within these extant taxa must be known before the fossils can be correctly placed within the hierarchy. That is to say, the crown groups must be clearly defined on the basis of shared derived characters before it is possible to place fossils into their correct stem group. Classification has been done on an *ad hoc* basis in the past by evolutionary systematists, who have ascribed fossils to extant classes on the basis of characters that they share. However, groups which lack some or all of those characters that are definitive of particular classes, have generally been hived off as new classes without considering their relationship within any hierarchical framework of extant taxa.

Only once the homologies that define individual taxa and those that act as synapomorphies at more inclusive levels (uniting two or more taxa) have been identified for the extant classes is it possible to be specific about the placement of fossil taxa within stem groups. We need to know the complement of attributive characters used to define extant groups in order to judge at what level a fossil with one or more of these attributive characters should be placed.

A second problem is that fossils only preserve a portion of the attributive characters which have been used to establish relationships amongst extant groups. Even when the skeletal parts of a fossil are entirely known (which is rare enough), crucial soft tissue characters are usually lost and may make placement of some fossils impossible at more than just a general level.

On the other hand, fossils do provide the only concrete evidence we have about character states that existed in the early history of echinoderms, and this can occasionally be illuminating. One of palaeontology's main strengths is the way in which it can provide evidence of homology and deter-

Echinoderm phylogeny and evolutionary biology (ed. C. R. C. Paul and A. B. Smith). Clarendon Press, Oxford, 1988.

mine character polarity. Some groups may be so highly derived that synapomorphic characters originally shared with a sister group have become modified out of certain recognition, or worse still, have been lost altogether. The fossil record may preserve 'snapshots' of a transformational series which aid the recognition of homology between end members. Alternatively, a fossil from the stem group of an extant taxon may have, in addition to at least one crown group autapomorphy, a character or characters that have subsequently been lost from the entire crown group. Such a fossil can then be used to establish that those characters have been secondarily lost. For example, holothurians are so derived and have reduced their skeleton to such an extent that it is very difficult to compare their morphology with that of other classes. By looking at stem group holothurians (assuming that they have been correctly identified as such), then certain characters such as the lantern apparatus can be shown to have been present in members of the stem group, but to have been lost prior to the divergence of crown group members.

Estimating divergence times

The divergence times for sister groups at class level and above cannot simply be read directly from the fossil record, but can only be estimated by trying to bracket the event as accurately as possible. The approach taken here is to try to identify the earliest member of each class, i.e. the oldest fossil which shows at least one of the autapomorphies of that class, and to try to identify fossils which are plesiomorphic sister taxa to clades of two or more classes combined, and which are thus potential ancestors. The earliest appearance of a member of either sister group obviously provides a definitive latest divergence time for those two taxa. Placing a lower limit on the divergence time is much more difficult and can never be definitive, because it is impossible to identify ancestors for certain. Fossil plesions that have all the characters of the crown group, but none of the autapomorphies of its constituent subgroups may include the latest common ancestor of the crown group, and provide an approximate date of divergence. Otherwise, the first appearance of the most derived plesion in the stem group of two or more classes can provide an approximate lower bracket to divergence timing.

CROWN GROUP ANALYSIS

The relationship between the five extant classes of the Echinodermata is not yet generally agreed, and recently proposed alternative schemes can be found in Smith (1984b), Smiley (1986, this volume), Raff (this volume), and Matsumura (this volume). One of the principal areas of disagreement lies in the placement of holothurians, which are treated either as the most derived group (Smith 1984b; Raff, this volume) or as the most primitive group (Smiley 1986, this volume). Smiley's analysis does a lot to clarify and correct errors that entered into my (1984b) cladistic treatment which resulted from mixing character comparisons of lecithotrophic and planktotrophic forms. However, there remain a number of problems with treating holothurians as a primitive sister group to other echinoderm classes.

First, although torsion in holothurian development may be absent (Smiley 1986), this does not prove that holothurians are primitively bilaterally symmetrical, as opposed to other echinoderms which undergo torsion. This is because in other classes a phase of asymmetrical development precedes the acquisition of pentamery and pentamery is an autapomorphy of echinoderms. If holothurians diverged prior to the acquisition of asymmetrical development then we have to invoke convergent evolution of pentamery. Furthermore, holothurians have only a single hydropore/stone canal, like other echinoderms with asymmetrical development and not the paired hydropore openings found in the outgroup hemichordates. It seems more parsimonious to assume that the condition seen in holothurians represents a derived state in which the asymmetrical phase has been drastically foreshortened or lost.

Secondly, the fact that the radial water vessel becomes internal in a somewhat different way to that seen in echinoids and ophiuroids weakens, but does not disprove their homology. The possession of internal radial water vessels remains a putative homology which is found in one of two states, one of which may be derived compared to the other. Thus, either the echinoid/ophiuroid or the holothurian mechanism of envelopment of the water vascular system may be primitive and the other derived.

Thirdly, the continuity of the axocoel/hydrocoel compartments in developing holothurians is not unique to that group since Gemmill (1914) also described an undivided axohydrocoel in the starfish *Asterias*. The separation of axocoel and hydrocoel is most marked in crinoids, but in other groups the division is less pronounced. Smiley (1986) argued that the undivided axohydrocoel in holothurians was primitive because structures that are directly induced by the separate axocoel in echinoids and asteroids, such as pentaradial gonads, are not found even vestigially in holothurians. Unfortunately, it is clear from Palaeozoic echinoids that pentaradial gonads are a derived feature, and that primitive echinoids have only a single gonopore (and presumably only a single genital rachis during development; Smith 1984a). Thus, either primitive echinoids had an undivided axohydrocoel or demarcation between the functions of axocoel and hydrocoel was not so rigidly established in the Lower Palaeozoic. In any case, the argument does not hold and other criteria must be used to establish the polarity of this character.

Finally, the buccal podia which are unique to holothurians are probably homologous with axocoelic attachment organs in the larvae of other echinoderms, as Smiley (1986) suggested. They could either be primitive, in which case their loss is a synapomorphy for non-holothurian echinoderm classes, or are derived larval characters retained into adulthood by holothurians. Since so much else of holothurian morphology, especially skeletal morphology, has larval characteristics, I prefer to treat buccal podia as an autapomorphy of the holothurians.

Holothurians pose considerable problems in placement within a phylogenetic scheme, largely because they have reduced and simplified their body skeleton to such a degree that there are so few characters with which to work. What fossil evidence there is suggests holothurians and echinoids are sister groups and is briefly reviewed below. The hierarchical relationship of the five classes adopted here is shown in Fig. 7.6.

THE CRINOID/ELEUTHEROZOAN DIVERGENCE

Divergence between the crinoids and other classes appears to have taken place before the end of the Lower Cambrian, some 550 Ma. A detailed analysis of the known Lower Cambrian fauna was given by Paul and Smith (1984) and need not be repeated here (see Fig. 7.1). Basically, we recognize helicoplacoids (Plate 7.1.(1)) to be so generalized in form as to belong to the stem group of the Echinodermata as a whole. They have at least three autapomorphies of living echinoderms, namely, the skeletal histology of stereom, the biserial ambulacra with coverplates, and the radial arrangement of the ambulacra (and radial water vessels by inference) around the mouth. However, with a laterally positioned mouth and ?apical anus, they lack pentaradial symmetry and possibly have only weakly developed visceral torsion both of which are characteristic of crown group echinoderms. As helicoplacoids appear in the fossil record prior to any member of the crown group, it is possible that they might be ancestral to the latest common ancestor of the extant echinoderm classes. A description of helicoplacoids was given by Durham and Caster (1963), and Durham (1967), while Derstler (1981) and Paul and Smith (1984) give a different interpretation.

Lepidocystis (Fig. 7.1) was described fully by Sprinkle (1973). It represents the earliest known fossil that has characters unique to crinoids, namely the elongated stem or holdfast and the presence of ambulacra which extend outside the theca to form a subvective system of arms. I take crinoids as originating somewhat earlier than most previous workers (but see Paul, this volume) because the Crinoidea, as conventionally conceived (viz. Moore and Teichert 1978) is only a subset of the stem group of the present day crinoids, the other part comprising cystoids *s.l.* (Paul, this volume). Thus, the stem group of the Crinoidea is coextensive with the Pelmatozoa.

Stromatocystites (Plate 7.1.(3)) (described by Smith 1986) represents a stem group member of the Eleutherozoa (asteroids + ophiuroids + echinoids ? + holothurians). It has a disc-shaped body with pentameral symmetry and appears to be without firm attachment as an adult. Unlike crown group eleutherozoans, it has a fixed mouth frame and no stone canal or calcified madreporite. I have interpreted this as a possible ancestor to all later eleutherozoan groups (Smith 1986).

Camptostroma (Plate 7.1.(2)) has been described briefly by Derstler (1981) and somewhat more fully by Paul and Smith (1984). This genus was originally placed in a trichotomy with *Stromatocystites* and the pelmatozoans by Paul and Smith (1984), since it seemed to be almost perfectly intermediate between these two groups. In the cladogram (Fig. 7.1) I have placed it as an

88 Andrew B. Smith

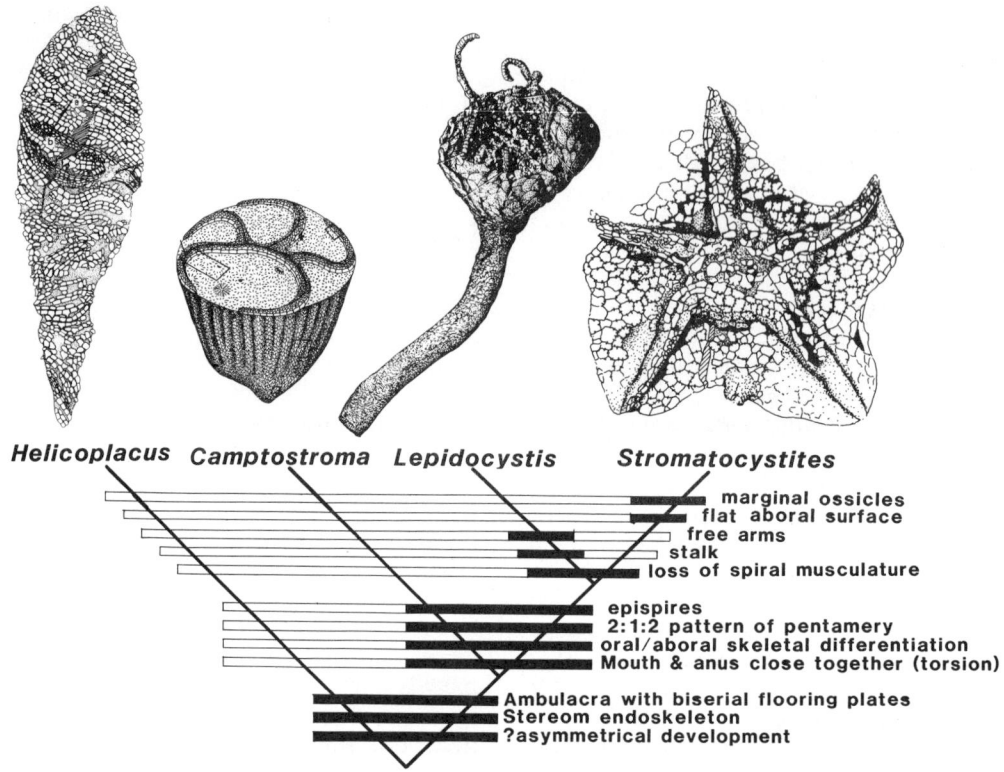

Fig. 7.1 Cladogram of four Lower Cambrian genera. *Lepidocystis* is a stem group crinoid and *Stromatocystites* a stem group eleutherozoan. Shared derived characters are shown as black bars.

advanced stem group member of the Echinodermata, because, unlike crown group echinoderms, but similar to *Helicoplacus*, it has spiral musculature aborally (seen in retracted specimens as spirally arranged zones of contraction).

Camptostroma, *Kinzercystis*, and *Stromatocystites* all occur in the upper part of the Lower Cambrian, whereas helicoplacoids are found in the middle Lower Cambrian. The age of *Kinzercystis* at about 550 Ma (on the scale of Harland *et al.* 1982) sets the youngest date by which divergence must have taken place. Similarly, if helicoplacoids have been correctly interpreted, then the earliest appearance of helicoplacoids at around 560 Ma is a conservative estimate for the lower date of divergence. The recognition of *Camptostroma* as the most crownward known member of the stem group of echinoderms (probably closely resembling the latest common ancestor of all extant echinoderms) supports the timing of divergence of crinoids and eleutherozoans shortly before 550 Ma.

THE ASTEROID/CRYPTOSYRINGID DIVERGENCE

The divergence between asteroids and other eleutherozoan echinoderms, which Smith (1984*b*) termed Cryptosyringida, is the next dichotomy. Cryptosyringids, comprising ophiuroids, echinoids, and holothurians, can be recognized in the fossil record by the autapomorphy of having an enclosed radial water vessel and (at least primitively) in having the first ambulacral ossicles in each ambulacral column modified into a jaw apparatus.

During the Middle Cambrian, *Stromatocystites* diversified to give rise to a number of forms, generally regarded as edrioasteroids, and here treated as stem group eleutherozoans. A cladistic analysis of this group has been recently given (Smith 1986). Amongst these forms is a disc-like animal with a prominent marginal ring of ossicles, called *Cambraster* (Plate 7.1.(5)) which has been thoroughly described by Ubaghs (1971) and Jell *et al.* (1985). *Cambraster* was originally mistaken for an asteroid by Jaekel (1923), and shares with *Archegonaster* and primitive asteroids a general similarity in body plan. However, *Cambraster* lacks certain crown group eleutherozoan autapomorphies, such as a calcified madreporite and stone canal, and still retains its periproct in the primitive position close to the mouth on the oral surface. It is, therefore, placed as a member of the stem group of all eleutherozoans, but more crownward than *Stromatocystites*. *Cambraster* is found through the Middle Cambrian.

The Upper Cambrian was a period with a very poor fossil record of echinoderms, and we know of only a handful of specimens from this period (see Sprinkle 1976 for a summary of the North American occurrences). The detailed history of *Cambraster* and its relatives during this period is, therefore, unknown and by the time we next find well preserved echinoderms, in the late Tremadoc and early Arenig (c. 490 Ma), asteroids and cryptosyringids had already diverged.

The first recognizable member of the asteroid stem group is *Uranaster* from the Lower Arenig of Ramsay Island (see Spencer 1918). This is rather poorly known, but does have a stellate body outline with a well marked marginal frame of ossicles. Slightly later species of *Uranaster* and *Petraster* are better known. In these, ambulacral ossicles are square and stout, abutting firmly against one another, and with adambulacral ossicles which are likewise stout and squarish. Adambulacrals are sutured to ambulacrals (which they oppose) and to marginals (with which they alternate) along most of the arm, but proximally there is an extensive zone of 'interambulacral' plating in a V-shaped interradial region (Fig. 7.2). The modified mouth angle plates and the squarish, sutured adambulacrals of this species are autapomorphies of the Asteroidea, and this form is without doubt an early member of the asteroid stem group.

Within about 20 million years (by the Caradocian, Upper Ordovician), starfish had diversified considerably and forms such as *Platanaster* (Plate 7.2.(1, 2)) are present. *Platanaster* shows additional asteroid autapomophies such as well developed paxillae aborally, and an odontophore.

Archegonaster (Plate 7.1.(4)), from the Llanvirn (Lower Ordovician) of Czechoslovakia is a very interesting form. It has a pentagonal outline and a well developed marginal frame of ossicles. The aboral surface is reduced to a uniform covering of minute spicules and there is no evidence of a periproct on either oral or aboral surface. The ambulacra have a straight perradial contact and imbricate slightly in the proximal/distal direction. There are no apparent podial basins, which distinguishes *Archegonaster* from all Lower Palaeozoic asteroids or ophiuroids. However, *Archegonaster* does have a well developed calcified madreporite and stone canal, and the passageway for the stone canal seen on the inner surface of the madreporite is scroll-shaped; both characters shared with asteroids, ophiuroids, and echinoids. A second important feature is that the first two ambulacral ossicles in each column are separated along the perradius to form buccal slits. Unfortunately, preservation in this region is not good, and the precise morphology of these mouth angle plates is difficult to decipher. From what can be seen, these mouth angle plates are only weakly differentiated from ambulacral plates.

Archegonaster has one peculiar feature of its adambulacral skeleton, shared with primitive cryptosyringids such as *Villebrunaster*. Adambulacral ossicles articulate on ambulacral ossicles and bear a row of spines. In overall morphology they are more like the slender lateral arm plates of ophiuroids than the blocky adambulacrals of asteroids. These adambulacral ossicles are also firmly sutured to a series of virgalia which run to the marginal ossicles (Plate 7.2.(6)). The position of *Archegonaster* in a phylogenetic scheme must await further work on its anatomy, but it is tentatively placed as an advanced stem group member of the eleutherozoans, intermediate in many ways between true asteroids and cryptosyringids on the one hand, and *Cambraster* on the other.

Several cryptosyringid echinoderms appear at the base of the Arenig in southern France (see Fell 1963; Courtessole *et al.* 1983). The most primitive of these is *Villebrunaster*, described by Fell (1963). This has a starfish-like body form, and a sequence of lateral plates between the ambulacra and marginals called virgalia. Virgalia seem to have no homology in later cryptosyringids. I have not studied this animal and have nothing to add to Fell's description. *Villebrunaster* has no autapomorphy shared with ophiuroids, echinoids or holothurians, so I treat it as a stem group member

90 Andrew B. Smith

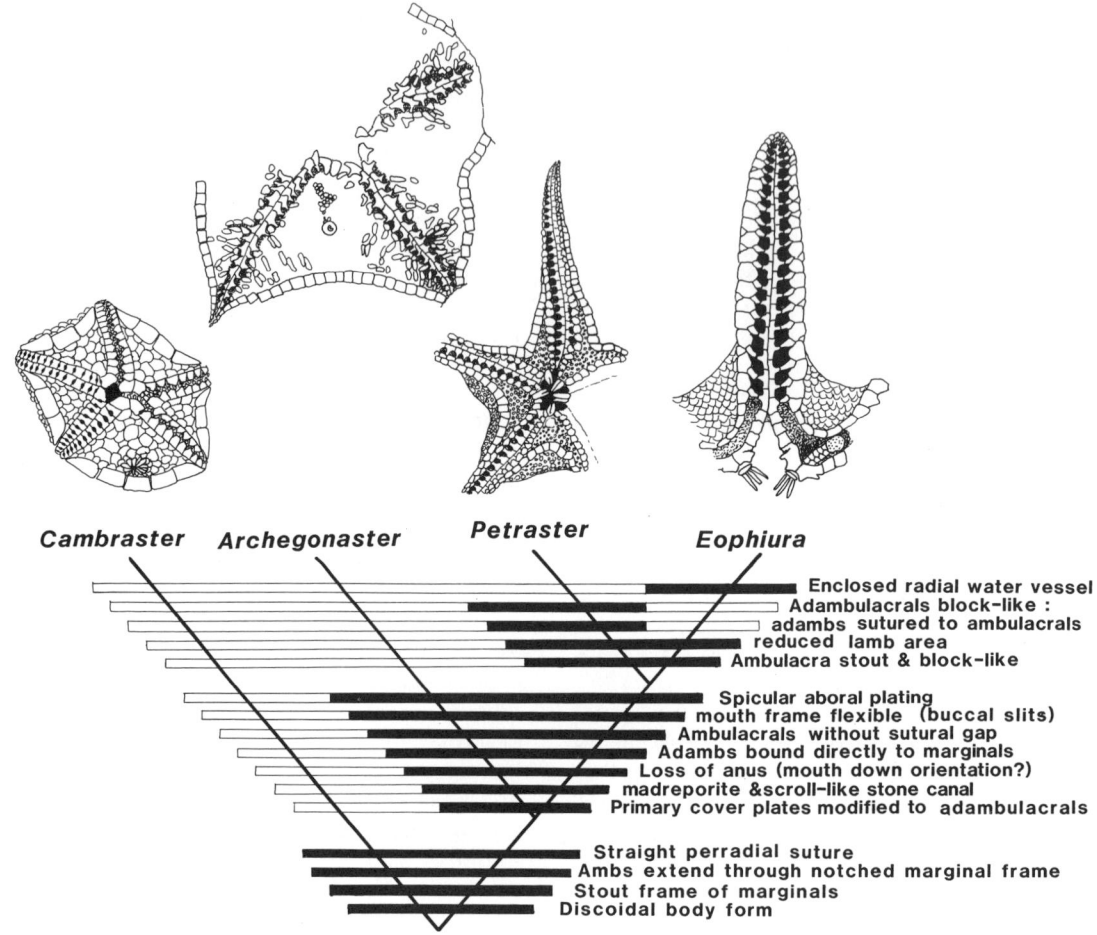

Fig. 7.2 Cladogram for four primitive eleutherozoan genera: *Cambraster* is a Middle Cambrian stem group eleutherozoan; *Archegonaster* is an advanced stem group eleutherozoan from the Llanvirn (Lower Ordovician); *Petraster* is a primitive asteroid from the Ordovician; and *Eophiura* is a stem group cryptosyringid from the Llanvirn (Lower Ordovician).

of the cryptosyringids. Along with *Villebrunaster* is found the earliest member of the ophiuroid stem group, *Pradesura* (Spencer 1951).

A slightly later member of the cryptosyringids that I have studied is *Eophiura*, from the Llanvirn of Czechoslovakia (see Spencer 1951). *Eophiura* is an ophiuroid-like echinoderm with a very large disc and short arms (Plate 7.3.(1, 2)). The aboral surface is covered in small ovoid plates which overlap slightly and leave sutural pores in places. These plates continue, without differentiation, right to the tip of the arms. On the oral surface there is a large calcified madreporite lying close to the mouth, and a plated stone canal. Ambulacral ossicles are large and block-like, resembling those of asteroids such as *Platanaster* except that the perradial groove that houses the radial water vessel is completely roofed over by thin flanges from each plate which meet centrally. As in asteroids, the podial basin is equally shared between the two adjacent ambulacral plates. Aborally, there are no notches for muscles between ambulacrals whereas in ophiuroids successive ambulacrals are separated by muscle blocks, giving the arms their characteristic flexibility. One of the most unusual features of *Eophiura* is its adambulacral skeleton (Plate 7.3.(2)). Here there are two sets of plates; an outer series of large, squarish plates which bear spines, and an inner series of smaller paddle-shaped plates which connect the outer series to the ambulacrals. There is one inner series plate to each ambulacral, and they articulate via a ridged and grooved surface that restricts movement to the vertical plane. The outer series of plates are more numerous than the inner series, with smaller and larger plates sometimes alternating, particularly towards the end of the arm. These smaller plates do not abut against the plates of the inner series.

The mouth frame of *Eophiura* (Plate 7.3.(2)) has well developed buccal slits, with the first five or so ambulacrals in each column separated from each other. The ossicles become twisted in orientation towards the mouth so that the podial basins come to face perradially into the mouth opening. The most proximal podial basin is shared equally between the first and second ambulacrals. The first ambulacral ossicle is fairly highly modified into a mouth angle plate (MAP) and adjacent MAPs from neighbouring ambulacra are united. This first ambulacral ossicle has a small distally-facing flange which is presumably for muscle attachment between it and the second ambulacral ossicle which has a complementary oral-facing flange. Each MAP has an oral-facing torus that carries a large number of mouth spines.

Eophiura has previously been classified as an ophiuroid (Spencer 1951; Spencer and Wright 1966), but seems to lack any clear-cut autapomorphies of the Ophiuroidea. In particular it lacks (1) the characteristic lateral arm plates, (2) musculature between ambulacrals that provide the arms with their distinctive whip-like flexibility, and (3) a V-shaped ambulacral jaw apparatus. Therefore, I place *Eophiura* as a stem group member of the cryptosyringids, more advanced than *Villebrunaster*.

A summary cladogram for *Cambraster*, *Archegonaster*, *Petraster*, and *Eophiura* is given in Fig. 7.2. The divergence of asteroids from cryptosyringids can be no later than *c*. 490 Ma, which is when the earliest cryptosyringids (*Villebrunaster*) are found. The interpretation of *Cambraster* as a moderately primitive stem group member in the Middle Cambrian, seems to provide a reasonable estimate for the earliest time of divergence at around 530 Ma.

THE OPHIUROID/ECHINOZOAN DIVERGENCE

The earliest fossils definitely attributable to the ophiuroid stem lineage are found in the same beds as *Villebrunaster*. These belong to the genus *Pradesura*, and were first described in detail by Spencer (1951). Slightly younger and contemporary with *Eophiura* is *Palaeura* (Plate 7.2.(3); Plate 7.3.(3)), also described by Spencer (1951) and more recently by Smith (1985). *Palaeura* comes from the Llanvirn of Czechoslovakia and Spain. Although its ambulacra are arranged as an alternating biseries, the individual ossicles are semi-circular in cross-section and there are large muscle gaps between successive ossicles which enlarge towards the arm tip. Like all modern ophiuroids, *Palaeura* had flexible, whip-like arms. A second important character shared with crown group ophiuroids is the structure of the lateral arm plates, which are single boot-shaped ossicles bearing a row of articulated spines. These ossicles articulate upon the ambularals and adjacent lateral arm plates imbricate upon one another. The proximal ambulacrals, though not yet forming as highly modified a jaw apparatus as is seen in later ophiuroids, have a first ambulacral ossicle that is slightly more differentiated than that in *Eophiura*. One noteworthy feature of *Palaeura* is that its disc is covered in flat plates each of which bears a single articulated spine.

Within a short period (by the Upper Ordovician) ophiuroids are found in which the ambulacrals are arranged opposite one another and, although not fused, are well on the way to becoming proper vertebrae (Plate 7.2.(4, 5)). These ophiuroids also have much more specialized jaw apparatuses in which the buccal slits are more or less absent and the first two ambulacral ossicles have become modified into a sophisticated biting apparatus.

Identifying a member of the echinozoan stem group lineage has proved to be very difficult. Here I use Echinozoa to include echinoids, holothurians, and the wholly extinct group ophiocistioids, characterized by their meridional growth pattern and primitively by their lantern apparatus. The oldest fossil member of this group appears to be *Unibothriocidaris* from the Llandeilo (Middle Ordovician) of North America (Kier 1982), but the relationship of bothriocidarids in general to other echinozoans is far from clear. Bothriocidarids have reduced aboral surface plating, such that the ambulacra extend almost to the apex, and they also have the first ambulacral ossicles highly modified and wholly internal, fully differentiated from the ambulacrals that make up the test. This jaw apparatus is very different from that found in echinoids or ophiocistioids.

The earliest members of the echinoid stem lineage that can be identified come from the late Ordovician (Ashgill) and were described by MacBride and Spencer (1938). These genera (*Aulechinus*, *Ectinechinus*, and *Eothuria*) have internal dental apparatuses (see Smith 1984a) which are rather simpler than those found in later echinoids as, for example, in the Lower Silurian *Aptilechinus* (see Kier 1973). Like ophiuroids, their ambulacral water vessels are fully enclosed by skeleton (Plate 7.2.(7)), since the ambulacral plates have upper and lower perradial flanges which surround the vessels. However they appear to have lost any remnant of the lateral arm plates, unless they are represented by the adambulacral spines that articulate upon ambulacral plates. Apart from the lantern, their overall shape is also characteristic of echinoids, with aboral plating reduced to only a small periproctal area and the ambulacra extending almost to the apex. By the Lower Silurian, echinoids had developed double perforations for tube feet which is another autapomorphic feature (seen in *Aptilechinus* Kier 1973).

Although a supposed ophiocistioid, *Volchovia*, has been reported from the Lower Ordovician (Hecker 1940; Regnéll 1945), this claim is based on very incompletely known material. Basically, these specimens show a slightly spinose marginal frame of broad plates which enclose somewhat smaller plates. In my opinion the material is so incomplete that it lacks any diagnostic features by which it might be placed. Similar body plans are known in, for example, mitrates. Therefore, I have treated these as indeterminate until more complete material comes to hand.

The first undoubted ophiocistioids are found in the Middle Silurian (c. 425 Ma) and were described by Ubaghs (1966). The best known forms, however, are slightly younger and have been described by Jell (1983). I have taken Jell's genus *Gillocystis* as an example for comparison here. *Gillocystis* shares with echinoids the large Aristotle's lantern, which is fully homologous in all its parts. Its ambulacra do not extend so far towards the apex as is the case in echinoids, which is probably a primitive feature, and ophiocistioids all have heavily plated tube feet. Heavily plated tube feet are also preserved in specimens of *Bothriocidaris*, *Eophiura*, *Villebrunaster*, and the starfish *Siluraster* that I have examined. Thus, this character may be shared by all primitive crown group eleutherozoans. The anus is lateral in *Gillocystis* and in other ophiocistioids where known.

A cladogram for the five genera considered in detail here is given in Fig. 7.3. The divergence between ophiuroids and echinozoans can be no later than about 490 Ma, when the first true ophiuroids are found. This is also the timing for the first appearance of cryptosyringids in general, so a conservative estimate for the lower limit for divergence can be taken at the first appearance of *Cambraster*, viz. at around 530 Ma.

THE ECHINOID/HOLOTHURIAN DIVERGENCE

The holothurians are the most difficult group to tackle, since they have a really appalling fossil record. There are only two Palaeozoic records of body fossils of holothurians, one from the Upper Carboniferous (as yet undescribed), the other from the Lower Devonian (*Palaeocucumaria*; Seilacher 1961). Characteristic body wall spicules are not uncommon in certain environments from the Triassic onwards (Frizzel and Exline 1955), but Lower Palaeozoic records are all unconvincing and could belong to any echinoderm group. The discovery that wheel spicules are not confined to holothurians *sensu stricto* (Haude and Langenstrassen 1976) means that these cannot be taken as diagnostic spicules for the group.

Clearly, the recognition of true echinoids in the latest Ordovician means that echinoids and holothurians must have diverged prior to this time. In considering the origin of holothurians, two genera have to be discussed, namely *Rotasaccus* and *Palaeocucumaria*.

Rotasaccus has been described in some detail by Haude and Langenstrassen (1976). It has several important features worth pointing out. First of all, its entire body wall skeleton, save for five double rows of internal wing-shaped ambulacral plates, is reduced to wheel-shaped spicules identical to those generally ascribed to the holothurian family Theeliidae. It has, in addition, five double rows of large plated tube feet that are identical in form and skeletization to those of Silurian ophiocistioids such as *Sollasina*. Possibly the most remarkable feature of *Rotasaccus*, however, is its Aristotle's lantern which has all the elements found in the echinoid lantern. Like other ophiocistioids, the teeth in this lantern are rather different in structure

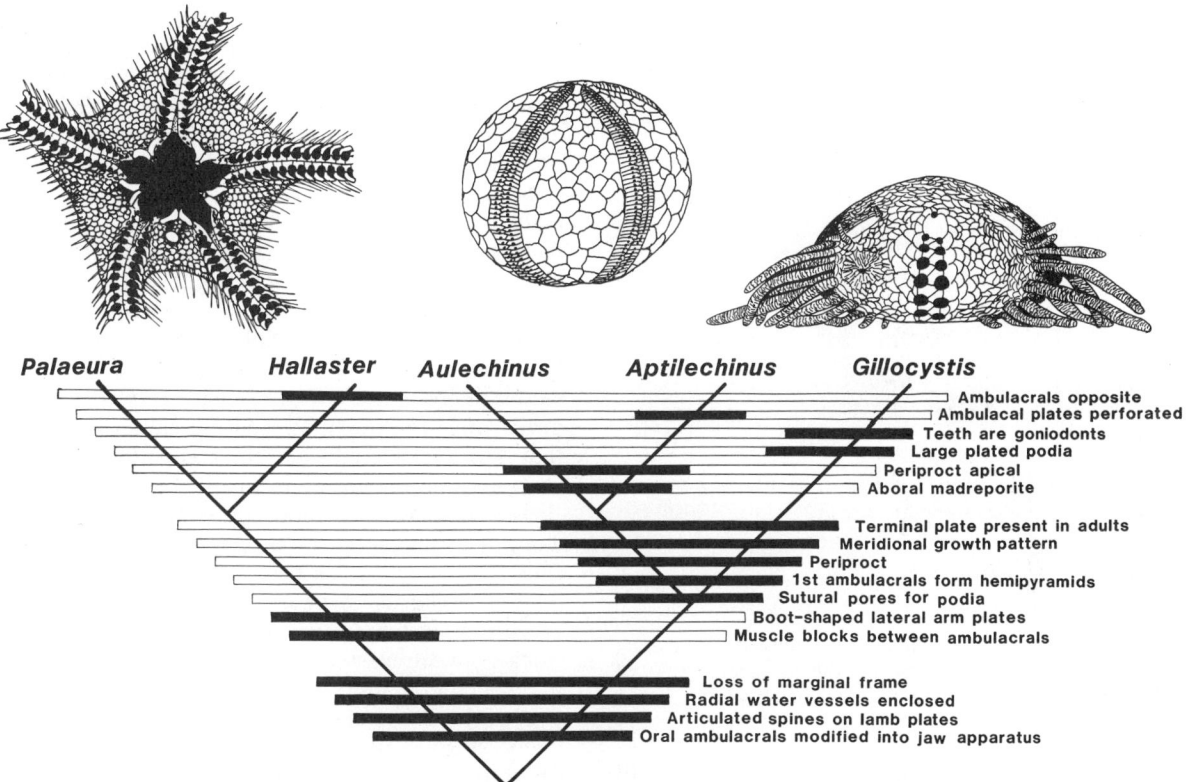

Fig. 7.3 Cladogram for five primitive cryptosyringid genera: *Palaeura* is a primitive ophiuroid from the Llanvirn (Lower Ordovician); *Hallaster* is a more derived ophiuroid from the Upper Ordovician (Ashgill); *Aulechinus* is a primitive Upper Ordovician echinoid; *Aptlechinus* is a slightly more derived echinoid from the Lower Silurian; *Gillocystis* is a Lower Devonian ophiocistioid.

to those of echinoids, being V-shaped goniodonts. Clearly then, *Rotasaccus* has characters that it shares with echinoids and ophiocistioids, but a type of body wall that is found only in holothurians.

Palaeocucumaria is the oldest holothurian for which we have any substantial morphological information (Seilacher 1961). However, because the skeleton is so reduced in this form, there are really very few attributes that can be recognized. *Palaeocucumaria* is undoubtedly a holothurian by anybody's reckoning since it possesses a calcareous circum-oesophageal ring which is an autapomorphy of extant holothurians. The rest of its body wall is reduced to spicules which Seilacher (1961) reported to be solid, but whether this is an artefact of preservation needs to be investigated. The only other feature of *Palaeocucumaria* is that it has large, plated, tube feet which are arranged into ambulacral rows, with at least four or five tube feet per row (Fig. 7.4). These tube feet are supported by curved spicules and may be digitate. The plated tube feet are a plesiomorphic feature shared with ophiocistioids, and it is noteworthy that *Palaeocucumaria* has rows of oral feeding tube feet rather than a ring of tube feet, as in all extant holothurians. The ring of buccal tube feet is thus a derived character in crown group holothurians and one which arose after the evolution of the calcareous circum-oesophageal ring. These characteristics of the tube feet at least hint at a close relationship with ophiocistioids.

A summary cladogram for *Palaeocucumaria*, *Rotasaccus*, and ophiocistioids is shown in Fig. 7.5. I would argue that the fossil record of holothurians, sparse as it is, points to ophiocistioids as belonging to their stem group. If this is so, then the presence of such a complex dental

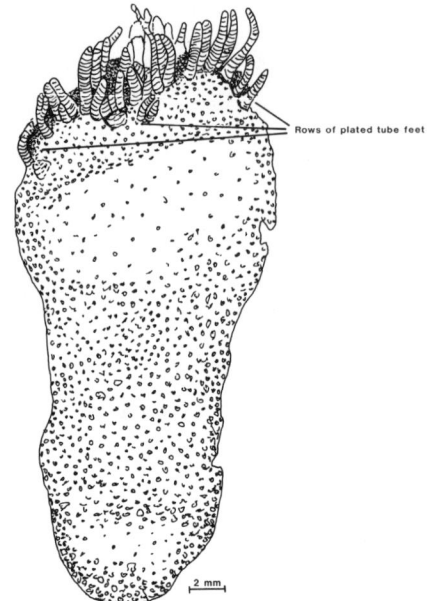

Fig. 7.4 Drawing of *Palaeocucumaria hunsruckiana* from Seilacher (1961) showing the apparent rows of ambulacral tube feet which are heavily plated.

apparatus in ophiocistioids homologous in all respects to that of echinoids, is convincing evidence that holothurians and echinoids are sister groups. The time of divergence for echinoids and holothurians can be no later than *c.* 440 Ma, with the first appearance of stem group echinoids. As a lower limit to the divergence time it seems reasonable to take the earliest appearance of a member of the Echinozoa, (*Unibothriocidaris*) at about 460 Ma.

IMPLICATIONS OF THIS HYPOTHESIS FOR
MOLECULAR ANALYSES OF PHYLOGENY

The data on divergence times for the five classes of echinoderms are summarized in Fig. 7.6. This is an hypothesis of relationships based on a cladistic analysis of comparative morphology where fossil taxa have been assigned to their appropriate stem group on the basis of the crown group characters that they possess. As such it should be used for test-like comparison with phylogenies and divergence times derived from independent sources of evidence (embryology, molecular biology). Clearly, I would hope that consensus exists between all lines of evidence, but where there is non-congruence this is useful in highlighting those relationships that need further investigation and corroboration. There are two points that come from this model of echinoderm phylogeny which have testable implications for those working on molecular based phylogenies.

1. In those groups which have a good fossil record (asteroids, echinoids, crinoids) then the divergence within the crown group that gave rise to all the extant forms, took place at around 250 Ma, near the Permo–Triassic boundary (see Simms this volume; Smith 1984a; Blake 1980; Gale 1987). From what little I know of the ophiuroids, the same seems likely to be true for this group also. This means that the maximum within-class divergence that is to be expected when comparing extant forms is approximately half that for divergences between classes (250 Ma as opposed to 550–450 Ma). It would be interesting to compare figures of within-class and between-class molecular divergence in relation to ideas of neutral models of molecular evolution and the comparison of the molecular clock between classes.

2. Secondly, it appears that the divergence of all five echinoderm classes occurred during a relatively short time interval, from *c.* 550–450 Ma, suggesting that on average there was a divergence event approximately every 20–30 Ma. Assuming molecular substitution takes place under a neutral model (Kimura 1983), and that at least one substitution is required between each divergence event in order to establish a synapomorphy in the derived sister group, then only those gene or amino acid sequences with a mutation rate of at least one substitution per 10–15 Ma are likely to be of use in fully resolving class relationships. However, because the divergences occurred so far back in geological time, a great number of additional mutations must have taken place after divergence in each line which are autapomorphies of each class, but which simply add 'noise' to any analysis of class relationships. If as an approximation we assume that divergence took place at 500 Ma and that one substitution has occurred every 10 Ma (for a fully resolved molecular cladogram), then 96 per cent of these substitutions are uninformative and simply add noise to the analysis at this level. With such a large signal to noise ratio, the probability of false signals being picked up seems high.

A more profitable approach to molecular analysis of class relationships may be to use analyses

Fossil evidence for class relationships 95

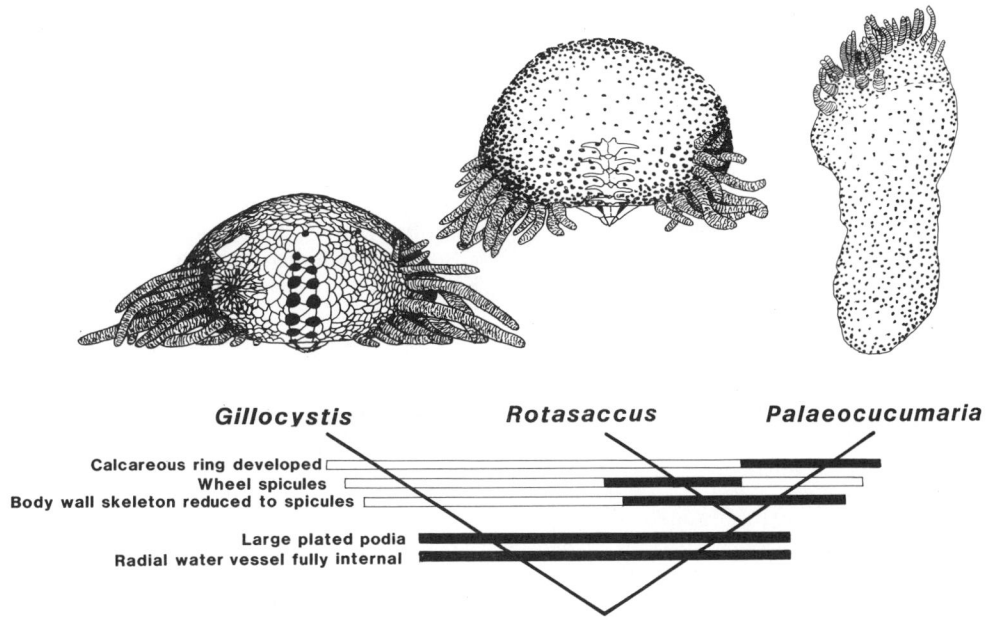

Fig. 7.5 Cladogram of Devonian ophiocistioids and holothurians; *Gillocystis* is a holothurian; *Rotasaccus* is an ophiocistioid/holothurian intermediate; *Palaeocucumaria* is a primitive holothurian.

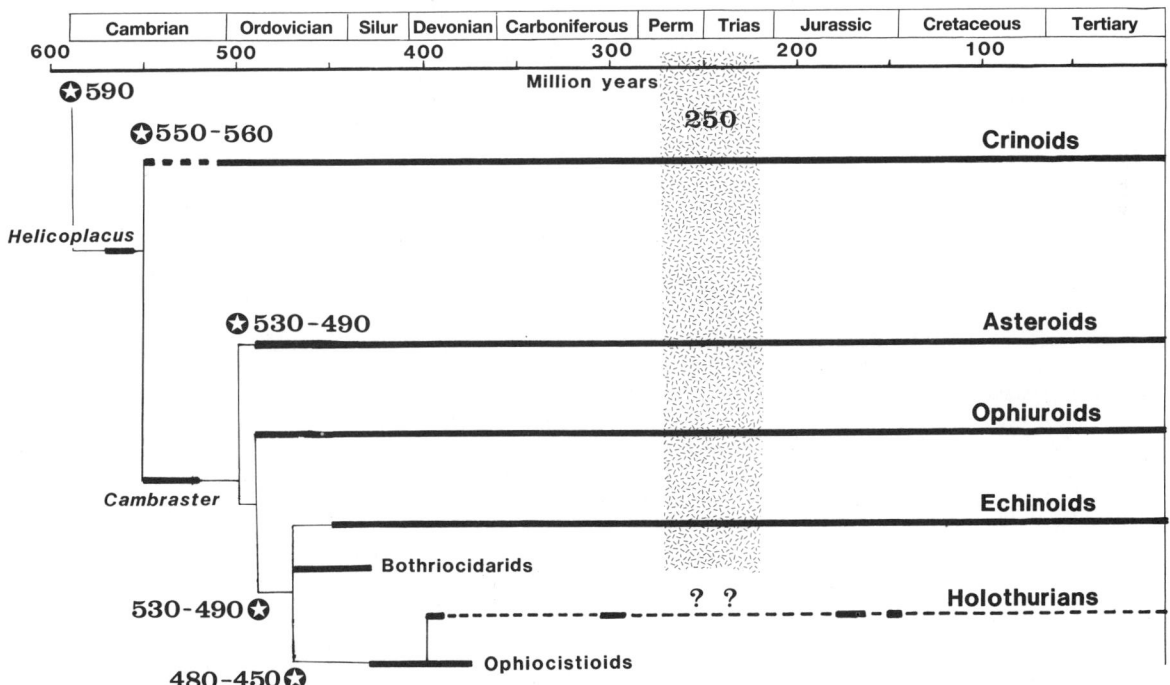

Fig. 7.6 Phylogram for the five classes of echinoderm showing known stratigraphic ranges (solid bars) and ranges of important stem group taxa. Times of divergence deduced from the fossil record are shown by stars, with definitive latest times of divergence and working estimates for the earliest time of divergence providing upper and lower limits.

of much more slowly mutating sequences (with changes in the range of every 100–150 Ma) which will give only partially resolved cladograms of relationship and then to combine several together to produce a fully resolved hypothesis of relationship.

REFERENCES

Blake, D. B. 1980. Post-Paleozoic Asterozoa. In *Echinoderms. Notes for a short course* (ed. T. W. Broadhead and J. A. Waters), *University of Tennessee, Department of Geological Sciences, Studies in Geology* **3**, pp. 200–14.

Courtessole, R., Marek, L., Pillet, J., Ubaghs, G., and Vizcaino, D. 1983. Calymenina, Echinodermata et Hyolitha de l'Ordovicien inférieur de la Montaigne Noire. *Mémoires de la Société d'Études Scientifiques de l'Aude* 1983, 62 pp, 13 pls.

Derstler, K. L. 1981. Morphological diversity of early Cambrian echinoderms. In *Short Papers for the Second International Symposium on the Cambrian System* 1981. U.S. Open File Report **81–743** (ed. M. E. Taylor), pp. 71–5.

Durham, J. W. 1967. Notes on the Helicoplacoidea and early echinoderms. *Journal of Paleontology* **41**, 97–102.

Durham, J. W. and Caster, K. E. 1963. Helicoplacoidea: a new class of echinoderms. *Science, New York* **140**, 820–2.

Fell, H. B. 1963. Phylogeny of sea-stars. *Philosophical Transactions of the Royal Society of London B*, **246**, 381–435.

Frizzell, D. L. and Exline, H. 1955. Monograph on fossil holothurian sclerites. *Missouri University School of Mines Metallurgy Bulletin Technical Series* **89**, 1–204.

Gale, A. S. 1987. Phylogeny and classification of the Asteroidea (Echinodermata). *Zoological Journal of the Linnean Society* **89**, 107–32.

Gemmill, J. F. 1914. The development and certain points in the adult structure of the starfish *Asterias rubens* L. *Quarterly Journal of Microscopical Science* **61**, 51–60.

Harland, W. B., et al. 1982. *A geological time scale*. Cambridge University Press, Cambridge.

Haude, R. and Langenstrassen, F. 1976. *Rotasaccus dentifer* n. g., n. sp., ein devonischer Ophiocistioide (Echinodermata) mit 'holothuroiden' Wandskleriten und 'echinoidem' Kauapparat. *Paläontologische Zeitschrift* **50**, 130–50.

Hecker, R. F. 1940. Carpoidea, Eocrinoidea und Ophiocistia des Ordoviziums des Leningrader Gebietes und Estlands. *Trudȳ Paleontologicheskogo Instituta, Akademia Nauk SSSR* **9**, 5–82. [In Russian with a German abstract.]

Jaekel, O. 1923. Zur Morphologie der Asterozoa. *Paläontologische Zeitschrift* **5**, 344–50.

Jell, P. A. 1983. Early Devonian echinoderms from Victoria (Rhombifera, Blastoidea and Ophiocistioidea). *Memoirs of the Association of Australasian Palaeontologists* **1**, 209–35.

Jell, P. A., Burrett, C. F., and Banks, M. R. 1985. Some Cambrian and Ordovician echinoderms from eastern Australia. *Alcheringa* **9**, 183–208.

Kier, P. M. 1973. A new Silurian echinoid genus from Scotland. *Palaeontology* **16**, 651–63.

Kier, P. M. 1982. Echinoids. In *Echinoderm Faunas from the Bromide Formation (Middle Ordovician) of Oklahoma*. (ed. J. Sprinkle), *The University of Kansas, Paleontological Contributions, Monograph* **1**, pp. 310–14.

Kimura, M. 1983. *The neutral theory of molecular evolution*. Cambridge University Press, Cambridge.

MacBride, E. W. and Spencer, W. 1938. Two new Echinoidea, *Aulechinus* and *Ectinechinus*, and an adult plated holothurian *Eothuria* from the Upper Ordovician of Girvan. *Philosophical Transactions of the Royal Society of London B*, **229**, 91–136.

Moore, R. C. and Teichert, C. (ed.) 1978. *Treatise on invertebrate paleontology: part T, Echinodermata 2*. Geological Society of America and University of Kansas Press, Lawrence, Kansas.

Paul, C. R. C. and Smith, A. B. 1984. The early radiation and phylogeny of echinoderms. *Biological Reviews* **59**, 443–81.

Regnéll, G. 1945. Non-crinoid Pelmatozoa from the Palaeozoic of Sweden: a taxonomic study. *Meddelanden från Lunds Geologisk-Mineralogiska Institution* **108**, 1–225.

Seilacher, A. 1961. Holothurien im Hunsruckschiefer (Unter-Devon). *Sonderdruck aus dem Notizblatt des Hessischen Landesamtes für Bodenforschung zu Wiesbaden* **89**, 66–72.

Smiley, S. 1986. Metamorphosis of *Stichopus californicus* (Echinodermata: Holothuroidea) and its phylogenetic implications. *Biological Bulletin of the Marine Biology Laboratory, Woods Hole*, **171**, 671–91.

Smith, A. B. 1984a. *Echinoid palaeobiology*. George Allen & Unwin, London.

Smith, A. B. 1984b. Classification of the Echinodermata. *Palaeontology* **27**, 431–59.

Smith, A. B. 1985. Ophiuroidea. In J. C. Guttierrez-Marcos, J. Chauvel, B. Melendez, and A. B. Smith, Los echinodermos (Cystoidea, Homalozoa, Stelleroidea, Crinoidea) del Paleozoico Inferior de los Montes de Toledo y Sierra Morena (Espana). *Estudios geologicos* **40**, pp. 440–2.

Smith, A. B. 1986. Cambrian eleutherozoan echinoderms and the early diversification of edrioasteroids. *Palaeontology* **28**, 715–56.

Spencer, W. K. 1918. A monograph of the British Palaeozoic Asterozoa. *Palaeontographical Society Monographs*, Part III, 109–68.

Spencer, W. K. 1951. Early Palaeozoic starfish. *Philosophical Transactions of the Royal Society of London B* **235**, 87–129.

Spencer, W. K. and Wright, C. W. 1966. Asterozoa. In *Treatise on invertebrate paleontology: Part U, Echinodermata* 3 (ed. R. C. Moore), pp. 4–107. Geological Society of America and University of Kansas Press, Lawrence, Kansas.

Sprinkle, J. 1973. Morphology and evolution of blastozoan echinoderms. *Special Publication of the Museum of Comparative Zoology, Harvard University*, 284 pp.

Sprinkle, J. 1976. Biostratigraphy and paleoecology of Cambrian echinoderms from the Rocky Mountains. *Brigham Young University Geology Studies* **23**, 61–73.

Ubaghs, G. 1966. Ophiocistioids. In: *Treatise on invertebrate paleontology: Part U, Echinodermata* 3 (ed. R. C. Moore), pp. 174–88. Geological Society of America and University of Kansas Press, Lawrence, Kansas.

Ubaghs, G. 1971. Diversité et specialisation des plus anciens échinodermes que l'on connaisse. *Biological Reviews* **46**, 157–200.

PART III
Molecules and relationships

8

DNA evolution and echinoderm systematics

ROY J. BRITTEN

California Institute of Technology and Carnegie Institution of Washington, USA

SEA URCHIN GENOME ORGANIZATION

When the total DNA of the sea urchin *Strongylocentrotus purpuratus* (Sp) is denatured and incubated (Cot 40 so that most of the repeats of more than 100 copies reassociate) and then treated with a single strand specific nuclease (S1), it is found that 21 per cent of the DNA remains as undigested duplexes (Britten *et al.* 1976). This is our best estimate of the amount of repeated DNA sequence present. The 79 per cent which was digested in this test is the 'single copy fraction' which is usually utilized for interspecies hybridization. It certainly contains contaminant high frequency repeats in small quantity which, however, have little effect on such measurements. There is evidence that it contains low frequency repeats as well (Klein *et al.* 1978; Moore *et al.* 1981). It is not known what part of the 'single copy fraction' is made up of sequences in a few copies but the kinetics of reassociation suggest that it is primarily single copy.

The sets of repeats of Sp have been fairly well examined, on a par with those of the *Drosophila* and human genome where recent interest in mobile elements has led to a large number of studies. Sp is the only species for which a large number of randomly cloned repeats of different kinds have been examined to obtain an overall picture of the characteristics of the families of repeats. Klein *et al.* (1978) studied frequencies and divergence characteristics of more than two dozen families and Moore *et al.* (1981) studied the evolution of 13 of these. On average, repeats show about half the rate of evolutionary sequence change as single copy sequences, while many of them show dramatic evolutionary changes in copy number.

The distribution throughout the genome was studied for a set of these families (Anderson *et al.* 1981; Scheller *et al.* 1981) and sequences were determined for a fraction of them (Posakony *et al.* 1981). It was shown that many of them are scattered throughout the DNA, interspersed with single copy DNA. Recent evidence suggests that some of them have the characteristics of mobile genetic elements (Posakony *et al.* 1983; Calzone *et al.* 1984; Johnson *et al.* 1984). Earlier measurements (Graham *et al.* 1974) had shown that about 80 per cent of single copy DNA sequences are interrupted about every kilobase by repeated sequences. The 'interspersion' of repeats and single copy requires that the DNA be sheared to small fragments in order to fractionate adequately repeats and single copy sequences for relationship measurements. If the fragments are a few kilobases long most of the single copy DNA will be linked to repeats and removed in the fractionation. There is reason to believe that single copy sequences which are in regions with few repeats do not exhibit the same evolutionary characteristics as those associated with repeats. Thus, for consistent results it is important to use short fragments and make measurements with the majority of the single copy DNA.

THE MAJORITY OF THE DNA

The coding sequences of the genes of eukaryotes make up only a small part of the total DNA and the function of the majority of the genome is unknown. The well known *C*-value paradox is clearly exemplified by comparing the haploid DNA content of *Drosophila melanogaster* (Dm) [150 million base pairs (bp)] with that of the amphibian *Triturus cristatus* (Tc; 42 000 million bp). The sea urchin *Strongylocentrotus purpuratus* (Sp) is intermediate in genome size (810 million bp) and is typical of echinoderms

Echinoderm phylogeny and evolutionary biology (ed. C. R. C. Paul and A. B. Smith). Clarendon Press, Oxford, 1988.

which may range from 600 to 1300 million bp (Thomas et al. 1981).

The amount of different sequence which appears in the RNA of mature eggs (RNA complexity) is a measure of the amount of coding sequence transcribed and for these three species is 40 (Tc), 37 (Sp), and 12 (Dm) million bases (Hough-Evans et al. 1980; Davidson 1986). Note that even for *Drosophila* with its relatively small DNA content, less than 10 per cent of the genome appears to be transcribed as mature egg RNA and the majority of the single copy DNA appears to be non-coding.

Evidence summarized below indicates that single copy DNA sequences are subject to neutral evolutionary drift and suggests that the actual sequence of most of the DNA matters little. However, the fact that most mammals have almost the same quantity of DNA, even though they diverged about 100 Ma, suggests that the quantity of their DNA is under natural selection and thus the majority of the DNA must have some function, however auxiliary and unexpected.

RATES OF DNA EVOLUTION

Few measurements have been made of interspecies hybridization of echinoderm DNA (Hall et al. 1980; Roberts et al. 1985; Smith et al. 1982, Busslinger 1980; Busslinger et al. 1982) and the fossil record is quite uncertain for the species that have been compared. Nevertheless, it appears that the rate of evolutionary change is more rapid for echinoderms than it is for birds and primates, where better rate estimates exist (reviewed in Britten 1986a). Figure 1 of that paper shows the results of three sea urchin comparisons for total single copy DNA. A line drawn through these values and rodent and *Drosophila* measurements has a slope of 0.66 per cent change in each lineage per million years. This estimate is not far different from one that could be made for the sea urchin data alone, but the echinoderm estimate is quite uncertain, perhaps by more than a factor of two.

Extensive future measurements will be required with echinoderm species having a better fossil record to obtain a good estimate. It appears that, in general, there is a range of rates among different systematic groups and some evidence suggests rate changes during the evolution of the primates (reviewed in Britten 1986a). Thus, the obvious question is whether the rate of evolution of echinoderm DNA will differ among different groups of echinoderms. Unfortunately, no answer can be given to this question as yet.

Differences in rate do not seriously interfere with the use of interspecies hybridization data in systematics. Furthermore, the so-called relative rate test may be used to determine if rate differences exist. In this test a more distant reference species is chosen and all of the more closely related species' DNAs are hybridized to it as well as to each other. If the rates are equal all of the species' DNA will show the same degree of divergence relative to the reference species' DNA.

DNA INTRASPECIES SEQUENCE VARIATION

The DNA of several sea urchins shows a large intraspecies sequence variation (Britten et al. 1978; Grula et al. 1982; Britten 1986b). It ranges from 3 per cent up to about 5 per cent, meaning that 1 out of 20 nucleotides may differ between two individuals in the same species. Members of the same population show as much difference as those from distant locations. In the case of *Strongylocentrotus droebachiensis* (Sd) the DNA of animals from the east and west North American coasts showed no more difference than those from the same location. Obviously the large variation must be considered when measurements are made between closely related species. Commonly mixtures of the DNA of many individuals are present in the tracer and driver DNA used in interspecies hybridizations. Then tracer length effects on the melting temperature and intraspecies variation are automatically subtracted. However, as argued below, it is better to measure the variation.

Roberts et al. (1985) considered how the variation might be taken into account in order to calculate systematic relationships from interspecies divergence measurements. Two extreme models are considered. In model A it is appropriate to subtract the intraspecies divergence as would be automatically done with mixed DNA preparations. In the much more likely model B, it is inappropriate to subtract the variation. The difference between the cases depends on how many of the actual nucleotide sequence variants are shared between the two species. In case B (no correction) a speciation event is proposed such that two populations pass through a bottleneck and variation is lost. They are then presumed to become successful species with large populations. Intraspecies variation and divergence would increase together and no variants would be shared (except by statistical accident).

In case A sequence variants are mostly shared since it is proposed that two large populations separated and retained their variation and speciation then occurred (mechanism unspecified). In this case while the species were closely related it would be appropriate to subtract the variation in estimating their relationship. However, after the two species have diverged for a few million years new variants would occur and the variation would slowly become independent.

Thus, only in the case of closely related species with a special (and perhaps unexpected) evolutionary history is correction by subtracting the variation appropriate. For this reason we have not applied the correction in any of our measurements. Although in some early measurements (Angerer et al. 1976), before we were aware of the large variation, mixed DNAs were used for tracer and driver and the subtraction occurred automatically and erroneously. Since that time we have attempted to refer all divergence measurements to precisely related duplexes (Hall et al. 1980).

MEDIAN SEQUENCE DIVERGENCE

For the comparison of DNA hybridization results from different laboratories a standard method of reporting is required which is independent of the precise conditions of hybridization. For example, in a measurement of divergence between species which are moderately distant, if the temperature of incubation is raised (without a compensating increase in salt concentration) then the amount of hybridization will fall as more divergent duplexes are prevented from forming. At the same time those sequences that do form are more precisely related on the average for the same reason and a higher melting temperature is observed (Smith et al. 1982). Therefore if only the melting temperature is taken into account measurements made for the same species pair may differ.

However, it is a simple matter to combine the two measurements and calculate the 'median sequence divergence' (Hall et al. 1980), where the term median has its usual meaning; half of the single copy DNA shows more divergence, and half less, than the median. This calculation can be made accurately only if the extent of hybridization is greater than 50 per cent, although extrapolation permits estimates where the hybridization is a little less. The procedure derives from D. Kohne and was first applied to higher primate DNA measurements. The result has been termed T50R (Kohne et al. 1972) or T50H (Sibley and Ahlquist 1984). Tracers usually contain short or damaged fragments and do not hybridize to 100 per cent even when reassociated with DNA from the same individual. Correction is made by dividing the amount of interspecies duplex formation by the control or 'homologous' hybridization and the result is called the normalized percent hybridization (NPH). The median divergence is measured by the temperature at which the percent remaining in duplexes (normalized) equals 50 per cent.

Single copy DNA hybridization results are clearly interpretable over a wide range of divergences. Closely related species with less than 1–2 per cent divergence require care and high precision. The limit for distantly related species is less well defined. The limitation in the use of the median sequence divergence is of course its inapplicability much below 50 per cent NPH. No standard method has been established for dealing with more distant relationships. It is worth considering the use of the upper quartile which would extend the range significantly, while presumably maintaining the independence from incubation conditions. However, by closely controlling the incubation conditions it should be possible to determine a relative measure of relatedness from the few per cent of hybridization that is observed, although this has not yet been attempted for systematic purposes among higher organisms.

DRIFT OR SELECTION?

As mentioned above, the bulk of the single copy DNA has no recognized function. There is evidence that it is freely drifting in sequence (Britten 1986a). The conclusion rests primarily on the comparison between the median divergence measured by single copy DNA hybridization and silent substitution divergence observed in gene coding regions. In the case of sea urchins fairly close agreement was obtained between the silent substitutions in histone genes and total single copy (Busslinger et al. 1982). Another comparison has been made between single copy DNA divergence measurements in the primates and substitutions in the eta-globin pseudo-gene (Koop et al. 1986). The percent substitutions are accurately equal for all of the comparisons over a wide range of divergence up to about 30 per cent.

The most extensive and probably most accurate comparison is between the genomes of rodents. W. S. Li has calculated the average silent substitution difference for 14 genes that have been sequenced in mouse and rat as 21 per cent (personal communication). The median single copy DNA divergence is 20 ± 2 per cent as measured in several different laboratories (reviewed in Britten 1986a), including an as yet unpublished measurement (C. C. Sibley and J. E. Ahlquist, personal communication). The degree of silent substitution in coding regions is clearly equal to the single copy DNA divergence measurements.

There is evidence that not all of the positions in coding regions which could accept silent substitutions actually do so freely. It is supposed that secondary structure of the message is not free to change. However, these positions must be in a minority. In addition the total single copy DNA fraction includes the coding regions and various regions which are auxiliary parts of genes. Thus, a minority of the single copy DNA is subject to selection. Presumably, in the case of the eta-globin pseudo-gene there is some selection to maintain the organization of the globin cluster. No large insertion deletion events (greater than 25 base pairs) have accumulated since the divergence of the lineages leading to the New World and Old World monkeys (Koop et al. 1986).

In summary, the rate of change of DNA sequences is equal within error as estimated in the following three ways: by silent substitution in coding regions; by the events that have occurred in a primate pseudogene; and by the median single copy DNA divergence. The simplest explanation is that all three measure free drift and show only minor effects of selection. This leads to the conclusion that the great majority of the single copy DNA is probably drifting freely in sequence.

APPLICATION TO EVOLUTIONARY STUDIES

Obviously, the comparison of the genomes of species is a valuable tool for systematics and evolution and will become more useful as more evidence is accumulated, whether whole genomes are compared by hybridization or regions are compared by cloning and sequencing. While valuable data have been obtained by sequencing specific genes, the focus in this review is entirely on estimations of the divergence of the total genome. I believe that the total DNA comparisons will ultimately be more informative since individual genes are under different degrees of selection and may be idiosyncratic in their history, while the average of many regions can be expected to be less so. The indication that free drift of sequence is occurring gives the DNA divergence measurements a special fundamental role as a basis for comparison with other measures of relationship which may show strong selection, such as morphology and behaviour.

It is now practical to sequence large regions of DNA and this will improve as automatic sequencing systems become available. However, not enough sequence data yet exist to compare with the hybridization data for systematic purposes, and we can only estimate what their relative value will be. One important issue is that it will very likely require much more time and effort to make sequence comparisons. Even if the sequencing itself required no effort, it is necessary to clone regions of DNA from various species, although some tricks can be used to reduce the effort. Always a number of DNA regions must be compared to get an overall view of genomic evolution. Even to get statistical accuracy that compares with the best thermal stability measurements (± 0.3 per cent) will require sequencing 10 000–50 000 nucleotides in each species, depending on their distance. However, at great evolutionary distance, where the divergence far exceeds 50 per cent, sequence comparison may be preferable. The few nucleotides held in common and the shared derived substitutions will be powerful tools, but again large regions will have to be sequenced and these should not be restricted to gene coding sequences.

It will ultimately be valuable to link together measurements from a variety of sources and establish a network of DNA relationships among echinoderms. In this way it may become possible to detect discordances between systematic relationships based on other data, such as morphology and behaviour on one hand and DNA divergence on the other. Each discordance or even apparent rate difference may supply clues regarding mechanisms of evolution. Transiently, it may be necessary to have two sets of discordant systematic groupings, as is evident among the birds, where large numbers of DNA relationship measurements have been made and a significant number of discordances exist. The ultimate systematic and evolutionary knowledge that will be derived from the resolution of the expected paradoxes and discordances may approach the actual historical patterns.

ACKNOWLEDGEMENTS

Research from this laboratory was supported by an NIH grant (GM34031) and by an Alfred P. Sloan Foundation grant (85-12-4).

REFERENCES

Anderson, D. M. *et al.* 1981. Repetitive sequences of the sea urchin genome: distribution of members of specific repetitive families. *Journal of Molecular Biology* **145**, 5–28.

Angerer, R. C., Davidson, E. H., and Britten, R. J. 1976. Single copy DNA and structural gene sequence relationships among four sea urchin species. *Chromosoma* **56**, 213–26.

Britten, R. J. 1986a. Rates of DNA sequence evolution differ between taxonomic groups. *Science, New York* **231**, 1393–8.

Britten, R. J. 1986b. Intraspecies genomic variation. In *Genetics, development, and evolution* (ed. J. P. Gustafson, G. L. Stebbins, and F. J. Ayala), pp. 289–306. Plenum, New York.

Britten, R. J., Graham, D. E., Eden, F. C., Painchaud, D. M., and Davidson, E. H. 1976. Evolutionary divergence and length of repetitive sequences in sea urchin DNA. *Journal of Molecular Evolution* **9**, 1–23.

Britten, R. J., Cetta, A. and Davidson, E. H. 1978. The single-copy DNA sequence polymorphism of the sea urchin *Strongylocentrotus purpuratus*. *Cell* **15**, 1175–86.

Busslinger, M. 1980. Identifikation möglicher Regulationsequenzen und Studium der Evolution varianter Histongene des Seeigels *Psammechinus miliaris*. PhD Thesis, University of Zurich.

Busslinger, M., Rusconi, S., and Birnstiel, M. L. 1982. An unusual evolutionary behaviour of a sea urchin histone gene cluster. *The EMBO Journal* **1**, 27–33.

Calzone, F. J., Jacobs, H. T., Flytzanis, C. N., Posakony, J. W., and Davidson, E. H. 1984. Interspersed maternal RNA of sea urchin and amphibian eggs. In *Biology of fertilization, Volume 3* (ed. C. B. Metz and A. Monroy), pp. 347–66. Academic Press, New York.

Davidson, E. H. 1986. *Gene activity in early development*, 3rd edn. Academic Press, Orlando, Florida.

Graham, D. E., Neufeld, B. R., Davidson, E. H., and Britten, R. J. 1974. Interspersion of repetitive and non-repetitive DNA sequences in the sea urchin genome. *Cell* **1**, 127–37.

Grula, J. W., *et al.* 1982. Sea urchin DNA sequence variation and reduced interspecies differences of the less variable DNA sequences. *Evolution* **36**, 665–76.

Hall, T. J., Grula, J. W., Davidson, E. H., and Britten, R. J. 1980. Evolution of sea urchin non-repetitive DNA. *Journal of Molecular Evolution* **16**, 95–110.

Hough-Evans, B. R., Jacobs-Lorena, M., Cummings, M. R., Britten, R. J., and Davidson, E. H. 1980. Complexity of RNA in eggs of *Drosophila melanogaster* and *Musca domestica*. *Genetics* **95**, 81–94.

Johnson, S. A., Davidson, E. H., and Britten, R. J. 1984. Insertion of a short repetitive sequence (D88I) in a sea urchin gene: a typical interspersed repeat? *Journal of Molecular Evolution* **20**, 195–201.

Klein, W. H., Thomas, T. L., Lai, C., Scheller, R. H., Britten, R. J., and Davidson, E. H. 1978. Characteristics of individual repetitive sequence families in the sea urchin genome studied with cloned repeats. *Cell* **14**, 889–900.

Kohne, D. E., Chiscon, J. A., and Hoyer, B. H. 1972. Evolution of primate DNA sequences. *Journal of Human Evolution* **1**, 627–44.

Koop, B. F., Goodman, M., Xu, P., Chan, K., and Slightom, J. L. 1986. Primate η-globin DNA sequences and man's place among the great apes. *Nature, London* **319**, 234–8.

Moore, G. P., Pearson, W. R., Davidson, E. H. and Britten, R. J. 1981. Long and short repeats of sea urchin DNA and their evolution. *Chromosoma* **84**, 19–32.

Posakony, J. W., Scheller, R. H., Anderson, D. M., Britten, R. J., and Davidson, E. H. 1981. Repetitive sequences of the sea urchin genome III. Nucleotide sequences of cloned repeat elements. *Journal of Molecular Biology* **149**, 41–67.

Posakony, J. W., Flytzanis, C. N., Britten, R. J., and Davidson, E. H. 1983. Interspersed sequence organization and developmental representation of cloned poly(A) RNAs from sea urchin eggs. *Journal of Molecular Biology* **167**, 361–89.

Roberts, J. W., Johnson, S. A., Kier, P., Hall, T. J., Davidson, E. H., and Britten, R. J. 1985. Evolutionary conservation of DNA sequences expressed in sea urchin eggs and early embryos. *Journal of Molecular Evolution* **22**, 99–107.

Scheller, R. H., *et al.* 1981. Organization and expression of multiple actin genes in the sea urchin. *Molecular and Cellular Biology* **1**, 609–28.

Sibley, C. G. and Ahlquist, J. E. 1984. The phylogeny of the hominoid primates, as indicated by DNA-DNA hybridization. *Journal of Molecular Evolution* **20**, 2–15.

Smith, M. J., Nicholson, R., Stuerzl, M. and Lui, A. 1982. Single copy DNA homology in sea stars. *Journal of Molecular Evolution* **18**, 92–101.

Thomas, T. L., Posakony, J. W., Anderson, D. M., Britten, R. J., and Davidson, E. H. 1981. Molecular structure of maternal RNA. *Chromosoma* **84**, 319–35.

9

DNA–DNA hybridization, the fossil record, phylogenetic reconstruction, and the evolution of the clypeasteroid echinoids

CHARLES R. MARSHALL

Committee on Evolutionary Biology, c/o Department of Geophysical Sciences, University of Chicago, 5734 South Ellis Avenue, Chicago, Illinois, 60637, USA

Ever since the proposal that many macromolecules, particularly DNA and proteins, might evolve in a clock-like fashion (Zuckerkandl and Pauling 1962), a great deal of interest has been focused on the rate of evolution of molecular characters and the potential contribution these characters might make to the construction of phylogenies (Wilson *et al.* 1977). Early enthusiasm for the clock hypothesis was dampened by the discovery that the rates of molecular evolution have been highly variable through time (Romero-Herrera *et al.* 1979) and interest shifted towards explaining the variation in rates in terms of the functional constraints on the molecules and the comparison between molecular and organismal rates of evolution (Goodman 1981). However, molecular characters continue to be an important source of phylogenetic information. One of the most powerful molecular techniques for phylogenetic purposes is the use of DNA–DNA hybridization to measure the single copy nuclear DNA sequence divergences between species.

The phylogenies of a wide range of organisms have been studied using DNA–DNA hybridization data, including several species of *Drosophila* (Hunt *et al.* 1981; Hunt and Carson 1983); four species of regular sea urchins (Angerer *et al.* 1976; Hall *et al.* 1980; Harpold and Craig 1978); five species of sea star (Smith *et al.* 1982); nine species of echinoderm, including a holothurian, asteroids, and echinoids (Poltaraus 1981); most of the birds (see Sibley and Ahlquist 1983 for a review) and a large number of the higher vertebrates (Brownell 1983; Benveniste 1985; O'Brien *et al.* 1985) including most of the primates (Kohne *et al.* 1972; Benveniste and Todaro 1976; Bonner *et al.* 1980; Sibley and Ahlquist 1984; O'Brien *et al.* 1985).

The first section of this paper explains what DNA–DNA hybridization data are and discusses the reasons for believing the data give robust phylogenies. In the second section I discuss how data from the fossil record and DNA–DNA hybridization studies can be combined to help understand the phylogenetic relationships of extinct taxa, and to identify paraphyletic groups and homoplasious characters. The fossil record can be used to estimate divergence times between lineages. Using these dates and DNA–DNA hybridization data the rate of single copy DNA evolution can be calculated, providing an interesting comparison between rates of morphological and molecular change.

At present I am undertaking a DNA–DNA hybridization study of the clypeasteroid echinoids (sand-dollars and sea-biscuits), and will use them to illustrate the points made above.

DNA-DNA HYBRIDIZATION DATA AND PHYLOGENETIC RECONSTRUCTION

There are five reasons for believing that DNA–DNA hybridization data will produce reliable phylogenies.

First, in the majority of DNA–DNA hybridization studies the phylogeny produced has found overwhelming support from other morphological,

Echinoderm phylogeny and evolutionary biology (ed. C. R. C. Paul and A. B. Smith). Clarendon Press, Oxford, 1988.

biochemical, and biogeographic data. Generally speaking, the only cases where DNA-DNA phylogenies differ from the phylogenies based on other data are where the relationships within the study group have always been unclear.

Secondly, in cases where the phylogeny of a group has been equivocal, the DNA-DNA hybridization data have sometimes added a crucial insight that has resolved the difficulties. For example, the long-standing problem of understanding the evolution of oscine passerines (true song-birds) has been elegantly clarified using DNA-DNA hybridization data (Sibley and Ahlquist 1983). See Diamond (1983) for a number of other bird examples.

Thirdly, DNA-DNA hybridization data show a remarkably clock-like rate of evolution in many of the groups studied, though the clock appears to run faster in some groups than others. Thus, the single-copy DNA sequence divergence measured does appear to reflect the pattern, if not the precise time, of divergence.

Fourthly, the power of DNA-DNA hybridization data can be attributed to the enormous complexity of the single copy DNA. There are in the order of 10^9 base pairs in the (haploid) single-copy DNA of most deuterostomes (Hinegardiner 1976), 10^5–10^6 times the base pairs in the 'average' gene. Most of the genome has no sequence-specific selection pressure acting on it. Thus, the chances of convergent or parallel evolution in the single-copy DNA nucleotide sequence, this the most complex of characters, is so overwhelmingly small that any significant similarities in the sequence difference between species must be due to shared ancestry (Gould 1985, 1986).

Fifthly, since all heritable phenotypic differences between species will be reflected in the structure or organization of the DNA, the measured difference in DNAs is, in some sense, a more fundamental and comprehensive measure of evolutionary distance than measured differences in any single character.

DNA-DNA hybridization—what does it measure?

When a double-stranded DNA molecule is heated sufficiently, the hydrogen bonds between the adenine (A)-thymine (T) and guanine (G)-cytosine (C) pairs break, and the two strands separate, melt or denature. Cooling results in the reforming of the hydrogen bonds and the DNA strands are said to reassociate or hybridize. The study of the reassociation kinetics of DNAs was pioneered by Britten and Kohne (1968), and provides information about the structure of the genome.

The eukaryote genome consists of a number of different components, each characterized by the number of times the included sequences occur in the genome. For ease of discussion there are often said to be four components; a highly repetitive (10^3–10^6 copies per sequence per genome), an intermediate repetitive (10^2–10^3 copies), a slow repetitive (10–100 copies) and single copy (one copy) components. Although structural analysis shows these DNA classes to possess distinct differences, both in sequence structure and chromosome location, for practical purposes the distribution approaches a continuum of sequences within the genome. Table 9.1 shows the sizes and number of copies of selected repeated sequences from *Strongylocentrotus purpuratus* (Klein *et al.* 1978) and indicates the range of variation in the repetitive DNA. The isolation of the single copy DNA, relies on the fact that the repetitive DNA is the first to hybridize during reassociation experiments. It is difficult to obtain pure single copy DNA and what is called single copy DNA may include up to ten or so copies of some sequences.

Genome structure of Strongylocentrotus purpuratus and clypeasteroids

The haploid genome size of three clypeasteroids has been determined; *Dendraster excentricus* and

TABLE 9.1 Lengths and number of copies per genome of eleven sequences of repetitive DNA in the genome of *Strongylocentrotus purpuratus* (After Klein *et al.* 1978)

Length (base pairs)	Approximate number of copies per genome
240	2100
1100	400
560	1000
325	900
290	90
165	130
420	3
220	140
380	6
990	40
155	12 500

Echinarachnius parma, 1.1 pg (1.06×10^9 bp) and *Mellita quinquiesperforata*, 0.98 pg (9.5×10^8 bp) (Hinegardiner 1974). The genome size of the *Strongylocentrotus purpuratus* is 0.89 pg (8.6×10^8 bp) (Hinegardiner 1974). Table 9.2 summarizes the genome structure of *Clypeaster japonicus* and *S. purpuratus*. Nothing else is known of the clypeasteroid genome organization. The percentage of single copy DNA in the genome of *S. purpuratus* has also been placed at 70 per cent (McColl and Aronson 1974) and 75 per cent (Galau et al. 1976b).

Most of the single copy DNA in *S. purpuratus* occurs in 10^3 bp (1 kb) segments interspersed with repeats a few hundred nucleotides in length (Graham et al. 1974). This type of organization is known as the short, or *Xenopus*, pattern of interspersion (Britten 1982), and is found in a wide range of organisms, including an oyster, surf clam, the horseshoe crab and a nemertean worm (Goldberg et al. 1975). The other major type of genome organization is the long or *Drosophila* pattern, where the single-copy stretches are much longer and the repeats average a few kilobases (Britten 1982). Seventy-five per cent of the total repetitive DNA in *S. purpuratus* is in 200–400 bp sequences, and more than 70 per cent of the single copy DNA is interspersed with repetitive DNA in 2000–3500 bp fragments (Galau et al. 1976b).

Measuring interspecies single-copy DNA sequence differences

Below, I present a simplified summary of how DNA-DNA hybridization techniques are used to measure single copy DNA sequence homology between species. Suppose we have three species, A, B, and C. Initially, the DNA is isolated from each species, sheared into 500 bp fragments and the repetitive DNA removed. Radioactively labelled single-copy DNA of species A (tracer or probe) is allowed to reassociate for 2–4 days with a 10^3–10^4 fold excess of unlabelled (driver) single-copy DNA from each of the three species. The excess driver prevents extensive self-hybridization between tracer fragments. The reassociated DNA from each experiment is isolated, the fraction of labelled DNA (tracer) hybridized is determined and the extent of reassociation calculated. The more distantly related the species, the smaller the extent of hybridization or reactivities of the DNAs. In each case, the results are corrected for self-reaction of the tracer (usually less than 2 per cent). The reactivity between the labelled and unlabelled DNA of species A (homologous hybrids) is set to 1.0 and the normalized reactivities for the heterologous duplexes (hybrids between labelled A DNA and unlabelled B or C DNA) are then calculated. The Cot curve of the reassociation (a graph that follows the proportion of DNA hybridized with time of re-association) between the species, and the reactivities and normalized reactivities are shown in Fig. 9.1.

The thermal stabilities of the duplexes formed between each pair of species are now determined. The more distantly related the species, the more divergent their single-copy DNA should be, and, therefore, the less tightly bound or thermally stable the hybrid duplexes. The duplexes for each species pair are heated incrementally until all the duplex has melted. At each temperature the

TABLE 9.2 Genome organization of: (a) *Clypeaster japonicus* (after Yanagisawa 1984). The genome size of *C. japonicus* was assumed to be 10^9 bp. (b) *Strongylocentrotus purpuratus* (after Britten et al. 1972). For both species the percentage repetition DNA is over-estimated

Component	%Genome	Complexity (bp)	Repetition frequency
(a) *Clypeaster japonicus*			
Moderate repetitive DNA	20%	3.9×10^6	515
Slow repetitive DNA	20%	6.1×10^6	35
Single copy DNA	46%	4.6×10^8	1
(b) *Strongylocentrotus purpuratus*			
Fast repetitive DNA	10%	1.3×10^4	6000
Moderate repetitive DNA	27%	10^6	250
Slow repetitive DNA	25%	10^7	20–50
Single copy DNA	38%	3.3×10^8	1

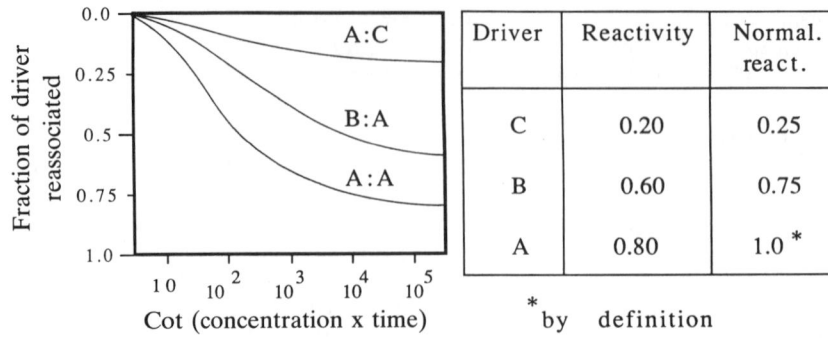

Fig. 9.1 Cot curve for the homologous (A hybridized with A) and heterologous (A hybridized with B and C) reactions. The reactivities and normalized reactivities for the hypothetical reactions are given.

proportion of duplex remaining is calculated, by measuring the relative radioactivities of duplex to single stranded DNA.

The cumulative proportion (fraction eluted) of denatured duplex is plotted against temperature and the melting temperature (T_m) (temperature at which 50 per cent of the duplexes are denatured) is determined. Figure 9.2 shows the thermal elution curves for the DNA hybrids formed between the three species, the T_m values and the difference between the T_m values (ΔT_m). Usually T_m values have to be corrected for differences in fragment length between tracer and driver (Britten *et al.* 1974) since it is difficult technically to provide fragments of exactly the same length.

The best measure of single copy DNA sequence divergence is based on the difference in median melting temperatures (T_{med}) between homologous duplexes and heterologous duplexes and is given the symbol $\Delta T_m R$ (Benveniste 1985) or $T_{50}H$ or $T_{50}R$ (Sibley and Ahlquist 1981). This measure takes into account both the reactivity and the thermal stability of the hybrids (T_m). The $\Delta T_m R$ values obtained using different experimental conditions for the same group of animals are usually in good agreement (Benveniste 1985; Britten 1986). Figure 9.3 shows graphically the relationship between T_m, ΔT_m, T_{med}, $\Delta T_m R$, and the normalized reactivity. The difference between Figs. 9.2 and 9.3 is that heterologous thermal stability curves are plotted so that the duplex remaining is normalized for the extent of reaction.

The melting temperature (T_m) alone is not a good measure of single-copy DNA sequence divergences because the measured T_m is often dependent on the particular experimental conditions used. If the hybridizations are only run for a short period, then only the most nearly similar DNA strands held in common between two species will hybridize and the duplexes will have a high

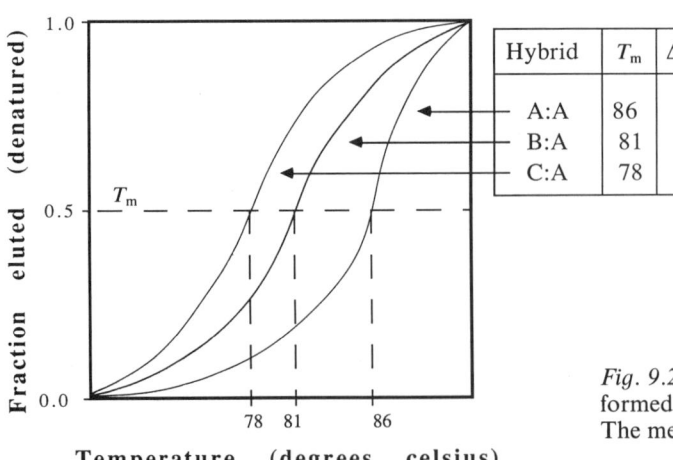

Fig. 9.2 Melting curves for the hybrids formed in the reactions shown in Fig. 9.1. The melting temperatures (T_m) and ΔT_ms are given.

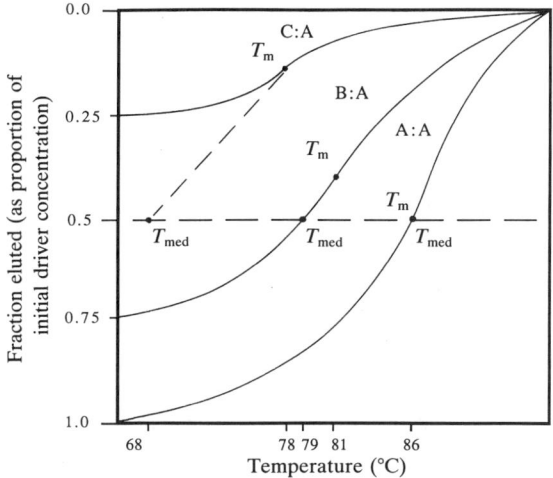

Fig. 9.3 Graph shows the relationship between the normalized reactivities (normalized fraction reassociated), melting temperatures (T_m and ΔT_m), and median melting temperatures (T_{med}, $\Delta T_m R$). See Hall *et al.* (1980) for an equation to calculate $\Delta T_m R$ when the normalized reactivity is less than 50 per cent (as in the case of A hybridized with C).

average melting temperature. For experiments run over a long period of time, less well matched strands will form hybrids, giving a lower average melting temperature.

Each species should be used as the reference species in turn and at least duplicates or triplicates of each run should be done. It is important to do the reciprocal hybrids to check that differences such as genome size are not large enough to invalidate the use of the results for phylogenetic reconstruction (Brownell 1983). For example, if genome A is far smaller than genome B, then most genome A may find homologous sequences in genome B, while a considerably smaller proportion of genome B will find homologous sequences in genome A.

Typical standard deviations for $\Delta T_m R$ values are $\pm 0.4°C$ for values of less than 5°C and $\pm 0.7°C$ for $\Delta T_m R$ values greater than 5°C (Benveniste 1985). In cases where the normalized reactivity is less than 50 per cent the actual value of $\Delta T_m R$ becomes only approximate. In DNA-DNA hybridization experiments a $\Delta T_m R$ of 1°C is equivalent to approximately a 1 per cent difference in nucleotide sequence (Hutton and Wetmur 1973; Britten *et al.* 1974). The experimental procedures outlined above can be used to measure DNA single copy sequence differences of between 0.5–20 per cent.

There is one important caveat. The degree of sequence divergence between individuals from the same species can be alarmingly high. Values of 4°C have been obtained for *S. purpuratus* (Britten *et al.* 1978), 3°C for *S. intermedius* and *S. franciscanus*, 2°C for *S. droebachiensis* (Grula *et al.* 1982) and 1°C for *Homo sapiens* (Britten 1982). Britten (this volume) discusses the problems these large intra-specific variations present for the use of $\Delta T_m R$ values for systematic purposes.

Mutational events detected by DNA-DNA hybridization

The sequence differences measured by DNA-DNA hybridization are due primarily to point mutations (base substitution at single locations on the DNA molecule). A plot of thermal melting point difference (ΔT_m or $\Delta T_m R$) against percentage hybrid detected with S1 nuclease, an enzyme that digests single-stranded, but not double-stranded DNA, yields the evidence for this conclusion (Benveniste 1985). An example of such a plot is shown in Fig. 9.4a. The slope of the line depends on the experimental conditions used. No points should be above the line since fully homologous hybrids could not have a low thermal stability. Points can only occur significantly below the line if the sequences contain major deletions, recombinations, gene transfers, or insertions. If an insertion, gene transfer, recombination, or deletion has occurred in a strand of DNA, then its matching strand would have perfect homology with the unaltered sequence and have no homology with the mutated section. The melting temperature would be the same between the strands before and after the mutation since in both cases large regions of perfectly paired regions would have to denature. However, S1 nuclease would digest the region of unpaired sequences in the hybrid between the unmutated and mutated strands, and hence, the percentage of hybrid detected would be reduced over the percentage hybrid measured between two

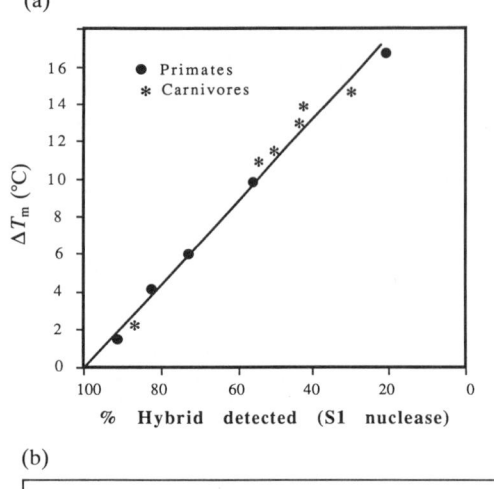

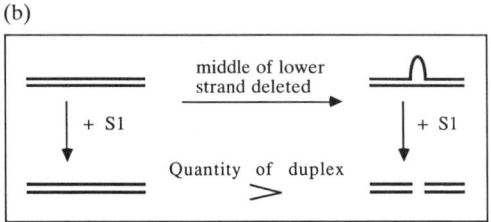

Fig. 9.4(a) Graph of per cent hybrid detected with S1 nuclease versus the difference in melting temperatures of DNA-DNA hybrids (ΔT_m). Graph shows that DNA-DNA hybridization data primarily detect point mutations (see text for explanation) (after Benveniste 1985). (b) A mutational event will cause a departure from the line in Fig. 9.4a. All molecules shown have identical melting points, but after a deletion and S1 nuclease treatment there is less duplex in the mutated molecule.

unmutated strands. This is shown diagrammatically in Fig. 9.4b. In strands where predominantly point mutations occur, the thermal stability is decreased because the density of bonds decreases as does the amount of hybrid detected since the S1 snips the strands at each site of point mutation.

Are the data good enough?

DNA-DNA hybridization data are most useful for phylogenetic reconstruction if the rate of base substitutions in the single copy DNA is relatively uniform with time. The validity of this assumption can be checked by testing the consistency of the data with the assumed rate and by seeing if a clock-like rate of evolution is consistent with what we know about the mechanisms of genome evolution. These two points are discussed below.

O'Brien et al. (1985) suggest four tests for determining if a set of $\Delta T_m R$ values is adequate for making phylogenetic inferences. First, the standard deviation of the $\Delta T_m R$s must be small. This ensures that a real signal is being measured and not just noise. Secondly, reciprocal measurements must be in good agreement. Cracraft (1987) has demonstrated how misleading DNA-DNA hybridization data can be if the reciprocal hybrids are not done. Thirdly, the internal consistency of the $\Delta T_m R$s is tested by checking the metricity of the values. All three-way comparisons must comply with the triangle inequality. Metricity alone does not establish a clock-like rate of evolution because the measured distances between taxa may be metric, but still show a variation of rates. Fourthly, the relative rate test is applied, which requires that the distances from an outgroup to all members of an internal group be equal, regardless of the pattern of branching within the internal group. The test has shown a uniform rate of genome evolution in the majority of primates (Benveniste and Todaro 1976), and the giant pandas and closest relatives (O'Brien et al. 1985), and a number of bird groups including the New Zealand wrens, Hawaiian honey creepers and Australo-Papuan fairy wrens (Sibley and Ahlquist 1983). If all these tests are met, then, avoiding cases where intraspecies variation is large with respect to internode length, the derived phylogeny should be reliable and the assumption of a clock-like rate of change is sound.

Not all groups show a clock-like evolution of the genome. For example, Bonner et al. (1980) have shown that rate of DNA evolution in the primates of Madagascar is approximately half of the rate seen in other primates and Sheldon (1987) has shown a similar variation in the rates of single copy DNA evolution in herons. However, even in cases such as these, where the rate of DNA evolution departs from a clock-like behaviour, the data can still provide sound cladistic information. Sibley et al. (1987) review some of the evidence that generation length or number of DNA replications or cell cycles per unit time in the germ line may effect the rate of DNA evolution.

To establish the topology of the tree any one of a number of distance algorithms can be used. O'Brien et al. (1985) used five different algorithms on their data for the primates and pandas. Each gave the same topology though internode lengths varied. While DNA-DNA hybridization data may give the correct branching order, the lengths of the internodes can also vary due to differences in experimental conditions. However, the variation is often small (Britten 1986). For example, several independent measurements between man and the Old World monkeys from different labs, using different techniques, are in close agreement; $\Delta T_m R = 7.4 \pm 0.5°C$ (Benveniste and Todaro 1976; Bonner et al. 1980; Sibley and Ahlquist 1984; O'Brien et al. 1985; Britten 1986).

Molecular clocks and genome evolution

Strongylocentrotus purpuratus has one of the best studied genomes. The single copy DNA constitutes about 75 per cent of the genome (6.45×10^8 bp). What proportion of the single copy DNA codes for protein? This is important because the rate of amino acid substitutions varies greatly from protein to protein (by at least an order of magnitude), presumably because of variation in selection pressure, and the rate of evolution over extended periods of time can be highly variable (Goodman 1981). Thus, a clock-like behaviour over extended periods of time would not be expected if most of the non-repetitive DNA codes for protein. If most of the non-repetitive DNA does not code for protein, then there should be little or no selection pressure on most of the base sequences and the non-repetitive DNA should evolve at some neutral rate.

Using a variation of the DNA-DNA hybridization technique described above, Galau et al. (1974) have shown the complexity of gastrula mRNA, an intermediate molecule in the translation of genes to proteins, of *S. purpuratus* is 1.7×10^7 bp. This is sufficient to code for some 14 000 different structural genes of median length 1200 bp and is 2.7 per cent of the complexity of the total genome or just 4 per cent of the single-copy DNA. Galau et al. (1976a) showed that the total complexity of the different mRNAs active in the gastrula and other development stages (ovary, oocyte, blastula, and pluteus) and some adult tissues (tubefoot, intestine, and coelomocyte) of *S. purpuratus* is approximately 10 per cent of the single copy DNA, some 6.5×10^7 bp, or enough to code for 35 000 different proteins.

These data suggest that some 90 per cent of the single-copy DNA has no *sequence-specific* selection pressure acting on it. It seems reasonable to assume that the non-coding sequences evolve at some neutral rate. Except for a few bases concerned with promoter sequences, we do not know the function, if any, of this non-coding DNA, thus generalizations concerning its rate of evolution must remain provisional. Support for a neutral rate of evolution comes from the good agreement between the rate of base substitution at the silent (degenerate) positions in the coding sequences, and in the short and long introns in regions of the histone gene cluster, where there is no selection pressure, and the rate of single copy DNA evolution measured in the same groups of animals (Hayashida and Miyata 1983; Britten 1986).

There is no universal intrinsic rate of evolution of DNA. The rate of single-copy DNA evolution varies by as much as a factor of five to ten between different groups of organisms (see Britten 1986 for a review). Even within the genome of individual organisms the single copy DNA shows a large range of rates of evolution. Roberts et al. (1985) have shown that expressed sequences in the eggs and early embryos of *S. purpuratus* are evolutionarily conserved with respect to non-expressed sequences. Grula et al. (1982) have shown that the single copy DNA of *S. purpuratus* can be separated into a 'more' and 'less' variable portions, each representing a third of the single copy DNA. The 'less' variable portion shows an average sequence variation of 2 per cent, including some completely conserved regions. The 'more' variable regions show an average sequence variation of 7 per cent, including a few percent that shows a 20 per cent variation. The reason for these variations is either due to some unknown selection pressure on the non-coding regions of the DNA, or due to differences in the efficiencies of the repair mechanisms on different parts of the genome. The crucial point is that the non-coding regions are not simply sitting in the nucleus mutating away at some intrinsic rate. The DNA is bathed in a panoply of enzymes many of which are responsible for repairing damage. If the genome displays an overall uniform rate of evolution it is because it is being held at that rate by the enzymatic machinery of the cell. Variations of rate are to be expected since the relative efficiency of this machinery doubtless varies between species.

Horizontal transfer—does it make a difference?

Does the discovery that DNA sequences can be transferred across species boundaries invalidate the use of measured DNA similarities between species to construct phylogenies? There is strong evidence of the transfer of a 500 bp fragment of mitochondrial DNA to the nuclear genome of *S. purpuratus* (Jacobs et al. 1983) (500 bp represents 0.0001 per cent of the single copy DNA). There is excellent evidence for the transfer of 10 kb sequences, via infectious retroviruses, from baboon ancestors to the domestic cat ancestor, from the ancestors of rodents to cats, from rodents to pigs, and from New World monkeys to Old World carnivores (reviewed in Benveniste 1985). The retroviral sequences may occur from 10 to 500–600 copies within the genome, and while as much as 0.1 per cent of the total primate genome may be virogene copies (Benveniste et al. 1977), only 0.01 per cent of the single copy DNA is of viral origin. DNA-DNA hybridization techniques have a resolution of approximately 0.2 per cent. The quantities that have been transferred are, to our present knowledge, far too small to be detected when comparing single-copy DNAs of species, and do not invalidate phylogenies determined by DNA hybridization.

The incorporation of viral genes into the genome and subsequent duplication must have some destabilizing effect on the genome. Horizontal transfer may have occurred frequently enough in geological time that destabilization of the genome upon incorporation could be a significant factor during evolution. For example, if repair mechanisms are disrupted, rapid sequence evolution could occur and the genome could show a punctuated style of evolution. We are still too ignorant of the mechanisms of genome evolution to be secure in any assumptions concerning the rate of genome evolution.

CLYPEASTEROIDS: DNA-DNA HYBRIDIZATION DATA AND THE FOSSIL RECORD

The clypeasteroids

The clypeasteroid echinoids arose in the Palaeocene, radiated in the Eocene (Kier 1982) and again, to a lesser extent, in the Miocene (Smith 1984). They have the best fossil record of any echinoid group (being infaunal and possessing reinforced tests) including 408 fossil species (Kier 1977). There are 129 extant species (Kier 1977) in 24 genera (Durham 1966) and nine extant families (Smith 1984), 22 of the 24 extant genera and all the extant families have a fossil record (Mortensen 1948; Sepkoski 1982).

There have only been three published phylogenies of the clypeasteroid families. Pre-cladistic phylogenies are those of Durham (1955, redrawn in 1966 with extended ranges) and Seilacher (1979). Smith (1984, see Fig. 9.5; and in Ghiold and Hoffman 1986) has the only published cladistic analysis to date, though he gives no characters to support his cladograms. Smith's (1984) phylogeny differs from the Ghiold and Hoffman cladogram in that in the former the Dendrasteridae and Mellitidae are sister groups. The phylogenies of Durham and Seilacher, with the extinct families omitted, are presented in the form of cladograms (Figs 9.6a and 9.6b, respectively). Figure 9.6c shows Smith's cladogram with the extinct families omitted. The strict consensus tree, which summarizes the monophyletic groups held in common between the cladograms, is given in Fig. 9.6d. No characters have been explicitly given to support any of the four phylogenies, though Seilacher justifies himself where he differs from Durham; hence, the strict consensus tree (Fig. 9.6d) is the favoured phylogeny. DNA-DNA hybridization data should certainly help resolve some of the nodes on the strict consensus tree.

Measuring absolute rates of genome evolution

Measuring the absolute rates of genome evolution is not only interesting in its own right, but is often necessary if the rates of genome change are to be compared between clades. DNA-DNA hybridization techniques have a maximum sequence divergence they can measure accurately. The relationship between $\Delta T_m R$ and time is generally only linear up to $\Delta T_m R$ values of 20°C, since the probability that a significant number of sites will have suffered multiple substitutions becomes large for distantly related species, and the method cannot distinguish between single and multiple events. There are methods for correcting for multiple hits (Sibley and Ahlquist 1983), but it is difficult to test the appropriateness of the corrections. It is also difficult to measure melting temperatures any more than 20°C below the homologous melting temperature in any given experiment. For mammals, birds, and sea urchins the DNA-DNA hybridization technique starts to become ineffective for measuring DNA sequence divergences at the inter-ordinal level. The only way to compare the rates of genome evolution between groups that show very large divergences in their single copy DNA base sequences is by

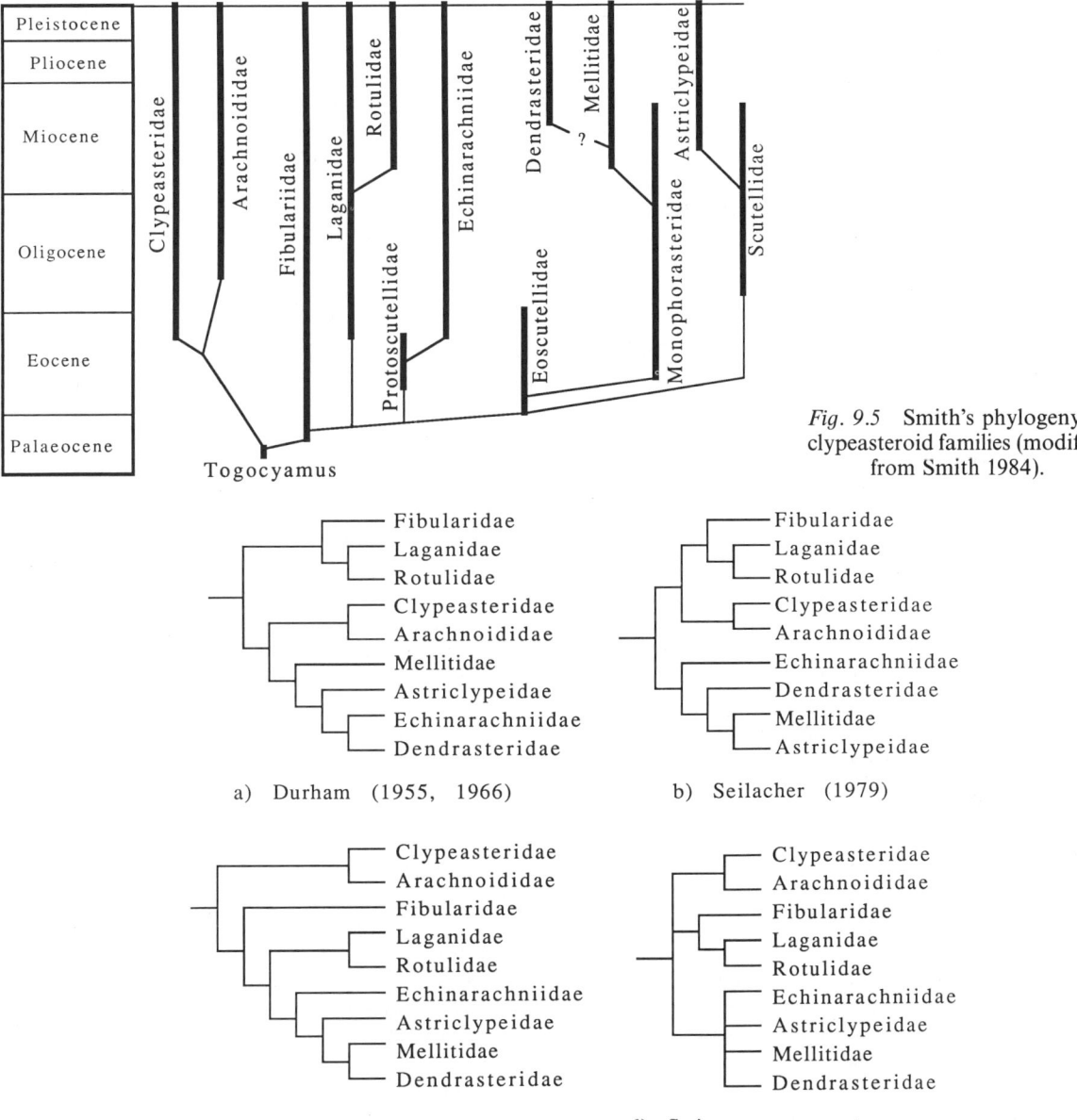

Fig. 9.5 Smith's phylogeny of clypeasteroid families (modified from Smith 1984).

Fig. 9.6 Cladograms of the extant clypeasteroid families derived from the phylogenies of: (a) Durham (1955, 1966). (b) Seilacher (1979). (c) Smith (1984) identical to the cladogram presented in Ghiold and Hoffman (1986). (d) Strict consensus tree derived from cladograms given in Figs 9.6a–c.

comparing the absolute rates of change of their base sequences. To calculate these absolute rates, divergence times must be obtained from the fossil record or biogeographic data.

None of the groups studied to date with DNA-DNA hybridization techniques has a good fossil record and good estimates of divergence times are virtually non-existent. For example, Sibley and Ahlquist (1983) use only two crude data, the opening of the Atlantic Ocean and the Tasman sea in the Late Cretaceous to scale the rate of evolution in birds. Angerer *et al.* (1976) resolve the divergence between two of the species in their study of regular echinoids to somewhere between 7 and 20 million years. Benveniste (1985) gives a range of possible dates for the divergence of man from chimpanzee and gorilla of between 4 and 20 million years ago; calculated rates vary by a factor of five! Surprisingly, the best controlled dates of divergence between lineages come from the Hawaiian *Drosophila* species, where the time of origin of the Hawaiian islands, and thus the time of (inferred) migration and speciation of the flies are well known (Hunt and Carson 1983; Easteal and Oakeshott 1985). However, the *Drosophila* species have diverged so recently (approximately 1–2 million years) that the experimental errors associated with the DNA-DNA hybridization data are large with respect to the actual distances measured and, thus, the calculated rates carry large uncertainties.

Moreover, there is no general method for assessing confidence limits on divergence times obtained from the fossil record. Lack of awareness of the inherent errors in dating the origin of lineages has led some workers to give rates of genome evolution with too many significant figures. For example, Hayashida and Miyata (1983) give a rate of base substitution per site per year (v) for the histone III gene, based on differences measured between *S. purpuratus* and *S. droebachiensis*, of 4.10×10^{-9}, assuming a divergence time of 5 Ma. Yet the two species could have diverged as much as 20 Ma ago (Angerer *et al.* 1976), which would give a rate of 1×10^{-9}. Clearly, for this datum Hayashida and Miyata have overestimated the precision of v by 100-fold! An extreme example of how incomplete the fossil record can be is the recent discovery of a fossil monotreme in Australia dated at about 110 Ma (Archer *et al.* 1985). The previously *oldest* find is only 22 Ma.

For groups with a good fossil record, such as the clypeasteroids, the first known occurrence in the fossil record is probably not too much later than the actual time of origin of the group. The method of stratigraphic gap analysis, first suggested by Shaw (1964) and more fully developed by Paul (1982), can be used to estimate confidence limits on the ranges for given species. The stratigraphic gap method will always give earlier times of origin for groups than a literal reading of the fossil record. This could be highly significant for the calculation of rates of evolution of molecular characters, since many workers use the first occurrence of a group to calculate such rates (for example see Busslinger *et al.* 1982).

Rates of divergence of DNA sequences vary from about 0.13 to 0.66 per cent per million years, sea urchins showing one of the fastest rates (Britten 1986). Given that $\Delta T_m R$ is linear with time in the range 0–20°C, clypeasteroids, with their excellent fossil record and Tertiary radiation, are ideal for studying the rate of genome evolution and promise to give one of the first well-controlled empirical rates of base substitution in single copy DNA.

DNA-DNA hybridization and the identification of paraphyletic groups

DNA-DNA hybridization data, assuming constant rates of base substitution, may be used to identify paraphyletic groups in two ways. The first method works only if there is at least one surviving member of the stem group. If two clades are monophyletic, then the measured distances between any species drawn from one clade to any species of the other clade, should always be the same. On the other hand, if one of the families is paraphyletic and at least one member of the stem group survives, then there should be one set of identical, small distances and at least one larger distance which corresponds to the stem group.

The second method relies on the fact that DNA-DNA hybridization data can be used to predict the time of origin of a clade. Consider Smith's 1984 phylogeny (Fig. 9.5). The way the branches are drawn between families implies the existence of a large number of paraphyletic groups (Durham's and Sielacher's phylogenies also imply a large number of paraphyletic groups). If the laganids really arose in the Eocene, then they are paraphyletic since they do not include the Rotulids. In this case, the divergence time between the lineages could be as late as the first Rotulid or sometime in the Early Miocene (20–25 Ma). If the Laganidae are monophyletic, then the divergence time of the lineages must predate both lineages (Paul 1982) and they probably diverged sometime in the Early Eocene (50–55 Ma). Given a rate of single copy DNA evolution, based on the analysis of other clypeasteroids, the expected divergence time of the Laganidae and Rotulidae can be estimated and compared with the divergence times implied by a monophyletic and paraphyletic Laganidae. This approach clearly works best when the difference between the implied divergence times under the two competing hypotheses is large.

In addition, if confidence limits can be placed on the divergence times of lineages using some variation of Paul's (1982) method mentioned above, the two alternative positions of the Laganidae can be decided by seeing which hypothesis is most consistent with the expected time of origin of the Rotulidae. A monophyletic Laganidae implies that we are missing about half the fossil record of the Rotulidae (from the earliest Miocene, to the Early Eocene), whereas a paraphyletic Laganidae is consistent with any time of origin of the Rotulidae between the Early Eocene and Early Miocene.

DNA-DNA hybridization data and the phylogenetic position of extinct taxa

DNA-DNA hybridization data can, in some cases, give insight into the correct phylogenetic position of extinct taxa. Let us consider the divergence time of the Dendrasteridae and the Mellitidae. I will assume both groups are monophyletic. According to Smith's phylogeny (Fig. 9.5) the Dendrasteridae and Mellitidae are closest relatives and, thus, the divergence time of the lineages must have been before the first representatives of each clade, or sometime in earliest Miocene of latest Oligocene (25 Ma). However, in the phylogenies of Durham (1966) and Seilacher (1979) the two groups are related via a number of extinct families, including the Eocene Eoscutellidae. The most recent common ancestor must predate all the families involved and, therefore, must have occurred in the lowermost Eocene, or latest Palaeocene (55 Ma) (see Fig. 9.5 for the stratigraphic position of the Eoscutellidae).

DNA-DNA hybridization data can help select between these two phylogenies. The rate of genomic evolution can be determined based on the divergence time between lineages whose relationship is clear. This rate can then be used to estimate the divergence time between the Dendrasteridae and Mellitidae. Clearly, if the divergence time of the two groups, as predicted by the DNA-DNA hybridization data, pre-dates the last known member of the Eoscutellidae, then that extinct group could not have been in the direct ancestry of the Dendrasteridae and Mellitidae.

DNA-DNA hybridization data can be used to predict the time of divergence between clades. Any phylogenetic scheme that implies a divergence time significantly different from the predicted value should probably be rejected.

DNA-DNA hybridization data and the search for homoplasies

There is, at present, no *a priori* way of identifying homoplasies (characters that evolve in parallel and/or convergently). Homoplasies are usually found during the course of a cladistic analysis. The particular characteristics identified as homoplasies will often vary with the method of analysis or with the particular set of characters used in the analysis. Phylogenies based on DNA-DNA hybridization data provide a way of searching for homoplasies. The homoplasious characters can be found by plotting morphological features onto the, assumed true, phylogeny of the group being studied based on the DNA-DNA hybridization data (a compatibility analysis applied in reverse).

Having identified those morphological features that have evolved in parallel or convergently, similarities between the homoplasious characters can be searched for. Do the homoplasious characters all develop at a characteristic time during embryogenesis, are they all related to structures associated with any particular function? By asking questions such as these, it may be possible to make generalizations about which morphologies are prone to parallel and convergent evolution. This information, in turn, may help in the classification of extinct taxa, by suggesting phylogenetically reliable characters.

ACKNOWLEDGEMENTS

I thank Raoul Benveniste, David Raup, and Hewson Swift for providing many useful comments on an early draft of the manuscript. I also thank the people listed above, and many of the staff and students associated with the Committee on Evolutionary Biology and the Department of Geophysical Sciences, University of Chicago, for many fruitful and enjoyable discussions. Hewson Swift reviewed the final manuscript, providing a number of useful suggestions.

REFERENCES

Angerer, R. C., Davidson, E. H., and Britten, R. J. 1976. Single copy DNA and structural gene sequence relationships among four sea urchin species. *Chromosoma* **56**, 213–26.

Archer, M., Flannery, T. F., Ritchie, A., and Molnar, R. E. 1985. First Mesozoic mammal from Australia—an early Cretaceous monotreme. *Nature, London* **318**, 363–6.

Benveniste, R. E. 1985. The contributions of retroviruses to the study of mammalian evolution. In *Molecular evolutionary genetics* (ed. R. J. MacIntyre), pp. 359–417. Plenum Press, London.

Benveniste, R. E. and Todaro, G. J. 1976. Evolution of type C viral genes: evidence for an Asian origin of man. *Nature, London* **261**, 101–8.

Benveniste, R. E., Callahan, R., Sherr, C. J., Chapman, V., and Todaro, G. J. 1977. Two distinct endogenus type C viruses isolated from the Asian rodent *Mus cervicolor*: conservation of virogene sequence in related rodent species. *Journal of Virology* **21**, 849–62.

Bonner, T. I., Heinemann, R., and Todaro, G. J. 1980. Evolution of DNA sequences has been retarded in Malagasy primates. *Nature, London* **286**, 420–3.

Britten, R. J. 1982. Genomic alterations in evolution. In: *Evolution and development. Dahlem Konferenzen* (ed. J. T. Bonner), pp. 41–64. Springer-Verlag, Berlin, Heidelberg, New York.

Britten, R. J. 1986. Rates of DNA sequence evolution differ between taxonomic groups. *Science, New York* **231**, 1393–8.

Britten, R. J. and Kohne, D. E. 1968. Repeated sequences in DNA. *Science, New York* **161**, 529–40.

Britten, R. J., Graham, D. E., and Henerey, M. 1972. Sea urchin repetitive and single copy DNA. *Carnegie Institution Washington Yearbook* **71**, 270–3.

Britten, R. J., Graham, D. E., and Neufeld, B. R. 1974. Analysis of repeating DNA sequences by reassociation. In: *Methods in Enzymology* **29**, Part E (ed. L. Grossman and K. Moldave), pp. 363–418. Academic Press, New York.

Britten, R. J., Cetta, A., and Davidson, E. H. 1978. The single-copy DNA sequence polymorphism of the sea urchin *Strongylocentrotus purpuratus*. *Cell* **15**, 1175–86.

Brownell, E. 1983. DNA/DNA hybridization studies of muroid rodents: symmetry and rates of molecular evolution. *Evolution* **37**, 1034–51.

Busslinger, M., Rusconi, S., and Birnstiel, M. 1982. An unusual evolutionary behaviour of a sea urchin histone gene cluster. *The EMBO Journal* **1**, 27–33.

Cracraft, J. 1987. DNA hybridization and avian phylogenetics. *Evolutionary Biology* **21**, 47–96.

Diamond, J. M. 1983. Taxonomy by nucleotides. *Nature, London* **305**, 17–18.

Durham, J. W. 1955. *Classification of clypeasteroid echinoids*. University of California Press, Berkeley and Los Angeles.

Durham, J. W. 1966. Clypeasteroids. In: Moore, R. C. (ed.) *Treatise on invertebrate paleontology* U, *Echinodermata* **3**(2), pp. 450–91. Geological Society of America and University of Kansas Press, Lawrence, Kansas.

Easteal, S. and Oakeshott, J. G. 1985. Estimating divergence times of *Drosophila* species from DNA sequence comparisons. *Molecular Biology and Evolution* **2**, 87–91.

Galau, G. A., Britten, R. J., and Davidson, E. H. 1974. A measure of the sequence complexity of polysomal messenger RNA in sea urchin embryos. *Cell* **2**, 9–20.

Galau, G. A., Klein, W. H., Davis, M. M., Wold, B. J., Britten, R. J., and Davidson, E. H. 1976a. Structural gene sets active in embryos and adult tissues of the sea urchin. *Cell* **7**, 487–505.

Galau, G. A., Chamberlin, M. E., Hough, B. R., Britten, R. J., and Davidson, E. H. 1976b. Evolution of repetitive and nonrepetitive DNA. In: *Molecular evolution* (ed. F. J. Ayala), pp. 200–24. Sinauer Press, Sunderland, Mass.

Ghiold, J. and Hoffman, A. (1986). Biogeography and biogeographic history of clypeasteroid echinoids. *Journal of Biogeography* **13**, 183–306.

Goldberg, R. B., *et al.* 1975. DNA sequence organization in the genomes of five marine invertebrates. *Chromosoma* **51**, 225–51.

Goodman, M. 1981. Decoding the pattern of protein evolution. *Progress in Biophysics and Molecular Biology* **37**, 105–64.

Gould, S. J. 1985. A clock of evolution. *Natural History* **4/85**, 12–25.

Gould, S. J. 1986. Fuzzy Wuzzy was a bear. Andy Panda, too. *Discover*, Feb., 40–8.

Graham, D. E., Neufeld, B. R., Davidson, E. H., and Britten, R. J. 1974. Interspersion of repetitive and non-repetitive DNA sequences in the sea urchin genome. *Cell* **1**, 127–37.

Grula, J. W. *et al.* 1982. Sea urchin DNA sequence variation and reduced interspecies differences of the less variable DNA sequences. *Evolution* **36**, 665–76.

Hall, T. J., Grula, J. W., Davidson, E. H., and Britten, R. J. 1980. Evolution of sea urchin non-repetitive DNA. *Journal of Molecular Evolution* **16**, 95–110.

Harpold, M. M. and Craig, S. P. 1978. The evolution of nonrepetitive DNA in sea urchins. *Differentiation* **10**, 7–11.

Hayashida, H. and Miyata, T. 1983. Unusual evolutionary conservation and frequent DNA segment exchange in class I genes of the major histocompatibility complex. *Proceedings of the National Academy of Sciences, USA* **80**, 2671–5.

Hinegardiner, R. 1974. Cellular DNA content of the Echinodermata. *Comparative Biochemistry and Physiology* **49B**, 219–26.

Hinegardiner, R. 1976. Evolution of genome size. In *Molecular Evolution* (ed. F. J. Ayala), pp. 179–99. Sinauer Press, Sunderland, Mass.

Hunt, J. A. and Carson, H. L. 1983. Evolutionary

relationships of four species of Hawaiian *Drosophila* as measured by DNA reassociation. *Genetics* **104**, 353–64.

Hunt, J. A., Hall, T. J., and Britten, R. J. 1981. Evolutionary distances in Hawaiian *Drosophila* measured by DNA reassociation. *Journal of Molecular Evolution* **17**, 361–7.

Hutton, J. R. and Wetmur, J. G. 1973. Effect of chemical modification on the rate of renaturation of deoxyribonucleic acid, deaminated and glyoxalated deoxyribonucleic acid. *Biochemistry, New York* **12**, 558–63.

Jacobs, H. T., et al. 1983. Mitochondrial DNA sequences in the nuclear genome of *Strongylocentrotus purpuratus*. *Journal of Molecular Biology* **165**, 609–32.

Kier, P. M. 1977. The poor fossil record of the regular echinoid. *Paleobiology* **3**, 168–74.

Kier, P. M. 1982. Rapid evolution in echinoids. *Palaeontology*, **25**, 1–9.

Klein, W. H., Thomas, T. L., Lai, C., Scheller, R. H., Britten, R. J., and Davidson, E. H. 1978. Characteristics of individual repetitive sequence families in the sea urchin genome studied with cloned repeats. *Cell*, **14**, 889–900.

Kohne, D. E., Chiscon, J. A., and Hoyer, B. H. 1972. Evolution of primate DNA sequences. *Journal of Human Evolution* **1**, 627–44.

McColl, R. S. and Aronson, A. I. 1974. Transcription from unique and redundant DNA sequences in sea urchin embryos. *Biochemistry and Biophysics Research Communications* **56**, 47–51.

Mortensen, T. 1948. *Clypeasteroida. A monograph of the Echinoidea.* Vol. IV. 2. Reitzel, Copenhagen.

O'Brien, S. J., Nash, W. G., Wildt, D. E., Bush, M. E., and Benveniste, R. E. 1985. A molecular solution to the riddle of the giant panda's phylogeny. *Nature, London* **317**, 140–4.

Paul, C. R. C. 1982. The adequacy of the fossil record. In *Problems of phylogenetic reconstruction* (ed. K. A. Joysey and A. E. Friday, pp. 75–117). Systematics Association Special Publication, Academic Press, London.

Poltaraus, A. B. 1981. The estimation of relative connections between nine species of echinodermata by molecular hybridization of their DNA. *Zhurnal Obshchei Biologii* **42**, 55–9.

Roberts, J. W., Johnson, S. A., Kier, P., Hall, T. J., Davidson, E. H., and Britten, R. J. 1985. Evolutionary conservation of DNA sequences expressed in sea urchin eggs and early embryos. *Journal of Molecular Evolution* **22**, 99–107.

Romero-Herrera, A. E., Lieska, N., Goodman, M., and Simons, E. I. 1979. The use of Amino acid sequence analysis in assessing evolution. *Biochimie* **61**, 767–79.

Seilacher, A. 1979. Constructional morphology of sand dollars. *Paleobiology* **5**, 191–221.

Sepkoski, J. J., Jr. 1982. *A compendium of fossil marine families*. Milwaukee Public Museum.

Shaw, A. B. 1964. *Time in stratigraphy*. McGraw-Hill, New York.

Sibley, G. C. and Ahlquist, J. E. 1981. The phylogeny and relationships of the ratite birds as indicated by DNA-DNA hybridization. In: *Evolution Today, Proceedings of the 2nd International Congress in Systems in Evolutionary Biology* (ed. G. G. E. Scudder and J. L. Reveal), pp. 301–35. Carnegie-Mellon University, Hunt Institute for Botanical Documentation, Pittsburgh, Pennsylvania.

Sibley, G. C. and Ahlquist, J. E. 1983. Phylogeny and classification of birds based on the data of DNA-DNA hybridization. In: *Current Ornithology* (ed. R. F. Johnston), pp. 245–92. Plenum Press, New York.

Sibley, G. C. and Ahlquist, J. E. 1984. The phylogeny of the hominoid primates, as indicated by DNA-DNA hybridization. *Journal of Molecular Evolution* **20**, 2–15.

Sibley, G. C., Ahlquist, J. E., and Sheldon, F. H. 1987. DNA hybridization and avian phylogenetics. Reply to Cracraft. *Evolutionary Biology* **21**, 97–125.

Sheldon, F. H. 1987. Rates of single-copy DNA evolution in herons. *Molecular Biology and Evolution* **4**, 56–69.

Smith, A. B. 1984. *Echinoid palaeobiology*. Allen & Unwin, London.

Smith, M. J., Nicholson, R., Stuerzul, M., and Lui, A. 1982. Single copy DNA homology in sea stars. *Journal of Molecular Evolution* **18**, 92–101.

Wilson, A. C., Carlson, S. S., and White, T. J. 1977. Biochemical Evolution. *Annual Reviews in Biochemistry* **46**, 573–639.

Yanagisawa, T. 1984. Base sequence complexity of *Clypeaster* and *Echinocardium* DNA. *Proceedings of the 5th International Echinoderm Conference, Galway* (ed. B. F. Keegan and B. D. S. O'Conner), p. 396. Balkema, Rotterdam.

Zuckerkandl, E. and Pauling, L. 1962. Molecular disease, evolution, and genetic heterogeneity. In *Horizons in Biochemistry* (ed. M. Kasha and B. Pullman), pp. 189–225, Academic Press, New York.

10

Phylogenetic implications of genome rearrangement and sequence evolution in echinoderm mitochondrial DNA

HOWARD T. JACOBS, PETER BALFE, BERNARD L. COHEN, ANDREW FARQUHARSON, AND LOREDANA COMITO

Department of Genetics, University of Glasgow, Church Street, Glasgow G11 5JS, Scotland, UK

FOREWORD

This paper presents a brief review of some of the difficulties associated with the use of molecular data to reconstruct phylogeny and evaluates a new approach to this task based upon mitochondrial DNA. The need for additional approaches arises from a number of problems: these include the difficulty of distinguishing orthologous from paralogous genes in multigene families, uncertainty whether sequences have evolved at constant rates in different lineages, and the unknown contributions of gene conversion, 'molecular drive' and promiscuous exchanges to observed sequence differences. All of these indicate the unreliability of phylogenetic relationships deduced solely from one type of sequence data. This paper explores the application of novel parameters for the elucidation of echinoderm phylogeny, namely the order of genes in the mitochondrial genome and mitochondrial DNA sequence divergence. It is suggested that these provide a valuable complement to the results of studies of nuclear genes.

ONTOGENETIC V. PHYLOGENETIC INFORMATION IN THE GENOME

The genomes of present day organisms not only provide the blueprint for individual development; they also constitute an historical record of the development of the species. Only a small proportion of the information contained in the eukaryotic genome appears to perform a definable function in the life-cycle of the organism (David 1986). Typically, cells express only a few thousand different mRNAs: studies of mRNA sequence complexity in different developmental stages and in adult tissues of the sea urchin have shown that the aggregate usage of protein-coding information probably represents no more than 5 per cent of the genome (Galau *et al.* 1976). Whereas this ontogenetic information is readily interpretable in terms of its function, the phylogenetic information is cryptic. Our ability to understand it depends on our knowledge of the processes which have shaped it, which remains fragmentary.

Fundamental to studies of molecular evolution is the idea that molecular information in the genome changes at a (roughly) uniform rate as a result of random processes, irrespective of the amount or type of phenotypic selection pressure to which the organism is subjected: the 'molecular clock' hypothesis (Margoliash 1963; Wilson *et al.* 1977). Recent results indicate that the eukaryotic nuclear genome is subject to various types of recombinational promiscuities (Petes and Fink 1982; Lewin 1983; Dover 1986), which would seem to undermine this hypothesis to some degree. At the very least, these phenomena introduce several potentially serious complications into attempts to reconstruct phylogeny on the basis of nuclear gene sequences.

Three such phenomena will be considered. First, it is normally assumed that the genes being studied are, and always have been, single-copy sequences in the genomes in which they reside.

Echinoderm phylogeny and evolutionary biology (ed. C. R. C. Paul and A. B. Smith). Clarendon Press, Oxford, 1988.

This assumption is clearly invalid for many commonly analysed genes (e.g. cytochrome c, Margoliash 1963; Schwartz and Dayhoff 1978) and may not apply *in extremis* to any eukaryotic gene. If multiple genes for one or more closely related functions co-exist for long periods, but are also subject to periodic loss, reduplication, and functional diversification, then molecular studies of extant gene sequences can, at best, establish the ancestry of the genes. This may be different from that of the organisms in whose genomes they are found. Even genes which are single-copy in all modern organisms and therefore assumed to be *orthologous* may in fact derive from *paralogous* members of an ancestral multigene family (Wilson *et al.* 1977; Schwartz and Dayhoff 1978; Fitch and Margoliash 1970). In addition, the precise history of a multigene family may be reconstructed only with some ambiguity, as the rate of change of some family members may show pronounced acceleration following gene duplication (Li and Gojorbi 1983), presumably as a result of relaxed functional constraints. Therefore, reliable phylogenies can only be inferred from data relating to more than one, and preferably to a large number, of different genes.

Secondly, the phenomena which may be grouped under the term 'molecular drive' bias the molecular evolution of multigene families (and perhaps all sequences, to some degree) as a result of the inherent properties of DNA and the enzymes involved in its replication and repair (Dover 1986). Homologous sequences which co-exist in a genome are subject to periodic gene conversions which can replace all or part of the sequence information of any one copy with that of another (Petes and Fink 1982). Estimation of the true ages of multigene family members may therefore prove impossible. Tandem repeats may also suffer periodic correction from dispersed homologues in the genome ('orphons', Childs *et al.* 1981) which themselves may be functional genes or, alternatively, pseudogenes evolving independently of phenotypic selection pressures. Different regions of the same genome appear to change at distinct rates, with mutationally active or hyperactive sequences commonly observed (Wyman and White 1980). One consequence of this is that when genes are mobilized by transposable elements, or are moved to new chromosomal locations by other means, their rates of evolution may alter considerably.

A third problem in extracting phylogenetic information from nuclear gene sequence data is that the molecular clocks of different species may not be identical (Jacobs and Pilbeam 1980). The molecular clock is an empirical observation: the processes which give rise to it are not understood. Therefore, it cannot be assumed that the rate of change is constant between lineages, or even within a lineage over the course of time.

MITOCHONDRIAL DNA AND ECHINODERM EVOLUTION

The use of mitochondrial DNA as a phylogenetic probe overcomes a number of these problems. The animal mitochondrial genome is a supercoiled, circular DNA of 15–20 kb, which always (except in nematodes) encodes the same thirteen polypeptide subunits of the mitochondrial inner membrane respiratory complexes, as well as the rRNAs and tRNAs required for their synthesis inside mitochondria (Anderson *et al.* 1981, 1982; Bibb *et al.* 1981; Roe *et al.* 1985; Saccone *et al.* 1983; Clary and Wolstenholme 1985). Mitochondrial DNA replication, transcription and translation are semi-autonomous processes within the cell, which nevertheless require proteins and some RNA moieties (Wong and Clayton 1986) encoded by nuclear genes. In animals, nuclear and mitochondrial genomes are physically and functionally separate, operating on principles as different, if not more so, as those of prokaryotes from eukaryotes. The only genetic exchange between them which has been demonstrated is the rare incorporation of sequences of mitochondrial origin into nuclear DNA (Lewin 1983; Jacobs *et al.* 1983; Gellissen 1983). This is best regarded as an illustration of the fluidity of the nuclear genome. By contrast, no animal mitochondrial DNA has been shown to have incorporated foreign sequences.

In animals, mtDNA is inherited uniparentally through the maternal line (Lansman *et al.* 1983). In each generation the mtDNA of an individual is believed to derive clonally from just a few mitochondria present in the developing oocyte, (Hauswirth and Laipis 1985). Moreover, and most importantly with respect to 'molecular drive', intermolecular recombination has *not* been demonstrated to occur in animal mtDNA. It seems reasonable, therefore, to discount the possibility that variant mitochondrial genomes might co-exist in a lineage over appreciable lengths of time and suffer periodic quantal change as a result of gene conversion phenomena such as occur in nuclear gene families. Mitochondrial DNA may, therefore, be regarded as if it behaved, over long evolutionary periods, as a set of true single-copy genes, even though it is physically present at hundreds of copies per cell.

Mutational change accumulates in mtDNA in a manner largely independent of processes occurring in nuclear DNA, although this independence is not absolute since mitochondrial gene products interact functionally with nuclear-coded polypeptides (Cann et al. 1984). Where comparable categories of mutational change have been examined, the mitochondrial 'clock rate' appears to be distinct from that of nuclear DNA, commonly an order of magnitude faster (Brown 1983; Ferris et al. 1983).

Against this background of relatively rapid change must be set two properties of the mitochondrial genome which exhibit extreme conservatism. First, the polypeptides encoded by mtDNA all participate in membrane-bound multi-subunit complexes (Capaldi 1982), some of which, such as NADH dehydrogenase, are multifunctional in nature. Each subunit must interact with a large number of other subunits, as well as several cofactors, substrates, and products. Many mitochondrial genes respond to these multiple constraints by exhibiting highly conserved amino acid sequences over much of their length. Secondly, the overall gene content and mode of organization of mtDNA appear to be universally conserved across the animal kingdom, with rearrangements in gene order occurring only with extreme infrequency. These conservative features permit useful phylogenetic inferences to be drawn from studies of the mitochondrial genome in different taxa.

The predictive usefulness of a new marker for phylogenetic relationships may only be judged by reference to groups whose ancestry is already well established either from the fossil record or by other means. The reasonably comprehensive fossil record of the echinoderms makes the phylum a particularly favourable subject for an exploratory study of mitochondrial genome rearrangements. In addition, it may be possible to draw from these studies further inferences relevant to echinoderm evolution, notably (i) the ancestral relationships between the five living classes; (ii) the derivation of the phylum itself, in relation both to other deuterostomes and to other invertebrates; and (iii) the precise relationships of specific groups which have diversified during much more recent evolutionary time, such as the Echinidae and related families.

MOLECULAR ANALYSIS OF SEA URCHIN mtDNA

The point of departure for this study was the molecular cloning of sea urchin mitochondrial DNA, which has permitted a detailed analysis of gene sequences and their arrangement in the genome. Clone λmt1 (Fig. 10.1a) was originally isolated as a λ phage recombinant from a *Stronglyocentrotus purpuratus* genomic library, on the basis of its hybridization with a previously characterized plasmid recombinant containing a cDNA copy of the large (16S) mitochondrial ribosomal RNA (Jacobs and Grimes 1986). λmt1 contains a 15.8 kb insert of sea urchin DNA which is, in fact, a linearized copy of the entire mitochondrial genome, as has been shown by sequencing from the two EcoRI sites at the borders of the insertion. The sequences at the two sites were found to encode contiguous segments of the N-terminal region of apocytochrome b, on the basis of (computer matched) homology with the corresponding vertebrate and *Drosophila* genes (Anderson et al. 1981, 1982; Bibb et al. 1981; Roe et al. 1985; Saccone et al. 1983; Clary and Wolstenholme 1985).

By generating a series of subclones in plasmid and M13 vectors, based upon the restriction map of λmt1 shown in Fig. 10.1a, each region of the genome was mapped in detail and sequenced. Sequences were identified on the basis of their homology, either at nucleotide or amino acid level, with their counterparts from other animal mtDNA sequences (human, mouse, rat, cow, *Xenopus*, and *Drosophila*: Anderson et al. 1981, 1982; Bibb et al. 1981; Roe et al. 1985; Saccone et al. 1983; Clary and Wolstenholme 1985) using search programmes provided by the University of Wisconsin Genetics Computer Group. The map locations of the sequences were determined by blot hybridizing the various subclones used for sequencing to restriction digests of λmt1 or its subclones. This procedure has resulted in the unambiguous identification, mapping, and orientation of all thirteen protein-coding genes and both rRNAs in *S. purpuratus* mtDNA (see Fig. 10.1b) as well as all of the 22 tRNAs (although for clarity only four of them are shown in Fig. 10.1b).

(a)

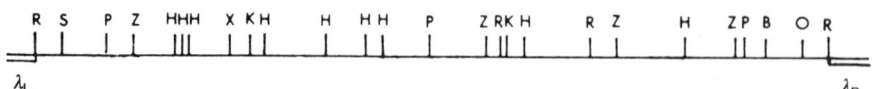

(b)

SEA URCHINS

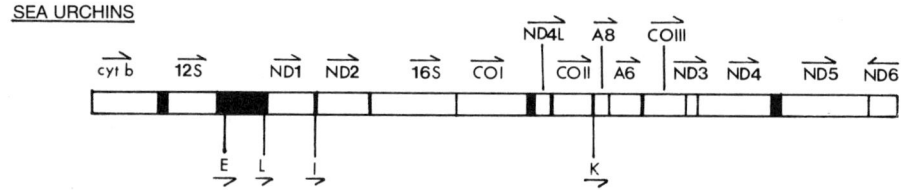

VERTEBRATES

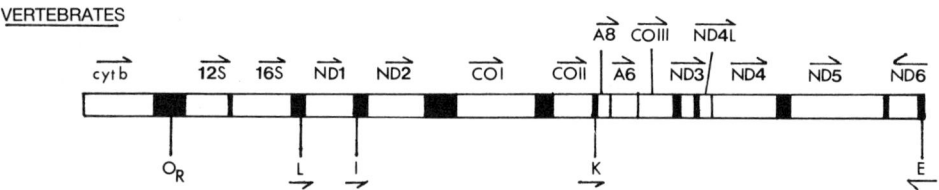

DROSOPHILA

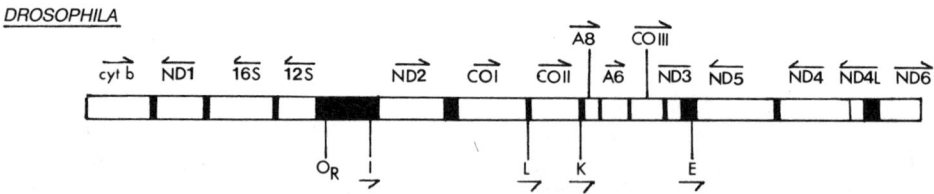

Fig. 10.1 (a) Restriction map of cloned *S. purpuratus* mtDNA (λmt1) showing sites for EcoRI (R), HindIII (H), SalI (S), SacI (Z), XbaI (X), KpnI (K), XhoI (O), Bg1II (B), and PstI (P). λ_L and λ_R denote, respectively, the left and right arms of the vector, λ Charon 4. (b) Schematic maps of the arbitrarily linearized mitochondrial genomes of sea urchins (aligned with the restriction map of λmt1 shown in (a), and based upon nucleotide sequencing of λmt1), humans (based on Anderson *et al.* 1981, aligned with the map for sea urchins at its left-hand side) and *Drosophila* (based on Clary and Wolstenholme 1985, and similarly aligned). Structural genes indicated are for the large (16S) and small (12S) rRNAs, subunits I, II, and III of cytochrome oxidase (COI, COII and COIII), subunits 6 and 8 of ATPase (A6 and A8), seven subunits of NADH dehydrogenase (ND1 etc.), and four tRNA genes decoding isoleucine (I), glutamate (E), lysine (K), and leucine UUR (L) codons. Arrows indicate the sense strand direction ($5' \to 3'$) for each gene. Regions shaded in black contain no open reading frames and include the 'control' region, containing the replication origin O_R (not mapped in sea urchins), tRNA genes, and non-coding segments.

As shown in Fig. 10.1b, two major rearrangements have occurred in the mitochondrial genome since the divergence of the vertebrate and echinoderm lineages, relocating the genes for the large (16S) ribosomal RNA and for NADH dehydrogenase subunit 4L (ND4L). Both of these genes have been transposed to new sites without inversion, preserving the enormous preponderance of coding information on one strand of the genome seen in vertebrates, though not in *Drosophila* (Clary and Wolstenholme 1985). Both 16S rRNA and ND4L are located upstream of the same genes (ND1 and ND4, respectively) in vertebrates and in *Drosophila*, suggesting that this arrangement, rather than that of the echinoids, may be ancestral. There are grounds for believing, therefore, that vertebrate mtDNA is organized more like that of the metazoan common ancestor than is the mtDNA of either of the other two groups considered. Since major rearrangements in mtDNA appear to be so rare it may never prove possible to determine the mechanism by which they occur. However, this should not detract from their usefulness as phylogenetic markers.

As regards the more variable positions of tRNA genes, only 5 of the 22 mitochondrial tRNAs lie in the same positions with respect to their adjacent genes on the 3' side in vertebrate and echinoid mtDNAs. *Drosophila* and vertebrate mtDNAs show nine conserved tRNA positions, but *Drosophila* and sea urchin only three. This variability may be a secondary consequence of the much greater clustering of tRNA genes in sea urchin mtDNA, 15 of which are located between the genes for 12S rRNA and ND1. It is hard to judge whether a clustered or dispersed arrangement should be regarded as ancestral. Although yeast tRNA genes are clustered (Borst and Grivell 1978), it would seem more reasonable to regard this feature as a case of convergent evolution with echinoids rather than view the relatively large number of equivalent tRNA positions in insects and vertebrates as coincidence. The possible role of tRNAs as occasional 'illicit' primers of DNA replication (Varmus 1982) may result in a trend towards their accumulation in the region of the replication origin in a circular genome such as mtDNA. The replication origin remains to be mapped definitively in *S. purpuratus* mtDNA; a 121 base-pair region within the tRNA cluster seems the likeliest candidate as its sequence has no other obvious function, and is the only extended unassigned sequence in the genome.

DEDUCING ANCESTRY FROM MITOCHONDRIAL GENE ORDER

Because of all the complications in the use of nuclear gene sequence data, gene order in mtDNA may prove to be a more reliable and objective parameter for establishing the interrelationships of classes and phyla. The stability of mitochondrial gene order is illustrated by the maps of *Xenopus* and mammalian mtDNAs (Roe *et al.* 1985; Anderson *et al.* 1981, 1982; Bibb *et al.* 1981), which are identical, despite the 320 Ma or so separating these taxa. As we show below, a similar degree of conservation appears to characterize the echinoid lineage. Disregarding the more variable positions of tRNA genes which, for reasons stated above, may relocate via a distinct mechanism, a small number of topological transformations is sufficient to interconvert the mtDNA maps of the different phyla. In pairwise comparisons, the numbers of transformations required to effect the topological interconversion of *Drosophila*, vertebrate, and echinoid mtDNAs are 3, 5, and 2, respectively, suggesting such events occur approximately once per 200 Ma. These rearrangements are so rare, in fact, that one cannot estimate a reliable rate of their occurrence for calculating actual divergence times. However, the extreme conservatism of this feature allows common ancestry to be deduced even without reference to clock rate, provided that successive taxonomic divergences did not all occur within too short a period of time.

As a preliminary illustration of the way this approach may be applied to specific questions of echinoderm phylogeny, we have analysed a number of mitochondrial genomes for the relative positions of the genes for 12S and 16S rRNAs, and for cytochrome oxidase subunit I (COI). Four echinoid species examined by blot hybridization to probes for 16S and COI show these genes to be adjacent (as already shown for *S. purpuratus*), and in the three cases where 12S has also been mapped, it lies upstream of 16S by approximately the same distance in each case, at least within the limits of accuracy of restriction mapping (Fig. 10.2). The species considered include examples of both suborders of regular echinoids, namely Stirodonta (*Arbacia punctulata*) and Camarodonta, of which

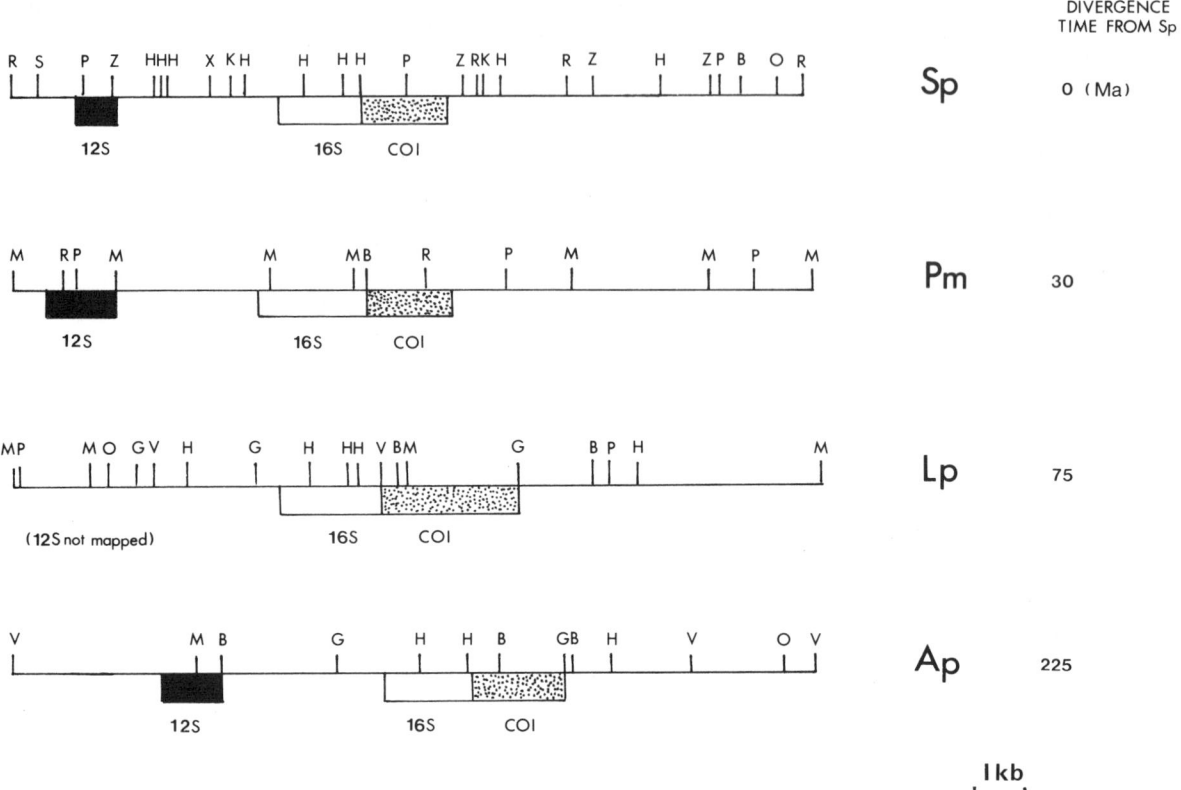

Fig. 10.2 Mitochondrial gene order in four species of echinoid. The positions of genes for the large (16S) rRNA (open bars), the small (12S) rRNA (solid bars) and cytochrome oxidase subunit I (dotted bars) are shown in arbitrarily linearized restriction maps of mtDNA from *Strongylocentrotus purpuratus* (Sp), *Psammechinus miliaris* (Pm), *Lytechinus pictus* (Lp), and *Arbacia punctulata* (Ap), whose approximate divergence times are as shown (Smith 1984). Restriction sites are as indicated in legend to Fig. 10.1, plus, additionally, BamHI (M), BglI (G), and PvuII (V). Maps have been approximately aligned to show the common arrangements of these genes, and are based on mapping and nucleotide sequencing of λmt1 (see Fig. 10.1), and blot hybridization to digests of uncloned mtDNA from the remaining three species, using probes SpP144 for 16S (Jacobs and Grimes 1986), SpG30 for COI (Jacobs *et al.* 1986) and mQ4 (HTJ unpublished) for 12S. Hybridizations were in 5 × SET at 55 °C with final wash in 1 × SET at 55 °C.

three families are represented (Strongylocentrotidae: *S. purpuratus*; Toxopneustidae: *Lytechinus pictus* and Echinidae: *Psammechinus miliaris*). The arrangement of these genes in *S. purpuratus* mtDNA may therefore be taken to be diagnostic for the echinoid lineage, providing further evidence that alterations to mitochondrial gene order occur less than once per 200 Ma or so.

This method has been applied to questions at class and phylum level by examining the organization of asteroid mtDNA. Using mtDNA from one species of starfish, *Asterias forbesi*, a combination of molecular cloning and blot hybridization has demonstrated that the 3' ends of 16S rRNA and COI are separated by at least 5 kb and probably rather more. Unfortunately, blot hybridization could not be used to map these sequences unambiguously in uncloned mtDNA for trivial reasons (the presence of too many inter-individual polymorphisms, plus a difficulty in obtaining cleanly digestible mtDNA from this one particular source). Blot hybridization and nucleotide sequencing of cloned restriction fragments of *A. forbesi* mtDNA have shown clearly that the region immediately downstream of the 3' end of 16S rRNA is unrelated to COI (see Fig. 10.3). The absence of open reading frames and tRNA genes suggests this may be the control region. It is

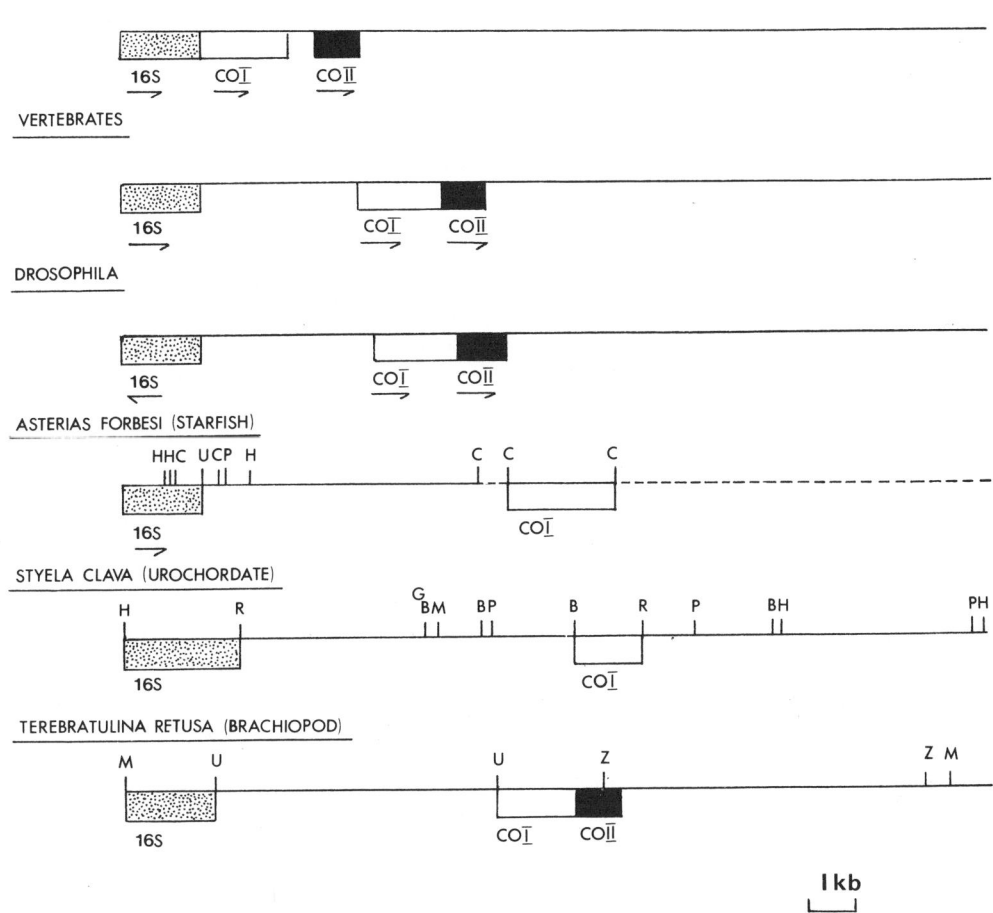

Fig. 10.3 Mitochondrial gene order in species from two classes of echinoderms plus four other phyla. The positions of genes for 16S rRNA (dotted bars), cytochrome oxidase subunits I (open bars) and II (solid bars), are shown in arbitrarily linearized maps of mtDNA from sea urchins, vertebrates and *Drosophila*, based on Figs 10.1 and 10.2, and from one species each of starfish, urochordate, and brachiopod. Maps have been aligned at the 16S rRNA gene. Restriction sites shown are as in Figs 10.1 and 10.2 plus, HincII (C) and Sau3AI (U), not all sites for which are shown. Maps are based on blot hybridization to digests of cloned and uncloned mtDNA of *A. forbesi*, nucleotide sequencing of cloned segments of *A. forbesi* mtDNA and blot hybridization to digests of uncloned *S. clava* and cloned *T. retusa* mtDNAs. Probes used were SpP144 for 16S (Jacobs and Grimes 1986), SpG30 for COI (Jacobs *et al*. 1983), mN6 (HTJ unpublished) for COII, and, additionally, mAfH30 (starfish 16S) and mAfH91 (starfish COI) (both unpublished). Hybridizations were carried out in 5 × SET at 50 or 55 °C with final wash in 1 × SET at the same temperature. Dotted lines indicate segments of unknown length separating the HincII fragments on which lie the 3′ end of COI and the sequence downstream of 16S rRNA in *A. forbesi* mtDNA.

apparent that asteroid mtDNA shows an organization distinct from that of echinoids or vertebrates. The logical extension of this work is to complete the mapping of the *A. forbesi* mitochondrial genome using the same techniques as for *S. purpuratus*, then to demonstrate that this is a conserved arrangement amongst asteroids by blot hybridization with the mtDNAs of other species. A still longer term aim is to define what mtDNA rearrangements have occurred during the evolutionary diversification of the entire phylum, by examining the mtDNAs of the remaining classes and of the cidaroids, the most anciently diverged order of echinoids (Smith 1984).

Amongst the phyla normally considered most closely related to the echinoderms, comparisons have been made using mtDNA from a urochordate (*Styela clava*), a phylum whose position is relatively well established, and from an articulate brachiopod (*Terebratulina retusa*), representing a phylum which occupies a particularly uncertain and intriguing phylogenetic position, being protostomes which exhibit a radial cleavage pattern and other features more suggestive of deuterostome embryology. The relative positions of 16S and COI sequences in each mtDNA have been determined using restriction mapping of cloned and uncloned DNA and blot hybridization to *S. purpuratus* probes (Fig. 10.3). The minimum distance between the 3′ ends of 16S and COI in the urochordate appears to be of the order of 7 kb and in the brachiopod 6 kb, both significantly more than the 4.5 kb observed in vertebrates, the 5.0 kb in *Drosophila* and the 1.6 kb in sea urchins. The results indicate that the genomes of each phylum considered are all differently arranged from one another with respect to the location of these genes, given that intergenic spacers in animal mtDNA are extremely short. The gene for COII has also been mapped in the brachiopod, where it appears adjacent to COI, as in vertebrates and *Drosophila*.

SEQUENCE COMPARISONS OF mtDNA IN DIFFERENT SPECIES

Because it is free from many of the perturbations associated with nuclear sequence evolution, molecular change in mtDNA is essentially dependent on two variables: the inherent error-proneness of mtDNA synthesis (or repair) and the degree of evolutionary constraint applying to the various gene sequences which it encodes. In other words, it fulfils the requirements for a true molecular clock. Another attractive aspect of mtDNA for phylogenetic studies is the fact that within it are to be found regions of DNA sequence under an enormous range of different selective pressures. For example, an extended sequence near the 3′ end of the large ribosomal RNA, which appears to function in peptidyl transferase activity (Kearsay and Craig 1981; Dujon 1980) is almost completely conserved amongst all the Metazoa. By contrast, other equally extended domains of the mitochondrial ribosomal RNAs appear to show no significant primary sequence homology at all between phyla, although the molecules fold to similar overall secondary structures. Protein-coding genes located in mtDNA also vary considerably in their degree of divergence. Figure 10.4 (a–e) displays some amino acid sequence data for five mitochondrial genes in *S. purpuratus*, alongside the corresponding sequences from man, mouse, *Xenopus*, and *Drosophila*. The data include representatives of polypeptides belonging to each of the four respiratory complexes containing mitochondrially encoded subunits (NADH dehydrogenase (Chomyn et al. 1985), cytochrome bc_1 complex, cytochrome c oxidase, and ATP synthase (Mariottini et al. 1983). Inspection of the data shows that variation in degree of divergence applies even to genes encoding subunits of the same enzyme such as ND1 and ND4L. The most variable mitochondrial gene, ATPase subunit 8 (see Fig. 10.4d) exhibits as little as 16 per cent absolute amino acid homology between animal phyla. Domains of other polypeptides, such as those shown from ND1 (Fig. 10.4b), or cytochrome oxidase subunit II (COII, Fig. 10.4e), commonly include runs of up to 15 consecutive amino acids invariant across all phyla analysed, with overall absolute homologies of up to 70 per cent. The ways in which such data may be used to deduce phylogenies and the limitations of this approach will be discussed in a later section.

Sequences are also shown for two tRNA genes (Figs 10.4f,g), in one of which ($tRNA_{ser-ucn}$) the sea urchin sequence is clearly more related to that of vertebrates than to that of *Drosophila*. In the case of $tRNA_{thr}$ the converse is true. This illustrates the danger of placing too much reliance on one type of sequence data in reconstructing phylogenies.

CALCULATING DIVERGENCE TIMES BASED ON THE RATE OF SILENT NUCLEOTIDE SUBSTITUTIONS IN MITOCHONDRIAL GENES

Superimposed upon the nucleotide sequence changes which lead to amino acid substitutions are a distinct class of silent point mutations. These accumulate at a much faster rate, as the degree of phenotypic selection pressure which constrains them is much less. However, this pressure is not zero, as can be inferred from considering the relative usage of different synonymous codons in mitochondrial genes (Table 10.1; Wolstenholme and Clary 1985). In sea urchin, mtDNA some sets

TABLE 10.1 Codon usage in five mitochondrial genes in *S. purpuratus*: ND1, ND4L, cytochrome b, ATPase 8, and COII. Amino acids are indicated using the standard one-letter code

Amino acid	Usage of synonymous codons (%)				Amino acid	Usage of synonymous codons (%)		
M	ATA[1] 5	ATG 95	GTG[1] 5		I	ATC 12	ATT 47	ATA 41
V	GTA 30	GTC 13	GTG 19	GTT 38	F	TTC 45	TTT 55	
G	GGA 47	GGC 15	GGG 21	GGT 18	N	AAC 35	AAT 21	AAA 42
R	CGA 65	CGC 18	CGG 12	CGT 6	K	AAG 100		
T	ACA 41	ACC 23	ACG 5	ACT 31	Q	CAA 63	CAG 37	
P	CCA 28	CCC 34	CCG 4	CCT 34	D	GAC 47	GAT 53	
A	GCA 41	GCC 11	GCG 3	GCT 46	E	GAA 54	GAG 46	
S	TCA 9	TCC 41	TCG 3	TCT 26	W	TGA 89	TGG 11	
	AGA 7	AGC 7	AGG 6	AGT 1	C	TGC 88	TGT 12	
L	CTA 27	CTC 13	CTG 3	CTT 18	Y	TAC 37	TAT 53	
	TTA 25	TTG 14						

[1] ATA and GTG used as start codons are presumed to encode methionine.

of synonymous codons are used with approximately equal frequency, such as GAA or GAG to specify glutamate. In other cases a considerable bias is evident, such as for tryptophan, where TGA predominates nine-fold over TGG, and for alanine where 87 per cent of codons are GCA or GCT. Some codons, such as AGT for serine or CCG for proline, are hardly ever used. These biases are usually assumed to be the result of selection for efficient (or in some specific cases, inefficient) translation (Grosjean and Fiers 1982). Nevertheless, the rate of silent substitutions is probably the parameter closest to a true molecular clock, in considering protein-coding genes.

This parameter has been used to estimate the divergence time of *Paracentrotus lividus* (family Echinidae) from *S. purpuratus* (family Strongylocentrotidae), subject to two caveats. First, correction must be made for saturation effects ('multiple hits'), taking account, if possible, of the likely relative frequencies of different types of substitutions, in part dependent on the degeneracy of the genetic code. In this example, an approximate formula based on Kimura and Ohta (1972) has been used, rather than the more elaborate method of Li *et al.* (1985) which would need adaptation to the properties of mitochondrial DNA. Secondly, the clock rate for silent substitution in sea urchin mtDNA remains somewhat uncertain (see below) due to the poor fossil record of the species studied. Given this uncertainty we do not consider a more rigorous analysis is justified at present.

Based on sequences of the genes for cytochrome oxidase subunits I and II and ATPase subunits 6 and 8 from both *S. purpuratus* and *P. lividus* (data which will be published in full elsewhere), the proportion of silent nucleotide differences between these species may be estimated at 38 per cent. Applying the formula of Kimura and Ohta (1972), this represents a substitution of 53 per cent of silent nucleotides since the divergence of these species. It can be noted that the frequencies of transitions and transversions in sea urchin mtDNA appear to be close to random expectation at this degree of divergence, taking account of the distribution of two-choice and four-choice amino

(a) ND4L

```
Sp    M---ALLIVILSMFYLGLMGILLNRLHFLSILLCLELLLISLFIGIALWNNNTGVPQNTTF
Hs    M--PLIYMNIMLAFTISLLGMLVYRSHLMSSLLCLEGMMLSLFIMATLMTLNTHSLLANIV
Mm    M--PSTFFNLTMAFSLSLLGTLMFRSHLMSTLLCLEGMVLSLFIMTSVTSLNSNSMSSMPI
Xl    M--TLIHFSFCSAFILGLTGLALNRSPILSILLCLELMLLMSMDGIVLTFLHLTIYLSSMM
Dy    MIMILYWSLPMILFILGLFCFVSNRKHLLSMLLSLEFIVLMLFFMLFIY-LNMLNYENYFS

Sp    NLFVLTLVACEASIGLSLMVGLSRTHSSNLVGSLSLLQY*
Hs    PIAMLVFAACEAAVGLALLVSISNTYGLDYVHNLNLLQC*
Mm    PI-TLVFAACEAAVGLALLVKVSNTYGTDYVQNLNLLQC*
Xl    LYIMLPFAAPEAATGLSLNSDHYTTHGTDKLFSLNLLEC*
Dy    MM-FLTFSVCEGALGLSILVSMIRTHGNDYFQSFSIM--*
```

(b) ND1

```
Sp    DGMKVFIKEELKPVNSSPYLFFFSPLLFLALALLLWNFMPVHTPTLDLQLSLLLVLGLSSL
Hs    DAMKLFTKEPLKPATSTITLYITAPTLALTIALLLWTPLPMPNPLVNLNLGLLLILATSSL
Mm    DAMKLFMKEPMRPLTTSMSLFIIAPTLSLTLALSLWVPLPMPHPLINLNLGILFILALSSL
Xl    DGVKLFIKEPVRPSTSSQTMFLIAPTMALALAMSIWAPLPMPFSLADLNLGILFILALSSL
Dy    DAIKLFTKEQTYPLLSNYLSYYISPIFSLFLSLFVWMCMPFFVKLYSFNLGGLFFLCCTSL

Sp    SVYAILGSGWASNSKYSLLGAIRAVAQTISYEISLALILLSLIIFSSSFNLTYIMNTQEFS
Hs    AVYSILWSGWASNSNYALIGALRAVAQTISYEVTLAIILLSTLLMSGSFNLSTLITTQEHL
Mm    SVYSILWSGWASNSKYSLFGALRAVAQTISYEVTMAIILLSVLLMNGSYSLQTLITTQEHM
Xl    AVYTILGSGWSSNSKYALIGALRAVAQTISYEVTLGLILLCMIMLAGGFTYTTLMTTQEQM
Dy    GVYTVMVAGWSSNSNYALLGALRAVAQTISYEVSLALIMLSFIFLIGSYNMIYFFYYQIYM

Sp    WFSLSCLPLFYIWFVSTLAETNRAPFDLTEGESEIVSGYNVEYAGG
Hs    WLLLPSWPLAMMWFISTLAETNRAPFDLAEGESELVSGFNIEYAAG
Mm    WLLLPAWPMAMMWFISTLAETNRAPFDLTEGESELVSGFNVEYAAG
Xl    WLIIPGWPMAAMWYISTLAETNRAPFDLTEGESELVSGFNVEYAGG
Dy    WFLIILFPMSLVWLTISLAETNRTPFDFAEGESELVSGFNVEYSSG
```

(c) cyt b

```
Sp    M-AAPLRKEHPIFRILNSTFVDLPLPSNLSIWWNSGSLLGLCLVVQILTGIFLAMHYTADI
Hs    M--TPMRKINPLMKLINHSFIDLPTPSNISAWWNFGSLLGACLILQITTGLFLAMHYSPDA
Mm    M--TNMRKTHPLFKIINHSFIDLPAPSNISSWWNFGSLLGVCLMVQIITGLFLAMHYTSDT
Xl    M-APNIRKSHPLIKIINNSFIDLPTPSNISSLWNFGSLLGVCLIAQIITGLFLAMHYTADT
Dy    MHKP-LRNSHPLFKIANNALVDLPAPINISSWWNFGSLLGLCLIIQILTGLFLANHYTADV

Sp    TLAFSSVMHILRDVNYGWFLRYV
Hs    STAFSSIAHITRDVNYGWIIRYL
Mm    MTAFSSVTHICRDVNYGWLIRYM
Xl    SMAFSSVAHICFDVNYGLLIRNL
Dy    NLAFYSVNHICRDVNYGWLLRTL
```

(d) ATPase 8

```
Sp   MPQLEFAWWIVNFSLIWASVLIVISLLLNSFPPNSAGQSSSSLTLN-KTTTNWQWL*
Hs   MPQLNTTVWPTNITPMLLTLFLITQLKMLNTNYHLPPSPKPMKMKNYNKPWEPKWTK...*
Mm   MPQLDTSTWFITIISSMITLFILFQLKVSSQTFPLAPSPKSLTTMKVKTPWELKWTK...*
Xl   MPQLNPGPWFLILIFSWLVLLTFIPPKVLKHKAFNEPTTQTTEKS-KPNPWNWPWT*
Dy   MPQMAPIRWLLLFIVFSITFILFCSINYYSYMPTS-PKSNE-LK-NINYNMNWKW*
```

(e) COII

```
Sp   EFDSYMVPTSDVSFGNPRLLEVDNRLVLPMQNPIRVLVSSADVLHSWAVPSLGTKMDAVPG
Hs   IFNSYMLPPLFLEPGDLRLLDVDNRVVLPIEAPIRMMITSQDVLHSWAVPTLGLKTDAIPG
Mm   CFDSYMIPTNDLKPGELRLLEVDNRVVLPMELPIRMLISSEDVLHSWAVPSLGLKTDAIPG
Xl   SFDSYMIPTNDLTPGQFRLLEVDNRMVVPMESPTRLLVTAEDVLHSWAVPSLGVKTDAIPG
Dy   EFDSYMIPTNEALIDGFRLLDVDNRVILPMNSQIRILVTAADVIHSWTVPALGVKVDGTPG

Sp   RLNQTTFFAARTGVFYGQCSEICGANHSFMPIVIESVPFNTFENWVTQYLEE-*
Hs   RLNQTTFTATRPGVYYGQCSEICGANHSFMPIVLELIPLKIFEMGPVFTL---*
Mm   RLNQATVTSNRPGLFYGQCSEICGSNHSFMPIVLEMVPLKYFENWSASMI---*
Xl   RLHQTSFIATRPGVFYGQCSEICGANHSFMPIVVEAVPLTDFENWSSSMLEAS*
Dy   RLNQTNFFINRPGLFYGQCSEICGANHSFMPIVIESVPVNNFIKWISRNNS--*
```

(f) tRNA$_{thr}$

```
Sp   G C C T T G A A A G C T C A A C A A - G A G C T T T G G T C C T T G T
Hs   G T C C T T G T A G T A T A A A C T A A A G C A C C A G T C C T T G T
Mm   G T C T T G A T A G T A T A A A A C A T T A C T C T G G T C C T T G T
Xl   G T C C T G A T A G C T T A A T T T A A A G C A T C G G T C C T T G T
Dy   G T T T T A A T A G T T T A A T A A - A A A C A T T G G T C C T T G T

Sp   A A A C C A G G A G A G A G G G T A A C T - C C C T C T C A A G G C T
Hs   A A A C C G G A G A T G A A A A C - - C T - T T T T C C A A G G A C A
Mm   A A A C C T G A A A T G A A G A T C T T C - T C T T C T C A A G A C A
Xl   A A G C C G A A G A T T G A G G C T A A A C C C T C C T C A A G A C T
Dy   A A A T C A A A A A T - A A G A T T A T T - T C T T - T T A A A A C T
```

(g) tRNA$_{ser-UCN}$

```
Sp   G G A G A A G T G G C A C G A T A G G A A T G C A T G C G G C T T G A A
Hs   T A G A A A A G T C A T G G A G G C C A T G G G G T T G G C T T G A A
Mm   T G A G A A A G A C A T - A T A G G A T A T G A G A T T G G C T T G A A
Xl   A A G A A A A T G G C A G A G T G G T G A T G C A A C T G A C T T G A A
Dy   A G T T A A T G A G C T T G A - A C - - A A G C G T A T G T T T T G A A

Sp   A C C G T T T G A T A G A G G T T T C - T T C C T C - T C T T C T C T T
Hs   A C C A G C T T T G G G G G G T T C G A T T C C T T C C T T T T T T G T
Mm   A C C A A T T T T A G G G G G T T C G A T T C C T T C C T T T C T T A T
Xl   A T C A G A G T A G G G G G G T T C G A T T C C T C - T T T T T C T C G
Dy   A A C A A A - G A T A G A A - T T T A A T T T T C T - - A T T A A C T T
```

Fig. 10.4 (a)–(e) Amino acid sequences, using the one letter amino-acid code, for segments of five mitochondrial genes deduced from mtDNA nucleotide sequences of *S. purpuratus* (*Sp*, H. T. Jacobs, unpublished data), human (*Hs*, Anderson et al. 1981), mouse (*Mm*, Bibb et al. 1981), *Xenopus* (*Xl*, Roe et al. 1985) and *Drosophila yakuba* (*Dy*, Clary and Wolstenholme 1985). (a) NADH dehydrogenase subunit 4L (entire gene); (b) NADH dehydrogenase subunit I (a region in the middle of the gene); (c) cytochrome b (N-terminal region); (d) ATPase subunit 8 (entire gene); and (e) cytochrome oxidase subunit II (C-terminal half). Dashes indicate gaps introduced to maximize homology. (f)–(g) Nucleotide sequences for tRNA genes from these same species, decoding (f) threonine (ACN) and (g) serine (UCN) codons. Dashes indicate gaps introduced to maximize homology.

acids in the sequence, a result similar to that observed for *Drosophila* (Wolstenholme and Clary 1985) and contradicting that found for mammals seen, admittedly, over briefer divergence times (Cann *et al*. 1984; Brown 1983; Ferris *et al*. 1983).

Jacobs and Grimes (1986) estimated the rate of change of sea urchin mtDNA at silent sites as 9.2 per cent *per nucleotide* (i.e. 25.6 per cent per silent nucleotide) over 37 Ma, the time during which they estimated the nuclear pseudogene of COI to have been evolving separately from its mitochondrial counterpart. Applying the formula of Kimura and Ohta (1972) to take account of multiple hits, the silent substitution rate in *S. purpuratus* mtDNA is therefore implied to be 0.84 per cent per Ma. Since this estimate is for the sequence change in mtDNA in a single species (i.e. comparing the extant with the inferred ancestral version of the sequence) it must be doubled to express the rate at which the mtDNAs of two species of sea urchin will be seen to diverge, assuming the molecular clocks of the two species advance at this same rate. The 53 per cent of silent substitution between *S. purpuratus* and *P. lividus* implies, therefore, that they last shared a common ancestor 53/1.68, i.e. 32 Ma, a figure in reasonable agreement with the fossil record (Smith 1984). The reliability of this value is compromised not only by the caveats already indicated, but also by the fact that the clock rate inferred by Jacobs and Grimes (1986) is dependent upon judging the age of a particular nuclear DNA sequence, using an empirical value for the rate of change of total single-copy nuclear DNA in sea urchins. This is expressed in terms of ΔT_m, the change in duplex melting temperature resulting from sequence mismatch (Angerer *et al*. 1976) and does not take account of mismatch due to intra-specific divergence. The sequence studied had, moreover, undergone a small number of insertion and deletion events which also contribute considerably to ΔT_m (Britten *et al*. 1974). The deduced silent divergence rate in mtDNA of 1.68 per cent/Ma may therefore be taken to be accurate only to within ± 50 per cent. It may be noted that the deduced rate of divergence in echinoid mtDNA is of a similar order of magnitude to that inferred for *Drosophila* (Solignac *et al*. 1986). However, for similar reasons as here, i.e. a poor fossil record, this rate is also somewhat uncertain. To clarify the point it will be necessary to sequence mitochondrial genes from a number of echinoid species whose divergence times are known from the fossil record with greater precision. Despite this, the potential value of the method, in elucidating the relationships of families and genera diverged over tens of millions of years is apparent.

CALCULATING DIVERGENCE TIMES FROM THE RATE OF AMINO ACID SUBSTITUTIONS IN MITOCHONDRIAL PROTEIN CODING GENES

As has already been indicated, nucleotide substitutions which alter the amino acid sequence of protein-coding genes occur much more slowly than silent substitutions. One may consider that two, largely independent molecular clocks are operating in the same DNA sequence. As was already indicated, the rate of change in different proteins, different subunits of the same enzyme and even different domains of the same polypeptide, varies enormously. Although free from the direct effects of gene conversion, duplication and deletion which occur in nuclear families, the evolution of individual mitochondrial genes may nevertheless be driven, in part, by events of this type occurring in nuclear genes encoding polypeptides with which they intimately interact e.g. other subunits of the same enzyme (Cann *et al*. 1984). Specific effects may also be produced in a particular polypeptide due to features of the environment or lifestyle of the organism. For example, adaptations to life within a particular temperature range or exposure to new metabolites may have highly selective effects on protein evolution. For these reasons it is probably unwise to place too much phylogenetic reliance on data from a single mitochondrial gene.

In the example shown (Figs 10.4, 10.5 and Tables 10.2 and 10.3) data have been combined from five of the mitochondrial protein-coding genes: two of the seven subunits of NADH dehydrogenase and one representative from each of the other three complexes providing a broad, though possibly not unbiased sampling of the mitochondrial genome. To take data from the entire genome may, in fact, lead to an even worse bias by accentuating possible distortion due to the specific concerted evolution of the NADH dehydrogenase genes, which make up more than half of the polypeptides encoded.

Two simple methods can be used to score for changes in protein sequences: absolute homology

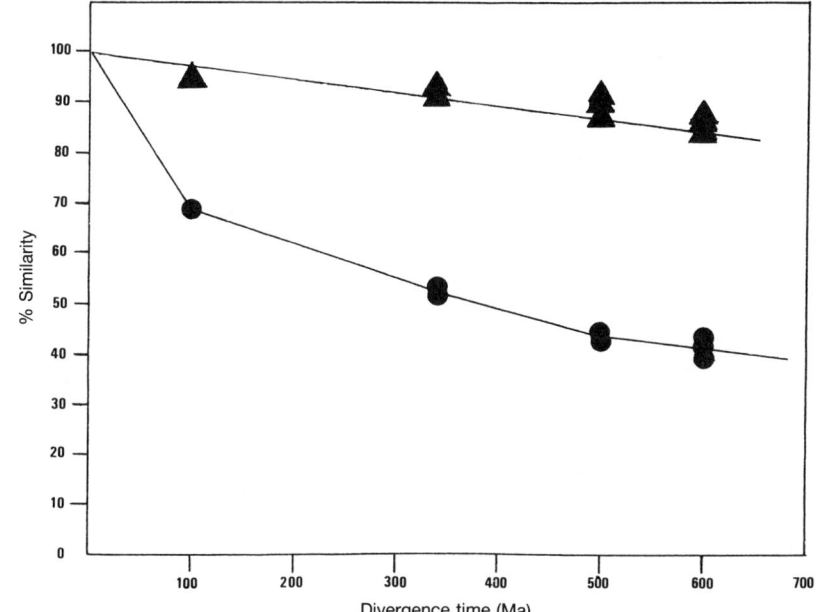

Fig. 10.5 Percentage of amino acid sequence similarities (as shown in Table 10.3, being based on the data of Fig. 10.4 analysed as indicated in Table 10.2), of the conserved domains (triangles) and unconserved domains (circles), as defined in the text, of five mitochondrial genes in five species, plotted against approximate species divergence time.

TABLE 10.2 An arbitrary method for scoring similarity of amino acid sequences. Amino acids are indicated using the standard one-letter code

Assigned numerical score	Feature scored
3	Absolute homology
2	Similarity of functional groups: Acidic—D, E Basic—K, H, R Hydroxyl—Y, S, T Aromatic—Y, F, W Kinky—G, P Small hydrophobic—L, M, I, F, V Amine—N, Q
1	Similar charge character: Charged—E, K, H, R, D Neutral polar—C, S, Y, T, A, G, P, N, Q Hydrophobic—W vs. L, M, I, F, V
0	All other substitutions
−3	Gap

TABLE 10.3 Amino acid similarity between five mitochondrial genes in five species (based on sequence data for ND1, ND4L, cyt b, ATPase 8, and COII shown in Fig. 10.2)

Species compared	% Similarity in		
	Highly conserved domains	Other residues	All residues
Human/Mouse	95	69	80
Human/Xenopus	92	52	69
Mouse/Xenopus	93	54	71
Human/Sea urchin	88	43	61
Mouse/Sea urchin	91	44	64
Xenopus/Sea urchin	90	44	63
Human/Drosophila	88	39	61
Mouse/Drosophila	88	42	62
Xenopus/Drosophila	87	44	62
Sea Urchin/Drosophila	84	42	60
Mean Inter-phylum	88.0 ± 1.9	42.6 ± 1.6	61.9 ± 1.2

(which is self-explanatory) and 'sequence similarity', a somewhat arbitrary measure based either upon theoretical consideration of the role of functional groups or upon an empirical determination of the relative frequencies with which particular residues substitute for each other in actual proteins. There are no universally agreed rules for carrying out such a computation, and a particular problem arises in attributing an appropriate weighting to the sequence gaps that are required to align sequences for maximal homology. However, these imperfections diminish as larger samples are analysed.

The method adopted (see Table 10.2) is a simpler version of a scheme used by Li et al. (1984). The similarity between residues at each particular position is assigned a value on a scale of 0–3, based on actual homology, equivalence of functional groups or the similarity of charge characteristics. A weighting of -3 (i.e. equivalent to one homologous residue being subtracted) is ascribed to each gap in the sequence. The deduced similarity values (as a percentage) for various pairwise comparisons of mitochondrial genes from different organisms are within an acceptable range of variation when organisms diverged by the same length of time are compared, e.g. mouse/Xenopus, human/Xenopus (see Table 10.3). Sea urchins, insects, and vertebrates are all observed to differ from each other by approximately the same degree (62 per cent overall similarity), and the standard errors for equivalent pairwise comparisons are remarkably small. Despite this, the slightly lower mean value obtained in comparing Drosophila with all four deuterostomes (61.3 per cent) as against sea urchins with all three vertebrates (62.7 per cent) is not statistically significant. This method is unable, therefore, to show unambiguously that the presumed common deuterostome ancestor diverged from the line leading to the protostomes significantly before the deuterostome phyla diverged from one another.

One reason for this apparent insensitivity of the method is that, within the coding regions analysed, the relative divergence rates of different domains are not equivalent. Whilst the amount of change in highly conserved regions remains minimal, other segments have effectively reached saturation with (allowable) mutations. The former should still be accumulating mutations as an approximately linear function of time, whereas changes amongst the latter group have begun experiencing multiple hits long before the 500–600 Ma which separate the phyla. The effect of the latter may be regarded as 'background noise' against the signal, deriving from the former. To examine whether the perception of 'signal' may be improved by filtering out this 'noise', we have arbitrarily reconsidered the data by dividing the regions analysed into highly conserved domains (which we define as stretches of residues where four out of each five consecutive residues are held in common amongst at least four of the five species analysed), and the remaining, less conserved segments. In pairwise comparisons for similarity, the highly conserved domains now do show a small but statistically significant difference between the *Drosophila* and deuterostome sequences, albeit with rather higher proportionate standard errors, presumably due to the inadequacies of the rather simplistic scoring method. Deuterostome sequences show a similarity of 89.7 ± 1.2 per cent, whereas the *Drosophila* sequence is 86.7 ± 1.5 per cent similar to the deuterostome sequences.

When the calculated values are plotted against (approximate) divergence time (Fig. 10.5) it is clear that the arbitrary division into highly conserved and less conserved domains does indeed distinguish two components of change, the one showing an approximately linear relationship with time even at the longest divergences plotted, the other showing a saturation effect at divergence times considerably below that for distinct phyla.

Thus, it can be seen that a single method, the sequencing of mtDNA, can be used to deduce phylogenetic relationships at genus/family level (using silent nucleotide substitutions) at family/order level (using the rate of amino acid replacement in the relatively less well conserved regions of proteins), and at class (and perhaps phylum) level (using the amino acid replacement rate in highly conserved domains).

One potential flaw in this approach arises from the strictly maternal inheritance of mtDNA. Attempts to detect paternal transmission of mtDNA in animals have proved totally negative, even where the progeny were repeatedly backcrossed to the paternal line over 91 generations (Lansman *et al.* 1983). This suggests that a freak inter-specific mating, followed by repeated paternal backcrossing would eliminate all trace of nuclear genes from the original maternal parent, but leave the maternal mtDNA genotype intact, resulting in an interspecific flow of mtDNA. This has indeed been demonstrated to have occurred in wild mice (Ferris *et al.* 1983). Since, in the laboratory, successful, though not necessarily fertile, hybrids have occasionally been raised between sea urchin genera diverged by at least 20 Ma, e.g. *Psammechinus* and *Paracentrotus* (Horstadius 1974), the possibility that such events occur in the wild cannot be entirely discounted.

Such events could invalidate estimates of mtDNA sequence divergence derived from some intergeneric comparisons, though not from comparisons between higher taxa.

Overall, the results of our studies have led us to conclude that the elucidation of mitochondrial gene order, especially when used in conjunction with mitochondrial gene sequence data, provides a highly promising new tool to assist in deriving an objective phylogeny of the echinoderms and ultimately of the whole animal kingdom.

ACKNOWLEDGEMENTS

HTJ thanks Dave Sherratt and Richard Wilson for useful discussions. HTJ is the Royal Society Shell Research Fellow. Both HTJ and LC were in receipt of EMBO short term fellowships which contributed to this work. BLC and PB acknowledge the assistance of Moyra Cohen and financial support from NERC and the University of Glasgow. We are also grateful for the contributions of undergraduate project students Carin Campbell, Susan Tweedie, Jacqueline Wilkinson, Mark Allan, David Matthews, and Karen Russell, the graphics of Roger van Wyk, and the secretarial assistance of Linda Kerr and Margaret White.

REFERENCES

Anderson, S., *et al.* 1981. Sequence and organisation of the human mitochondrial genome. *Nature, London* **290**, 457–65.

Anderson, S., de Bruijn, M. H. L., Coulson, A. R., Eperon, I. C., Sanger, F., and Young, I. G. 1982. Complete sequence of bovine mitochondrial DNA. Conserved features of the mammalian mitochondrial genome. *Journal of Molecular Biology* **156**, 683–717.

Angerer, R. C., Davidson, E. H., and Britten, R. J. 1976. Single-copy DNA and structural gene sequence relationships among four sea urchin species. *Chromosoma* **56**, 213–26.

Bibb, M. J., Van Etten, R. A., Wright, C. T., Walberg, M. W., and Clayton, D. A. 1981. Sequence and gene organisation of mouse mitochondrial DNA. *Cell* **26**, 167–80.

Borst, P. and Grivell, L. A. 1978. The mitochondrial genome of yeast. *Cell* **15**, 705–21.

Britten, R. J., Graham, D. E., and Neufeld, D. R. 1974. Analysis of repeating DNA sequences by reassociation. In *Methods in enzymology*, Vol. 29E (ed. L. Grossman and K. Moldave), pp. 363–406. Academic Press, New York.

Brown, W. M. 1983. Evolution of animal mitochondrial DNA. In *Evolution of genes and proteins* (ed. R. K. Koehn and M. Nei), pp. 62–8. Sinauer Associates, Sutherland, Massachusetts.

Cann, R. L., Brown, W. M., and Wilson, A. C. 1984. Polymorphic sites and the mechanism of evolution of human mitochondrial DNA. *Genetics* **106**, 479–99.

Capaldi, R. A. 1982. Arrangement of proteins in the mitochondrial inner membrane. *Biochimica et Biophysica Acta* **694**, 291–306.

Childs, G., Maxson, R., Cohn, R., and Kedes, L. H. 1981. Orphons: dispersed genetic elements derived from tandem repetitive genes of eukaryotes. *Cell* **23**, 651–63.

Chomyn, A., *et al.* 1985. Six unidentified reading frames of human mitochondrial DNA encode components of the respiratory chain NADH-dehydrogenase. *Nature, London* **314**, 592–7.

Clary, D. O. and Wolstenholme, D. R. 1985. The mitochondrial DNA molecule of *Drosophila yakuba*: nucleotide sequence, gene organisation and genetic code. *Journal of Molecular Evolution* **22**, 252–72.

Davidson, E. H. 1986. *Gene activity in early development*, 3rd edn. Academic Press, New York.

Dover, G. 1986. Molecular drive in multigene families: how biological novelties arise, spread and are assimilated. *Trends in Genetics* **2**, 159–65.

Dujon, B. 1980. Sequence of the intron and flanking exons of the mitochondrial 21S rRNA gene of yeast strains having different alleles at the ω and *rib*-1 loci. *Cell* **20**, 185–97.

Ferris, S. D., Sage, R. D., Huang, R-C., Nielsen, J. T., Ritte, U., and Wilson, A. C. 1983. Flow of mitochondrial DNA across a species boundary. *Proceedings of the National Academy of Sciences, USA* **80**, 2290–4.

Ferris, S. D., Sage, R. D., Prager, E. M., Ritte, U., and Wilson, A. C. 1983. Mitochondrial DNA evolution in mice. *Genetics* **105**, 681–721.

Fitch, W. M. and Margoliash, E. 1970. The usefulness of amino acid and nucleotide sequences in evolutionary studies. *Evolutionary Biology* **4**, 67–109.

Galau, G. A., Klein, W. H., Davis, M. M., Wold, B. J., Britten, R. J., and Davidson, E. H. 1976. Structural gene sets active in embryos and adult tissues of the sea urchin. *Cell* **7**, 487–505.

Gellissen, G., Bradfield, J. Y., White, B. N., and Wyatt, G. R. 1983. Mitochondrial DNA sequences in the nuclear genome of a locust. *Nature, London* **301**, 631–4.

Grosjean, H. and Fiers, W. 1982. Preferential codon usage in prokaryotic genes: the optimal codon-anticodon interaction energy and the selective codon usage in efficiently expressed genes. *Gene* **18**, 199–209.

Hauswirth, W. W. and Laipis, M. J. 1985. Transmission genetics of mammalian mitochondria: a molecular model and experimental evidence. In *Achievements and perspectives of mitochondrial research*, Vol. 2, (ed. E. Quagliariello *et al.*) pp. 49–59. Elsevier, North Holland, Amsterdam.

Horstadius, S. 1974. *Experimental embryology of echinoderms*. Clarendon Press, Oxford.

Jacobs, H. T. and Grimes, B. 1986. Complete nucleotide sequences of the nuclear pseudogenes for cytochrome oxidase subunit I and the large mitochondrial ribosomal RNA in the sea urchin *Strongylocentrotus purpuratus*. *Journal of Molecular Biology* **187**, 509–27.

Jacobs, H. T., *et al.* 1983. Mitochondrial DNA sequences in the nuclear genome of *Strongylocentrotus purpuratus*. *Journal of Molecular Biology* **165**, 609–32.

Jacobs, L. L. and Pilbeam, D. 1980. Of mice and men: fossil based divergence dates and molecular 'clocks'. *Journal of Human Evolution* **9**, 551–5.

Kearsey, S. E. and Craig, I. W. 1981. Altered ribosomal RNA genes in mitochondria from mammalian cells with chloramphenicol resistance. *Nature, London* **290**, 607–8.

Kimura, M. and Ohta, T. 1972. On the stochastic model for estimation of mutational distance between homologous proteins. *Journal of Molecular Evolution* **2**, 87–90.

Lansman, R. A., Avise, J. C., and Huettel, M. D. 1983. Critical test of the possibility of 'paternal leakage' of mitochondrial DNA. *Proceedings of the National Academy of Sciences, USA* **80**, 1969–71.

Lewin, R. 1983. Promiscuous DNA leaps all barriers. *Science, London* **219**, 478–79.

Li, W. H. and Gojorbi, T. 1983. Rapid evolution of goat and sheep globin genes following gene duplication. *Molecular Biology and Evolution* **1**, 94–108.

Li, W.-H., Wu, C.-I., and Luo, C.-C. 1984. Nonrandomness of point mutation as reflected in nucleotide substitutions and its evolutionary implications. *Journal of Molecular Evolution* **21**, 58–71.

Li, W.-H., Wu C.-I., and Luo, C.-C. 1985. A new method for estimating synonymous and nonsynonymous rates of nucleotide substitution considering the relative likelihood of nucleotide and codon changes. *Molecular Biology and Evolution* **2**, 150–74.

Margoliash, E. 1963. Primary structure and evolution of cytochrome c. *Proceedings of the National Academy of Sciences, USA* **50**, 672–9.

Mariottini, P., Chomyn, A., Attardi, G., Trovato, D., Strong, D. D., and Doolittle, R. F. 1983. Antibodies against synthetic peptides reveal that the unidentified reading frame A6L, overlapping the ATPase 6 gene, is expressed in human mitochondria. *Cell* **32**, 1269–77.

Petes, T. and Fink, G. R. 1982. Gene conversion between repeated genes. *Nature, London* **300**, 216–17.

Roe, B. A., Ma, D.-P., Wilson, R. K., and Wong, J. F-H. 1985. The complete nucleotide sequence of the *Xenopus laevis* mitochondrial genome. *Journal of Biological Chemistry* **260**, 9759–74.

Saccone, C., *et al.* 1983. Rat mitochondrial DNA:

evolutionary considerations based on nucleotide sequence analysis. In *Mitochondrial genes* (ed. P. Slonimski, *et al.*), pp. 121–8. Cold Spring Harbor Laboratory.

Schwartz, R. M. and Dayhoff, M. O. 1978. Cytochromes. In *Atlas of protein sequence and structure*, Vol. 5, Suppl. 3 (ed. M. O. Dayhoff), pp. 29–44. National Biomedical Research Foundation, Washington DC.

Smith, A. B. 1984. *Echinoid palaeobiology*. George Allen and Unwin, London.

Solignac, M., Monnerot, M., and Mounoulou, J-C. 1986. Mitochondrial DNA evolution in the *melanogaster* species subgroup of *Drosophila*. *Journal of Molecular Evolution* **23**, 31–40.

Varmus, H. E. 1982. Form and function of retroviral proviruses. *Science, New York* **216**, 812–20.

Wilson, A. C., Carlson, S. S., and White, T. J. 1977. Biochemical evolution. *Annual Review of Biochemistry* **46**, 573–639.

Wolstenholme, D. R. and Clary, D. O. 1985. Sequence evolution of *Drosophila* mitochondrial DNA. *Genetics* **109**, 725–44.

Wong, T. W. and Clayton, D. A. 1986. DNA primase of human mitochondria is associated with structural RNA that is essential for enzymatic activity. *Cell* **45**, 817–25.

Wyman, A. R. and White, R. 1980. A highly polymorphic locus in human DNA. *Proceedings of the National Academy of Sciences, USA* **77**, 6754–8.

11

What molecular biology tells us about the genomic programme for development

ERIC H. DAVIDSON

*Division of Biology, California Institute of Technology,
Pasadena, CA 91125*

INTRODUCTION: EVOLUTION AND
DEVELOPMENTAL GENE REGULATION

Novel biological forms must arise in phylogeny primarily through alterations in the genome that affect developmental programmes of gene regulation. This proposition appears, at present, altogether self-evident, although despite its simple logical force it did not seem so 15 years ago when first considered specifically from molecular and genomic points of view (Britten and Davidson 1971). Differential patterns of gene activity in various specialized cell types that have come into being during development have now been extensively described, and the field is converging on some general insights into the molecular processes by which developmental gene regulation occurs. With *Drosophila*, the sea urchin is at present among the organisms for which knowledge of the molecular basis of early embryonic development is most advanced. The following brief, and rather selective review of some of the salient points that have emerged may serve to improve our focus on the kinds of genomic change required for the echinoderm evolution described in this conference to have taken place. Unfortunately, most fossil evidence concerns adult echinoderm forms, while most molecular evidence concerns the initial organization of the embryo, the direct product of which (in most species) is the larva. As discussed below, early embryonic development and the morphogenesis of the adult form in the larval imaginal rudiment are, in certain respects, very different processes.

Developmental gene regulation

The essential subject of this essay refers in the following specifically to those genomic control mechanisms by which differential programmes of gene expression are instituted and maintained in appropriate cell lineages or cell types during ontogeny. It is now beyond reasonable debate that differential gene activity is directly responsible for the diversification of function displayed by the various kinds of cells of which an animal organism is constituted. The sea urchin embryo provides a number of excellent examples in which particular structural genes have been identified that endow given cell types with their specific functional capacities. Except that they exist, and except for some molecular evidence regarding the mode of action of their products, we know much less about the genes responsible for the regulation of gene expression in embryonic time and space. The initial stage of the process by which a spatially diverse pattern of differential gene function appears early in development is called *specification*. By this event the progenitors of given cell lineages, or regional patches of previously undifferentiated cells, are first set onto the course that will ultimately result in transcriptional activation of those genes required for the particular functional behavior of their progeny. The state of specification may often remain for a time plastic and reversible, e.g. as demonstrated if embryonic cells are placed in ectopic positions. However, under normal circumstances these cells (or their progeny) ultimately become *committed* or *determined*, even though transcription of the genes to be activated may not yet have occurred. This paradigm implies that spatial differentiation results from a process in which sequence-specific, selective

---

Echinoderm phylogeny and evolutionary biology (ed. C. R. C. Paul and A. B. Smith). Clarendon Press, Oxford, 1988.

activation rescues given genes in given cells from a general, default state of repression. The mechanism of activation is, in the cases so far analysed, interaction with diffusible, i.e., *trans*-acting proteins, that bind to specific regulatory 'receptor' sequences in the immediate vicinity of their target genes (i.e. the *cis*-regulatory sequences). It is to be stressed, however, that only a minor fraction of active genes are spatially regulated in the early embryo, in the sense that their products appear in some regions and not in others. At least in the sea urchin embryo, these genes all belong to the set that is initially quiescent, and is activated as development proceeds (Angerer and Davidson 1984; Davidson 1986). Such genes account for only about 10 per cent of the total complexity of the set of genes expressed in the embryo (Davidson et al. 1982). The great majority of expressed genes are initially transcribed during oogenesis, are represented in stored maternal RNA, and are also expressed by the embryo genomes, until development has advanced to the stage where their products are no longer required.

In considering later stages of ontogeny it becomes particularly clear that developmental gene regulation is at least a two-stage process. Thus, though interaction with *trans*-activators in terminally differentiated cell types is necessary for the expression of their respective tissue specific genes, it is often not sufficient. An additional and prior process, occurring during the ontogeny of each cell lineage, is required to place these genes in a state such that they are able to respond to their terminal stage *trans*-activators. This accounts for the fact, observed in many gene transfer experiments, that exogenous genes or gene fragments introduced into various cells may be activated for transcription, while the homologous endogenous genes of the same cells, which have undergone an 'ontogenic experience', remain silent until the cell has achieved its fully differentiated state (see Davidson 1986, for general review of these aspects of developmental gene regulation).

Developmental gene regulation clearly includes features beyond those required for what might be termed *physiological* gene regulation, as opposed to *ontogenic* or developmental gene regulation. Many genes are needed for maintenance of homeostasis, e.g. genes that respond to extracellular signals such as heat shock; the advent of toxic heavy metals or organic compounds; genes that respond to nutrient depletion or excess; and genes that are controlled directly by hormonal signals or other triggers of external origin, often acting at the cell surface. When the external signal is removed these genes are shut off (or turned on, depending on whether their control is positive, negative, or both), and the system shortly reverts to its initial state. This is rarely true of the differential, cell type specific patterns of gene activity that arise during animal ontogeny. Physiological gene regulation occurs at all grades of cellular organization, including bacteria, yeasts, animals, and plants, and in some relatively simple organisms, such as cellular slime molds, it is utilized for short term differentiation-like processes. Thus, for example, in the process of aggregation and fruiting body formation in *Dictyostelium* a set of genes is activated in response to external cAMP that is not expressed in the vegetative cells (e.g. Gomer et al. 1986). The developmental gene regulation mechanisms of animals and plants go beyond physiological modulation, in part because they result in commitment, and thus may be utilized to programme unidirectional and usually irreversible changes in structure, form, and function. The appearance of the underlying molecular mechanisms early in evolution can be regarded as a fundamental step required for the ontogeny of complex multicellular organisms.

THE EMBRYO-TO-LARVA AND RUDIMENT-TO-JUVENILE PROGRAMMES FOR ECHINODERM DEVELOPMENT

An initial consideration is the *size* of the genomic programme for development, i.e. the overall amount of protein coding genomic information it includes. The first molecular measurements that approach this point directly were carried out on sea urchin embryos. The *complexity* of the mRNA undergoing translation on polysomes of embryos at various stages of development were derived from mRNA excess single-copy DNA hybridizations by Galau et al. (1974, 1976; Hough-Evans et al. 1977). These complexity measurements yielded the result that about 10^4 different messages are required, or at least are being utilized, in the developing embryo. All but about 10 per cent of these mRNA species are represented at similar levels in the maternal RNA stored in the unfertilized egg. The embryo genomes in general transcribe the same genes as are utilized to provide the maternal mRNA, but also activate about 10^3 other genes, the products of which are absent or present only at very low levels in the maternal RNA. This conclusion has been confirmed in studies carried out on embryo cDNA clone populations (Lasky *et al.* 1980; Flytzanis *et al.* 1982). A similar pattern of events obtains in other species, including *Drosophila* and *Xenopus* (reviewed by Davidson 1986). Genetic and molecular evidence converge on a slightly lower value of about 6000 genes required for the embryonic development of *Drosophila* (see calculation of Davidson 1986). Figure 11.1 shows that over a range >30 in genome size, the complexity of the maternal RNA is more or less a constant for an assemblage of unrelated creatures, only *Drosophila* displaying a slightly smaller value.

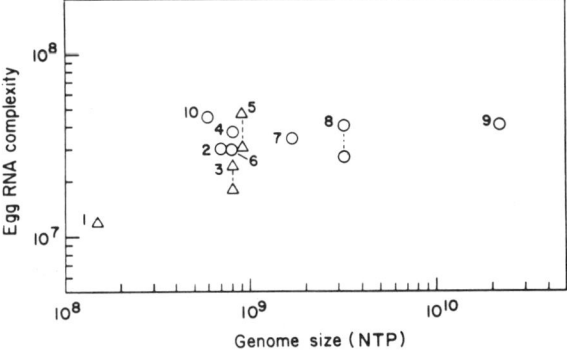

Fig. 11.1 Complexity of egg RNA in various species, as a function of genome size. Genome sizes (haploid) are given in Hough-Evans *et al.* (1980). Two points connected by a dashed line indicate independent determinations. Both protostomes (△) and deuterostomes (○) are represented. 1, *Drosophila melanogaster*; 2, *Arbacia punctulata*; 3, *Musca domestica*; 4, *Strongylocentrotus purpuratus*; 5, *Urechis caupo*; 6, *Lytechinus pictus*; 7, *Tripneustes gratilla*; 8, *Xenopus laevis*; 9, *Triturus cristatus*; 10, *Pisaster ochraceus*. [From Davidson (1986).]

The most general conclusions that can be drawn from all the measurements of embryonic complexity that have been made are that development from the egg is a process that is extremely expensive in terms of genomic information, that always requires many thousands of genes; and that is controlled by a complex regulatory programme, the function of which is to organize the transcriptional activity of this large number of genes in developmental time and space. In whatever way these genes are divided up into regulatory sets, or 'gene batteries' (Morgan 1934; Britten and Davidson 1969), there must be many highly pleiotropic control elements. These elements would constitute a large group of genomic targets at which certain kinds of evolutionary alterations could result in potent developmental effects (Britten and Davidson 1971).

The functional significance of the relatively enormous absolute values for embryo mRNA complexity remains obscure. Why are the products of $\sim 10^4$ different genes needed to construct an embryo? An hypothesis put forth a decade ago (Hough-Evans *et al.* 1977) suggests that the explanation might lie in the large amount of genetic information required to build three-dimensional morphological structures. This is illustrated by several well-studied cases, e.g. the chorion of silk moth eggs, which has been shown to contain at least 180 different proteins (Regier *et al.* 1980), and the ubiquitous flagellum, which is composed of at least 120 different proteins (Piperno *et al.* 1977).

Relatively little is known of the genomic programme called into operation in the development of the imaginal rudiment from which the juvenile sea urchin is formed. Rudiment development probably occurs very differently from embryonic development, though of course both ultimately rely on differential gene expression. In the embryo spatial differentiation is set up in the *absence of any net growth* within a cytoplasmic matrix preformed during oogenesis. At least one axis of this matrix, the vertical, or animal-vegetal axis, is definitely imposed during oogenesis. The initial set of spatial differentiations in the embryo probably depends both on inheritance of given sectors of topologically localized egg cytoplasm, and on inductive interactions occurring amongst the early blastomeres. In contrast, the imaginal rudiment develops within the larva from pouches of actively dividing cells that can transfer to it no preformulated spatial information, since these pouches form from progeny of single, free-wandering mesenchyme cells, and of a few other cells carried into the anterior region of the embryo on the tip of the archenteron. Thus, rudiment morphogenesis must depend mainly on cellular interactions that set up *de novo* the initial radial polarizations of its structure, and then subdivide it into regions of different fate. We can imagine that somehow these regional distinctions select among various possible patterns of differential genomic function. A similar mechanism is probably engaged in the initial development of the mammalian embryo from the pluripotential cells of the inner cell mass. No self-organizing system of this nature has yet been analysed at the gene level, and it is impossible to more than guess at what the genomic requirements might be.

The echinoderm regulatory programme for rudiment morphogenesis and that for embryonic and larval development appear to be separate. Thus, sea urchin species that develop directly never produce a feeding larva, instead forming the rudiment immediately from embryonic mesenchyme cells (Mortensen 1921; Williams and Anderson 1975; Raff et al. 1984). These species must lack functional versions of the later portions of the programme of embryonic development that normally operate in their indirectly developing congeners, and in most other sea urchin groups. Another kind of evidence concerns the structural genes that are included by the embryo-to-larva and the rudiment-to-juvenile programmes. We found, for example, that the cytoskeletal actin gene CyIIIa is utilized only in the embryo and larva, and not in any tissue of the juvenile (Shott et al. 1984). Similarly, the well known 'early histone' genes, the 'cleavage stage' histone genes, and several of the 'late' histone genes are utilized only in the embryo (reviewed by Hentschel and Birnstiel 1981; Davidson 1986; see e.g. Lieber et al. 1986). The CyIIIa actin gene codes for a protein that differs by only 1–2 per cent in sequence from other cytoskeletal actins, and the H3 and H4 early histones differ not at all in sequence from the respective late histones utilized in adult tissues. The main distinguishing features of these genes can be regarded as their inclusion in the embryonic regulatory programmes. Conversely, other genes are utilized exclusively after metamorphosis, in tissues of the juvenile and adult, e.g. the sperm histone genes (Busslinger and Barberis 1985; Lieber et al. 1986); the gene that codes for the sperm protein binding (Gao et al. 1986); and the gene that codes for yolk proteins (Shyu et al. 1986).

Many genes are of course called on in both sets of programmes (Galau et al. 1976). Some specific examples include others of the actin gene family (Shott et al. 1984); and a gene utilized in the production of embryonic spicules (Benson et al. 1987; Sucov et al. 1987), which is also utilized in adult sea urchin spines (F. Wilt, personal communication).

SPATIAL SPECIFICATION OF GENE
EXPRESSION IN THE EMBRYO

Many experiments, both classical and modern, that cannot be reviewed here, indicate that at the beginning of development the nuclei of the sea urchin embryo are mutually equivalent (see Davidson 1986). By the 100–200 cell early blastula stage, however, the embryo is already a regionally determined mosaic of differential patterns of gene expression. The spatial location of these regions can be understood directly in terms of their cell lineage. Thus, each region consists of the clonal progeny of a small number of founder cells, that arise early in cleavage. There is no cell migration at the early stages, and the location of each region of the blastula follows simply from the position of its founder cells in the spherical late cleavage morula. The relation between the initial cell lineage and the fate and function of the progeny of the different lineage elements is diagrammed in Fig. 11.2. This figure is based both upon classical sources (Selenka 1883; Boveri 1901; Hörstadius 1939), and upon recent experiments performed in our laboratory, in which a fluorescent lineage tracer was injected into individual blastomeres of the eight-cell stage and the disposition of their progeny was determined later in development (Cameron et al. 1987). The fundamental issue in understanding this process is, of course, the molecular mechanism by which the initial lineage-specific spatial patterns of gene regulation are established.

Portions of the genomic programme responsible are evidently 'read out' during oogenesis. Thus, species hybrid and other experiments show that the timing and the position of the early cleavages

Fig. 11.2 A diagram of the first four cleavages in the sea urchin embryo illustrating cell fates at the mesenchyme blastula and pluteus stages. Although the timing differs among species, the patterns of the early lineage, the sequence of division cycles and the cell fates are the same among those species that have been studied (reviewed in Okazaki 1975). This is a simplified version of the diagram published in Davidson (1986) wherein can be found a complete discussion of the derivation of the diagram. The symbols used are as follows: A, a, aboral; L, lateral; M, macromere; *M*, micromere; N, animal; O, o, oral; V, vegetal. In the nomenclature used wholly capital designations indicate founder cells that will give rise to cell lineages of several diverse fates, as perceived at the late (mesenchyme) blastula stage, while cell names including a lower case letter designate a progenitor of a clone of cells of identical fate (so far as is

Genomic programme for development   143

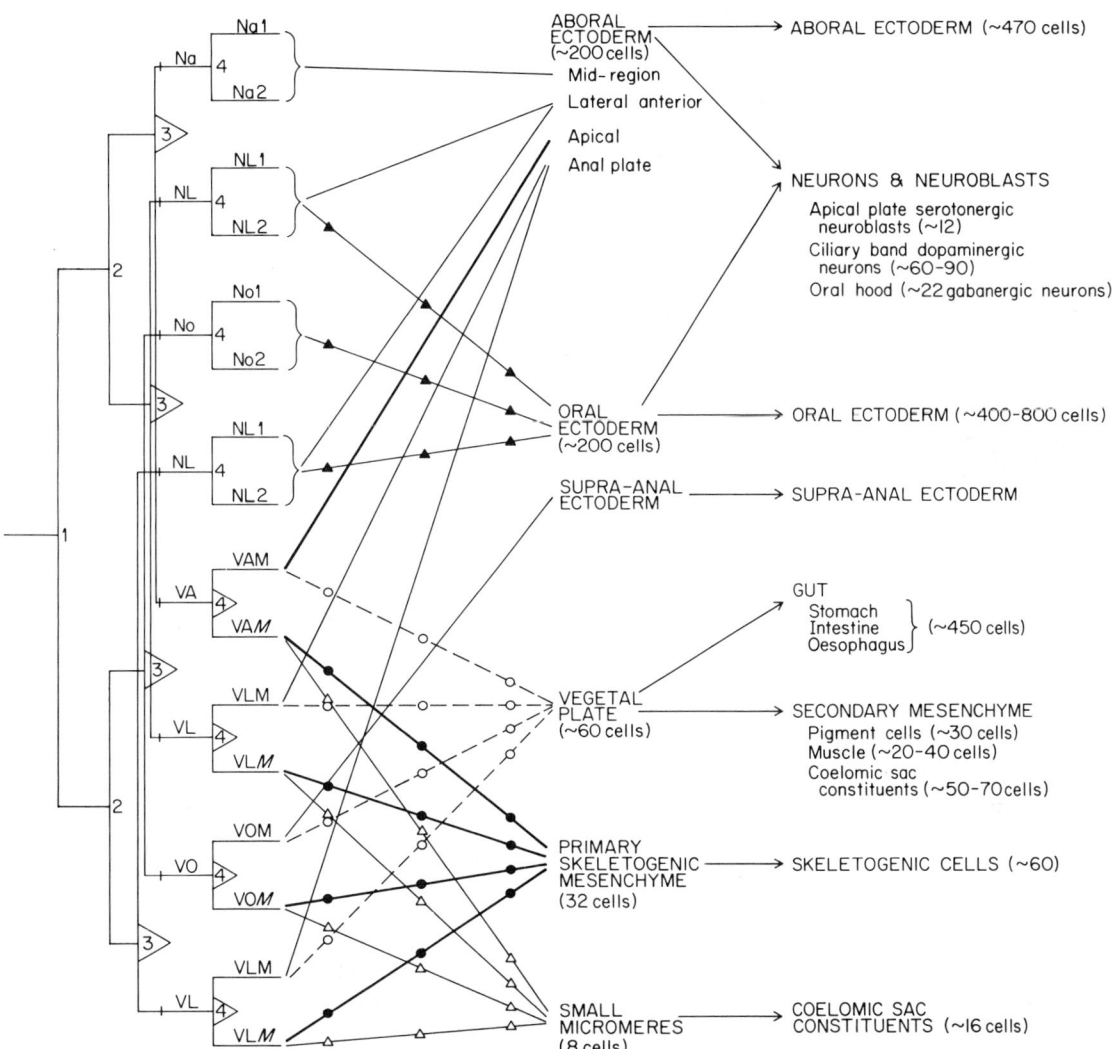

known). Thus for example all the descendants of the Na blastomere give rise exclusively to aboral ectoderm cells (see Fig. 4.5 of Davidson 1986). Each horizontal line represents a cell and each vertical line a division. The number of each cleavage (i.e. first-fourth) is given, and those divisions in which the cleavage plane is horizontal are marked with a triangle around the cleavage number. The first two cleavage planes are vertical (i.e. parallel to the animal-vegetal axis, or meridional). The third cleavage plane is horizontal (perpendicular to the animal-vegetal axis, or equatorial). In the animal quartet the fourth cleavage is vertical, producing the mesomeres (Na1, Na2, the two NL1 and the two NL2 cells, and No1 and No2). In the vegetal quartet, the fourth cleavage is horizontal and unequal, producing the macromeres (VAM, the two VLM cells, VOM) and the micromeres (VAM, the two VLM cells, and VOM). In the next two cleavages the macromeres first divide vertically and then horizontally, so that at the sixth division clones of the lower tier of cells contribute exclusively to the vegetal plate [named the veg$_2$ lineage by Hörstadius (1939, 1973)] and clones of the upper tier of cells contribute exclusively to the ectoderm [called the veg$_1$ lineage by Hörstadius (1939, 1973)]. Approximate cell numbers for each cell type later in development are indicated in parentheses. [From Cameron *et al.* (1987).]

by which the lineage elements displayed in Fig. 11.2 are established are a property of the egg cytoplasm (reviewed in Davidson 1986). Furthermore, the animal-vegetal polarization of the egg, which as noted above is clearly also maternal, has direct consequences in respect to the patterns of gene expression. The simplest example of this is the set of blastomeres that give rise to the primary skeletogenic mesenchyme (VAM, VLM, VLM, and VOM, the 'micromeres'). These cells derive from the vegetal pole of the eggs, and their skeletogenic progeny express a certain specific set of genes, some of which have been identified. An example is shown in Plate 11.1. Thus, it is an attractive idea that the maternal cytoplasm at or near the vegetal pole contains maternal regulatory factors that specifically activate those structural genes required for the initial stages of skeletogenic mesenchyme differentiation.

A different model is provided by some recent studies in which we have been engaged that concern the CyIIIa cytoskeletal actin gene, mentioned above. This gene functions in aboral ectoderm cells of the embryo and larva, as shown, for example, in Plate 11.1b. CyIIIa is one of at least ten different genes (two coding for cytoskeletal actins, and eight for Ca^{2+} binding proteins) that function exclusively in these cells (reviewed in Davidson 1986). The hypothetical concept of 'developmental gene batteries' is thus in this case clearly a reality. The aboral ectoderm cells descend from six clonal progenitors (see Fig. 11.2). The oral-aboral axis of the embryo may not be specified until the eight-cell stage, and the process of specification could depend on a polarizing inductive interaction occurring horizontally between blastomeres on opposite sides of the embryo (Davidson 1986; Cameron et al. 1987). Though the regulatory factors that control CyIIIa gene activation may also be maternal in origin, their distribution, activation, and/or presentation must be affected by the events which lead to specification of this axis. In any case, the existence of sequence specific regulatory factors that are responsible for spatial and temporal CyIIIa gene expression in the embryo is directly implied by the results of recent experiments. We constructed a fusion gene consisting of a bacterial 'reporter' gene coding for chloramphenicol acetyl transferase (CAT) ligated to upstream CyIIIa DNA sequences that we presumed should include its *cis*-regulatory elements. After injection of this fusion gene into eggs, CAT gene expression is activated in the early blastula, almost exactly when the endogenous CyIIIa gene is activated (Flytzanis et al. 1987). The CAT mRNA appears only in aboral ectoderm cells. This is shown by *in situ* hybridization using a probe complementary to CAT message, as illustrated in Plate 11.1c. These experiments in themselves indicate that *spatial* and *temporal* control of the CyIIIa gene in early development depends on interactions of *cis*-regulatory sequences with diffusible *trans*-acting regulatory factors. The *cis*-regulatory sequences have been further identified by deletion and other experiments (Flytzanis et al. 1987). In addition we have identified eleven different sites where interactions have been detected in DNA-binding proteins that display 10^4–10^5-fold specificity for particular CyIIIa sequences located within the upstream regulatory regions (Calzone et al. 1987; submitted for publication).

What conclusions may be drawn regarding the genomic programme for development? Generalization seems warranted, since although there is yet little comparable information for other genes that are regionally active at the very beginning of the life cycle, at a basic level the mechanisms emerging seem in essence similar to those operating at the final stages of gene activation in terminally differentiating cells. Thus, the *cis* genomic information required to activate given genes is encoded in their immediate vicinity, while the sequence-specific *trans*-activators to which the gene responds are encoded elsewhere. What is special to the embryo, and not yet understood is the mechanism by which the products of these regulatory genes are *spatially* distributed or presented in the early embryo. Gene activation is seen to be a pleiotropic, battery level process, and it still seems reasonable to imagine, as in our earlier speculations (Britten and Davidson 1969, 1971; Davidson and Britten 1971; Davidson and Posakony 1982) that the molecular basis for developmental co-ordination at the battery level lies in homologies among the *cis* receptor sequences of genes that are to be activated in concert.

As these aspects of developmental regulatory mechanisms are clarified, hypothetical speculations on the nature of evolutionary genomic changes that might alter embryonic forms can be sharpened. Given that there are *cis*-regulatory sequences that confer response to spatially localized *trans*-acting factors, it is possible, perhaps inevitable, that such sequences will be tested during evolution by random translocation, in positions where they could bring other structural genes into play in particular spatial domains or cell lineages of the embryo. A pleasing aspect that could not have been conceived a decade ago is that it should soon be possible to demonstrate this kind of process experimentally, by deliberately positioning a spatial control sequence conferring expression in a given region near an ontogenically regulated gene not normally active in that region. This is essentially the experiment shown in Plate 11.1c, except that CAT is not a sea urchin gene normally utilized elsewhere during development. The most powerful forms of genomic change are of course potentially those that would affect the genes coding for the pleiotropic regulatory factors themselves, their level and time of expression, and the localization of their products. Such genes will soon be cloned and characterized. I harbour the optimistic view that the venerable problem of genomic control of cell specification in the embryo will be solved in principle in the not too distant future. There will then be opened far reaching pathways to understanding evolutionary change in developmental programmes, ultimately including the recreation, at least of certain examples, in the laboratory.

ACKNOWLEDGEMENTS

Research from this laboratory was supported by NIH grant HD-05753.

REFERENCES

Angerer, R. C. and Davidson, E. H. 1984. Molecular indices of cell lineage specification in sea urchin embryos. *Science, New York* **226**, 1153–60.

Benson, S., Sucov, H. M., Stephens, L., Davidson, E. H., and Wilt, F. 1987. Lineage-specific gene encoding a major matrix protein of the sea urchin embryo spicule. I. Authentication of the cloned gene and its developmental expression. *Developmental Biology* **120**, 499–506.

Boveri, T. 1901. Die Polarität von Oocyte, Ei und Larve des *Stronglyocentrotus lividus*. *Zoologische Jahrbücher Abteilung für Anatomie und Ontogenie* **14**, 630–53.

Britten, R. J. and Davidson, E. H. 1969. Gene regulation for higher cells: A theory. *Science, New York* **165**, 349–58.

Britten, R. J. and Davidson, E. H. 1971. Repetitive and nonrepetitive DNA sequences and a speculation on the origins of evolutionary novelty. *Quarterly Review of Biology* **46**, 111–38.

Busslinger, M. and Barberis, A. 1985. Synthesis of sperm and late histone cDNAs of the sea urchin with a primer complementary to the conserved 3' terminal palindrome: Evidence for tissue-specific and more general histone gene variants. *Proceedings of the National Academy of Sciences, USA* **82**, 5676–80.

Calzone, F. J., Flytzanis, C. N., Fromson, D. R., Britten, R. J., and Davidson, E. H. 1987. Protein-DNA interactions within regulatory regions required for embryonic activation of the sea urchin CyIIIa actin gene. In *Molecular Approaches to Developmental Biology* (ed. R. A. Firtel and E. H. Davidson), pp. 205–21. A. R. Liss, New York.

Cameron, R. A., Hough-Evans, B. R., Britten, R. J., and Davidson, E. H. 1987. Lineage and fate of each blastomere of the eight-cell sea urchin embryo. *Genes and Development* **1**, 75–85.

Davidson, E. H. 1986. *Gene activity in early development*, 3rd edn. Academic Press, Orlando, Florida.

Davidson, E. H. and Britten, R. J. 1971. Note on the control of gene expression during development. *Journal of Theoretical Biology* **32**, 123–30.

Davidson, E. H. and Posakony, J. W. 1982. Repetitive sequence transcripts in development. *Nature, London* **297**, 633–6.

Davidson, E. H., Hough-Evans, B. R., and Britten, R. J. 1982. Molecular biology of the sea urchin embryo. *Science, New York* **217**, 17–26.

Flytzanis, C. N., Brandhorst, B. P., Britten, R. J., and Davidson, E. H. 1982. Developmental patterns of cytoplasmic transcript prevalence in sea urchin embryos. *Developmental Biology* **91**, 27–35.

Flytzanis, C. N., Britten, R. J., and Davidson, E. H. 1987. Ontogenic activation of a fusion gene introduced into sea urchin eggs. *Proceedings of the National Academy of Sciences USA* **84**, 151–5.

Galau, G. A., Britten, R. J., and Davidson, E. H. 1974. A measurement of the sequence complexity of polysomal messenger RNA in sea urchin embryos. *Cell* **2**, 9–21.

Galau, G. A., Klein, W. H., Davis, M. M., Wold, B. J., Britten, R. J., and Davidson, E. H. 1976. Structural gene sites active in embryos and adult tissues of the sea urchin. *Cell* **7**, 487–505.

Gao, B., Klein, L. E., Britten, R. J., and Davidson, E. H. 1986. Sequence of mRNA coding for bindin, a species-specific sea urchin sperm protein required for fertilization. *Proceedings of the National Academy of Sciences, USA* **83**, 8634–8.

Gomer, R. H., Armstrong, D., Leichtling, B. H., and Firtel, R. A. 1986. cAMP induction of prespore and prestalk gene expression in *Dictyostelium* is mediated by the cell-surface cAMP receptor. *Proceedings of the National Academy of Sciences, USA* **83**, 8624–8.

Hentschel, C. C. and Birnstiel, M. L. 1981. The

organization and expression of histone gene families. *Cell* **25**, 301–13.

Hörstadius, S. 1939. The mechanics of sea urchin development studied by operative methods. *Biological Reviews* **14**, 132–79.

Hörstadius, S. 1973. *Experimental embryology of echinoderms*. Oxford University Press (Clarendon), London and New York.

Hough-Evans, B. R. and Davidson, E. H. 1987. Gene transfer in the sea urchin. *Genetic Engineering* **9**, 1–25.

Hough-Evans, B. R., Wold, B. J., Ernst, S. G., Britten, R. J., and Davidson, E. H. 1977. Appearance and persistence of maternal RNA sequences in sea urchin development. *Developmental Biology* **60**, 258–77.

Hough-Evans, B. R., Jacobs-Lorena, M., Cummings, M. R., Britten, R. J., and Davidson, E. H. 1980. Complexity of eggs of *Drosphila melanogaster* and *Musca domestica*. *Genetics* **95**, 81–94.

Lasky, L. A., Lev, A., Xin, J.-H., Britten, R. J., and Davidson, E. H. 1980. Messenger RNA prevalence in sea urchin embryos measured with cloned cDNAs. *Proceedings of the National Academy of Sciences, USA* **77**, 5317–21.

Lieber, T., Weisser, K., and Childs, G. 1986. Analysis of histone gene expression in adult tissues of the sea urchin *Strongylocentrotus purpuratus* and *Lytechinus pictus*: Tissue-specific expression of sperm histone genes. *Molecular and Cellular Biology* **6**, 2602–12.

Morgan, T. H. 1934. *Embryology and genetics*. Columbia University Press, New York.

Mortensen, T. 1921. *Studies of the development and larval forms of echinoderms*. E. C. Gad, Copenhagen.

Okazaki, K. 1975. Normal development to metamorphosis. In *The Sea urchin embryo: biochemistry and morphogenesis* (ed. G. Czihak), pp. 177–232. Springer Verlag, Berlin and New York.

Piperno, G., Huang, B., and Luck, D. J. L. 1977. Two-dimensional analysis of flagellar proteins from wild-type and paralyzed mutants of *Chlamydomonas reinhardtii*. *Proceedings of the National Academy of Sciences, USA* **74**, 1600–4.

Raff, R. A., et al. 1984. Origin of a gene regulatory mechanism in the evolution of echinoderms. *Nature, London* **310**, 312–14.

Regier, J. C., Mazur, G. D., and Kafatos, F. C. 1980. The silkmoth chorion: Morphological and biochemical characterization of four surface regions. *Developmental Biology* **76**, 286–304.

Selenka, E. 1883. *Die Keimblatter der Echinodermen. Studien über die Entwicklungsgeschichte der Tiere.* 1:2. Wiesbaden.

Shott, R. J., Lee, J. J., Britten, R. J., and Davidson, E. H. 1984. Differential expression of the actin gene family of *Strongylocentrotus purpuratus*. *Developmental Biology* **101**, 295–306.

Shyu, A.-B., Raff, R. A., and Blumenthal, T. 1986. Expression of the vitellogenin gene in female and male sea urchin. *Proceedings of the National Academy of Sciences, USA* **83**, 3856–69.

Sucov, H. M., Benson, S., Robinson, J. J., Britten, R. J., Wilt, F., and Davidson, E. H. 1987. Lineage-specific gene encoding a major matrix protein of the sea urchin embryo spicule. II. Structure of the gene and derived sequence of the protein. *Developmental Biology* **120**, 507–19.

Williams, D. H. C. and Anderson, D. T. 1975. The reproductive system, embryonic development, larval development, and metamorphosis of the sea urchin *Heliocidaris erythrogramma* (Val.) (Echinoidea: Echinometridae). *Australian Journal of Zoology* **23**, 371–403.

PART IV
Ontogeny and phylogeny

12

Heterochrony and the evolution of echinoids

KENNETH J. MCNAMARA

*Western Australian Museum, Francis Street, Perth,
Western Australia*

INTRODUCTION

The relationship between ontogeny and phylogeny, after many years of neglect, is once again attracting interest from biologists and palaeontologists. Historically, the fossil record has played an important role in explaining the causal relationship between the life history of organisms and their evolutionary development. Of early echinoid workers, only Jackson (1912) attempted to assess the nature of the complex interplay between development and evolution. However, in keeping with the predominant view of his day, Jackson saw heterochrony (as the ontogeny/phylogeny relationship has since become known) largely in terms of 'progressive variation' by acceleration in development. In other words, descendant species, by terminal addition, passed through adult growth stages of their ancestors during their own ontogenies. Jackson did recognize that the opposite situation, which he called 'arrested variation', also occurred, but much less frequently.

Although Durham (1955, 1966) and Kier (1965, 1974), have investigated the patterns of evolution in echinoids, there has been little understanding of the processes which might have caused the patterns. None of these general studies of echinoid evolution has investigated the significance of heterochrony in the evolution of the group. The principal reason for this probably lies, to a large degree, in the general disfavour into which heterochrony as an explanation for evolutionary changes had, until recent times, fallen.

The resurgence of interest in heterochrony has focused on a limited number of groups of organisms (see McNamara 1982a, 1986, 1988, for recent examples). Documentation of heterochrony in echinoids has been largely restricted to studies

Echinoderm phylogeny and evolutionary biology (ed. C. R. C. Paul and A. B. Smith). Clarendon Press, Oxford, 1988.

in spatangoids (McNamara 1982b, 1984, 1985, 1987a, 1987b, in press; McNamara and Philip 1980a, 1984) and oligopygoids (McKinney 1984), apart from a recent general analysis of the possible ecological causation of heterochrony in echinoids (McKinney 1986).

To understand the influence of heterochrony in echinoid evolution, it is important to examine the basic mechanisms of coronal growth, in particular the pattern of plate production and the nature of subsequent plate growth. The effect of changes in coronal ontogenetic development over time involves not only structural changes to plates, but also heterochronic changes to the morphologies of the structures which they bear, such as tubercles and pore pairs. A major part of this study will involve analysing the impact of such heterochronic changes in coronal structures on echinoid evolution.

This study will also, hopefully, serve as a guide to the identification of heterochrony in echinoid lineages, by outlining many of the potential changes which may possibly arise by changes to developmental regulation. It will focus on examples from both living species and the fossil record, and will consider only morphological effects of changes to developmental regulation in post-larval echinoids. Developmental changes at the molecular and embryonic level have been more fully dealt with elsewhere (Raff *et al.* 1984; Raff in press; Raff *et al.* this volume).

THE CONCEPT OF HETEROCHRONY

Most recent studies of heterochrony have accepted, with minor emendations, the nomenclatural scheme proposed by Alberch *et al.* (1979; see also McNamara 1986; Domergues *et al.* 1986). Heterochrony, which has been defined (de Beer

1930; Gould 1977) as 'changes in relative time of appearance and rate of development of characters already present in ancestors', may involve extension or contraction of development, termed *peramorphosis* and *paedomorphosis*. These are not themselves processes, but merely the morphological expression of processes.

Three processes may cause paedomorphosis: *progenesis*, which is the phenomenon of precocious sexual maturation resulting in the development of a descendant which is, in many characters morphologically juvenile in appearance and, generally, smaller in size than its ancestor; *neoteny*, where rate of morphological development is reduced throughout ontogeny; and *post-displacement*, where onset of growth is delayed. The peramorphic corollaries of these terms, which result in descendants being morphologically more 'advanced' than their ancestors, are: *hypermorphosis*, where onset of maturation is delayed; *acceleration*, where rate of morphological development is increased; and *pre-displacement*, where onset of growth is advanced.

Without information on the actual timing of size and shape changes it is not always possible to state with certainty that the relationship between size increase and time will always be the same in ancestors and descendants. With this in mind, McKinney (1986) has recently applied the heterochronic terminology to the comparison of size and shape alone, so removing the time component. While heterochrony involves the dissociation of not only shape and time, but also size, the overwhelming importance of size and shape to organisms make it methodologically valid to use the heterochronic terminology in terms of size and shape alone. Future research on the relationship between size and time in echinoids, perhaps using growth line analysis, may allow the three parameters to become fully integrated.

THE NATURE OF CORONAL GROWTH

Plate generation and peripheral accretion

In echinoids, post-metamorphic skeletal growth has a two-tiered pattern. New plates are formed at the apical system and added throughout the life of individual echinoids in all groups, except some spatangoids (Kier 1974). These plates, once incorporated into coronal columns, undergo subsequent plate growth by peripheral accretion.

Raup (1968) has shown how in the regular echinoid *Strongylocentrotus pallidus* peripheral accretion is greater equatorially, while more stereom is deposited proximally than distally. As the test adds new plates which then, to varying degrees, undergo peripheral accretion, the test increases in overall size. The number of plates added during ontogeny varies enormously, from as few as 85 in *Lysechinus incongruens* Gregory, up to 3000 in *Tripneustes ventricosus* (Lamarck) (Kier 1974).

An important aspect to understand when trying to evaluate heterochrony in echinoids is that test plates are of different ages, having been formed at different periods in the life history of the individual, becoming relatively younger toward the apical system. The record of growth of individual plates is preserved in the growth rings. Comparison of growth rings between plates (Deutler 1926) reveals differential growth rates between different plates in an individual. It is therefore possible to assess whether changes in the morphology of successive plates are a function of allometric shape change alone, or whether there has been a real increase in the rate of plate growth; the time component being added to the size and shape factors.

Plate translocation

The number of plates produced in ambulacral and interambulacral columns generally varies substantially. Thus, for instance cidaroids may have only six to eight interambulacral plates, but up to 60 ambulacral plates in each column. Although conventional wisdom has it that ambulacral and interambulacral plates always remain in the same position relative to one another during ontogeny (Deutler 1926; Melville and Durham 1966; Durham 1966; Kier 1974; Märkel 1981; Smith 1984), a recent study (McNamara 1987b) has shown that often, particularly in irregular echinoids, ambulacral and interambulacral plate growth is dissociated, resulting in effective migration of plates in adjacent columns. This has been termed plate translocation (McNamara 1987b). As will be discussed below, not only does plate translocation occur equatorially between adjacent columns, but also meridionally, plates from one column translocating between pairs of plates in adjacent columns, so bisecting them. It is the complex interplay of heterochronic changes to rate of plate production, plate allometries, and plate translocation which

Heterochrony in echinoids 151

are the controlling factors in the diversification of echinoid morphological evolution.

Plate size

The size of plates varies markedly within an individual as well as between species. The youngest, most recently developed plates, are the smallest and least developed morphologically. The smallest plate is the last generated from the ocular plates. However, the oldest plate in each column is not necessarily the largest. Such differential plate growth is perhaps one of the most significant features which has led to the occurrence of heterochrony in echinoids.

In regular echinoids the largest plates are situated at the ambitus (Fig. 12.1A); in spatangoids the largest occur on the adoral surface (Fig. 12.1B); in clypeasteroids, however, the smallest plates occur at the ambitus (Fig. 12.2). It is this

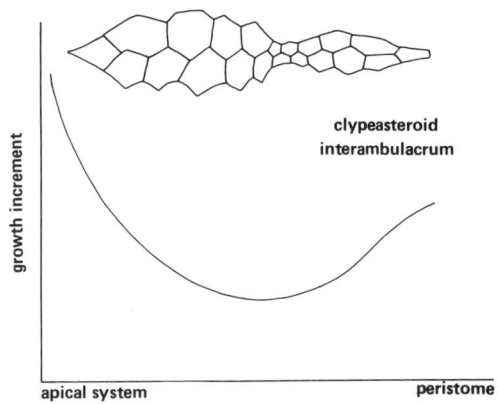

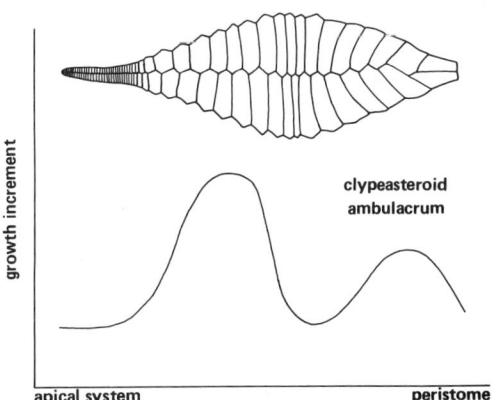

Fig. 12.2 Growth increments in clypeasteroids.

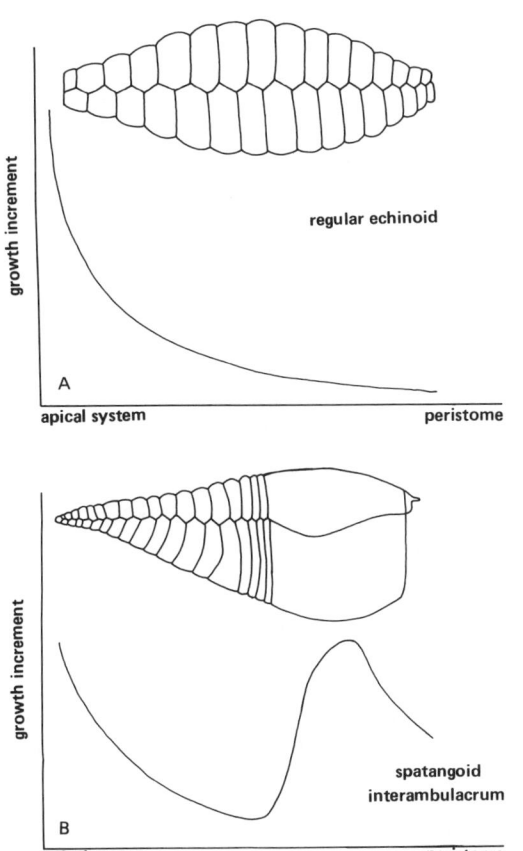

Fig. 12.1 Graph of growth increments in (A) regular echinoids and (B) spatangoids.

heterochronic variation in growth rates of plates in particular parts of the test and at different stages in ontogenetic development, which has influenced the wide range of coronal shapes that have evolved. It is important to remember that the small size of adoral plates in regular echinoids, or of ambital plates in clypeasteroids, is not a function of resorption (Raup 1968). It reflects variation between plates in their growth rates. Each plate may be viewed as having its own ontogenetic trajectory. It may share the same allometry as an adjoining plate, i.e. shape and size changes are the same, but if one plate began its period of growth earlier, or later, depending on its position on the test, it can reach a different ultimate size and shape from its neighbour. If allometries vary between plates, this, combined with age differences, can result in substantial differences in the shapes and sizes of plates within single coronal columns.

Ontogenetic changes in rate of peripheral accretion

The significance of changing interplate growth rates along columns on test shape can be great. Raup (1968) has described how in the regular echinoid *Strongylocentrotus pallidus* the rate of plate growth (expressed by Raup as growth increment) decreases away from the apical system, rapidly at first, then slower toward the peristome, until plate growth there has almost stopped (Fig. 12.1A). As a consequence of this growth pattern, the first formed plates, with little peripheral accretion, are naturally smaller than later formed plates. The small size of plates closest to the apical system is a function of their short life span. Ambital plates, in this case, are the largest, due to a combination of their older age than more adapical plates, even though their rate of peripheral accretion will be less, and their faster rate of accretion than the older plates on the adoral surface. The shape of the test is a function of the two interacting factors of rate of plate addition and rate of decline in peripheral accretion away from the apical system. The faster the decline in rate of accretion, the more globose the test.

Thus, the size and shape of any plate depends on a combination of a number of factors: the relative age of the plate and its rate of growth by peripheral accretion; and the allometric growth of structures within or upon the plate. Growth of a given plate boundary will depend, to a large degree, upon its distance from the apical system. The rate of migration of the plate from the apical system will, therefore, also affect the growth history of the plate. A plate will migrate quicker if new plates are added at a high rate. It will also undergo less growth by peripheral accretion.

HETEROCHRONY AND TEST SHAPE

Spherical tests

I have outlined above how the shape of the echinoid test is strongly influenced by both the rate of plate production and, perhaps more importantly, by differential peripheral accretion of the plates. Many Palaeozoic echinoids are characterized by their spherical shape, as well as multicolumned tests. The multicolumnal state occurred because rate of plate production was relatively high compared with plate growth. Consequently, the growth gradient was so steep (Raup 1968) that plates reached their maximum meridional size very early. The ambitus in these near spherical forms was high.

Flat tests

During the Palaeozoic, a flattened test shape evolved a number of times. Kier (1965) has considered that genera such as *Proterocidaris*, *Meekechinus*, and *Pronechinus* possessed flat tests because of the presence of a zone of very small plates in the region of the ambitus. The same pattern is apparent in many Tertiary clypeasteroids where, in adults, interambulacral plates become larger away from the apical system at first, then generally reduce in size toward the ambitus (Fig. 12.2) only then to increase in size on the adoral surface, before decreasing slightly close to the peristome. This complex system of intracolumnal plate allometries, with negative allometry in the ambital plates, and positive allometry in the growth of many adoral and some aboral interambulacral plates, results in a flattened test shape. Post-metamorphic clypeasteroids have a spherical test. With test expansion the ambital plates undergo a paedomorphic reduction in growth rate. These plates end up resembling the homologous plates of the ancestor at a very early growth stage.

Phelan (1977) has suggested that the earliest clypeasteroid, the Palaeocene *Togocyamus* arose from a cassiduloid ancestor by a paedomorphic process. Its small size suggested to Dommergues *et al.* (1986) that the process was progenesis. However, this progenetic appearance of a small adult which retained ancestral cassiduloid juvenile characters, such as a lantern, was not, in itself, a sufficient cause of the evolution of an entire order. Probably, the most significant overall morphological feature which sets clypeasteroids apart from their cassiduloid ancestors is the flat test. This occurs because of paedomorphosis in a few critical plates by cessation of their growth early in post-metamorphic development. The full morphological and functional potential of the ambital paedomorphosis was not realized until later clypeasteroids increased in size. This also occurred by a heterochronic process, in this case hypermorphosis. As a consequence of the larger test size, the stronger growth of the non-ambital plates, caused by delay in the onset of maturity, accentuated the difference in plate size, and led to the development of the flattened test.

The pattern of the plot of peripheral accretionary plate growth rate against distance from the apical system (Fig. 12.2) for clypeasteroids is quite different from that for regular echinoids (Fig. 12.1A). Because of the dissociation of growth patterns between interambulacra and ambulacra in irregular echinoids, this pattern only applies to the interambulacra. The growth pattern for the ambulacra of clypeasteroids is even more complicated in that on the aboral surface the plates which comprise the petals indicate a rapid production of plates. The same is true for most irregular echinoids which have petals. The way in which ambulacral plate growth in irregulars differs from that in regulars is that the first, perhaps 15, plates are generated at a lower rate than in regulars and hence they undergo greater peripheral accretion (apart from the shorter ambital plates). Following that, plate growth reverts to the fast rate of initial production seen in regulars. This neotenous early ambulacral plate development results in the production of ambulacra which show striking changes in width. The slow growing early plates are much wider than those in the petals. Thus, in clypeasteroids the non-petaloid ambulacral plates are as wide or wider than the interambulacral plates. Heterochronic changes in the sudden increase in rate of ambulacral plate production affect the relative lengths and widths of the petals.

Domed tests

Other irregular echinoids, apart from clypeasteroids, also evolved a test with a flattened adoral surface. However, these other groups, such as spatangoids, holasteroids, and cassiduloids, have a domed aboral surface and thus show a different pattern of incremental growth (Fig. 12.1B). Many spatangoids evolved relatively large adoral plates, such as the plastron and adjacent lateral interambulacra. Thus, whereas in regular echinoids accretionary plate growth declined in the adoral plates as the test increased in size, a number of interambulacral plates in spatangoids retained the juvenile pattern of rapid growth into the adult. In allometric terms there was a change from negative to positive allometry. The importance of plate translocation in facilitating the expression of these allometric changes was critical. Without the ability to 'slide' effectively past adjacent ambulacral columns the strongly positive allometry of the interambulacral plates would just not have evolved.

The ontogenetic and phylogenetic effects of such plate translocation can be seen in species of the spatangoid *Breynia* from Western Australia (McNamara 1987b). For example, in the living *Breynia desorii* (Gray) the second plate of interambulacrum 1a on the adoral surface is relatively short, its length being 38 per cent of test length (TL) in juveniles about 10 mm long. With a tenfold increase in test size the plate grows with positive allometry to occupy about 50 per cent TL. However, the adjoining ambulacral plates (3–11) do not all undergo similar positive allometric increase in size. As a consequence, the junction between the second and third plates of 1a translocates from being opposite plate 8 or 9 in the smallest specimens, to being opposite plate 17 in the largest (McNamara in press). Such strong dissociation of growth patterns between adjacent columns is characteristic of spatangoids and, combined with plate translocation, is one of the reasons for the group's high morphological diversity.

Heterochrony in this particular character can be documented between fossil and living species of *Breynia*. In a Middle Miocene species from north western Australia, *B*. aff. *carinata* d'Archiac and Haime, the extent of positive allometry of plate 2 of interambulacrum 1a was less than in the living species, such that the junction of plates 2 and 3 never reached more than plate 13 of the adjacent ambulacrum. Similarities in slopes of the plot of test length against plate junction reached between the two species suggests that for this particular character the living species has undergone predisplacement resulting in peramorphosis; the second plate of 1a has developed beyond the stage reached by the ancestral species.

Evolution of the holasteroid rostrum

Plate translocation and the retention of high juvenile growth rates of particular adoral plates in some holasteroids has been instrumental in the development of elongate test shapes. This is most spectacularly shown by the late Cretaceous holasteroids *Infulaster* and *Hagenowia*, and the pourtalesiids *Echinosigra* and *Pourtalesia*.

Gale and Smith (1982) noted how there was a trend in the evolution of *Hagenowia* from *Infulaster* for apical elongation of the test. The apical system in *Infulaster* is 'uninterrupted'. In other words, there was no plate translocation within the apical system. Consequently, apical elongation was restricted in extent in *Infulaster*. The evolution of *Hagenowia* from *Infulaster* by the formation of a very long, thin rostrum, composed only of plates of ambulacra IIb and IVa, and interambulacra 1b, 2a, 3b, and 4a, was able to occur because of plate translocation in the apical system in the earliest species, *H. rostrata* (Forbes). This took the form of translocation of plates of interambulacra 1a and 4b between the two posterior oculars. Plates of interambulacra 1a and 4b, were themselves translocated by plates of adjacent columns 1b and 4a. In the descendant *H. anterior* Ernst and Schulz, and *H. blackmorei* Wright and Wright this translocation probably occurred earlier in ontogeny; in other words there was pre-displacement, allowing the development of a relatively longer rostrum in later species of *Hagenowia*. Rostral development thus occurred in *Hagenowia* both as a result of the pre-displacement of plate translocation and the positive allometric meridional growth of plates between the disjunct apical system. The progressive increase in extent of this peramorphosis along the lineage resulted in the evolution of a peramorphocline (Dommergues *et al.* 1986).

Recent work by David (1985, 1987, in press) on plate development in *Echinosigra* and *Pourtalesia*, and other pourtalesiids has shown the influence of heterochronic changes to the degree of plate translocation in adoral plates of ambulacrum I and V between the labrum and plastron. Strong positive allometric growth of ambulacral plates, almost entirely meridionally, resulted in the evolution of many forms with bizarre, very elongate tests. Repeated overlapping translocation allowed the rapidly elongating plates to replace much of the area occupied by the labrum and plastron early in ontogeny and phylogeny. David (in press) has suggested that an increasing degree of plate translocation between a number of genera led to the formation of a peramorphocline in this character.

A similar pattern of meridional plate translocation of adoral ambulacral plates between the labrum and plastron can also be seen in Miocene species of *Breynia*, such as *B. carinata* from India and *B.* aff. *carinata* from north-western Australia. In juveniles of *B.* aff. *carinata* the labrum and plastron are in contact. Between 30 and 40 mm test length positive allometric growth of plate 3 of ambulacra Ia and Vb resulted in plate translocation between the labrum and plastron (Fig. 12.3). These plates grow with negative allometry. A consequence of this pattern of growth was the development of a relatively elongate test. This same character shows paedomorphosis by post-displacement in living species of *Breynia* (McNamara 1982*b*). These paedomorphic species have broader tests than the ancestral Miocene species.

Lunules

Variations in plate allometries along different axes of individual plates can also have a marked affect on coronal morphology, in particular the development of structures such as lunules, both enclosed and marginal. These structures, apart perhaps from the anal lunule (Smith and Ghiold 1982), do not form by resorption. They form by changes in growth allometries along different plate axes. Lunule formation results from strongly negative allometric growth along the interambulacral suture (Seilacher 1979, fig. 8C). As a result, there is virtually no growth along this suture. Phylogenetic changes to these intraplate allometries by heterochrony can result in a rapid and distinctive degree of lunule development.

Heterochronic changes in marginal lunules can be traced in a lineage of rotulids from the Miocene to Recent of north and west Africa (Fig. 12.4). The heterochronic changes form a peramorphocline as the marginal lunules become deeper and increase in numbers. The earliest species, the late Miocene *Rotuloidea vieirai* Darteville, develops very late in ontogeny, slight marginal indentations in the posterior and lateral margins of the test. In the succeeding early Pliocene, *R. fimbriata* Etheridge, juvenile specimens compare with the adults of *R. vieirai* in lacking, or only having faint, marginal indentations. With test growth, the nine indentations became more pronounced in *R. fimbriata*, reaching up to about 8 per cent TL in depth. In the succeeding Pliocene species, *R. fonti* Lambert, the lunules are a little deeper and increased in number to 11 (Roman 1963). In the descendant early Pleistocene to Recent *Heliophora orbiculus* (Linnaeus) onset of lunule formation occurs at an earlier stage of ontogeny. The rate of increase in marginal plate peripheral accretion accelerated, with the consequence that the lunules attain a depth of up to almost 25 per cent TL in large species. The extreme development of the lunules occurred during the Pleistocene, when an offshoot from the main stock of *H. orbiculus*, a form called *H. orbiculus angolensis* by Gonçalves and Roman (1963), developed deep lunules around the entire perimeter of the test. Some were extremely deep,

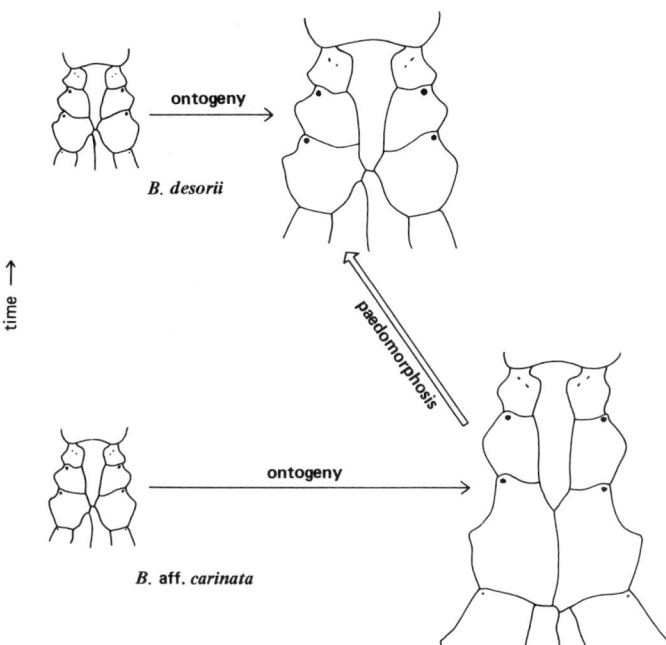

Fig. 12.3 Paedomorphic plate translocation in labrum/plastron in *Breynia*.

reaching up to one-third of the length of the test. The peramorphocline in test outline, therefore, evolved by a combination of pre-displacement and acceleration of peripheral acccretion in the plates immediately above and below the ambitus.

It is likely that the evolution of marginal and enclosed lunules in other clypeasteroid lineages occurred by the selection of such peramorphic morphotypes. Smith and Ghiold (1982), and Ghiold (1986) have suggested that the development of lunules improves feeding by effectively increasing the test perimeter and expediting transport of food from the aboral surface to the mouth. The allometric growth of lunules is related to the fact that the animal's food requirement increases faster than its linear dimension (Smith and Ghiold 1982). However, Telford (1983) considers that lunules serve a hydrodynamic function.

HETEROCHRONY AND AMBULACRAL STRUCTURES

Structural and functional significance of changes to rate of plate production

Raup's (1968) model of the effect of variations in rate of plate generation to shape and size of coronal plates readily accommodates the patterns of ambulacral plate generation seen in many echinoid lineages. For instance, in the *Schizaster* lineage in the Tertiary of southern Australia, the rate of plate generation in ambulacrum III accelerated, with the consequence that plate number, and hence number of pore pairs, in each ambulacrum increased (McNamara and Philip 1980a, fig. 8). As plate generation accelerated, peripheral accretion, particularly equatorially, diminished causing close packing of the plates. The stage was reached in species of the highly peramorphic subgenus *Ova* where plate occlusion in adults occurred to allow accommodation of as many pore pairs as possible. In other words slight differences in intracolumnal plate growth caused localized plate translocation and occlusion (see McNamara 1987b). The functional significance of the acceleration in plate generation in ambulacrum III was that more mucus-secreting funnel-building tube feet were produced. This factor, combined with other morphological changes (see McNamara and Philip 1980a) allowed the occupation of finer-grained sediments by the peramorphic descendants.

Within the *Schizaster* lineage the acceleration in ambulacral plate generation was accompanied by a progressive increase in depth of the petals. Such a trend has been documented in a number of irregular echinoid lineages by Kier (1974). Sunken

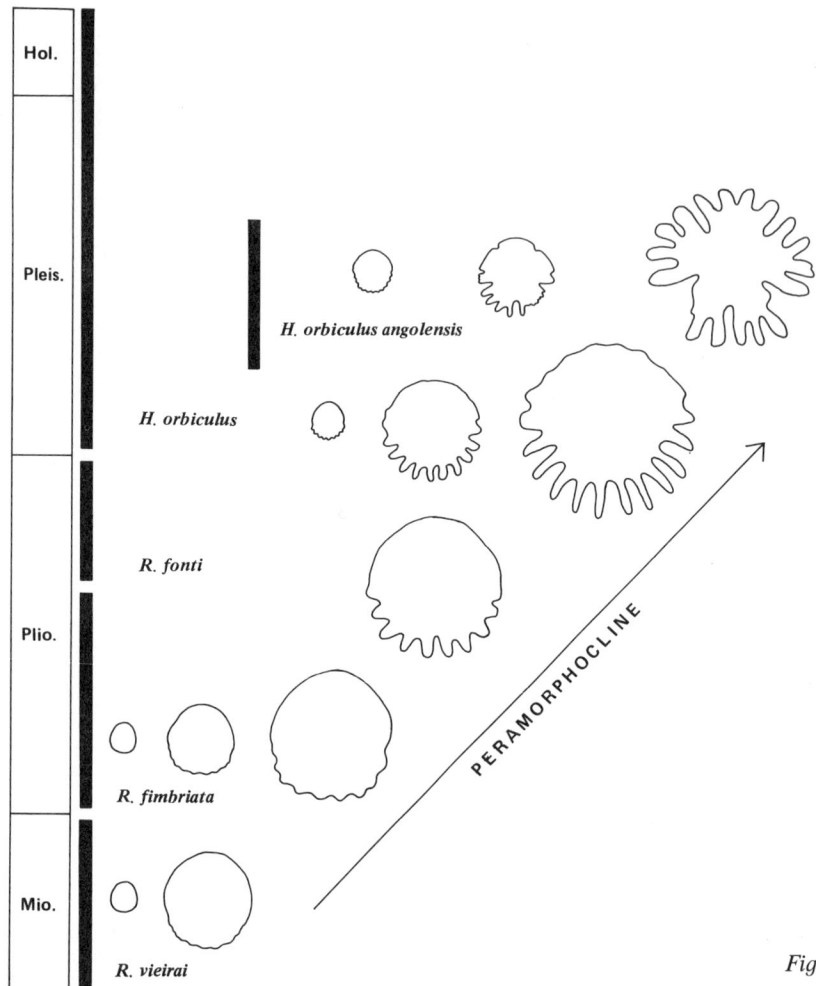

Fig. 12.4 Peramorphic evolution of the rotulids.

petals first appeared in the early Cretaceous. Deepening of petals in these other lineages is also likely to have occurred by acceleration. The formation of such sunken areas on the test may have occurred as a mechanical consequence of increase in meridional allometry of adjoining interambulacral plates. If such an increase in allometry, itself by acceleration, was restricted to interambulacral and not ambulacral plates, a natural result would have been for the ambulacral plates to have been forced downward as they grew, so as to allow the test to accommodate the faster growing interambulacral plates. Selection of these forms will have been favoured, because the deepening of the petals was functionally advantageous to forms burrowing in fine grained sediments,

improving water flow over the respiratory tube feet in the petals. The pattern of ambulacral plate generation and ultimate plate occlusion which occurs in *Schizater* (*Ova*) parallels the development seen in many lineages of regular echinoids. This is shown by plate compounding and, in many cases, an accompanying occlusion of plates.

Evolution of petals

A significant evolutionary development in irregular echinoids during the Cretaceous was the evolution of petals. As Kier (1974) has noted, pore pairs in species of the holasteroid *Pygorhytis* are small and circular in all aboral ambulacra. By the late Cretaceous, forms such as *Holaster* had evolved differentiated pore pairs, with pores in ambulacrum III retaining the ancestral morphology, whereas those in the paired ambulacra became larger, elongate, and slit-like. This reflects the evolution of functional partitioning of tube feet in irregular echinoids, those in the paired petals being restricted to a respiratory function.

This change in the character of pore pairs was a function of allometric changes to the morphology of the tube feet and pore pairs. In most petals the first formed pore pairs undergo restricted development and retain the ancestral circular pore morphology, as do the last formed, which have undergone a shortened growth period. Most pores on fully developed ambulacral plates undergo greater morphological change for the same increase in plate size in non-petaloid ancestors, attaining a larger, elongate shape. There has once again been acceleration in ancestral allometries. Rate of plate production and onset of development of the normal allometry will be factors affecting the lengths to which the petals grow. Interspecific variation in this character within lineages reflects changes in either of these two parameters. Although the initial trend in development of these structures is one of peramorphosis, within some lineages there may be a reversal of the trend, resulting in paedomorphosis, such as in the southern Australian Tertiary *Hemiaster* lineage (McNamara 1987a), where pores become increasingly circular in descendant species.

Evolution of tube feet

Many of the changes in tube feet morphology, such as the increase in papillae in feeding tube feet in spatangoids (Smith 1980) may be a function of acceleration of development. Smith (1980) has suggested that the simpler aboral tube feet in ambulacrum III in some spatangoids reflect paedomorphosis. However, it is more likely that these tube feet have simply changed less from the ancestral type found in regular echinoids. It is therefore probably more appropriate to regard petaloid and phyllodal tube feet in spatangoids as being peramorphic.

Evolution of the phyllode

Phylogenetic changes in the character of the phyllode are generally likely to have resulted from heterochronic changes. Kier (1974) has recorded an overall dominant trend in reduction in the number of phyllodal tube feet in cassiduloid, holasteroid and spatangoid lineages. For example, in the early Bajocian holasteroid *Pygorhytis* there are 22 pore pairs in each phyllode, whereas in Valanginian species of *Toxaster* there may be no more than 12. Such a paedomorphic reduction in the number of phyllodal pores is largely a consequence of a progressively earlier reduction in allometric development of those pore pairs during ontogeny.

Specific changes to the nature of the phyllodal plates may be quite complicated, involving changes to growth patterns of peripodial structures, which in turn affect the number of pores on each plate. Reduction in pore number, from two to one has been documented in a number of spatangoid lineages, such as *Linthia* (Kier 1974), *Schizaster* (McNamara and Philip 1980a), and *Protenaster* (McNamara 1985). Although it might seem that such a reductive change reflects paedomorphosis, analysis of these changes in the *Protenaster* lineage has shown that, on the contrary, complicated peramorphic structural changes resulted in the reduction in pore number. The earliest species, the late Eocene *P. preaustralis* McNamara has a pair of pores separated by a raised interporal partition ('bridged' condition). The descendant late Oligocene *P. philipi* McNamara lost this bridged condition by resorption, attaining the 'breached' stage. The early Miocene *P. antiaustralis* (Tate) has breached first formed phyllodes, but later plates continued their growth for a longer period, resulting in the stumps rejoining to form a reniform ridge adjacent to a single pore. Lastly, the living species *P. australis* (Gray) passes through all the preceding morphological stages of development during its ontogeny, passing beyond the reniform stage to the 'platform' stage. Here the rest of the plate develops a raised, swollen platform in large adults. Phyllodal plates in this lineage, therefore, form a peramorphocline by acceleration.

A principal feature of these changes in morphology of the site of muscle attachment for the feeding tube foot, was a general decrease in surface area, both phylogenetically and ontogenetically in later species. Accompanying these morphological changes were changes in the nature of the sediment, from coarse to fine grained, in which successive species lived, and from which they fed. The trend for reduced muscle attachment area along the lineage probably reflects specialization in feeding from progressively finer grained sediments.

Clypeasteroid food grooves

Heterochrony has also been an important factor in the evolution of food grooves in clypeasteroids. This can be demonstrated by a lineage of four species of *Peronella* from western Australia (Fig. 12.5). The earliest species is an undescribed late Pliocene form from the Roe Plains. This is succeeded by a mid Pleistocene species, *P. rictum* (Gregory), the late Pleistocene *P. orbicularis* (Leske), and the Recent *P. tuberculata* Clark, all of which occur in the Shark Bay region. During the ontogeny of the oldest species the food grooves reach only 2–3 per cent TL at a test length of 30 mm. The grooves are hardly depressed, but occur as narrow zones free of tubercles. Small juveniles of *P. rictum* also have very faintly developed food grooves. Their ontogenetic development is greater than in the older Pliocene species, attaining a maximum length of about 8 per cent TL at a test length of 63 mm. The grooves are slightly depressed, and in large individuals weak bourrelets are developed. This trend continues in *P. orbicularis* and *P. tuberculata*, ontogenetic deepening and lengthening of the food grooves resulting in them attaining about 12 per cent TL and 23 per cent TL, repectively, in these two youngest species, at a test length of 30 mm. This peramorphocline in food groove development probably occurred by a combination of pre-displacement and acceleration.

Adoral ambulacral plate growth in clypeasteroids

The adoral plates of clypeasteroids themselves can undergo increased allometric growth and result in peramorphosis. Durham (1955) has shown how in small individuals of *Dendraster excentricus* (Eschscholtz) ambulacral and interambulacral columns are both in contact with the peristome. During ontogeny the second ambulacral plates show positive allometry compared with adjacent interambulacral plates and translocate equatorially between the first and second interambulacral plates in each column. With continued allometric growth of the ambulacral plates the area occupied by the ambulacra relatively increases during ontogeny. In some genera, such as *Arachnoides*, two pairs of ambulacral plates translocate, with the result that ambulacral plates occupy almost the entire adoral surface (Seilacher 1979, fig. 16B).

Accelerating this plate translocation can produce peramorphosis. Thus, in the *Rotuloidea-Heliophora* lineage plate translocation only occurs in the peramorphic *Heliophora* (Roman 1973). In the earliest clypeasteroids, such as the Fibulariidae, Laganidae, Neolaganidae and Eoscutellidae, adoral plate translocation does not occur. How-

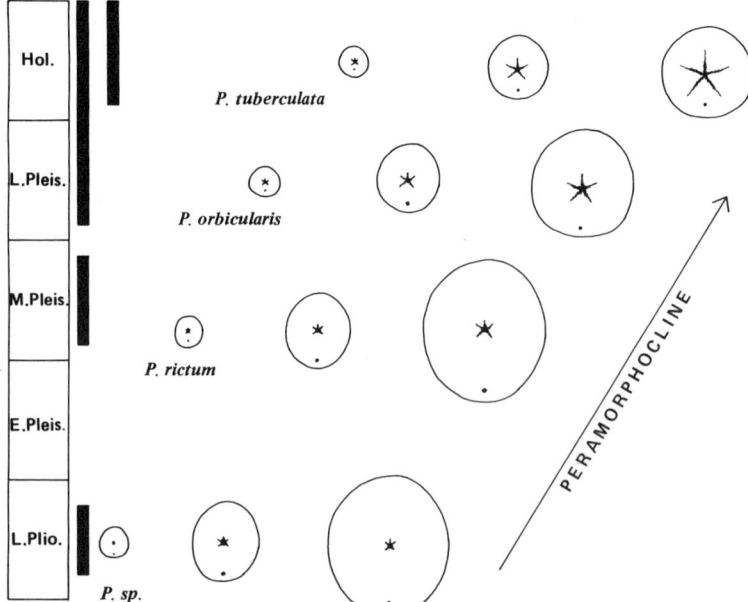

Fig. 12.5 Peramorphic evolution of food grooves in *Peronella*.

ever, many later clypeasteroids, including the Arachnoididae, Clypeasteridae, Mellitidae, Astriclypeidae, Echinarachniidae, Dendrasteridae, and Scutasteridae, show varying degrees of equatorial ambulacral/interambulacral plate translocation. In this regard all these families are peramorphic.

The increase in adoral surface area of the ambulacra in many clypeasteroids was accompanied by the evolution of extensive food grooves. The importance of plate translocation to the increase in food grooves can be seen by comparing species of *Echinarachnius*. Thus, in *E. parma* (Lamarck), the interambulacra are often continuous. Food grooves are less well developed than in *E. laganolithinus* Nisiyama, which has undergone plate translocation and developed a larger adoral ambulacral surface area.

HETEROCHRONY AND INTERAMBULACRAL STRUCTURES

Evolution of tubercles

During ontogeny tubercle development can occur in two ways, analagous, in many respects, to plate development on the test. Firstly the number of tubercles on the interambulacral plate will be affected by the size of the plate. If the plate undergoes little peripheral accretion there is little room for tubercles to develop. Increase in interambulacral plate size will result in an increase in tubercle number. On many irregular echinoids in particular the first formed tubercles and spines are relatively large (Gordon 1926a, fig. 7; 1929, fig. 18), much the same relative size as in regular echinoids (Gordon, 1926b, fig. 17). As the plate increases in size the tubercles and spines in irregular echinoids undergo little actual growth, so developing with negative allometry. As the plate surface expands it is filled by more, small tubercles and spines. Increasing the rate of tubercle production need have little effect on the tubercle size (although, as I note below, there are exceptions), but it will increase the density of tuberculation.

Such acceleration in rate of tubercle production, producing peramorphic descendants with more tubercles, has been documented in a number of spatangoid lineages, particularly affecting the density of tubercle concentration on the adoral surface, as in *Schizaster* (McNamara and Philip 1980a) and *Protenaster* (McNamara 1985), and on the aboral surface, as in *Hemiaster* (McNamara 1987a). Increasing the concentration of lateral adoral tubercles reflects an increase in density of burrowing spines, making for more effective burrowing in finer grained sediments.

Phylogenetic changes in tubercle, and hence spine, size can also occur by a reduction in the degree of negative allometry. In other words the rate of tubercle growth increases. Variation in the timing of this allometric change will cause either pre- or post-displacement and will affect the final number of tubercles which develop. The evolution of primary tubercles in irregular echinoids, such as *Lovenia*, *Breynia*, and *Eupatagus*, is a function of a combination of heterochronic changes to the initial allometries of tubercle and spine growth, and to the subsequent timing of onset of development of each tubercle.

Once the allometric change has occurred, variation to rate of tubercle production can be traced along lineages. For instance, in the three living species of *Breynia* from Australia, which form a paedomorphocline (McNamara 1982b) from the western Australian *B. desorii* through the eastern Australian *B. australasiae* (Leach) to the northern Australian *B. neanika* McNamara, adoral primary tubercle numbers decrease along the lineage. During ontogeny of all three species, tubercle number increases, but at different rates and with different times of onset of development. Along the paedomorphocline onset of tubercle formation is delayed between successive species, and rate of development is reduced. These two heterochronic processes are post-displacement and neoteny. Changes in primary tubercle concentration also occur in a lineage of *Lovenia* from the Oligocene to Pliocene of southern Australia. Not only is there variation in rate of development of primary tubercles resulting in peramorphosis, but onset of development in different areas of the test may be delayed, so much so that tubercles fail to develop altogether in descendant species, resulting in paedomorphosis (McNamara in press).

Evolution of bourrelets

Many cassiduloids have strongly developed bourrelets. These are swollen interambulacral areas surrounding the peristome. Species of *Echinolampas* from southern Australia show both ontogenetic and phylogenetic changes in the intensity of development of these structures (McNamara and Philip 1980b). In the main lineage, from the late

Eocene *E. posterocrassa posterocrassa* Gregory, through the Oligocene *E. posterocrassa curtata* McNamara and Philip, and early Miocene *E. gambierensis* Tenison Woods and *E. ovulum* Laube, the intensity of bourrelet development increases more and more through ontogeny. This occurred partly by acceleration, but also by hypermorphosis, as later species are larger, having attained maturity at a larger size. The progenetic early Miocene *E. morgani* Cotteau has very weakly developed bourrelets. Changes in intensity of bourrelet development in other cassiduloids were probably also controlled by such heterochronic changes.

Evolution of fascioles

Many spatangoids have fascioles, bands of minute tubercles which bear tiny spines called clavulae. Their function is both mucus secretion and current formation. In many spatangoids the most prominent fasiole is the peripetalous. The clavulae here serve an important function in secreting mucus to cover the petaliferous area of the aboral surface of the test. This tent of mucus effectively prevents sediment from falling onto the respiratory tube feet, which might otherwise affect respiratory functioning.

In some lineages, in particular *Schizaster*, *Hemiaster*, and *Pericosmus*, there is a steady increase in fasciole width in species inhabiting progressively finer grained sediments. This increase is, once again, due to heterochrony and reflects a change from a low rate of clavula production in ancestral species to a higher rate in descendants. Such acceleration produces a peramorphic descendant with proportionately more clavulae and thus a wider fasciole, often covering more ambulacral plates (McNamara 1987a, fig. 5).

Interambulacral plate number

Variation in interambulacral plate number occurs in all echinoid lineages by acceleration, producing peramorphs, or neoteny, producing paedomorphs. In spatangoids timing of cessation of plate production is another factor. It occurs as a result of separation from the interambulacral column of the ocular plate, which is the generative region for plates. This separation occurs by plate translocation, either by adjacent ambulacral plates translocating equatorially between the ocular and interambulacral plates (Kier 1956; McNamara 1987b) or by translocation of a genital plate between the ocular and most recently formed interambulacral plate (Kier 1956, fig. 2). Timing of this translocation will be crucial phylogenetically in affecting not only the number of plates produced, but also the degree of peripheral accretion of existing plates. It is likely that precocious translocation and cessation of plate production, resulting in the evolution of a paedomorphic descendant, will be accompanied by increased peripheral accretion in pre-existing plates. In some instances this had profound morphological and functional significance, where the early formed plate which grew larger carried, for example, burrowing spines.

HETEROCHRONY AND OTHER CORONAL STRUCTURES

Peristome and periproct

A characteristic feature of many echinoids is that the peristome and periproct are proportionately larger in juveniles than adults (see for example McNamara 1982a,c, 1985; McNamara and Philip 1980c). In other words, these structures grow with negative allometry. McKinney (1984) has demonstrated how in an oligopygoid lineage from the Eocene of Florida changes to allometry and timing of maturation of the test can affect the resultant size of these structures in descendant adults. Paedomorphosis or peramorphosis may eventuate.

Apical system

A number of phylogenetic changes in the character of the apical system have occurred in many lineages and may, in part, be attributed to heterochrony. Kier (1974) has discussed the trend for a change from a tetrabasal to a monobasal condition in most cassiduloids, apart from *Apatopygus*, which retains the tetrabasal system. Kier has demonstrated that the individual genital plates have not fused together, as crystallographically the sole plate is a single crystal. The monobasal condition arose by a reduction in development of genital plates 1, 3 and 4. Only genital 2 developed. Significantly, this is the plate which carries the madreporite. The result of this change was the evolution of a larger madreporite. The same trend also occurs in some oligopygoids, clypeasteroids, holasteroids and spatangoids.

The mechanism for the tetrabasal to monobasal transition may be analogous to the ontogenetic changes which have been demonstrated in the clypeasteroid *Echinarachnius* (Gordon 1929). In this genus all genital plates were present in immature forms, although three occurred only as remnants of larval spicules. These were resorbed and lost during development. In terms of the heterochronic model, the plate which failed to develop may be considered to have undergone extreme post-displacement, to such an extent that their growth never continued beyond the larval spicule stage. The area which they would otherwise have occupied was filled by a greatly enlarged genital 2.

It is possible that changes in plate allometries within the apical system, combined with plate translocation, may have played a significant role in release of the periproct from the apical system during the evolution of irregular echinoids (McNamara 1987b). The ontogenetic development of the apical system in *Echinocardium cordatum* shows some similarities to the phylogenetic development of irregular echinoids. In *E. cordatum* early in the history of post-larval development, genital 2 grows with stronger positive allometry than other genital plates. As a consequence, it translocates past genital 4 and ocular 5, before coming into contact with plates of interambulacrum 5. The periproct, because of the posterior migration of genital 2, was also moved posteriorly. Following equatorial plate translocation of plate 9 of interambulacrum 5 between the posterior margin of genital 5 and the adapical periproct plates, the periproct became separated from the apical system. Continued production of interambulacral plates resulted in migration of the periproct to its marginal position. It is tempting to speculate that cessation of production of interambulacral plates in spatangoids was a functional necessity to prevent the periproct from being forced onto the adoral surface.

The evolutionary progression of the periproct out of the apical system has been described in species of Disasteridae and Nucleolitidae (Jesionek Szymanska 1968). Smith (1984) considers it to have occurred in three separate lineages. Although the sequence of plate changes may vary in different lineages, there are similarities in the mechanism in the ontogeny of the living spatangoids and the fossil species which suggest that a combination of increased specific plate allometries and plate translocation was important in the evolution of irregular echinoids. The functional significance of the change was of secondary importance. Without the structural changes the irregular echinoids would not have evolved.

CONCLUSIONS

Variations over time in the rate of plate production and in the rate of individual plate growth have been commonplace in echinoid evolution. Likewise, changes to the rate of morphological development of peripodial structures, tubercles and spines appear to have been largely influenced by heterochrony. From the examples which have been documented to date, both paedomorphosis and peramorphosis appear to have been equally common. Perhaps one of the more interesting aspects to emerge from recent studies of heterochrony in echinoid evolution is the phenomenon of dissociated heterochrony: some traits in a species being paedomorphic, others peramorphic (McKinney 1984; McNamara 1987a). If such dissociated heterochrony is more common than hitherto has been appreciated, then it may help to explain the enormous range of morphological variation in many groups of echinoids.

Heterochrony has operated at many taxonomic levels, from the intraspecific to the ordinal. At this level, major innovations have appeared by progenesis, as in clypeasteroids and neolampadids. Although much research still needs to be done, there is an indication that perhaps progenesis was more common in the evolution of regular rather than irregular echinoids. The reason for this lies in the nature of the ontogenetic change in the two groups. In regulars, ontogenetic change involves, predominantly, rates of plate production. However, in many irregulars, such as spatangoids, holasteroids, and clypeasteroids, many of the morphological changes during ontogeny relate to variation in individual plate allometries. Changes to these will produce neotenous or accelerated descendants. Pre- and post-displacement are likely to have been influential in the evolution of all groups of echinoids.

Many of the heterochronic processes which have been documented in recent years, in particular in spatangoids (McNamara in press), show patterns of directed speciation: evolutionary trends producing morphological adaptations to the changing nature of the environment. Almost

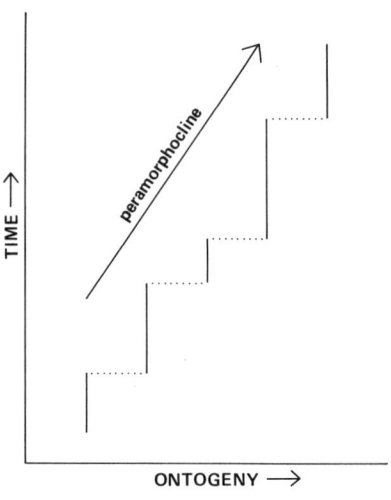
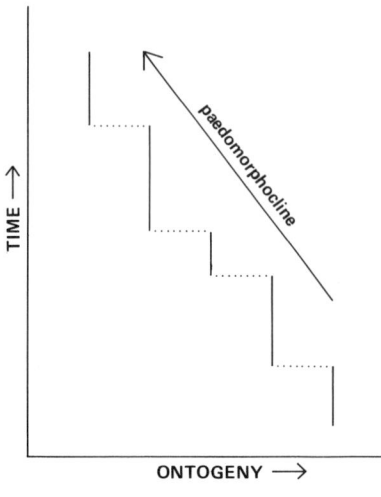

Fig. 12.6 Pattern of (A) stepped peramorphoclines and (B) stepped paedomorphoclines, shown by many echinoid lineages.

without exception the morphological changes are adaptations to living in, burrowing in, or feeding from, progressively finer grained sediments. In other words, heterochrony in spatangoid development allowed the successful colonization of a new niche: fine grained sediments. The patterns of these trends form either peramorphoclines or paedomorphoclines. Interestingly, although ancestral environments persisted, that is to say the relatively coarse grained sediments, with the evolution of a descendant paedomorph or peramorph into a finer grained sediment, the ancestral morphotype, in most cases, became extinct. The resultant pattern is one of stepped pera- and paedomorphoclines (Fig. 12.6).

Kier (1982) has commented on the rapid evolution of key groups of echinoids, such as the irregulars and clypeasteroids. It is quite possible that such rapid morphological changes occurred as the result of the operation of heterochronic processes by small scale genetic changes. Perhaps heterochrony was the 'unknown mechanism' of rapid evolution which Kier was searching for.

REFERENCES

Alberch, P., Gould, S. J., Oster, G. F., and Wake, D. B. 1979. Size and shape in ontogeny and phylogeny. *Paleobiology* **5**, 296–317.

David, B. 1985. La variation chez les échinides irréguliers. Unpublished Doctorates-Sciences, Université de Dijon.

David, B. 1986. Dynamics of plate growth in the deep-sea echinoid *Pourtalesia miranda* Agassiz: a new architectural interpretation. *Bulletin of Marine Science* **39**, 29–47.

David, B. in press. Jeu en mosaïque des hétérochronies: variation et diversité chez les Pourtalesiidae (echinides abyssaux). In *Ontogènese et Evolution* (ed. J. Chaline *et al.*). C.N.R.S., Paris.

de Beer, G. R. 1930. *Embryos and ancestors*. Clarendon, Oxford.

Deutler, F. 1926. Über das Wachstum des Seeigelskeletts. *Zoologische Jahrbücher* **48**, 120–200.

Dommergues, J-L., David, B., and Marchand, D. 1986. Les relations ontogenèse-phylogenèse: applications paléontologiques. *Geobios* **19**, 335–56.

Durham, J. W. 1955. Classification of clypeasteroid echinoids. *University of California Publications in Geological Sciences* **31**, 73–198.

Durham, J. W. 1966. Evolution among the Echinoidea. *Biological Reviews* **41**, 368–91.

Gale, A. S. and Smith, A. B. 1982. The palaeobiology of the Cretaceous irregular echinoids *Infulaster* and *Hagenowia*. *Palaeontology* **25**, 11–42.

Ghiold, J. 1986. The African sand dollar *Rotula*. In *Echinodermata*. Proceedings of the 5th Echinoderm Conference, Galway (ed. B. F. Keegan and B. D. S. O'Conner), pp. 269–73. A. A. Balkena, Rotterdam.

Gonçalves, F. and Roman, J. 1963. Une sous-espèce nouvelle de *Rotula orbiculus* Linné dans les formations Plio-Quaternaires de l'Angola. *Boletim do Museu e Laboratório Mineralogico e Geológico da (Faculdade de Ciêncas) Universidade di Lisboa* **9**, 99–106.

Gordon, I. 1926a. The development of the calcareous test of *Echinocardium cordatum*. *Philosophical Transactions of the Royal Society of London* **B215**, 255–313.

Gordon, I. 1926b. The development of the calcareous

test of *Echinus miliaris*. *Philosophical Transactions of the Royal Society of London* **B214**, 259–312.

Gordon, I. 1929. Skeletal development in *Arbacia*, *Echinarachnius* and *Leptasterias*. *Philosophical Transactions of the Royal Society of London* **B217**, 289–334.

Gould, S. J. 1977. *Ontogeny and phylogeny*. Belknap, Cambridge. Mass.

Jackson, R. T. 1912. Phylogeny of the Echini, with a review of Palaeozoic species. *Memoirs of the Boston Society for Natural History* **7**, 1–443.

Jesionek-Szymanska, W. 1968. Irregular echinoids—an insufficiently known group. *Lethaia* **1**, 50–65.

Kier, P. M. 1956. Separation of interambulacral columns from the apical system in the Echinoidea. *Journal of Paleontology* **30**, 971–4.

Kier, P. M. 1965. Evolutionary trends in Paleozoic echinoids. *Journal of Paleontology* **39**, 436–65.

Kier, P. M. 1974. Evolutionary trends and their functional significance in the post-Paleozoic echinoids. *Paleontological Society Memoirs* **5**, 1–95.

Kier, P. M. 1982. Rapid evolution in echinoids. *Palaeontology* **25**, 1–9.

Märkel, K. 1981. Experimental morphology of coronal growth in regular echinoids. *Zoomorphologie* **97**, 31–52.

McKinney, M. L. 1984. Allometry and heterochrony in an Eocene echinoid lineage: morphological change as a by-product of size selection. *Paleobiology* **10**, 207–19.

McKinney, M. L. 1986. Ecological causation of heterochrony: a test and implications for evolutionary theory. *Paleobiology* **12**, 282–9.

McNamara, K. J. 1982a. Heterochrony and phylogenetic trends. *Paleobiology* **8**, 130–42.

McNamara, K. J. 1982b. Taxonomy and evolution of living species of *Breynia* (Echinoidea: Spatangoida) from Australia. *Records of the Western Australian Museum* **10**, 167–97.

McNamara, K. J. 1982c. A new species of the echinoid *Rhynobrissus* (Spatangoida: Brissidae) from northwest Australia. *Records of the Western Australian Museum* **9**, 349–60.

McNamara, K. J. 1984. Observations on the light-sensitive tube feet of the burrowing echinoid *Protenaster australis* (Gray, 1851). *Records of the Western Australian Museum* **11**, 411–20.

McNamara, K. J. 1985. Taxonomy and evolution of the Cainozoic spatangoid echinoid *Protenaster*. *Palaeontology* **28**, 311–30.

McNamara, K. J. 1986. A guide to the nomenclature of heterochrony. *Journal of Paleontology* **60**, 4–13.

McNamara, K. J. 1987a. Taxonomy, evolution and functional morphology of southern Australian Tertiary hemiasterid echinoids. *Palaeontology* **30**, 109–42.

McNamara, K. J. 1987b. Plate translocation in spatangoid echinoids: its morphological, functional and phylogenetic significance. *Paleobiology* **13**, 312–25.

McNamara, K. J. 1988. The abundance of heterochrony in the fossil record. In *Heterochrony in evolution; a multidisciplinary approach* (ed. M. L. McKinney). Plenum, New York.

McNamara, K. J. in press. The role of heterochrony in the evolution of spatangoid echinoids. In: *Ontogenèse et Evolution* (ed. J. Chaline et al.). C.N.R.S., Paris.

McNamara, K. J. and Philip, G. M. 1980a. Australian Tertiary schizasterid echinoids. *Alcheringa* **4**, 47–65.

McNamara, K. J. and Philip, G. M. 1980b. Tertiary species of *Echinolampas* (Echinoidea) from southern Australia. *Memoirs of the National Museum of Victoria* **41**, 1–14.

McNamara, K. J. and Philip, G. M. 1980c. Living Australian schizasterid echinoids. *Proceedings of the Linnean Society of New South Wales* **104**, 127–46.

McNamara, K. J. and Philip, G. M. 1984. A revision of the spatangoid echinoid *Pericosmus* from the Tertiary of Australia. *Records of the Western Australian Museum* **11**, 319–56.

Melville, R. V. and Durham, J. W. 1966. Skeletal morphology. In *Treatise on invertebrate paleontology. Part U. Echinodermata 3* (ed. R. C. Moore), pp. 220–57. Geological Society of America and University of Kansas Press, Lawrence.

Phelan, T. F. 1977. Comments on the water vascular system, food grooves, and ancestry of the clypeasteroid echinoids. *Bulletin of Marine Science* **27**, 400–22.

Raff, R. A. et al. 1984. Origin of a gene regulatory mechanism in the evolution of echinoderms. *Nature, London* **310**, 312–4.

Raff, R. A. in press. Macroevolutionary changes in echinoid ontogeny: Mechanisms and phylogenetic implications. In *Ontogenèse et Evolution* (ed. J. Chaline et al.). C.N.R.S., Paris.

Raup, D. 1968. Theoretical morphology of echinoid growth. In: Paleobiological aspects of growth and development, a symposium. *Paleontological Society Memoirs* **2**, 50–63.

Roman, J. 1963. Les rotules du Sahara Espagnol. *Notas y Comunicaciones del Institute Geologica y Minero de España* **70**, 103–21.

Seilacher, A. 1979. Constructional morphology of sand dollars. *Paleobiology* **5**, 191–221.

Smith, A. B. 1980. The structure, function and evolution of tube feet and ambulacral pores in irregular echinoids. *Palaeontology* **23**, 39–83.

Smith, A. B. 1984. *Echinoid palaeobiology*. Allen and Unwin, London.

Smith, A. B. and Ghiold, J. 1982. Roles for holes in sand dollars (Echinoidea): a review of lunule function and evolution. *Paleobiology* **8**, 242–53.

Telford, M. 1983. An experimental analysis of lunule function in the sand dollar *Mellita quinquiesperforata*. *Marine Biology* **76**, 125–34.

13

Roles of allometry and ecology in echinoid evolution

MICHAEL L. McKINNEY

*Department of Geological Sciences, University of Tennessee,
Knoxville, Tennessee 37996-1410, USA*

Shape change is the heart of organic evolution and we study its effects in all areas of palaeontology from taxonomy to functional morphology. We have made much progress describing such changes, beginning with Thompson (1917) and Huxley (1932) on up through the multivariate techniques described by Shea (1985). However, we have had, until recently, little of substance to offer concerning the mechanisms and causes of such change. A growing awareness of heterochrony (see McKinney in press; McNamara 1986, this volume) may be changing all that. These changes in the timing of development, which can affect both local growth fields and the whole organism, may be the mechanism we have searched for. Furthermore, by tying heterochronic change to environmental change we can say much more about the ultimate (ecological) causes of phylogenetic shape change.

Echinoids make excellent subjects for such allometric-heterochronic studies. Because they have a rigid test and use external appendages for mobility and feeding (so that these activities are dependent on test surface area and location), echinoid shape is closely linked to function and, thus, environment. Here, a number of cases are reviewed where shape change has been described and I put forth some speculative models to relate form and function. Also, I will review the sometimes subtle relationship between heterochrony and its allometric expression. Finally, I will discuss how heterochrony may often be tied to life history tactics with the important implication that, in direct contrast to the first topic, some traits may not be optimally adaptive, but are instead by-products of life history changes such as body size.

SHAPE CHANGE AS MORPHOLOGICAL ADAPTATION

In this section, some of the many cases of shape change in echinoid evolution are noted in the context of environmental adaptation. However, it is of great heuristic interest to ask first, why is there so much shape change among the echinoids, notably the irregulars? We may take this for granted, but consider body shape in other kinds of burrowers which lack the rigid test and external appendages of echinoids. Worms and snakes, for example, are all very similar in shape even through a huge range of sizes. The reason, I propose, is that their flexible skeletons and internal musculature permit these latter to adjust behaviourally to a much greater variety of substrate conditions. For example, finer, more cohesive sediment can be dealt with simply by greater exertion with the internal musculature. In contrast, the echinoid must live with a fixed, rigid test which completely determines the location and number of appendages that can be applied to feeding and movement. Not surprisingly, the optimal placement of these appendages (and thus test shape) varies with the substrate, as will be shown in more detail. Thus, environmental changes should very quickly result in respondent morphological changes in future generations if a lineage is to survive. These changes may be either ecophenotypic, especially where minor environmental fluctuations are common, or genetic, if they are long-term. As Smith and Paul (1985, p. 34) noted: 'test shape is notoriously variable in extant echinoid species and

Echinoderm phylogeny and evolutionary biology (ed. C. R. C. Paul and A. B. Smith). Clarendon Press, Oxford, 1988.

has in a number of cases been correlated with environmental conditions'. The following discussion focuses on test shape, but similar variability can be shown for the shape of other organs as well.

One of the most common trends in test shape is that of 'streamlining' in many groups of irregular echinoids. Many amateur echinoid collectors have pointed this out to me, but mistakenly analogize it to aerodynamic and hydrodynamic designs which it closely resembles. This analogy fails because streamlining in the latter is for minimization of friction. This is not the case with echinoids because they excavate their medium (sediment) before they move into it rather than push directly into it. This means that anterior surface area should be tied to the way it affects the number of anterior, scraping spines.

Figure 13.1 shows that, because of this unique burrowing mode, the resistance of an echinoid's medium (sediment) scales at the same proportions as its ability to excavate its way through it. That is, for a constant shape, both resistance (sediment in the way) and spine number (ability to move sediment) scale up, at about the same rate, with increasing anterior area. Thus, on these considerations alone, it does not matter what size or shape the echinoid is. Of course, other considerations also apply and it is these that determine specific echinoid shape. Aside from those unique to each major group, such as flatness in the clypeasteroids for adapical feeding (to be discussed shortly), one

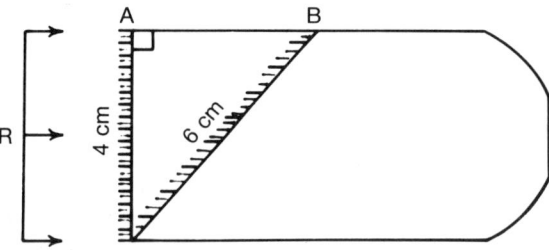

Fig. 13.1 Schematic illustration of how sediment resistance (arrows) scales at the same rate as burrowing ability (anterior area with 'tube feet'). Given frontal surface A, if the organism were reduced to half its original height, both resistance and burrowing ability would be halved. Figure also shows how anterior 'wedging' helps to increase the number of tube feet (anterior area applied to sediment) without increasing the cross-section (and thus resistance). Given frontal surface B, an approximate 50 per cent increase in area is gained from a slope of 45°.

general factor is test width. One simply does not see very narrow cylindrical echinoids. The constraining factor here, I think, is respiration. The ambulacra depend, of course, on surface area so that even though echinoids can become quite flat (diminishing greatly the vertical dimension), they are limited in how horizontally narrow they can become. In contrast, holothurians with respiratory trees around the cloaca are not so limited and are free to reduce the lateral dimension.

Spatangoids

Changes in test shape for this group have been amply described in the fossil record beginning with Rowe's (1899) classic on *Micraster*. The changes are often regular and directional in time and space so that McNamara (1982) has denoted them as morphoclines. While clinal changes occur in many traits, (see below) one of the most obvious is that the test becomes more wedge-shaped as grain size decreases (e.g. *Schizaster*, McNamara and Philip 1980; *Micraster*, Smith 1984). This is illustrated in Fig. 13.2 which shows the posterior migration of the apical system and the vertical and horizontal tapering of the anterior end.

This change can be tied to rather clear-cut functional design. As shown in Fig. 13.1, the wedge shape of the tapering anterior end results in more scraping spines (defined in Smith and Crimes 1983) being applied to the sediment in front. For instance, (to take an extreme case) the anterior test area with spines available for clearing sediment, is increased by about 50 per cent when the anterior test slope is increased to 45° compared to its being vertical. Sloping in the horizontal plane also occurs (Fig. 13.2) and has the same effect. The key point is that this change in anterior shape increases the frontal area in contact with the sediment *without* increasing the anterior cross-section, i.e. sediment to be moved. This shape change is the one way out of the simultaneous scaling of resistance and ability to move sediment. Not surprisingly, this 'wedging' occurs, as noted, in finer sediments where more particles per unit volume and greater cohesiveness make excavation more difficult.

Many other traits in this group also show regular shape change with environmental change. To name some of the more conspicuous, depth and length of the anterior ambulacrum increase with decreasing grain size (Smith 1984). Apparently, this is to accommodate funnel-building tube feet

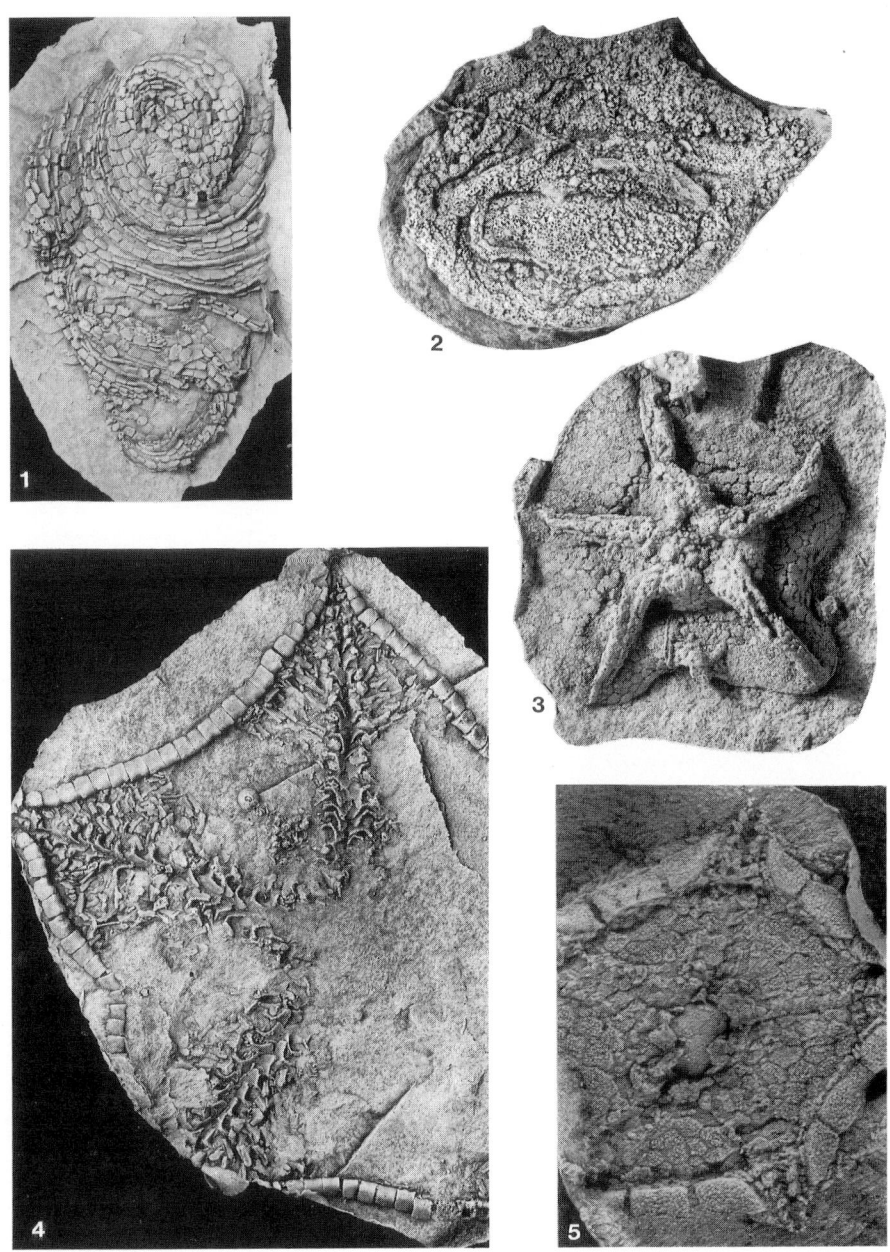

Plate 7.1. (1) *Helicoplacus everndeni* Durham (BMNH E27094), Poleta Formation, Lower Cambrian of Westgard Pass, California, USA, x1.2. **(2)** *Camptostroma roddyi* Ruedemann (USNM 85181), Kinzers Formation, Lower Cambrian Pennsylvania, USA; oral surface, x0.6. **(3)** *Stromatocystites pentangularis* Pompeckj (BMNH E16004), Middle Cambrian, Jince, Czechoslovakia; oral surface, x0.8. **(4)** *Archegonaster pentagonum* Spencer (EH30—latex in the BMNH), Lower Ordovician (Llanvirn) of Ozek, Czechoslovakia; oral surface, x1.2. **(5)** *Cambraster tastudorum* Jell *et al.* (NMVP 107060), Cateena Formation, Upper Middle Cambrian of Ulverstone, northern Tasmania, Australia; oral surface, x5.

Plate 7.2. **(1,2)** *Platanaster ordovicus* Spencer (GSM 8238), Upper Ordovician (Caradocian) of Madeley, Shropshire, UK; **(1)** oral; **(2)** aboral; both x0.8. **(3)** *Palaeura neglecta* Schuchert (specimen in the Národní Museum, Prague), Lower Ordovician (Llanvirn) of Ozek, Czechoslovakia, aboral surface, x1.6. Note spination. **(4,5)** *Hallaster cylindricus* (Billings) (BMNH E53165), Upper Ordovician (Ashgill) of Girvan, Scotland, x1.6. **(4)** aboral; **(5)** oral. **(6)** *Archegonaster pentagonum* Spencer (latex of a specimen in the Národní Museum, Prague), Lower Ordovician (Llanvirn) of Ozek, Czechoslovakia. Detail of arm tip showing virgalia structure, x1.6. **(7)** *Echinocystites pomum* Wyville Thomson (BMNH E75596), Upper Silurian (Ludlow) of Leintwardine, Herefordshire, UK; internal of oral area showing internal lantern apparatus and V-shaped tooth structure, x1.2.

Plate 7.3. (1,2) *Eophiura petaloides* Jaekel; Lower Ordovician (Llanvirn) of Ozek, Czechoslovakia; **(1)** (EH 94—latex in the BMNH) oral surface, x1.2. **(2)** (EH 101—latex in the BMNH) oral surface of arm tips—note the two series of plates in the adambulacral position. **(3)** *Palaeura neglecta* Schuchert (specimen in the Narodní Museum, Prague); Lower Ordovician (Llanvirn) of Ozek, Czechoslovakia, oral, x1.6.

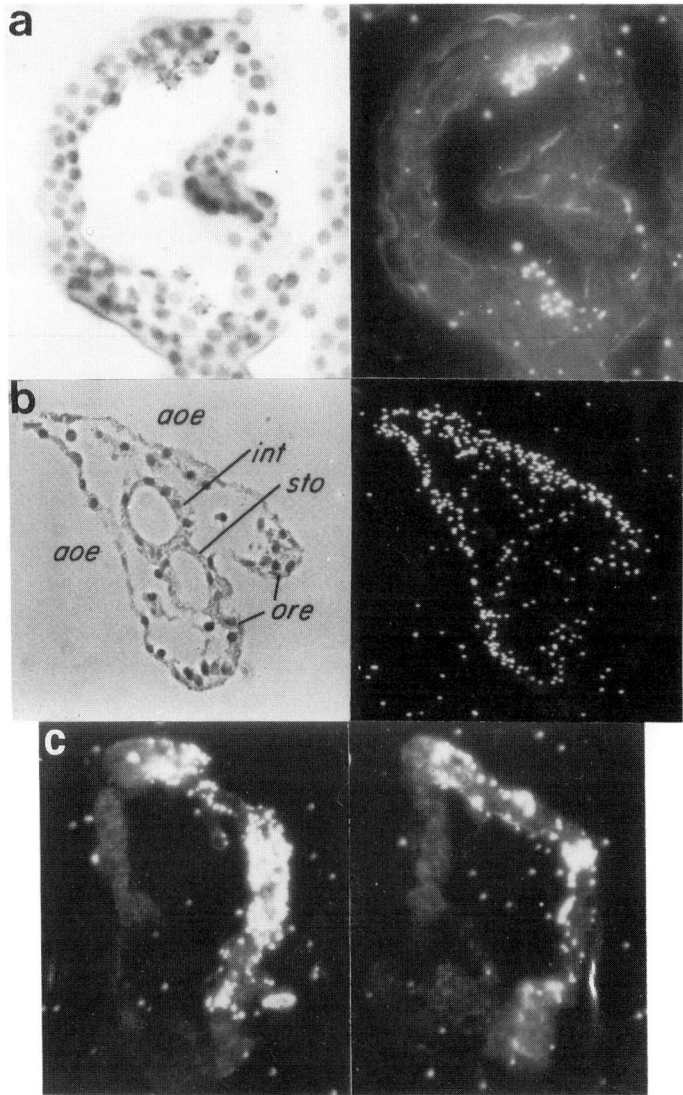

Plate 11.1. Localization of specific gene products by *in situ* hybridization in section of sea urchin embryos. (a) Localization of mRNA coding for a spicule matrix protein gene (SM50) in skeletogenic mesenchyme cells. *Strongylocentrotus purpuratus* embryos at gastrula stage were fixed, processed and hybridized overnight with a ^{3}H-labelled single stranded RNA probe complementary to the message. On the left is a phase photomicrograph and on the right a dark field view of the same section. The autoradiographic grains appear light. They are centred over bilateral aggregations of mesenchyme cells about to secrete the initial triradiate spicules. [From Benson *et al.* (1987) *Developmental Biology.* **120**, 499–506.] (b) Localization of mRNA coding for CyIIIa actin in aboral ectoderm cells of pluteus stage embryo. The sections were reacted with an antisense probe complementary to the 3′ region of CyIIIa mRNA, which is specific to this actin message. Only aboral ectoderm cells display the mRNA. Phase micrographs left, dark field, right. Abbreviations: aeo, aboral ectoderm; int, intestine; ore, oral ectoderm; sto, stomach. [From Angerer and Davidson (1984) *Science.* **226**, 1153–60.] (c) Localization of CAT mRNA in pluteus stage embryos injected with the CyIIIa-CAT fusion gene. As described in text in this construct the CAT gene is under control of the CyIIIa regulatory sequences. Two sections are shown in dark field. The CAT antisense mRNA probe reacts only with aboral ectoderm cells, and not with gut, or oral ectoderm that can be observed at the left of the section. [From Hough-Evans and Davidson (1987) *Genetic Engineering.* **9**, 1–25.]

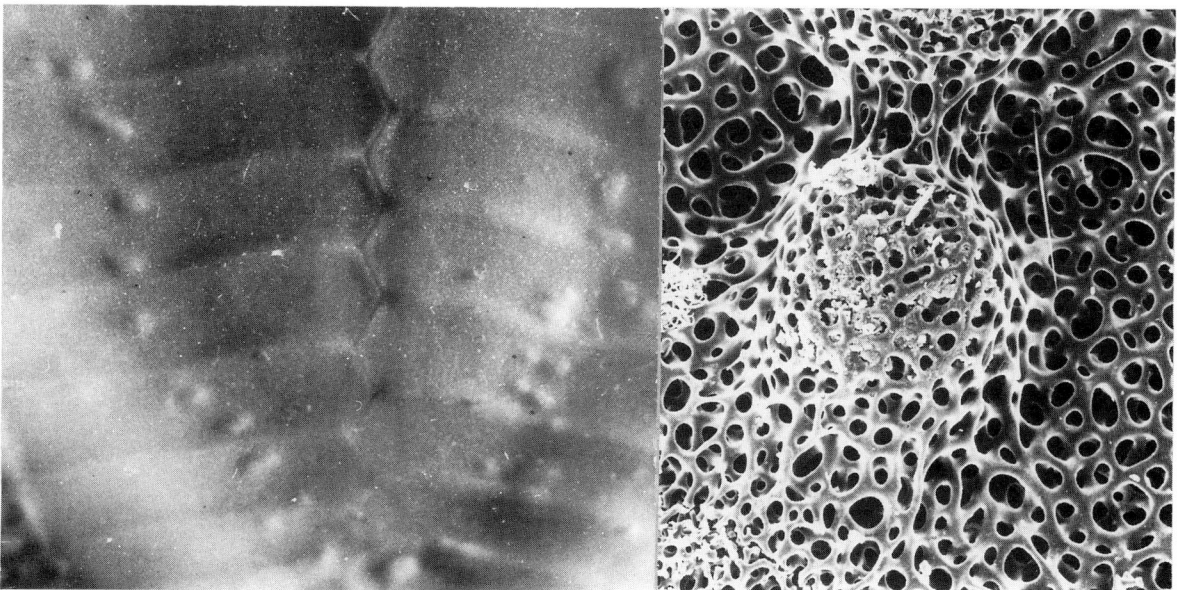

Plate 14.1 Left: The internal body wall of *Tripneustes gratilla elatensis*, showing large protuberances, serving for the attachment of mesenterial threads (x4.5). Right: The microstructure of such protuberances is enhanced by remodelling of surface trabeculae. Their orientation generally corresponds to the directions indicated by Fig. 14.2 (x90).

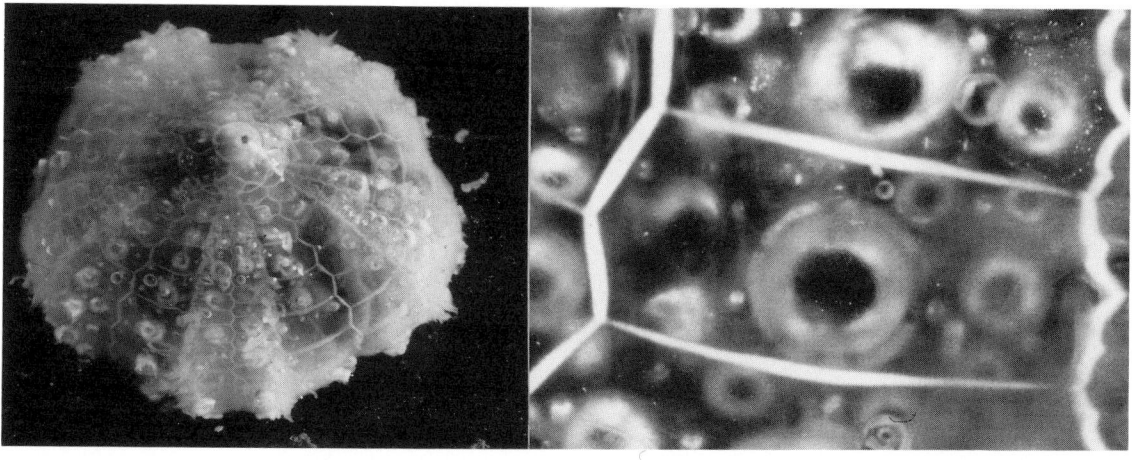

Plate 14.2. An EDTA treated regular echinoid *Psammechinus miliaris*, 25 mm horizontal diameter, submerged in water. Left: the decalcified test in a partly shrunken state following lowering of the water level (spines were removed to show the collagen sutures and internal organs). Right: interambulacral plate 'capsule' of the same, photographed under crossed nicols, showing the glowing light colour of sutural collagen bands (x8).

Plate 14.3. Two types of pollution-induced deformities in *Tripneustes gratilla elatensis*. Left: an abnormally tall specimen; right: aborally depressed urchin.

Plate 14.4. Deformed *Tripneustes gratilla elatensis* showing ambulacral 'pinches'.

Plate 14.5. Test configuration of a *Tripneustes gratilla elatensis*, regenerated from experimental ambulacrum V immobilization. A plane of bilateral symmetry has developed due to impaired growth of the treated ambulacrum and subsequent elongation of the apical complex. Note arching of the adjacent ambulacral rays towards the treated one (Dafni 1986a).

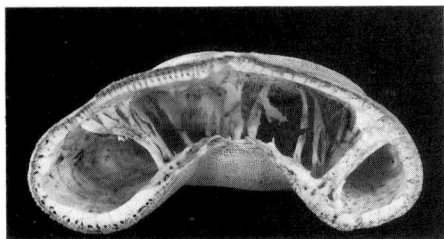

Plate 14.6. Section of *Clypeaster rosaceus*, showing the internal supporting pillars, connecting interambulacral plates of the adoral and aboral sides.

Plate 14.7. Left: abnormal 'petaloid' formed by ambulacral pinching in deformed *Tripneustes gratilla elatensis* (35mmHD), compared with, right, the true petal of *Clypeaster humilis*.

Plate 14.8. Left: brood pouches in the petal of *Abatus cavernosus* (female, 25 mmHD), compared with, right, abnormally formed pits in the aboral tips of the ambulacra of a 72 mmHD *Tripneustes gratilla elatensis*.

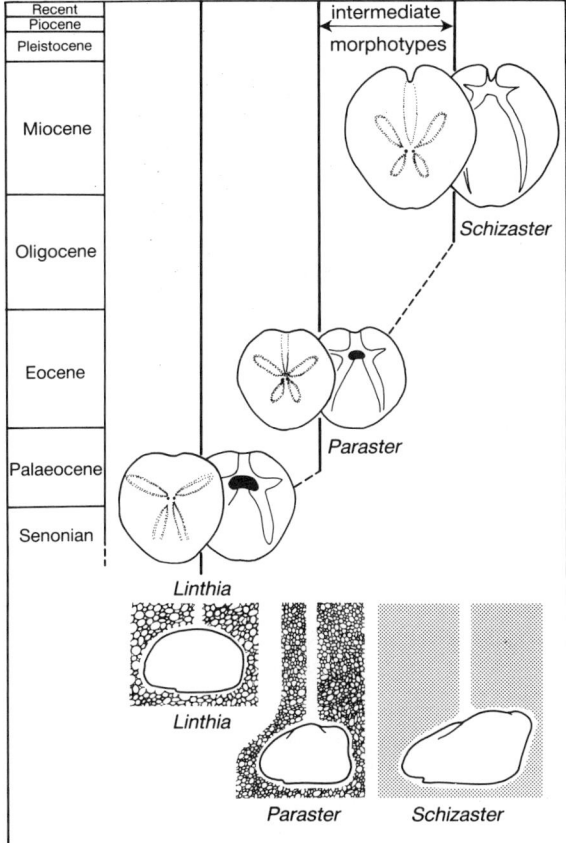

Fig. 13.2 A spatangoid morphocline from Australia showing the anterior 'wedging' which occurs with decreasing sediment size (modified from McNamara and Philip 1980).

and increased respiratory difficulties. Labrum elongation, plastron enlargement, and increase in tubercle density are other changes that occur with decreasing grain size (McNamara and Philip 1980). McNamara (1985) has also demonstrated some notable changes in phyllodal pores. An excellent study of polymorphism in extant sympatric spatangoids was done by Chesher (1968).

Clypeasteroids

Members of this group inhabit primarily coarse sediments and so do not show such adaptations as noted above. Instead, by flattening the vertical dimension and paedomorphically increasing the number of spines, they were able to exploit the relatively nutrient-rich sediment-water interface via adapical feeding (Seilacher 1979; Smith 1984; Ghiold 1984). Again, test design is seen in the usually broad lateral dimension which allows maximum respiratory and nutrient intake while the reduced vertical dimension eliminates unneeded body volume. Not surprisingly, species which are extremely small or continue to feed above the surface are not nearly as flat (Ghiold 1984; Seilacher 1979). Variations in test shape and lunule development (see below) do seem to correlate closely with water depth (energy levels) according to Seilacher (1979). In general, longitudinal eccentricity of the test decreases and lunule size increases as depth decreases. However, more study of this is badly needed before it can be regarded as a safe generalization.

Dendraster has developed a mode of suspension feeding (Timko 1976). Test eccentricity increases in areas of strong currents, probably to serve as anchorage (Raup 1956). Probably, an ecophenotypic change, this has been useful in palaeoecological studies (Dodd and Stanton 1981).

One important supplement to the flattened shape of deposit feeding sand dollars is the development of food grooves and lunules to assist in adapical feeding. In an elegant study, Alexander and Ghiold (1980) tested the importance of lunules in feeding via experiment. Allometric analysis showed that lunule length increased with a growth ratio of about 1.5 relative to test length. In fact, this is not quite what one would expect from scaling theory; nutrient needs should increase faster than 1.5 as test length increases. That is:

$$\text{body mass} = k_1 \text{ (test length)}^3$$
$$\text{daily energy needs} = k_2 \text{ (body mass)}^{0.67}$$

where k_1 and k_2 are shape constants; lower equation from Calder (1984). Therefore:

$$\text{daily energy needs} = k_2[k_1 \text{ (test length)}^3]^{0.67} = k_3 \text{ (test length)}^2.$$

Because lunule perimeter rather than lunule length is the main functional part in feeding one might question this argument. However, perimeter is a linear function of length if there is no shape change. So:

$$\text{Lunule perimeter length} = c_1 \text{ (lunule length)}$$

where c_1 is a constant.

From Alexander and Ghiold (1980):

$$\text{lunule length} = c_2 \text{ (test length)}^{1.5}$$

so

$$\text{lunule perimeter length} = c_3 \, (\text{test length})^{1.5},$$

where $c_3 = c_1 \times c_2$. Compare this to:

$$\text{daily energy needs} = k_3 \, (\text{test length})^2$$

and we see that there seems to be an increasing disparity between lunule food transferring ability and food needs with growth in these forms. The reason is unknown and warrants investigation.

Aside from test shape, a number of other traits show shape change in association with environmental changes. Durham (1978) discussed polymorphism in the sand dollar *Merriamaster* involving petal elevation and tubercle abundance. Unfortunately, these were related to rather uncertain salinity and temperature parameters as opposed to sedimentological ones so that environmental correlates are unclear. He concluded that the polymorphism was genetic, but given the commonness of ecophenotypic variation and his sparse evidence, this is not convincing.

Holectypoids, cassiduloids, and holasteroids

These more primitive groups are less specialized for burrowing than those aforementioned and reflect it in the allometry of their tests and other features. The adoral surface is, of course, flattened to maximize surface area in contact with the sediment base and the test is often mildly longitudinally elongated. However, most of the test variation is seen in the vertical dimension, i.e. height. The adaptive reasons for this variation are poorly understood. Nevertheless, it seems likely that non-flat tests in these shallow/epifaunal forms are adaptive in general. Increasing test height may be a better way to maximize respiratory (petal) area than increasing test width. Unlike the clypeasteroids, these forms lack the multitudinous spines useful in removing suspended particles. By growing taller they increase respiratory surface area without increasing the horizontal area available for suspended matter to fall upon. (For the same reason as anterior spatangoid tapering mentioned above, this allows a change in the area of one vector without changing cross-sectional area.) It also allows materials to be more easily swept off. Just how high the tests get can vary quite a bit, even within one species. In many cases, test height seems to increase with decreasing grain size.

Height increase in shallow burrowers is most clearly demonstrated by Smith and Paul (1985) in *Discoides* which becomes more conical as grain size decreases. They suggest that this was an ecophenotypic process and that the function of this increased relative height was to maintain burrow walls in the muddy sediment. The steeper walls might prevent collapse or cascading of the walls.

In addition, I have noted (in preparation) much height variation in fossil cassiduloids (Fig. 13.3) of Florida which seem to have been only partly buried in life (Kier 1965). This variation is not only

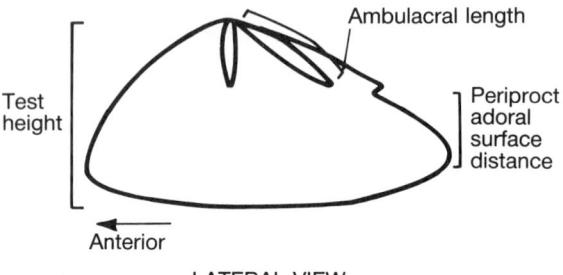

Fig. 13.3 A typical fossil cassiduloid from Florida, *Rhyncholampas*.

intraspecific but interspecific as well (Fig. 13.4). Again, taller forms seem to correlate with finer grain size. In this case, the reason may be that, as noted, increased suspended matter in fine-grained sediment is more easily shed by relatively taller tests. In addition, increased height will increase petal length (McNamara and Philip 1980), creating more surface area for respiration in the muddier water. Much test height variation has also been reported by McNamara and Philip (1980) in *Echinolampas*. More precise ecological correlates of such height alteration are badly needed to test these ideas.

Finally, I note the most drastic shape change yet discussed. The holasteroid *Hagenowia* developed a very elongated rostrum in the late Cretaceous (Gale and Smith 1982). This seems to have been a response to deepening conditions. Fewer nutrients led to the development of this sulcus as a feeding mechanism to take in what was available.

RELATING ALLOMETRY TO HETEROCHRONY

Having discussed how common allometric changes are in echinoid evolution, and how they may be related to environment, I turn now to the developmental growth processes that carried them

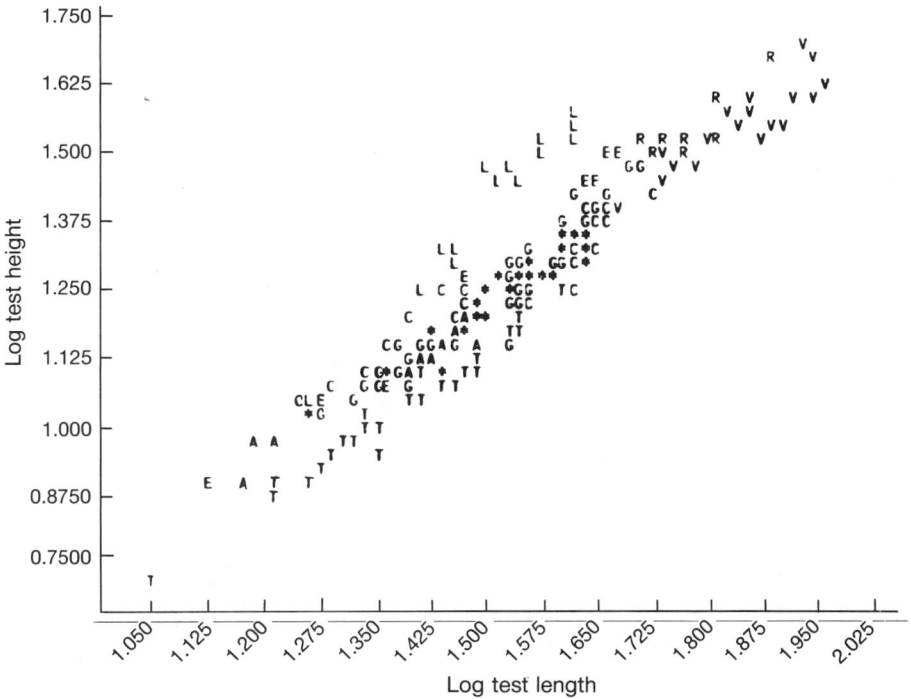

Fig. 13.4 Plot showing test height variation at the same length, among a number of fossil cassiduloids from Florida. Each letter is a different species.

out. Heterochrony, or change in the timing and/or rate of development, is emerging as a common theme of evolution (e.g. McKinney in press; McNamara 1986). And no wonder, it is easy to carry out (change only one or a few regulatory genes) and safe (it uses pre-existing developmental pathways). McNamara (this volume) discusses how heterochronic changes in echinoids are manifested in individual plate development. I wish to focus on shape change at the coarser, organismic level.

Heterochronic change is expressed by shape change. Sometimes all traits are affected at once (i.e. the change is 'global'), but often only some traits change shape as only local growth fields are affected (i.e. they are 'dissociated'). McNamara (1986) has published a qualitative method of comparing adult and juvenile morphologies to classify heterochrony, while I (1986) have published the quantitative, bivariate allometric analogs. However, it is critical to note that both of these methods rely on size and shape alone, omitting the third heterochronic factor, ontogenetic timing. It can be shown (McKinney in press)

that this lack introduces some serious problems for the correct assignment of heterochrony. Happily knowledge of ontogenetic timing is available in some fossil organisms (e.g. molluscs) including perhaps echinoids. Smith (1980) discussed growth lines in plates which may be annual or some other regular increment. In addition, Ebert (1982) has noted that fast-growing species should have relatively thinner plates, and this alone contributes to resolving the timing problem (McKinney 1984). In any case, further work to refine these approaches, especially in irregular echinoids where shape change is best preserved, is crucial. Indeed, I would venture that it is the most important task facing fossil echinoid workers at this time.

In the meantime it is worth noting that knowledge of only size and shape can still tell us a great deal. After all, the size and shape traits of organisms essentially tell what the organisms do in life, as demonstrated in the first section of this paper. They not only form the whole of fossil taxonomy, but of functional morphology as well. With this in mind, I present the allometric classification scheme in Fig. 13.5. Note, however,

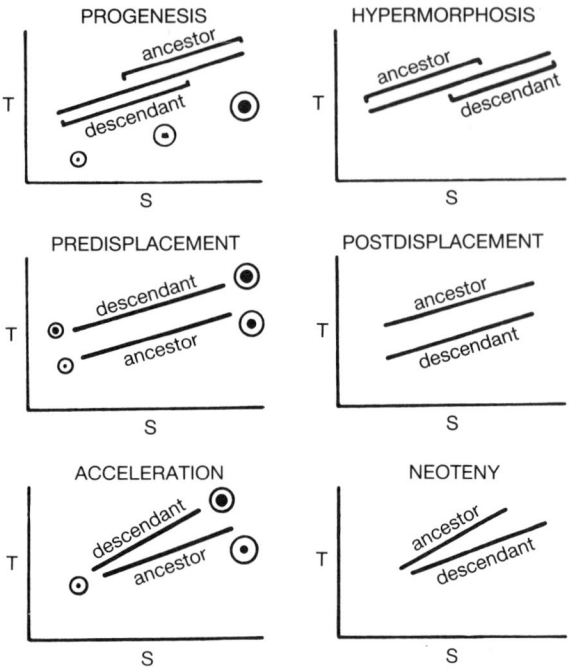

Fig. 13.5 Allometric expressions of heterochrony. S = body size, T = trait shape. Modified from McKinney (1986).

that because only size and shape are used, the terms given may not reflect the 'true' heterochronic processes defined by ontogenetic timing (Alberch *et al.* 1979). Therefore, they should be qualified as 'allometric' processes when obtained in this way (and McNamara's 1986). Thus, growth along the same allometric trajectory but to smaller size in the descendant is 'allometric progenesis' while an increased slope (growth ratio) in the descendant is 'allometric acceleration'. In essence, these terms refer to shape change relative to body size rather than relative to time. We can see then that allometrically accelerated test height in a cassiduloid (e.g. Fig. 13.4) means that the descendant has grown a taller test for a given test length relative to its ancestor. However, we cannot say whether test height truly grew at a faster rate (per unit time) as true acceleration would have done, because the species may not be the same age at a given length; the 'accelerated' form may have taken much longer to reach that length so that the higher test was reached by growing for a longer time (at the same or an even slower rate than the ancestor).

Although these allometric terms may not correspond to the true heterochronic processes, they provide a useful terminology for comparing fossil ontogenies when timing data are not available. This even applies to the ubiquitous bivariate plots used so often in taxonomic descriptions. For further discussion, see McKinney (in press).

ALLOMETRY AND ECOLOGY

The association of allometry and heterochrony has major implications for shape change in echinoids far beyond simply providing a mechanism for that change. Consider that heterochrony, by affecting echinoid growth rate and timing of reproduction, maturation, longevity, body size, and other such traits must also be closely allied to life history tactics in the echinoid. For example, Gould (1977) proposed that progenesis would be fostered in relatively unstable environments because it leads to smaller organisms with shorter generation times and other associated life history traits (see Calder 1984). More stable regimes would tend to foster hypermorphosis and neoteny for converse reasons.

I recently (1986) carried out a test of this generalization with fossil echinoids which seems to give it some validity. Thirty-one species from the circum-Caribbean area were analysed for the types of allometric heterochrony that occurred between pairs of closely related species. Clypeasteroids, spatangoids, and cassiduloids were all included. Of fifteen pairs of closely related species, thirteen had the relatively hypermorphic or neotenous one living in a deeper-water environment. This proportion was shown to be statistically significant at the 0.05 level of confidence indicating that it represents a non-random preference of these processes in deeper environments. While these are only allometrically derived heterochronic assignments (see above), in each case the relatively hypermorphic or neotenous species was significantly larger, indicating that body size, and hence life history, was indeed involved. Because environment was compared to a relative state of heterochrony, determining which species was descendant and which was ancestral was not necessary. Thus, given that increasing water depth is usually correlated with increasing stability (from less random nutrient input, decreased energy levels, and so on, e.g. Levinton 1982), the results tentatively substantiate a relationship between types of heterochrony and environmental stability.

Allometry and ecology in echinoid evolution   171

The key ramification here is that heterochrony is not just a process for attaining certain shape proportions such as test height and so on. It must be adaptive in its effects on a whole suite of life history variables as well. These two effects, shape and life history, are not necessarily exclusive, but it is probable that selective pressures on each may often conflict. For instance, what if increasing environmental stability selects for hypermorphic forms but some of the various organ shapes (e.g. tubercle density) are not maximally functional at that size if the ancestral allometric trajectory is simply extrapolated? The implication is that some trait proportions are allometric by-products of life history related 'size' selection. Consider the width of the peristome, a species-diagnostic trait among three species of oligopygoids (Fig. 13.6).

In some cases, such a situation must result in a trait being suboptimal in function. Yet this is nothing really new. As Mayr (1983) pointed out, the 'struggle of parts' for conflicting needs has been discussed for over a hundred years. The point here is that traits must not only vie with one another, but with the needs of life history features as well.

Yet, having said this, two mitigating observations are important here. One, in many cases extrapolating the ancestral ontogenetic trajectory in fact does give optimal or near-optimal shape. This is because, as Bonner (1965) has emphasized, size increase during ontogeny must be allometrically adaptive throughout, into adulthood. Thus, any allometric proportions are already going to be 'aimed' in the direction of the needs of larger size. Two, even if the ancestral trajectory is inadequate upon extrapolation, 'dissociated' heterochrony can alter shapes of certain organs, as noted above. For example, for Fig. 13.6 I have hypothesized (1984) that the allometric trajectory of the smallest species was insufficient for the larger, so was altered. The point is that even if life history tactics are major targets of selection, traits need not be suboptimal allometric by-products of size change. The allometry may be 'preadaptive' or it can be 'adjusted'.

Returning to the initial topic of this paper, shape change, such as that of the test, is indeed probably highly adaptive in many cases. This is true in spite of the discussion here which shows how allometry can be 'piggy-backed' on the needs of life history tactics. It conforms to traditional interpretations

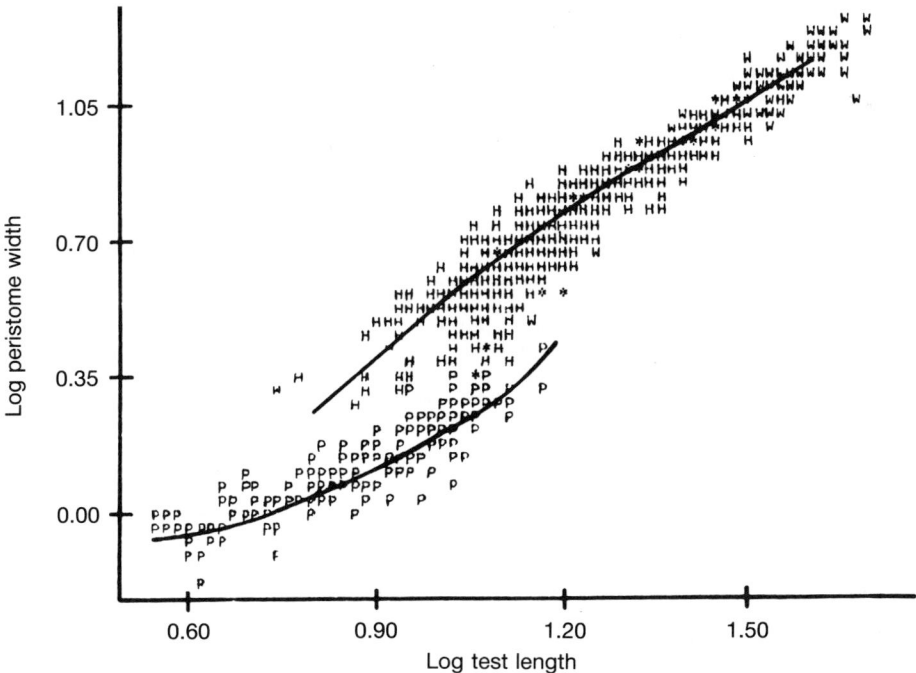

Fig. 13.6 Plot showing allometric relationships among three species of oligopygoid echinoids designated by a different letter. The two larger follow the same trajectory, but the smaller one is distinct. Modified from McKinney (1984).

of evolutionary trends by echinoidologists and such functional-morphological explanations.

To take a concrete example, consider the spatangoid examples of test shape and other changes which occur in lineages as grain size decreases. The strictly functional interpretation (e.g. McNamara 1982; Smith 1984) views the heterochronic changes as mechanisms for 'adjusting' trait shapes to deal with the changing substratum. An extreme view might even have size change as a way of 'getting at' a certain trait proportion. However, we see here that the demands of life history must also be considered. Changing body size is only one part of a suite of life history traits which are under their own selective pressures. Thus, the test and other shape changes in the spatangoids might also be associated with selection for larger body size, either to allow it or as a by-product of it. This would be consistent with the Gould hypothesis because decreasing grain size is indicative of lower energy levels, usually greater depth, and, thus, greater environmental stability.

Shape allometry and life history can accommodate one another. There may be trade-offs, but one suspects that few shape or life history traits are truly inadaptive. However, it should be noted that where substantial changes in body size occur, as they often do in echinoids, it is especially likely that life history considerations are involved. For instance, an increase in length from 50 to just 65 mm involves doubling an echinoid's food needs. Such an added burden is hard to visualize without substantial life history advantages.

REFERENCES

Alberch, P., Gould, S. J., Oster, G. F., and Wake, D. B. 1979. Size and shape in ontogeny and phylogeny. *Paleobiology* **5**, 296–317.

Alexander, D. E. and Ghiold, J. 1980. The functional significance of the lunules in the sand dollar *Mellita quinquiesperforata*. *Biological Bulletin of the Marine Biology Laboratory, Woods Hole* **159**, 561–70.

Bonner, J. T. 1965. *Size and cycle*. Princeton Univ. Press, Princeton, New Jersey.

Calder, W. A. 1984. *Size, function, and life history*. Harvard University Press, Cambridge, Mass.

Chesher, R. H. 1968. The systematics of sympatric species in West Indian spatangoids: a revision of the genera *Brissopsis*, *Plethotaenia*, *Paleopneustes*, and *Saviniaster*. *Studies in tropical Oceanography* **7**, 1–168.

Dodd, J. R. and Stanton, R. J. 1981. *Paleoecology: concepts and applications*. Wiley, New York.

Durham, J. W. 1978. Polymorphism in the Pliocene sand dollar *Merriamaster* (Echinoidea). *Journal of Paleontology* **52**, 275–86.

Ebert, T. A. 1982. Longevity, life history, and relative body wall size in sea urchins. *Ecological Monographs* **52**, 353–94.

Gale, A. S. and Smith, A. B. 1982. The palaeobiology of the Cretaceous irregular echinoids *Infulaster* and *Hagenowia*. *Palaeontology* **25**, 11–42.

Ghiold, J. 1984. Adaptive shifts in clypeasteroid evolution – feeding strategies in the soft-bottom realm. *Neues Jahrbüch für Geologie und Paläontologie Abhandlungen* **169**, 41–73.

Gould, S. J. 1977. *Ontogeny and phylogeny*. Harvard University Press, Cambridge, Mass.

Huxley, J. S. 1932. *Problems of relative growth*. Methuen, London.

Kier, P. M. 1965. Echinoid distribution and habits, Key Largo Coral Reef preserve, Florida. *Smithsonian Miscellaneous Collections* **149**, 1–68.

Levinton, J. S. 1982. *Marine ecology*. Prentice-Hall, Englewood Cliffs, New Jersey.

Mayr, E. 1983. How to carry out the adaptationist program? *American Naturalist* **121**, 324–34.

McKinney, M. L. 1984. Allometry and heterochrony in an Eocene echinoid lineage: morphological change as a by-product of size selection. *Paleobiology* **10**, 207–19.

McKinney, M. L. 1986. Ecological implications of heterochrony: a test and implications for evolutionary theory. *Paleobiology* **12**, 282–9.

McKinney, M. L. (ed.) in press. *Heterochrony in evolution: a multidisciplinary approach*. Plenum Press, New York.

McNamara, K. J. 1982. Heterochrony and phylogenetic trends. *Paleobiology* **8**, 130–42.

McNamara, K. J. 1985. Taxonomy and evolution of the Cainozoic spatangoid echinoid *Protenaster*. *Palaeontology* **28**, 311–30.

McNamara, K. J. 1986. A guide to the nomenclature of heterochrony. *Journal of Paleontology* **60**, 4–13.

McNamara, K. J. and Philip, G. M. 1980. Australian Tertiary schizasterid echinoids. *Alcheringa* **4**, 47–65.

Raup, D. M. 1956. *Dendraster*: a problem in echinoid taxonomy. *Journal of Paleontology* **30**, 685–94.

Rowe, A. W. 1899. An analysis of the genus *Micraster*, as determined by rigid zonal collecting from the zone of *Rhynchonella cuvieri* to that of *Micraster coranguinum*. *Quarterly Journal of the Geological Society of London* **55**, 494–547.

Seilacher, A. 1979. Constructional morphology of sand dollars. *Paleobiology* **5**, 191–221.

Shea, B. T. 1985. Bivariate and multivariate growth allometry: statistical and biological considerations. *Journal of Zoology, London (A)* **206**, 367–90.

Smith, A. B. 1980. Stereom microstructure of the echinoid test. *Special Papers in Palaeontology* **25**, 1–81.

Smith, A. B. 1984. *Echinoid palaeobiology*. Allen & Unwin, London.

Smith, A. B. and Crimes, T. P. 1983. Trace fossils formed by heart urchins (Echinoidea): a study of *Scolicia* and related traces. *Lethaia* **16**, 79–92.

Smith, A. B. and Paul, C. R. C. 1985. Variation in the irregular echinoid *Discoides* during the early Cenomanian, *Special Papers in Palaeontology* **33**, 29–37.

Thompson, D. W. 1917. *On growth and form*. Cambridge. University Press, Cambridge.

Timko, P. L. 1976. Sand dollars as suspension feeders: a new description of feeding in *Dendraster excentricus*. *Biological Bulletin of the Marine Biology Laboratory, Woods Hole* **151**, 247–59.

14

A biomechanical approach to the ontogeny and phylogeny of echinoids

JACOB DAFNI

Interuniversity Institute of Eilat, P.O. Box 469, Eilat 88103, Israel

INTRODUCTION

This paper reviews evidence concerning echinoid skeletal growth and morphology, reintroducing a 70-year-old approach, the biomechanical regulation of ontogenetic and phylogenetic skeletal morphogenesis.

The question whether shape is predetermined, selected to a well-defined set of environmental conditions; or plastic, responsive to a variable set of physical conditions by ontogenetic modification, is of cardinal importance in morphology. D'Arcy Thompson (1917, 1942, p. 15) was the first to show how,

> the forms of living things, and that of the parts of living things, can be explained by physical considerations ... and no organic forms exist save such as are in conformity with physical and mathematical laws

and

> ... it is not merely the nature of the motions ... which we must interpret in terms of force, but also the conformation of the organism itself, whose permanence or equilibrium is explained by the interaction or balance of forces, as described in statics.

These pioneering ideas remained dormant for several decades, possibly because they could not be accommodated easily within the prevailing doctrine, that morphological adaptation is genetically determined. Thompson's mechanical analogies had been considered, at the most, as illustrations of the effective way natural selection eliminates those forms that do not conform to mechanical and economical principles of design.

The rapid progress in the fields of genetics and molecular biology held the promise that all possible phenotypic variability, of which natural selection makes use in evolution, will be amply accommodated within the unlimited capacity of the genome. And yet, after so many years of enormous efforts invested in these disciplines, one still wonders

> what controls the process of morphogenesis, or ... how does a one-dimensional genetic code ultimately specify a tri-dimensional organism (Edelman 1984, p. 118).

A possible solution to this dilemma is that organisms yield to mechanical stimuli throughout their morphogenesis, and that the gene complement that favours such responsiveness is retained by selection (Bonner 1961). The expression of the genes is, according to this view, limited to the realm of constructive materials and regulative processes, whereas morphogenesis itself is self-organized, influenced by the internal constraints and the physico-mechanical external environment. Growing organisms follow mechanical principles and design, through which the nature of the constructional elements and materials affects their ultimate structural morphology (Stevens 1976; Wainwright *et al.* 1976), in processes that are largely independent of the direct control by the genome.

This is best illustrated by the prevalence of pneumatic structures in nature (Bach *et al.* 1976). Cells, organs, and frequently also whole organisms, assume the typical shape of the *pneu*—a globular or roundish inflated body, filled with fluid or gas, which is under higher than ambient hydrostatic pressure, and surrounded by a flexible strained membrane. The pneu paradigm is equally common among living and non-living forms, and is known for its being optimized by a mechanical force-balance. By adopting this principle organisms attain mechanical stability and save energy.

Echinoderms have a skeleton unlike that of any other invertebrate group. In contrast to the inert molluscan shell, or the ephemeral disposable arthropod carapace, they have a mesodermal endoskeleton, composed of skin-embedded plates, and an internal fluid-filled cavity, the coelom. Seilacher (1979) coined the term 'mineralized pneu' for permanently fixed pneus, such as eggshells and foraminiferal chambers. The echinoid test consists of a still more complex form of pneu—a *plastic mineralized pneu*. Thompson (1917, 1942, p. 945) was the first to describe its pneumatic nature:

> The sea-urchin shell consists of a membrane, stiffened into rigidity by calcareous deposits, which constitute a beautiful skeleton of separate, neatly fitting ossicles. The rigidity of the shell is more apparent than real, for the entire structure is, in a sluggish way, plastic; inasmuch as each little ossicle is capable of growth, and the entire shell grows by increments to each and all of these multitudinous elements, whose individual growth involves a certain amount of freedom to move relative to one another; in a few cases the ossicles are so little developed that the whole shell appears soft and flexible. The viscera of the animal occupy but a small part of the space within the shell, the cavity being mainly filled by a large quantity of watery fluid, whose density must be very near to that of external sea-water. . . . While the sea-urchin is alive, an immense number of delicate tube-feet, with suckers at their tips, pass through minute pores in the shell, and, like so many long cables, moor the animal to the ground. They constitute a symmetrical system of forces, with one resultant downward, in the direction of gravity, and another outwards in the radial direction; and if we look upon the shell as originally spherical, both will tend to depress the sphere into a flattened cake . . . This is precisely the condition which we have to deal with in a drop of liquid lying on a plate; the form of which is determined by its own uniform surface-tension, plus gravity, acting against the internal hydrostatic pressure.

ECHINOID SKELETAL MORPHOGENESIS RESEARCH

Thompson's biomechanical approach concerning echinoid morphology was adopted by a few echinoderm students, whose morphometric observations seemed to conform with his hypotheses, while most students of morphology ignored it altogether.

In the last two decades echinoid skeletal research progressed appreciably. Many studies dealt with skeletal growth and allometry (Moss and Meehan 1968; Raup 1968; Kobayashi and Taki 1969; Régis 1969, 1973; Jensen 1969; Ebert 1967, 1975, 1982; Seilacher 1979; Smith 1980, 1984; Dafni 1980, 1983b; Dafni and Erez 1982, 1987a,b). Jensen (1972), Pearse and Pearse (1975), Märkel (1975, 1976, 1981) and Smith (1978, 1980) used enhanced microscopy techniques to observe echinoid microstructure, while Weber *et al.* (1969), Emlet (1982) and Telford (1985) studied the mechanical design of the skeleton. Okazaki (1975) described the morphogenesis of normal and abnormal larval skeletons. Although all these students contributed immensely to our knowledge of both inherited and adaptable morphology, only few attempted to elucidate the regulatory mechanisms which determine echinoid skeletal morphogenesis.

Among the latter were Raup (1968), and Moss and Meehan (1967, 1968). Raup, an ardent supporter of Thompson's ideas, tried to formulate the growth process, using computer simulation. As to the factors which control skeletal growth, he believed that while rates of new plate addition and plate peripheral accretion are species-specific, skeletal morphology as a whole is a product of the ontogenetic development of the plate mosaic. He argued that

> growth (of plates) occurs around their periphery and is concentrated in those regions where pressure counter to this growth is lowest (p. 58).

In other words, the growth of echinoid plates is controlled by sutural compression. Moss and Meehan (1968), who found an analogy between the growth of echinoid tests and the growth of mammalian clavarial bones, believed that plate growth is triggered by surface tensional stresses rather than compression. They speculated that the individual echinoid plates, as well as the whole test, grow passively in response to the expansion of the enclosed living mass, which they termed 'functional matrix', a mechanically obligatory process.

The differential growth and downward shift of plates represent, according to Moss and Meehan (1968), ontogenetic changes in an internal reference register, rather than a real movement of plates.

Although Moss and Meehan's approach is more consistent with the pneu paradigm, close examination of test plates, and their design, indicates the existence of sutural compression in some areas of the test (Seilacher 1979; Telford 1985; Dafni 1986a), supporting Raup's proposition. It is argued below that sutural compression and tension are in fact two sides of the same coin.

Telford (1985) analysed the architectual design of the echinoid skeleton, comparing it to an arched dome, physically adapted to carry its own weight and sustain a variety of external forces. Loads, which weigh down the test dome are transformed, according to Telford's analysis, into compressional stresses in the radial (=longitudinal) direction and tensile stresses circumferentially. He argued that some allometric trends in growth of regular echinoids, namely growing taller with size and forming thicker radial ribs on the inner surface of the test, as well as the ample sutural collagen fibres lashed across the longitudinal sutures, all represent adaptations to minimize circumferential tension, which may tear the test at the sutures.

A somewhat different attitude is held by the author (Dafni 1986a). He views the echinoid test as a mineralized pneumatic structure, which is ontogenetically modified by means of differential growth. Growth, according to this hypothesis, is controlled by sutural stresses, tension, and to a lesser extent compression, brought about by either external or internal forces.

The plastic response of the coronal test to changes in the physico-mechanical environment, without endangering the structural stability, is enabled by the combination of compression-resistant mineral plates and flexibility of the sutural fibres, which are tensed during expansion. Hence, the articulated echinoid test combines the properties of a flexible pneu with those of a rigid mineralized pneu.

Environment influences echinoid growth inducing morphological plasticity, through resource allocation (Ebert 1980). Food deficiency was found to suppress growth rates, but at the same time cause exaggerated growth of food gathering apparatuses, possibly due to an energy trade-off between the various body components (Ebert 1980).

A BIOMECHANICAL MODEL OF ECHINOID GROWTH AND MORPHOGENESIS

Based on the above approach, a model has been presented, describing skeletal growth and morphogenesis of echinoid tests in terms of a biomechanical force-balance resulting from the pneu principle and the activity of various contractile and elastic tissue elements (Dafni 1986a). Essentially, this model integrates Moss and Meehan's (1968) and Raup's (1968) models. The principles of this model, as well as evidence supporting it, are summarized here to justify phylogenetic considerations presented later.

According to this model the echinoid test, characterized by an external plate armour, increases its volume owing to internal expansion, which forces the mineral plates apart, inducing growth and calcification along their free margins. During intensive growth, tension prevails throughout the coronal surface, whereas a non-growing mineralized pneu relies upon its skeletal framework for rigid support. Under the latter conditions the external static and loading forces are being absorbed by the combination of compression-resistant plates and tension-dispersing collagen fibres (Strathmann 1981; Telford 1985).

The various mechanical components of the model which affect the alleged 'stress-balance', controlling the peripheral growth of plates and, ultimately, the entire test morphology, are (Fig. 14.1): (1) the inner hydrostatic pressure; (2) the mechanical pull of the ambulacral tube-feet; (3) the tethering effect of mesenterial threads; and (4) the muscles of the Aristotle's lantern, which pull the lowermost part of the test centrally. A somewhat more passive role is played by the sutural collagen fibre system (Dafni 1986a), to keep the test plates together, and to enable plastic response.

The following aspects of the model have been observed or experimentally tested.

Tension-induced growth

Observations suggest that tensile stresses control the peripheral growth of test plates by forming sutural gaps, which are spanned by fast-growing

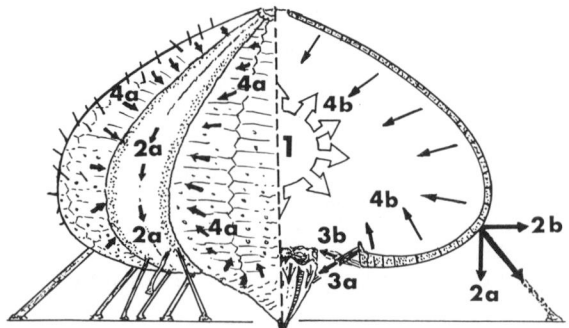

Fig. 14.1 Schematical presentation of the proposed model, showing the various components of force involved in echinoid morphogenesis, side view (left) and section (right): (1) expansion of the inner mass; (2) forces produced by the tubefeet; (3) lantern muscles (retractor and protractor); (4) mesenterial threads pulling at the plates in the plate level (4a) or centripetally (4b) (Dafni 1986a).

trabecular rods; and that trabecular growth ceases when compression develops along non-stretched sutures (Dafni 1986a). This has been tested experimentally by culturing skeletal explants in seawater and, using an artificial spring device, stretching their plates either along the horizontal (latitudinal) or the vertical (meridional) sutures. The pattern of uptake of ^{45}Ca of the stretched sutures, as well as the vertical/horizontal calcification ratio of vertically stretched explants were significantly greater than those of the non-stretched or horizontally stretched control, thus supporting this hypothesis (Dafni 1985).

Slow-growing, confined sutures are characterized by a denser trabecular mesh while rapid trabecular growth, in stretched sutures, usually yields a more spacious stereom (Dafni 1986a). It has also been shown that well-fed urchins produced a fast-growing well-aligned galleried stereom, whereas starved urchins produce a labyrinthic stereom with poorly-aligned trabeculae (Pearse and Pearse 1975). Since growth-lines are microstructurally defined as alternations between different stereom types, often showing different porosity (Smith 1980), and rapid growth is associated with sutural tensional stresses (Dafni 1985, 1986a), the various mesh sizes or alignment types may reflect different physico-mechanical sutural conditions.

Echinoid species with collagen embedded test plates, not confined to a tight mosaic, usually lack periodic growth-lines. This is consistent with Smith's (1980) observation that coronal plates of diadematoid and echinothurioid echinoids, with a more mouldable test, lack growth banding. It is also noteworthy that the peristomial plates of *Tripneustes*, fully enveloped by connective tissue, are roundish rather than polygonal, and have no growth-lines (J. Dafni, unpublished results). The scarcity of growth zones in the tightly-arranged cidaroid plates (Smith 1980), may be explained by their extremely slow growth (Ebert 1982).

Possible role of internal hydrostatic pressure

Internal higher-than-ambient pressure is an inherent property of the pneu paradigm. Furthermore, a flexible plate armour makes sense only if being hydrostatically supported (Seilacher 1986). It has been suggested (Dafni 1986b) that the flexibility of the sutural binding of the plates plays a decisive role during growth, when the internal mass expands, while their resistance to compression is exhibited mainly when the test is overwhelmed by external loads. The plate stereom seems to be well-adapted to these conditions, since porosity is known to reduce the difference between tensional and compressional strength (Emlet 1982). This dual function, a rapidly expanding structure interlocked by sutural fibres at one time and a dome-shaped solid armature of exceptional mechanical stability (Telford 1985) at other times, is perhaps the most remarkable structural property of the echinoid mineralized pneu. Hence, Telford's (1985) static model, implying morphological adaptation to external loads, and Dafni's (1986a) model, emphasizing the role of internal forces, do not conflict, since both apply to different states, or temporal phases, of the same structure.

Alternations from one state to another are associated with the echinoid's metabolism. Starvation, low temperature or physiological stress inhibit skeletal growth (Kobayashi and Taki 1969; Pearse and Pearse 1975; Dafni and Tobol in press), even causing 'negative growth' (Ebert 1967, 1968). Temporal growth fluctuations, recorded as growth-line patterns (Jensen 1969; Smith 1980), may well reflect fluctuations in the internal pressure. Moreover, since there is no conclusive evidence for peripheral resorption of plates (Märkel 1981), it is speculated that slight negative growth in the size order of <5 per cent in the test diameter, described by several authors (Ebert 1967, 1968; Régis 1979; Dafni 1984), may result from reduction of the internal pressure and test shrinkage due to contraction of the sutural gaps.

Specialized metabolic processes are employed by various invertebrates in an effort to control their internal fluid pressure, thus controlling their body form and size under the different osmotic conditions, or for special functions (e.g. carapace shedding in arthropods, Prosser 1973).

In analogy to these animals, Dafni (1986b) suggested that higher than ambient internal hydrostatic pressures are obtained either by muscles, which brace the body wall and confine the internal fluid, or by maintaining internally higher osmotic pressures, that attract water inflow through the semi-permeable body wall. Rapid expansion of the test can also be obtained by temporarily reducing its surface tension, possibly by 'softening' the tensility of the sutural collagen fibres (see below).

Proving the existence of internal pressures is rather tricky. First, we already suggested that during periods of non-growth the sutural gaps close, which corresponds to a state of zero internal pressure. Secondly, these pressures may well be small, and act for relatively shorter periods. Thirdly, the plates themselves are enclosed in pneu-like capsules (see below) which further complicates the model. Confirming this model and measuring the pressures involved may be a challenging task.

Tube-feet activity

Thompson (1917) suggested that test profiles of sea urchins are modified by the mechanical interaction with the substratum, via the ambulacral tube-feet, and that stronger adherence yields flatter profiles (i.e. lower height v. horizontal diameter, H/D). Observations by Moore (1935), McPherson (1965), and Lewis and Storey (1984) support this hypothesis, showing that regular echinoids exposed to surf conditions, where firm adherence is essential, tend to be flatter than those living in low energy habitats. Similarly, Nichols (1982) found, in a biometric study among different populations of *Echinus esculentus*, a significant increase in H/D ratio with increasing water depth, possibly due to calmer water conditions.

Dafni (1983a) showed that two sub-species of *Tripneustes gratilla*, *T. g. gratilla*, and *T. g. elatensis*, differ in their typical habitat. *T. g. gratilla*, which inhabits seagrass meadows and sandy areas, has a typically high H/D and domed shape, while *T. g. elatensis*, which lives mainly in rocky shores, is flatter.

Testing Thompson's hypothesis, Dafni (1986a) transferred live urchins of the latter sub-species from their natural hard substratum into a sand-bottomed artificial habitat. As expected, they grew taller by ~ 5 per cent, while those which were carried back to their original habitat, after this treatment, showed a reversed trend. Moreover, the differential calcification pattern of these urchins was changed appreciably (Dafni and Erez 1987b). The hypothesis that the test profile is affected by tube-feet activity is supported also by the observation that abnormal echinoid tetramers, with only four ambulacra and, accordingly fewer functional tube-feet, are on the average taller than their normal conspecifics (Dafni 1986a).

Role of internal tethering elements

Ideal 'mineralized pneus' are expected to grow by uniform increase of each individual plate, in response to the expansion of the internal fluid (Moss and Meehan 1968). The actual pattern is anything but uniform; some plates grow rapidly in one peripheral direction, while in other directions they grow slowly or do not grow at all. Save for some species-specific differences, similar patterns are exhibited by most regular echinoids (Deutler 1926; Raup 1968; Märkel 1975, 1976, 1981; Dafni 1984; Dafni and Erez 1982, 1987a). According to the proposed model the internal pressure is transformed into a differential tensile stress pattern which determines plate growth pattern. The same process is shared by most species of regular echinoid (Dafni 1986a).

Dafni and Erez (1982) compared the plate peripheral patterns with the activity and spatial distribution of the ambulacral tube-feet and various tethering tissue elements, which, as suggested by earlier observations, pull the plates and affect their confinement patterns. Among the latter were lantern muscles and the mesenterial threads, which fasten the alimentary system to the inner surface of the coronal test (Dafni 1983b, 1986a). Using the radioisotope ^{45}Ca, they measured the calcification patterns of individual *T. gratilla* plates, along a vertical column, in all four directions. Assuming that test plates, arranged in 'close-packing' patterns (Raup 1968), grow more in tensed sutures than in compressed sutures, they transformed the measured rates into growth vectors, and studied their patterns (Fig. 14.2). They found that for the resultant vector of each plate there is a contractile element (mesenterial

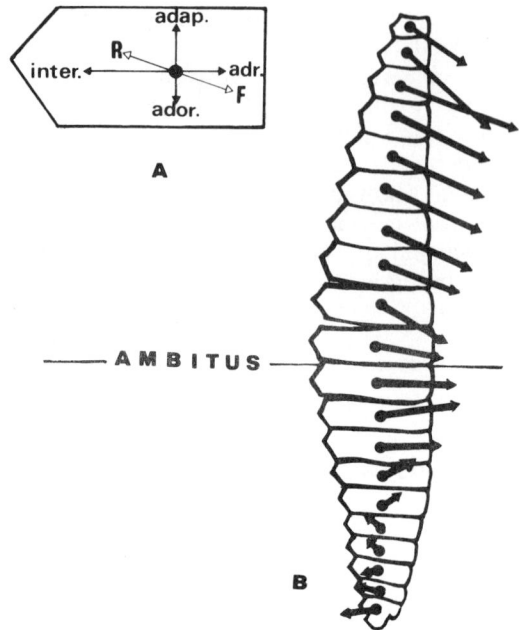

Fig. 14.2 (A) Vector addition of the differential calcification rates at the different directions in an interambulacral plate of *Tripneustes gratilla elatensis*. R, resultant growth vector, F–inferred tensional force vector. (B) Map of F vectors for plates along the interambulacral column (Dafni and Erez 1982).

thread or tube-foot), attached to the plate, which pulls the plate in a direction 180° to the growth direction. For example, an ambulacral plate, pulled downwards by tube-feet attached to it, will grow more in its upper margins. This analysis also confirmed observations, indicating a similar biomechanical effect caused by the mesenterial threads, owing to the presence of scattered muscle fibres in these threads (Dafni 1983b).

The homology between the mesenterial threads and the unique mesenterial muscles (Jensen 1985), that flex the elastic tests of various echinothuriid species, supports the hypothesis about the biomechanical role of the threads (Dafni 1983b). Indicative of the mesenterial pull are conspicuous protuberances on the internal surface of the interambulacral plates, where these threads are attached (Plate 14.1). Their orientation corresponds with that of the above-mentioned vectors (Dafni 1983b; Dafni and Erez 1987b).

These facts lead to the conclusion that these elements form an 'equilibrium of forces', which determines echinoid growth pattern. Hence, deformities in urchins exposed to pollution (Allain 1978; Dafni 1980, 1983b; Dafni and Erez 1987b), may well result from deviation from this equilibrium.

Although not treated in this analysis, lantern muscles also play a biomechanical role, pulling the lowermost part of the test, the peristome margin, internally. This pull is possibly counteracted by lantern growth, which limits peristome marginal growth, as indicated by the positive allometry of the lantern size, as compared to the peristome diameter (Dafni and Erez 1987a).

The soap-bubble analogy

This is another illustration of the biomechanical nature of echinoid plate growth. Thompson (1917) and Raup (1968) pointed out the resemblance between echinoid plate patterns and soap-bubble configurations: equal-sized plates share straight suture lines, whereas unequal plates have curved sutures, with the smaller plates arching into their larger neighbours. The typical plate patterns greatly change ontogenetically, while they move downwards and change their size relative to their neighbours.

Using the pneu paradigm, the soap-bubble analogy of the plates is easily explained (Dafni 1986a); the highly porous plate is contained in a membrane-surrounded 'capsule', containing the quasi-fluid stroma, which serves as a template for calcification (Moss and Meehan 1967). Thus, the echinoid plates are continuously growing densely-packed pneus, fixed by calcification within the boundaries of their capsule, and the test as a whole is best described as a complex mineralized pneu. It has also been demonstrated that whole tests, as well as their dead, decalcified plate capsules, tend to retain their typical bubble-like form, owing to the flexible membrane and fluid content (Dafni 1986b) (Plate 14.2).

Although plate growth is controlled by the physico-mechanical relations between them, observations suggest that the plate plasticity is constrained by the mineral phase. When two neighbouring plates grow at different rates, the curvature of their common sutures usually reverses, as predicted by the soap-bubble analogy. Lowermost, non-growing coronal plates often fail to change their plate curvature, since this must involve resorption, which is extremely rare in echinoid tests (Dafni 1986a).

Pollution induced deformities

Skeletal deformities may serve as 'natural experiments', demonstrating the effect of regulatory factors on the developing body. Three types of echinoid deformities, developed in severely polluted sites, affecting the majority of *T. gratilla elatensis* populations there (Dafni 1980, 1983b; Dafni and Erez 1987b), were interpreted using the above model

(1) The extremely flat tests with aboral depression apparently resulting from a mechanical collapse of the aboral 'dome' (Dafni 1983b) (Plate 14.3).
(2) Five symmetrically arranged pits, superimposed on the aboral depression, coincidental with the uppermost attachment of the mesenterial threads to the test plates, appeared in many of the deformed tests. They are believed to be caused by internal tethering (Dafni 1983b).
(3) Another deformity, excessively tall and irregular-shaped urchins (Dafni 1980), was often accompanied by abnormal narrowing of the ambulacra, forming 'ambulacral pinches' (Koehler 1924; Jackson 1927; Moore 1974), which occurred in 69 per cent of the abnormally tall urchins (Plate 14.4). It has been suggested that the pinches result from a 'strangling effect', imposed by internal tethering (Dafni 1984; Dafni and Erez 1987b).

Deviations from radial symmetry

It has been suggested that deviation from the round symmetry of regular echinoids is associated with imperfect development of the ambulacral system (Chadwick 1924). Since the latter component produces a biomechanical effect, Dafni (1986a) proposed that the perfect pentaradial symmetry of regular echinoids corresponds to a symmetrical arrangement of the ambulacral system. He tested this hypothesis by inactivating one ambulacral ray of normal *T. gratilla elatensis*, severing the radial vessel and the associated neural system. The normal development of this ambulacrum was arrested, and the apical plate system became strongly oval. Consequently, the entire test became bilaterally symmetrical, with the damaged ambulacrum and the elongated apical diameter in the plane of symmetry. This was followed also by considerable skewedness and curving of the non-damaged ambulacra towards the damaged one (Plate 14.5).

Similarly, it has been shown that 'imperfect' tetramers, sea-urchins having four intact ambulacral rays, with a normal five-jawed lantern, always show a rudimentary fifth ambulacrum near the peristome, the growth of which was apparently impeded by physical damage during early post-larval ontogeny (Dafni 1986a).

OTHER ENVIRONMENTALLY-ASSOCIATED TRAITS

A number of additional morphological traits seem to be linked with the physico-mechanical environment.

Thicker tests

Echinoid species living in surf-exposed habitats tend to have thicker tests (Ebert 1982). Moreover, populations of the same species living in more exposed habitats usually have more massive tests than those from sheltered habitats (Moore 1935; Lewis and Storey 1984).

Thicker spines

Echinoids living in wave exposed reef habitats have thicker spines than conspecific populations inhabiting sheltered areas (Dotan and Fishelson 1985).

Roundness

The relationship between the long and short diameter in the oval-shaped echinoid *Echinometra lucunter* is also linked to the mechanical interaction between the test and the habitat (Lewis and Storey 1984).

PHYLOGENETIC CONSIDERATIONS

Teratological phenomena and experimentally induced deformities have often been used to provide evidence for a phylogenetic relationship between distantly related taxa, offering insight into evolutionary processes.

It is tempting to speculate, observing the striking resemblance of the abnormal *T. gratilla elatensis* phenotypes to those of more specialized echinoid taxa (e.g. irregulars), that these deformities actually 'mimic' phylogenetic development occurring in this group.

Jackson (1927) suggested that abnormal variation may serve as a clue to developmental processes. He grouped the various deformities in five types, related to phylogenetic trends:

(1) *arrested variation*—characters appearing in juveniles of one species, which are shared by adults of allied primitive species;
(2) *progressive variation*—not typical of the species, but which occurs typically in more specialized species;
(3) *regressive variation*—bearing characters of primitive species of the same lineage;
(4) *parallel variation*—variation shared by taxonomically unrelated species;
(5) *aberrant variation* that bears no resemblance to any known phylogenetic trend.

Jackson argued that even the aberrant variation is apt to follow definite lines of its own.

The following discussion highlights ontogenetic 'progressive variations' in the test shape among anomalous regulars, showing their correspondence to recognized phylogenetic trends among the Echinoidea. The rationale for this deduction is that the same constructional 'repertoire' is available for both processes.

Phylogenetic trends within the Echinoidea are discussed in greater detail by Durham (1966), Kier (1974) and Smith (1984). The present discussion is restricted to changes which affected the whole test and other characteristics related to our model.

ADAPTIVE RADIATION OF TEST MORPHOLOGY IN THE REGULARIA

Since earliest Palaeozoic echinoids ('Perischoechinoidea') were almost spherical (Durham 1966; Smith 1984), rigid or flexible (e.g. *Bothriocidaris* and *Eothuria*), it is safe to assume that echinoid phylogeny started with a non-specialized spherical archetype, evolving into the varied forms which inhabit today's marine environments.

Deviation from the globular archetype

Incorporation of the pneu paradigm into the echinoid archetype was facilitated by the flexible articulation of up to 3000 plates (Kier 1974). Hence, a major constraint ruling echinoid evolution was the mechanical balance emanating from the morphology (relative size, thickness, imbrication), ontogeny (vertical shift), and spatial organization of the plate shield. Other constraints which may have influenced this process were the pentamery imposed by the ambulacral pull; the mechanical properties and spatial arrangement of the alimentary system, festooned from the inner surface of the test; and possibly also the amount and organization of the sutural collagenous fibres, which provide echinoid tests with a varied degree of flexibility (Wilkie 1984).

While whole test morphology and plate close-packing patterns are biomechanically determined, the sculpture of the external surface and the shape of other skeletal appendages (spines, pedicellaria, etc.), as well as structures emanating from functional adaptation, are evidently under direct genetic control.

A perfect globular shape is of no advantage for epibenthic echinoids, because they must orientate their body to the substratum for feeding, or stabilize it against external forces. It is, therefore, likely that the globular shape of Palaeozoic echinoids, several of which were probably elongated in the vertical axis (Kier 1966), may have been associated with a reclined posture, not unlike that of modern sea cucumbers.

Some globular species with long spines, such as the deep sea diadematoid *Plesiodiadema indicum*, display a stilt-like gait on the loose substratum (Durham 1966). Extant globular echinoids with short spines are known to cradle among seaweeds, clinging tenaciously to the algal tissue around them (Smith 1978). In both types there is no significant downward component of the ambulacral adherence. A similar tendency is shown by cidaroids, extant and extinct, whose feeble tube-feet emerge from a very narrow ambulacrum. They usually use their thick spines to fasten to rock crevices. Although both cradled and spine-supported urchins may become flattened owing to an inwards pull by mesenterial threads and other tethering tissues, the plane of their ambitus is usually at mid-height (Dafni 1986a).

Effect of ambulacral system development

A conspicuous trend in the phylogeny of the Regularia is a progressive increase of ambulacral adherence. The tube-feet pores of early Palaeozoic (Ordovician) species, located next to the perradial (mid-ambulacral) suture, shifted later to their present position, at the adradial edge of the ambulacral plates (Kier 1966). A further development was the appearance of compound plates, with many pores pairs for each plate (Kier 1974).

However, the tendency to increase the circumferential angle of the ambulacral system prevailed throughout echinoid evolution. A morphometric survey (Dafni 1986a,b), in which 150 modern and fossil post-Palaeozoic rigid echinoid species from four orders were analysed, representing progressive evolutionary stages showed (Fig. 14.3):

(1) the mechanical efficiency of the ambulacra, represented by the relative breadth of the ambulacral system, increased, from 0.2 of the interambulacral breadth in the cidaroids, up to 0.6 in the most advanced group (Camarodonta);
(2) cidaroid species display an almost median ambitus, while in the other groups the ambitus is progressively lowered.

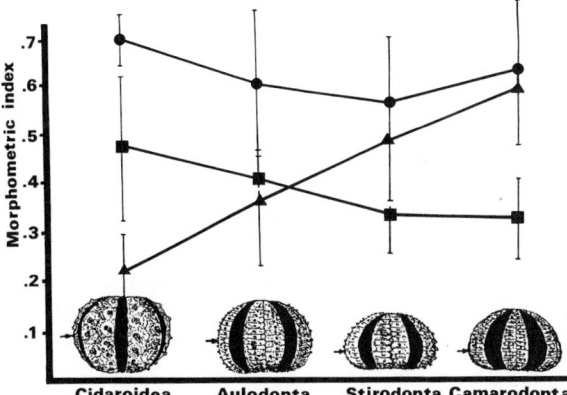

Fig. 14.3 Morphological indices among four regular echinoid orders: H/D ratio (●); ambulacral system relative breadth (▲); ambitus position (■). The breadth of the ambulacra in a typical representative of each order is emphasized by darkening. Vertical bars = SD (Dafni 1986a).

The morphometric survey failed, however, to show a significant trend in tallness (H/D ratio). This suggests that although lower H/D ratio generally reflects enhanced ambulacral adherence, this effect is probably restricted to ontogeny, whereas the related phylogenetic trend is lowering of the ambital line.

Adaptation to rock boring

An unusual test profile is shown by urchins of the tropical genus *Echinostrephus*. Their test is usually flat aborally and is markedly conical and tapering adorally (Fig. 14.4). The ambitus lies approximately at, or slightly above mid-height. This unique form is possibly an adaptation to infaunal life on a hard substratum. *Echinostrephus* live in perfectly round holes slightly larger than their size, which they bore deep into limestone rocks, using their short adoral spines and possibly also their jaws (Campbell *et al.* 1973). The longer aboral spines are apparently used for protection.

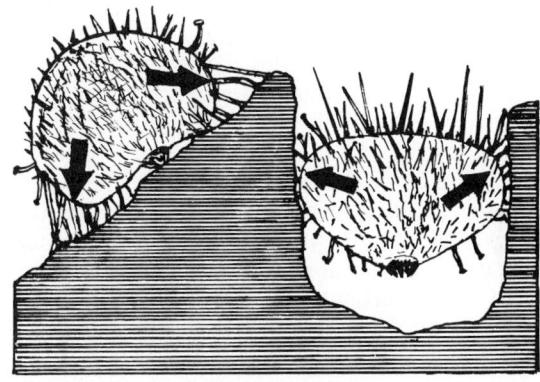

Fig. 14.4 A semidiagrammatic representation of a typical regular echinoid (left) and *Echinostrephus* shown in its natural habitat, bored holes in the rock. Arrows show the normal orientation of the ambulacral tube-feet, possibly affecting the test profile.

This unique profile reflects the continuous pull exerted by the ambulacral tube-feet sidewards, and toward the aperture, compared to the downwards orientation of most epifaunal echinoids. A similar profile is shown also by several other echinoid species such as *Heliocidaris erythrogramma armigera*, described by Mortensen (1943) as rock-boring, and the Jurassic cidaroid *Procidaris edwardsi*, whose habitat is unknown (Mortensen 1928).

Oral flattening in soft-bottom echinoids

Flattening of the subambital region is uncommon among regular echinoids. Yet it is shown by several Recent deep-sea regular genera, such as *Hebrocidaris*, *Baueria*, etc. (Mortensen 1935). These echinoids have a low ambitus, forming a sharp angle separating the domed aboral, from the flat adoral, side. This is evidently an adaptation to allow epibenthic locomotion on unconsolidated substrata.

This unusual flattening seems to be independent

of ambulacral activity, since tube-feet adherence to the loose sediment is feeble, and their only means of locomotion are the short adoral spines. A similar phenomenon, shown by the holectypoids and holasteroids, marks an early stage in the evolution of irregular echinoids (Smith 1984).

The smaller subambital plates of those early irregulars, unlike those of modern sand-dollars (Seilacher 1979), indicate that their sharp ambital bending did not interrupt plate migration beyond the ambitus. Hence, the sutural binding of these plates must have changed its elasticity, yielding to the effect of gravity to form the sharp ambital angle, which never became fixed.

The variable tensility of echinoderm collagen (Wilkie 1984; Motokawa 1985) is a possible mechanism accounting for reversible oral flattening and ambital bending. It is possible that imbrication, found in plates of *Holectypus* and other irregulars (Jesionek-Szymanska 1968), facilitated sutural articulation of bending sutures.

TRENDS IN THE EVOLUTION OF IRREGULAR ECHINOIDS

The emergence of the Irregularia marks a transition from epibenthic to infaunal habit, accomplished through profound morphogenetic changes which took place within a relatively short period in the Mesozoic (Kier 1982).

Five evolutionary trends are relevant in this development.

Bilateral symmetry

Common to all irregular echinoids, this trend was heralded by the departure, in the Holectypoida, of the periproct from the apical region, moving downwards along interambulacrum 5 (Lovén's plane), well below the ambitus. In advanced irregular orders this was associated with elongation of the interambulacrum 5–ambulacrum III axis. In the Spatangoida, ambulacrum III became anterior and adapted to serve as a food-groove, while the periproct marks the posterior end. In several taxa this process was accompanied by arching forward of the paired ambulacra (Smith 1984).

Although it seems as if this chain of events started by the backward migration of the anus, anal migration and true bilateral symmetry are not necessarily linked. The tendency to move the periproct away from the apex was shown in groups, which are otherwise classified as regular echinoids, such as acrosaleniid species (Jesionek-Szymanska 1968).

It is speculated that in phylogeny, the excentric position of the periproct and bilateral orientation impair the mechanical balance between the ambulacra, producing a mechanical constraint on the perfect pentameral symmetry. This is supported by the frequent occurrence of teratologically induced 'bilateral' symmetry, apparently by damaging one ambulacrum (Plate 14.5), among a majority of almost perfect pentaradial specimens (Koehler 1924; Jackson 1927; Mortensen 1943; Dafni 1986a). It is plausible that ontogenetic 'bilateria' results from the same mechanical constraints which dominated bilateral symmetry in the evolution of irregular echinoids.

The observation that the ontogeny of irregular echinoids (e.g. spatangoids) starts with a slightly oval test with pentameral symmetry, and that periproctal shift and the typical bilateral symmetry are obtained during subsequent ontogenetic development (Gordon 1926), enables a comparison between the phylogenetic and teratologically induced bilateral phenomena.

Test flattening

Adoral flattening, shown by the earliest irregulars, developed into the extremely flat tests shown by the Clypeasteroida. It is reasonable to assume that this trend was primarily an adaptation to epibenthic life on soft substrata, as suggested for modern deep-sea regular echinoids, but later shifted to infaunal burrowing (Seilacher 1979). This development was facilitated by both a streamlined hydrodynamic design (Telford 1981) and constructional modification, such as sutural interlocking (Seilacher 1979) and magnetite accumulation (Chia 1973).

Two apparently antagonistic trends are exhibited here: (1) flattening by inner tether combined with yielding to gravity; and (2) creation of a supportive structure to prevent the entire collapse of the test. It is suggested that variable tensility of collagenous tissues may account for the former trend (see above), whereas the latter effect is discussed in the following section.

Internal pillar support

The introduction of vertical pillars across the coelomic cavity in the clypeasteroids, stiffened the test and enabled its infaunal way of life. Aside from fixing the flat profile of the test, it also arrested plate migration around the ambitus. In the suborder Scutellina the large marginal plates are folded around the ambitus, remaining in this position throughout adult life, whereas in the Clypeasterina the ambital margins consist of several narrow plates which are nevertheless fixed by the internal pillars (Seilacher 1979).

The origin of these pillars is obscure. Observations suggest their association with mesenterial threads (Dafni 1986b). In the epibenthic *Clypeaster rosaceus*, with an inflated test and relatively few pillars (Poddubiuk 1985), these pillars span the aboral interambulacra to the adoral interambulacra next to the lantern auricles (Plate 14.6). The orientation of the pillars is thus similar to that of the intestinal mesenterial threads of the Regularia (Dafni 1983b). Although in other clypeasteroids the pillars form a more complex pattern, the main connection is nevertheless between interambulacral plates in both sides of the ambitus. It is noteworthy that the pillars have a median suture (Seilacher 1979; Smith 1980), where the horizontal mesenteria are stabilized, appearing as if they are pierced by the pillars.

The association of the pillars with the mesenterial threads suggests that the latter may serve as guiding tissue, directing the aboral calcareous 'stalactites' towards the adoral 'stalagmites'. It also suggests that the same tissue which might have produced the inner tether has turned into a mechanical supportive structure simply by its mineralization.

Formation of the petals

In irregular echinoids the aboral ambulacra are typically adapted for gaseous exchange, whereas the lower, adoral ambulacra mainly became part of a highly complex food-gathering system. The aboral ambulacra are constricted at some distance from the apex, forming the flower-like petals, with narrow plates and specialized pores. Petals vary from one group to another within the Irregularia, and some variability in their shape is associated with the environmental conditions prevailing in the natural habitat (Smith 1984).

A striking similarity shows between petal constrictions and 'pinches' appearing in deformed urchins (Plate 14.4). First, they are found at about the same 'latitude', halfway between the ambitus and the apex; and secondly, it has been noticed that these constrictions usually coincide with the point where the foregut crosses the ambulacral ray (J. Dafni, unpublished observations). This further supports the hypothesis that they are produced by a biomechanical effect, the inward pull by mesenterial threads on either side of the ambulacra, which squeezed them together. Furthermore, it is suggested that 'pseudopetals' (Plate 14.7), consisting of five concurring pinches, that form under the specific pathologic conditions which characterize *T. g. elatensis* tall deformity (Plate 14.4), are a 'progressive variation' in Jackson's (1927) terminology. This suggestion can potentially be tested by observing the ontogeny of irregulars, in particular by studying the relations between the gut mesenteries and the development of petals at the early post-larval stage.

Brooding chambers in regular and irregular echinoids

Brooding chambers, in the form of coronal depressions, are a special adaptation found in cold-water echinoids (Hyman 1955). Philip and Foster (1971) listed several types of brooding chambers, ranging from a huge aboral cavity to five equal-sized depressions developed either within the petal region, or in other species-specific locations.

From a morphogenetical point of view these brooding chambers are 'natural deformities', rendered useful for this particular requirement. It is not surprising, therefore, that several of these shapes are reminiscent of abnormal depressions shown in deformed regular echinoids:

(1) A large, single aboral depression was typical for larger specimens of *T. g. elatensis*, possibly afflicted by calcification inhibitors (Dafni 1983b);
(2) small cavities at each of the ambulacral tips—occurring less frequently in both types of deformity (Plate 14.8).

The formation of these cavities can possibly be induced either by internal tethering or by localized 'softening' of the collagen links between adjacent plates, which allows them to grow excessively at the sutures and sag inwards.

CONCLUSIONS

Recently experimental observations have added to the hitherto descriptive skeletal research on the role of biomechanics. This review tries to explain the ontogeny and phylogeny of the echinoid skeleton in terms of continuous interaction of the genetically controlled structure with the physico-mechanical environment. Although the proposed model is still sketchy and roughly defined, it provides possible clues to the morphogenesis and constructional morphology of echinoids and other invertebrates with hard skeletons.

First, it defines the sort of structural modifications which are more likely to occur, either in normal or abnormal growth conditions, using universal laws and constructional principles.

Secondly, it links ontogenetic and phylogenetic changes, by assuming that the same constructional repertoire applies to both processes. Moreover, it seems that the ontogeny of advanced forms 'recapitulates' their phylogeny, simply because the growth process itself is self-organizing, and mainly depends on, or is controlled by, the constructional materials, which are similar, yet subjected to evolutionary changes.

Thirdly, this model enables us to put forward fruitful hypotheses for further experimental research, aiming to gain deeper insight into the constructional morphology of fossil and living forms.

REFERENCES

Allain, J. Y. 1978. Déformations du test chez l'oursin *Lytechinus variegatus* (Lamarck) de la Baie de Carthagene. *Caldasia* **12**, 363–75.

Bach, K., et al. 1976. *Pneus in Natur und Technik.* Publication 9, Instituts für Leichte Flächentragwerke, University of Stuttgart.

Bonner, J. T. (ed.) 1961. *On growth and form* (abridged edn). Cambridge University Press, Cambridge.

Campbell, A. C., Dart, H. K. G., Head, S. M., and Ormond, R. F. G. 1973. The feeding activity of *Echinostrephus molaris* (de Blainville) in the central Red Sea. *Marine Behaviour and Physiology* **2**, 155–69.

Chadwick, H. C. 1924. On some abnormal and imperfectly developed specimens of the sea urchin *Echinus esculentus*. *Proceedings of the Zoological Society of London* **94**, 163–72.

Chia, F-S 1973. Sand dollar: a weight belt for the juvenile. *Science, New York* **181**, 73–4.

Dafni, J. 1980. Abnormal growth patterns in the sea urchin *Tripneustes* cf. *gratilla* (L.) under pollution (Echinodermata: Echinoidea). *Journal of Experimental Marine Biology and Ecology* **47**, 259–79.

Dafni, J. 1983a. A new sub-species of *Tripneustes gratilla* (L.) from the northern Red Sea (Echinodermata: Echinoidea: Toxopneustidae). *Israel Journal of Zoology* **32**, 1–12.

Dafni, J. 1983b. Aboral depressions in the tests of the sea urchin *Tripneustes* cf. *gratilla* (L.) in the Gulf of Eilat, Red Sea. *Journal of Experimental Marine Biology and Ecology* **67**, 1–15.

Dafni, J. 1984. *Skeletal growth and calcification in the short-spined sea urchin (Tripneustes gratilla elatensis).* PhD Dissertation, Hebrew University, Jerusalem. (In Hebrew with English summary.)

Dafni, J. 1985. Effect of mechanical stress on the calcification pattern in regular echinoids. In *Echinodermata: Proceedings of the fifth International Echinoderm Conference, Galway* (ed. B. F. Keegan and B. D. S. O'Connor), pp. 233–6. A. A. Balkema, Rotterdam.

Dafni, J. 1986a. A biomechanical model for the morphogenesis of echinoid tests. *Paleobiology* **12**, 143–60.

Dafni, J. 1986b. *Echinoid skeletons as pneu structures.* Universität Stuttgart und Tübingen, Stuttgart.

Dafni, J. and Erez, J. 1982. Differential growth in *Tripneustes gratilla* (Echinoidea). In *Echinoderms: Proceedings of the International Conference, Tampa* (ed. J. M. Lawrence) pp. 71–5, A. A. Balkema, Rotterdam.

Dafni, J. and Erez, J. 1987a. Skeletal calcification patterns in the sea urchin *Tripneustes gratilla elatensis*: I. Basic patterns. *Marine Biology* **95**, 275–87.

Dafni, J. and Erez, J. 1987b. Skeletal calcification patterns in the sea urchin *Tripneustes gratilla elatensis*: II. Effect of various treatments. *Marine Biology* **95**, 289–97.

Dafni, J. and Tobol, R. in press. Population structure patterns of a common Red Sea echinoid (*Tripneustes gratilla elatensis*). *Israel Journal of Zoology* **34**.

Deutler, F. 1926. Über das Wachstum des Seeigelskeletts. *Zoologische Jahrbücher* **48**, 119–200.

Dotan, A. and Fishelson, L. 1985. Morphology of *Heterocentrotus mammillatus* (Echinodermata, Echinoidea) and its ecological significance. In *Echinomata: Proceedings of the fifth International Echinoderm Conference, Galway* (ed. B. E. Keegan and B. D. S. O'Connor), pp. 253–60. A. A. Balkema, Rotterdam.

Durham, J. W. 1966. Phylogeny and evolution. In *Treatise on Invertebrate Paleontology* (U) Echinodermata 3 (ed. R. C. Moore), pp. U266–70. Geological Society of America and University of Kansas Press, Lawrence, Kansas.

Ebert, T. A. 1967. Negative growth and longevity in the purple sea urchin *Strongylocentrotus purpuratus* (Stimpson). *Science, New York* **157**, 557–8.

Ebert, T. A. 1968. Growth rates of the sea urchin *Strongylocentrotus purpuratus* related to food availability and spine abrasion. *Ecology* **49**, 1075–91.

Ebert, T. A. 1975. Growth and mortality of post-larval echinoids. *American Zoologist* **15**, 755–75.
Ebert, T. A. 1980. Relative growth of sea urchin jaws: an example of plastic resource allocation. *Bulletin of Marine Science* **30**, 461–74.
Ebert, T. A. 1982. Longevity, life history, and relative body wall size in sea urchins. *Ecological Monographs* **52**, 353–94.
Edelman, G. M. 1984. Cell-adhesion molecules: a molecular basis for animal form. *Scientific American* **250**, 80–91.
Emlet, R. B. 1982. Echinoderm calcite: a mechanical analysis from larval spicules. *Biological Bulletin of the Marine Biology Laboratory, Woods Hole*, **163**, 264–75.
Gordon, I. 1926. The development of the calcareous test of *Echinocardium cordatum*. *Philosophical Transactions of the Royal Society of London* **B215**, 255–313.
Hyman, L. H. 1955. *The Invertebrates*, Vol. IV. Echinodermata, the coelomate Bilateria. McGraw Hill, New York.
Jackson, R. T. 1927. Studies of *Arbacia punctulata* and allies and of non-pentamerous echinids. *Memoirs of the Boston Society of Natural History* **8**, 435–565.
Jensen, M. 1969. Age determination of echinoids. *Sarsia* **37**, 41–4.
Jensen, M. 1972. The ultrastructure of the echinoid skeleton. *Sarsia* **48**, 39–48.
Jensen, M. 1985. Functional morphology of test, lantern and tube feet ampullae system in flexible and rigid sea urchins (Echinoidea). In: *Echinodermata: Proceedings of the fifth International Echinoderm Conference, Galway* (ed. B. E. Keegan and B. D. S. O'Connor), pp. 281–8. A. A. Balkema, Rotterdam.
Jesionek-Szymanska, W. 1968. Irregular echinoids—an insufficiently known group. *Lethaia* **1**, 50–65.
Kier, P. M. 1966. Noncidaroid Paleozoic echinoids. In *Treatise on invertebrate paleontology* (U) *Echinodermata* 3 (ed. R. C. Moore), pp. 298–318. Geological Society of America and University of Kansas Press, Lawrence, Kansas.
Kier, P. M. 1974. Evolutionary trends and their functional significance in the post-Paleozoic echinoids. *Memoirs of the Paleontological Society* **5**, 1–95.
Kier, P. M. 1982. Rapid evolution in echinoids. *Palaeontology* **25**, 1–10.
Kobayashi S. and Taki, J. 1969. Calcification in sea urchins. I. A tetracyclin investigation of growth of the mature test in *Strongylocentrotus intermedius*. *Calcified Tissue Research* **4**, 210–23.
Koehler, R. 1924. Anomalies, irregularités et déformations au tests chez les echinids. *Annales de l'Institut Oceanographique, Monaco, Nouvelle serie* **1**, 159–480.
Lewis, J. B. and Storey, G. S. 1984. Differences in morphology and life history traits of the echinoid *Echinometra lacunter* from different habitats. *Marine Ecology—Progress Series* **15**, 207–11.
Märkel, K. 1975. Wachstum der Coronar skelettes von *Paracentrotus lividus* Lmk. *Zoomorphologie* **82**, 259–80.
Märkel, K. 1976. Struktur und Wachstum des Coronarskelettes von *Arbacia lixula* Linné (Echinodermata: Echinoidea). *Zoomorphologie* **84**, 279–99.
Märkel, K. 1981. Experimental morphology of coronal growth in regular echinoids. *Zoomorphologie* **97**, 31–52.
McPherson, B. F. 1965. Contribution to the biology of the sea urchin *Tripneustes ventricosus*. *Bulletin of Marine Science* **15**, 2298–44.
Moore, H. B. 1935. A comparative study of the biology of *Echinus esculentus* in different habitats. *Journal of the Marine Biology Association UK* **20**, 109–28.
Moore, H. B. 1974. Irregularities in the test of regular sea urchins. *Bulletin of Marine Science* **24**, 545–60.
Mortensen, T. 1928. *A monograph of the Echinoidea, I. Cidaroida.* C. A. Reitzel, Copenhagen.
Mortensen, T. 1935. *A monograph of the Echinoidea, II. Bothriocidaroida, Melonechinoida, Lepidocentroida, and Stirodonta.* C. A. Reitzel, Copenhagen.
Mortensen, T. 1943. *A monograph of the Echinoidea, III*(2), *Camarodonta*. C. A. Reitzel, Copenhagen.
Moss, M. L. and Meehan, M. 1967. Sutural connective tissues in the test of an echinoid, *Arbacia punctulata*. *Acta Anatomica* **66**, 279–304.
Moss, M. L. and Meehan, M. 1968. Growth of the echinoid test. *Acta Anatomica* **69**, 409–44.
Motokawa, T. 1985. Catch connective tissue: The connective tissue with adjustable mechanical properties. *Echinodermata: Proceedings of the fifth International Echinoderm Conference, Galway* (ed. B. E. Keegan and B. D. S. O'Connor), pp. 69–73. A. A. Balkema, Rotterdam.
Nichols, D. 1982. A biometrical study of populations of the European sea-urchin *Echinus esculentus* (Echinodermata: Echinoidea) from four areas of the British Isles. *Australian Museum Memoirs* **16**, 147–63.
Okazaki, K. 1975. Spicule formation by isolated micromeres of the sea urchin embryo. *American Zoologist* **15**, 567–81.
Pearse, J. S. and Pearse, V. B. 1975. Growth zones in the echinoid skeleton. *American Zoologist* **15**, 731–53.
Philip, G. M. and Foster, R. J. 1971. Marsupiate Tertiary echinoids from South-Eastern Australia and their zoogeographic significance. *Palaeontology* **14**, 666–95.
Poddubiuk, R. H. 1985. Evolution and adaptation in some Caribbean Oligo-Miocene Clypeasters. *Echinodermata: Proceedings of the fifth International Conference, Galway* (ed. B. E. Keegan and B. D. S. O'Connor), pp. 75–80. A. A. Balkema, Rotterdam.
Prosser, L. 1973. *Comparative Animal Physiology*. Saunders, Philadelphia.

Raup, D. M. 1968. Theoretical morphology of echinoid growth. *Memoirs of the Paleontological Society* **2**, 50–63.

Régis, M. B. 1969. Premières données sur la croissance de *Paracentrotus lividus* Lmk. *Tethys* **1**, 1049–56.

Régis, M. B. 1973. Premières données sur la croissance de l'échinid *Arbacia lixula* (L.). *Tethys* **5**, 167–72.

Régis, M. B. 1979. Croissance négative de l'oursin *Paracentrotus lividus* (Lamarck). *Compte Rendu Hebdomadaire des Séances de l'Académie des Sciences, Paris*, **288**, 355–8.

Seilacher, A. 1979. Constructional morphology of sand dollars. *Paleobiology* **5**, 191–221.

Seilacher, A. 1986. Introduction. In *Echinoid skeletons as pneu structures*. Konzepte Sonder forschungsbereich 230 Heft 13, Universität Stuttgart und Tübingen, Stuttgart.

Smith, A. B. 1978. A functional classification of the coronal pores of regular echinoids. *Palaeontology* **21**, 759–90.

Smith, A. B. 1980. Stereom microstructure of the echinoid test. *Special Papers in Palaeontology* **25**, 1–81.

Smith, A. B. 1984. *Echinoid palaeobiology*. Allen and Unwin, London.

Stevens, P. S. 1976. *Patterns in nature*. Penguin, Middlesex.

Strathmann, R. R. 1981. The role of spines in preventing structural damage to echinoid tests. *Paleobiology* **7**, 400–6.

Telford, M. 1981. A hydrodynamic interpretation of sand dollar morphology. *Bulletin of Marine Science* **31**, 605–22.

Telford, M. 1985. Domes, arches and urchins: The skeletal architecture of echinoids (Echinodermata). *Zoomorphology* **105**, 114–24.

Thompson, D. W. 1917. *On growth and form*, 1st edition. Cambridge University Press, Cambridge.

Thompson, D. W. 1942. *On growth and form*, 2nd edition. Cambridge University Press, Cambridge.

Wainwright, S. A., Biggs, W. D., Currey, J. D., and Gosline, J. M. 1976. *Mechanical design in organisms*. Edward Arnold, London.

Weber, J., Greer, R., Voight, B., White, E., and Roy, R. 1969. Unusual strength properties of echinoderm calcite related to structure. *Ultrastructural Research* **26**, 355–66.

Wilkie, I. C. 1984. Variable tensility in echinoderm collagenous tissues: a review. *Marine Behavioural Physiology* **11**, 1–34.

15

Experimental embryology as a tool for studying the evolution of echinoderm life histories

LARRY R. McEDWARD

Department of Zoology, University of Alberta, Edmonton, Alberta T6G 2E9, Canada

An understanding of life history evolution in echinoderms requires information on the diversity of life histories that have evolved and the processes by which they undergo evolutionary change. Over the past 50-odd years, descriptive and comparative studies have provided much information on what life history patterns exist, what traits characterize each pattern, and in which taxa and in which environments these patterns occur (e.g. Thorson 1936, 1946, 1950; Mileikovsky 1971, 1974; Chia 1974; Strathmann 1974; Hendler 1975; Emlet *et al.* 1987). Life history studies in echinoderms (and other marine invertebrates) have focused on the adaptive significance of the three most common modes of development: planktotrophy, pelagic lecithotrophy, and brooding (reviewed by Grahame and Branch 1985; Strathmann 1985).

Recently, interest has developed in understanding the mechanisms of life history evolution, i.e. how natural selection modifies reproduction and development to yield the patterns we observe in nature. Quantitative, theoretical models have been the primary tool used to explore the effects of selection on various life history traits (e.g. Vance 1973*a,b*; Christiansen and Fenchel 1979; Strathmann 1985). Future developments will depend on successful integration of the insights from theoretical research with the generalizations from empirical studies.

Caution must be exercised when attempting to integrate the data base obtained from empirical studies of echinoderm reproduction and development with the life history theory. The reason is that data relevant to describing and classifying life history patterns may not be appropriate for the development and testing of theoretical models which explore how evolutionary change occurs. To illustrate this point, I will concentrate on the relationship between parental investment and offspring fitness. It is instructive to contrast the information that we obtain by describing patterns of parental investment with the information we need to evaluate responses of parental investment to natural selection. The assumption in life history studies is that as parents allocate more materials and energy to each individual offspring, they produce fewer offspring. However, this is balanced by the fact that those offspring are of higher quality and are more likely to survive and reproduce (Fig. 15.1). In other words, it is assumed that parental investment per offspring directly determines offspring fitness (e.g. Smith and Fretwell 1974; Vance 1973*a,b*; Christiansen and Fenchel 1979). This is a central assumption of the theory that underlies most of the quantitative models.

Species comparisons have shown some striking correlations between egg size (an index of parental investment) and several traits likely to be important determinants of larval success (e.g. Thorson 1946). For example, there are two basic life history patterns in asteroids with pelagic development (Table 15.1, see Emlet *et al.* 1987). Species that free-spawn do not provide parental care after release of the gametes, therefore egg size (organic content) constitutes the entire investment by the parent. The planktotrophic pattern (e.g. *Pisaster ochraceus*, *Pycnopodia helianthoides*) is characterized by small eggs (approx. 100–200 μm diameter) that are produced in large numbers (approx. 10^5–10^7 per female per reproductive season). Offspring from small eggs typically spend several weeks to several months in the plankton and develop into planktotrophic (feeding) larvae. Pre-

Echinoderm phylogeny and evolutionary biology (ed. C. R. C. Paul and A. B. Smith). Clarendon Press, Oxford, 1988.

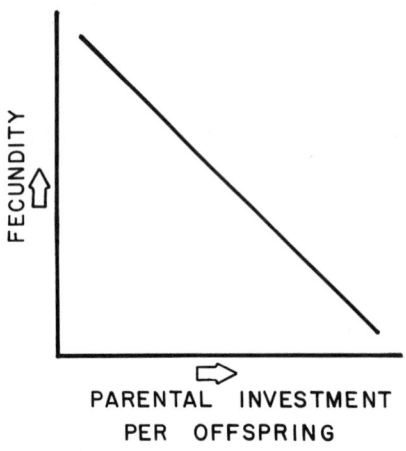 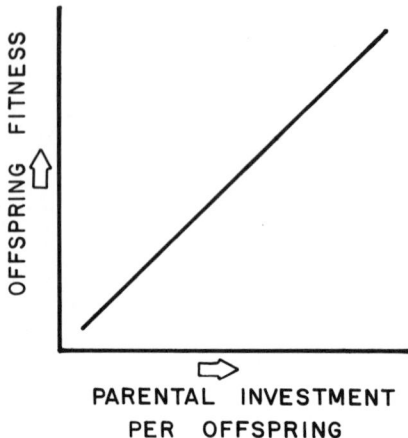

Fig. 15.1 Diagrammatic representation of the assumed relationships among parental investment per offspring, fecundity, and offspring fitness in the evolution of echinoderm life histories.

TABLE 15.1 Characteristics of life history patterns in asteroids with pelagic larvae. Data drawn from Emlet et al. 1987

Characteristic	Planktotrophy	Lecithotrophy
Egg size (μm diameter)	100–200	800–1400
Fecundity (annual)	10^5–10^7	10^3–10^5
Pelagic period	weeks to months	days to weeks
Larval type	feeding	nonfeeding
Dispersal capability	very high	moderate to high
Larval mortality	very high	high
Juvenile size (μm diameter)	500	1000
Examples	*Pisaster ochraceus*	*Mediaster aequalis*
	Pycnopodia helianthoides	*Solaster stimpsoni*

sumably, they suffer high mortality. Planktotrophic larvae recruit (via settlement and metamorphosis) into benthic populations as small juveniles (approximately 500 μm diameter). In the alternative pattern, pelagic lecithotrophy (e.g. *Mediaster aequalis, Solaster stimpsoni*), much larger eggs (approx. 1000 μm diameter) are produced, but in smaller numbers (10^3–10^5) per reproductive season). The pelagic period is only a few days to 1 or 2 weeks, and the larvae do not feed. Presumably, they suffer low mortality and recruit as large juveniles (approx. 1000 μm diameter). The features that distinguish these patterns are likely to influence larval energetics, dispersal, mortality by predation, and the success of the early benthic stages. Most asteroids either follow the small egg-high fecundity pattern or the large egg-low fecundity pattern. Few species are intermediate or follow mixed strategies (Fig. 15.2).

Comparative biology provides a powerful tool for elucidating the correlations among traits that yield these life history patterns. However, there are two important reasons why the comparative approach limits what we can learn about the evolution of parental investment. First, comparisons can identify correlations among characters, but cannot identify the causal relationships that produce the correlations. It is not clear whether egg size directly influences development time or juvenile size, or whether they vary together because of relationships with other important traits. Different life histories are characterized by non-random associations of traits, which represent co-adapted suites of characters that were

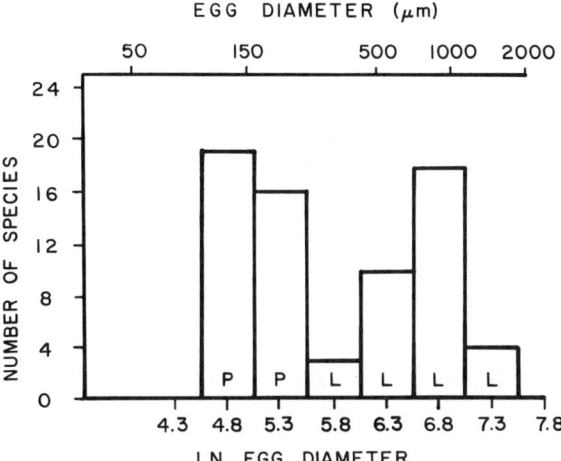

Fig. 15.2 Frequency histogram of the number of asteroid species and egg size. Only species with pelagic larvae are included. Egg size is reported as diameter. Modified from Emlet *et al.* 1987.

designed by natural selection to solve an ecological problem (Stearns 1977). We do not know, however, if those associations are the result of selection acting on relatively independent traits or if they represent selective compromises determined in large part by strong interdependence of traits.

The second reason is related to the fact that natural selection does not operate on differences among species, families, or classes, rather it operates on intraspecific variation. Although the quantitative models of marine invertebrate life history evolution (e.g. Vance 1973a,b; Christiansen and Fenchel 1979) provide insights into the adaptive features of different modes of development, they are designed to predict responses of different life history characters to natural selection. Clearly, 'crude interspecific comparisons are inadequate to reveal intraspecific evolutionary potential' (Vance 1974, p. 880). It is the relationship between life history traits, within any one species, that underlies the evolution of reproductive mode (Vance 1974).

It is possible to measure intraspecific variation in parental investment by comparing individuals within and among populations. Biologists routinely measure egg size and infer parental investment per offspring (i.e. egg organic content). A reliable rank correlation exists between egg size and organic content, when measured among species of echinoderms (Strathmann and Vedder 1977; Turner and Lawrence 1979). However, I have found that within species, egg size is a poor predictor of parental investment. Comparisons were made among females from different populations, within populations, and among the sibling offspring (i.e. within single spawns) of individual females. Measurements of the volume and the organic content of individual eggs failed to yield useful predictive relationships in the three species of asteroids examined: *Solaster stimpsoni*, *Pteraster tesselatus*, and *Henericia sp.* (*leviuscula*?) (McEdward and Carson 1987; McEdward and Coulter 1987). If, as these results suggest, parental investment cannot be inferred from egg size, then the organic content of an egg must be measured directly; but that destroys the egg. Therefore, it is not possible to measure parental investment and examine its consequences for larval development within any one species. These results call into question the validity of applying interspecific trends and correlations in life history characters to estimation of intraspecific variation.

An alternative to the comparative approach is experimental manipulation of life history traits. The regulative nature of echinoderm development provides a rare opportunity to manipulate parental investment and study its effects on subsequent stages of development. This is not possible in most phyla of marine invertebrates (Kume and Dan 1968; Reverberi 1971; M. Strathmann 1987). It is known from the work of Horstadius (1936, 1973, 1975) and other developmental biologists that blastomeres isolated from early embryos will develop into qualitatively normal larvae. Isolation of blastomeres is equivalent to a drastic reduction in the size of the egg. This allows observation of the effects of egg size without the genetic differences that complicate interpretation of specific comparisons. These experiments provide direct evidence of the causal relationships among life history traits.

Isolation of blastomeres requires removal of the fertilization membrane and, in echinoids, dissolution of the hyaline which binds the blastomeres together. Various techniques have been described for removal of fertilization membranes (Hinegardner 1975a). Note that chemical procedures are often effective only on one or a few species. Hyaline can be dissolved in calcium-free seawater (Hinegardner 1975a), Blastomeres can be isolated by various physical and chemical means (see Horstadius 1936, 1973; Harvey 1940; Okazaki and Dan 1954; Takahashi and Okazaki 1979 for echinoids; see Dan-Sohkawa and Satoh 1978; Mita 1983 for asteroids). Isolation of blastomeres at the two- and four-cell stages, produces 'eggs' of 1/2 and 1/4 of the normal size. These 'eggs' are totipotent cells capable of development through metamorphosis.

Sinervo and McEdward (1988) utilized blastomere isolation techniques to examine the effects of an experimental reduction in egg size (parental investment per offspring) on larval size, development rate, and size of the juvenile at metamorphosis in two species of sea urchins. The use of two species (*Strongylocentrotus droebachiensis* and *S. purpuratus*) allows comparison of experimental results with correlates of an evolved change in egg size. This provided a check on the results of the experimental treatments as well as indicating the extent of the differences in the life histories of these two species that were attributable to differences in egg size. *Strongylocentrotus droebachiensis* and *S. purpuratus* are closely related, co-occurring sea urchins with similar later larval stages but very different egg sizes and early larvae. *Strongylocentrotus purpuratus* has an egg of approximately 80 μm diameter and *S. droebachiensis* has an egg 150 μm diameter (Table 15.2). The eggs differ by six-fold in volume, and Strathmann and Vedder (1977) measured a five-fold difference in egg organic content. Larval sizes and developmental rates were compared among the following treatments: normal size *S. droebachiensis*, 1/2 size *S. droebachiensis*, 1/4 size *S. droebachiensis*, normal size *S. purpuratus*, and 1/2 size *S. purpuratus*. These treatments provided a series of egg sizes spanning an order of magnitude in volume.

Three clear results were obtained by Sinervo and McEdward (1988):

(1) development rate was correlated with egg size;
(2) larval size was correlated with egg size;
(3) size at metamorphosis was independent of egg size.

The effects of egg size were restricted to the early larval stages, prior to the six-armed pluteus stage. Rates of development and larval sizes were similar among treatments and between the two species during later stages of larval development.

These results confirm the importance of the level of parental investment per offspring in determining the correlations among early life history traits in strongylocentrotid echinoids (McEdward 1986a). Reduced larval size may be a general result of reduced egg size among echinoids. Experimental reduction in the volume of cytoplasm in the early embryonic stages does not influence blastomere size, nor cleavage schedules, but does reduce the total number of cells in the embryo and early larva (Hagstrom and Lonning 1964, 1965, on *Echinocyamus pusillus* and *Psammechinus miliaris*; Takahashi and Okazaki 1979, on *Clypeaster japonica*; Dan-Sohkawa and Satoh 1978, on *Asterina pectinifera*). The developmental basis of size regulation during the larval stages is not known, but may be a general phenomenon among echinoids (Harvey 1949, on *Arbacia punctulata*; Hinegardner 1975b, on *Lytechinus pictus*). Observations by McEdward (1986b), Kawamura (1970), and Naidenko (1983) indicate that initial differences in larval size among strongylocentrotid urchins did not persist throughout development. An important consequence of size regulation is constancy of juvenile size at metamorphosis. Metamorphic size is remarkably conservative within the class Echinoidea, especially compared to the Asteroidea (Emlet *et al.* 1987). The significance of metamorphic size for juvenile survival deserves investigation.

There are several potential advantages of larger body size in planktotrophic larvae. Larger larvae

TABLE 15.2 Comparison of egg sizes and organic contents between *Strongylocentrotus droebachiensis* and *Strongylocentrotus purpuratus*. Egg size data from Sinervo and McEdward (1988). Egg content data from Strathmann and Vedder (1977). Egg content has units of μg organic matter per egg

Species	Egg diameter	Egg volume		Egg content
Strongylocentrotus purpuratus	82 μm	0.30	$10^6 \mu m^3$	100 μg
Strongylocentrotus droebachiensis	152 μm	1.84	$10^6 \mu m^3$	500 μg

have more extensive feeding structures (McEdward 1984, 1986a) and greater maximal clearance rates (Strathmann 1971). They can utilize a wider size range of food particles (Strathmann 1971) and, therefore, obtain more food from any given volume of water. It has been assumed that larger larvae have greater body reserves and should be more resistant to starvation (Vance 1973b), but empirical data are lacking. Since larvae that are derived from larger eggs start larger and need to grow less to reach metamorphic size, they should require less food to complete development. The effects of larval size on susceptibility to planktonic predation are not known. It is not yet clear how these various factors combine to influence larval survival under natural conditions. However, appropriate data can be generated in the laboratory via blastomere isolations and comparative rearing studies.

Rates of cleavage are the same in embryos derived from isolated blastomeres as in whole embryos and developmental stage depends on the number of cleavages, rather than the size or the number of cells, during the embryonic stages (Takahashi and Okazaki 1979). Experimental increases in egg or embryo size result in giant embryos which cleave at normal rates and yield giant larvae (Horstadius 1957, 1970). Developmental rate is independent of egg size during embryonic and prefeeding larval stages. However, developmental rate is markedly dependent on egg size during the early feeding larval stages (Sinervo and McEdward, 1988). Since smaller larvae must grow more to reach any given size and since equivalence of larval size and developmental rate is achieved at the same stage of development (six-armed pluteus; Sinervo and McEdward 1988), egg size related differences in developmental rate probably arise entirely from differences in initial larval size. Marcus (1979) reported that total time to metamorphosis is increased in larvae of *Arbacia punctulata* derived from isolated blastomeres.

One of the likely advantages of increased developmental rate is a reduction in mortality from planktonic predation (Thorson 1946, 1950; Chia 1974; Strathmann 1985; Vance 1973a,b; Christiansen and Fenchel 1979). In echinoids, developmental stage (i.e. body form) has a greater influence on susceptibility to some planktonic predators than does body size (Rumrill et al. 1985). Embryos and larvae derived from larger eggs develop more rapidly during the early stages (Sinervo and McEdward 1988) and should reduce the mortality from predation during these vulnerable stages. Larger eggs also result in a shorter pelagic period and therefore less total time exposed to planktonic predators.

Strathmann (1985; Emlet et al. 1987) has developed a quantitative model for evaluating the importance of the pelagic period on larval mortality (similar to Vance 1973a, Equation 3). It assumes that parental investment per offspring is inversely proportional to fecundity and has a direct influence on the duration of the pelagic period. An obvious disadvantage of investing more energy in each offspring is that fewer offspring can be produced. However, some benefit is gained if they develop faster and have greater chances of surviving to metamorphosis. It is assumed that a constant mortality applies to all larvae, at all developmental stages. Using the model one can calculate the mortality rate necessary to yield sufficient differences in survival to balance differences in fecundity in animals with different levels of parental investment per offspring (Table 15.3). The number of offspring released into the plankton is given by fecundity. They decrease in number exponentially at a constant rate throughout the pelagic period to yield the final number that metamorphose. The mortality rate necessary to balance mortality and fecundity is given by the logarithm of the ratio of parental investment per offspring divided by the difference in development time (Table 15.3).

Blastomere isolations yield known ratios of parental investment and subsequent rearing of the larvae provides information on development time. For *S. droebachiensis* and *S. purpuratus*, egg size is not traded against size at metamorphosis. Therefore, any effects of differing levels of parental investment per offspring on offspring fitness should occur prior to metamorphosis, during the planktonic larval stages. Comparisons among experimental treatments from *S. droebachiensis* and *S. purpuratus* (Sinervo and McEdward 1988) indicate that mortality rates between approximately 5 and 35 per cent per day are sufficient to offset differences in fecundity. These mortality rates are rather low compared to many reported for planktonic larval stages of various taxa (reviewed by Strathmann 1985). The advantages of investing more energy in each offspring can apparently balance the reduction in fecundity.

Numerous additional problems in life history

TABLE 15.3 Model balancing fecundity and larval mortality.
Model from Strathmann (1985; Emlet *et al.* 1987). Data from B. R.
Sinervo and L. R. McEdward (unpublished observations)

Terms:
- F   fecundity
- a,b   subscripts denoting individuals with different levels of parental investment per offspring
- T   development time (=duration of the pelagic larval period)
- d   mortality rate (per day)
- PI   parental investment per offspring
- M   number of offspring that reach metamorphosis
- SD   *Strongylocentrotus droebachiensis*
- SP   *Strongylocentrotus purpuratus*

Equations:
(1) $F_a e^{-T_a d} = M_a = M_b = F_b e^{-T_b d}$
(2) $F_a/F_b = e^{-(T_b - T_a)d}$
(3) $\ln(PI_b/PI_a) = -(T_b - T_a)d$

Calculations:

Comparison	PI_b/PI_a	d
SD: SD 1/2	0.5	0.14–0.35
SD 1/2: SP	0.3	0.04–0.07
SD: SP	0.15	0.05–0.08

biology can be investigated using manipulations of egg size via blastomere isolations.

(1) Several theoretical models of life history evolution in marine invertebrates predict that only the extreme egg sizes within the planktotrophic and lecithotrophic strategies will be evolutionarily stable (e.g. Vance 1973a,b; Christiansen and Fenchel 1979). The fact that 1/2 or 1/4 size embryos and larvae of echinoderms derived from blastomere isolation can develop through metamorphosis argues that selection has not produced eggs of the minimum size compatible with development (reviewed by Emlet *et al.* 1987).

(2) There are numerous reasons why natural selection might favour different sizes of the juvenile at metamorphosis. It is entirely possible that in some taxa egg size is traded against metamorphic size rather than development time. Blastomere isolation experiments provide a powerful tool for testing this hypothesis when comparative studies suggest correlations between egg size and juvenile size. Okazaki and Dan (1954) reported that lecithotrophic larvae of the echinoid, *Peronella japonica*, derived from isolated blastomeres, metamorphose at the same time as normal larvae, but are smaller.

(3) The energetics of lecithotrophic echinoderm larvae are poorly understood. However, it is likely that the amount of material stored in the egg will determine larval and juvenile success. Manipulations of the amount of material, via blastomere isolations, will facilitate understanding the life history tradeoffs that underlie this adaptive strategy.

(4) Ecological consequences of larval size, for instance, susceptibility to predation, resistance to starvation, swimming capabilities, capabilities for delaying settlement, etc., can be investigated within species, thereby providing information relevant to understanding the responses of life histories to various selective pressures.

Stearns (1977) has argued that

> Conclusions reached from experimental tests of risky hypotheses can advance our knowledge by pointing out weaknesses in our models. It is much more difficult to falsify a prediction made at the interspecific level, where one can always argue that confounding effects obscure the main trend. But generalizations made from experiments carried out on a single species are subject to question, perhaps more so than generalizations based on correlations drawn from many

species. The more certain we are that we know what is causing an evolutionary trend, the less certain we are that the cause is general. And the more certain we are that a trend is in fact general, the less certain we can be about its cause or causes. In living within these limits, I feel strongly that we must do experiments wherever possible, and base our arguments on the best possible evidence available, rather than resort to appealing to the beauty of untested speculation.

Echinoderms offer a rare opportunity to do experiments on important life history traits. Experimental embryology, combined with traditional comparative studies, provides a powerful tool for generating data relevant to understanding the evolution of life histories in echinoderms.

REFERENCES

Chia, F. S. 1974. Classification and adaptive significance of developmental patterns in marine invertebrates. *Thalassia Jugoslavica* **10**, 121–30.

Christiansen, F. B. and Fenchel, T. M. 1979. Evolution of marine invertebrate reproductive patterns. *Theoretical Population Biology* **16**, 267–82.

Dan-Sohkawa, M. and Satoh, N. 1978. Studies on dwarf larvae developed from isolated blastomeres of the starfish, *Asterina pectinifera*. *Journal of Embryology and Experimental Morphology* **46**, 171–85.

Emlet, R. B., McEdward, L. R., and Strathmann, R. R. 1987. Larval ecology viewed from the egg. In *Echinoderm Studies* 2 (ed. M. Jangoux and J. M. Lawrence), pp. 55–136. A. A. Balkema, Rotterdam.

Grahame, J. and Branch, G. M. 1985. Reproductive patterns of marine invertebrates. *Oceanography and Marine Biology Annual Reviews* **23**, 373–98.

Hagstrom, B. E. and Lonning. S. 1964. The rate of development in isolated halves of sea urchin embryos. *Sarsia* **15**, 17–22.

Hagstrom, B. E. and Lonning, S. 1965. Studies of cleavage and development of isolated sea urchin blastomeres. *Sarsia* **18**, 1–9.

Harvey, E. B. 1940. A new method of producing twins, triplets and quadruplets in *Arbacia punctulata*, and their development. *Biological Bulletin of the Marine Biology Laboratory, Woods Hole* **78**, 202–16.

Harvey, E. B. 1949. The growth and metamorphosis of the *Arbacia punctulata* pluteus, and late development of the white halves of centrifuged eggs. *Biological Bulletin of the Marine Biology Laboratory, Woods Hole* **97**, 287–99.

Hendler, G. 1975. Adaptational significance of the patterns of ophiuroid development. *American Zoologist* **15**, 691–715.

Hinegardner, R. T. 1975a. Care and handling of sea urchin eggs, embryos, and adults (principally North American species). In *The sea urchin embryo: biochemistry and morphogenesis* (ed. G. Czihak), pp. 10–25. Springer-Verlag, New York.

Hinegardner, R. T. 1975b. Morphology and genetics of sea urchin development. *American Zoologist* **15**, 679–89.

Horstadius, S. 1936. The mechanics of sea urchin development, studies by operative methods. *Biological Reviews* **14**, 132–79.

Horstadius, S. 1957. On the regulation of bilateral symmetry in plutei with exchanged meridional halves and giant plutei. *Journal of Embryology and Experimental Morphology* **5**, 60–73.

Horstadius, S. 1970. Giant larvae of *Paracentrotus lividus*. *Arkiv für Zoologie* **23**, 417–22.

Horstadius, S. 1973. *Experimental embryology of echinoderms*. Clarendon Press, Oxford.

Horstadius, S. 1975. Isolation and transplantation experiments. In *The sea urchin embryo: biochemistry and morphogenesis* (ed. G. Czihak), pp. 364–406. Springer-Verlag, New York.

Kawamura, K. 1970. On the development of the planktonic larvae of Japanese sea urchins, *Strongylocentrotus intermedius* and *S. nudus*. *Scientific Reports of the Hokkaido Fisheries Experimental Station* **12**, 25–32.

Kume, M. and Dan, K. 1971. *Invertebrate embryology*. Reprinted by National Technical Information Service, Springfield, VA.

Marcus, N. 1979. Developmental aberrations associated with twinning in laboratory-reared sea urchins. *Developmental Biology* **70**, 274–7.

McEdward, L. R. 1984. Morphometric and metabolic analysis of the growth and form of an echinopluteus. *Journal of Experimental Marine Biology and Ecology* **82**, 259–87.

McEdward, L. R. 1986a. Comparative morphometrics of echinoderm larvae. I. Some relationships between egg size and initial larval form in echinoids. *Journal of Experimental Marine Biology and Ecology* **96**, 251–65.

McEdward, L. R. 1986b. Comparative morphometrics of echinoderm larvae. II. Larval size, shape, growth, and the scaling of feeding and metabolism in echinoplutei. *Journal of Experimental Marine Biology and Ecology* **96**, 267–86.

McEdward, L. R. and Carson, S. F. 1987. Variation in organic content and its relationship with egg size in the starfish, *Solaster stimpsoni*. *Marine Ecology: Progress Series* **37**, 159–69.

McEdward, L. R. and Coulter, L. K. 1987. Egg volume and energetic content are not correlated among sibling offspring of starfish: Implications for life history theory. *Evolution* **41**, 914–17.

Mileikovsky, S. A. 1971. Types of larval development in marine bottom invertebrates, their distribution and ecological significance: a re-evaluation. *Marine Biology* **10**, 193–213.

Mileikovsky, S. A. 1974. Types of larval development in marine bottom invertebrates: An integrated ecological scheme. *Thalassia Jugoslavica* **10**, 171–9.

Mita, I. 1983. Studies on factors affecting the timing of early morphogenetic events during starfish embryogenesis. *Journal of Experimental Zoology* **225**, 293–9.

Naidenko, T. K. 1983. Laboratory culture of the sea urchin *Stronglyocentrotus intermedius*. *Soviet Journal of Marine Biology* **9**, 46–50.

Okazaki, K. and Dan, K. 1954. The metamorphosis of partial larvae of *Peronella japonica* Mortensen, a sand dollar. *Biological Bulletin of the Marine Biology Laboratory, Woods Hole* **106**, 83–99.

Reverberi, G. 1971. *Experimental embryology of marine and freshwater invertebrates*. North Holland Publishing Co., Amsterdam.

Rumrill, S. S., Pennington, J. T., and Chia, F. S. 1985. Differential susceptibility of marine invertebrate larvae: Laboratory predation on sand dollar, *Dendraster excentricus* (Eschscholtz), embryos and larvae by zoeae of the red crab, *Cancer productus* Randall. *Journal of Experimental Marine Biology and Ecology* **90**, 193–208.

Sinervo, B. R. and McEdward, L. R. 1988. Development consequences of an evolutionary change in egg size: an experimental test. *Evolution* (in press).

Smith, C. C. and Fretwell, S. D. 1974. The optimal balance between size and number of offspring. *American Naturalist* **108**, 499–506.

Stearns, S. C. 1977. Life-history tactics: A review of the ideas. *Quarterly Reviews in Biology* **51**, 3–47.

Strathmann, M. 1987. *Reproduction and development of marine invertebrates of the northern Pacific cost*. University of Washington Press, Seattle.

Strathmann, R. R. 1971. The feeding behavior of planktotrophic echinoderm larvae: mechanisms, regulation, and rates of suspension-feeding. *Journal of Experimental Marine Biology and Ecology* **6**, 109–60.

Strathmann, R, R. 1974. Introduction to function and adaptation in echinoderm larvae. *Thalassia Jugoslavica* **10**, 321–39.

Strathmann, R. R. 1985. Feeding and nonfeeding larval development and life-history evolution in marine invertebrates. *Annual Review of Ecology and Systematics* **16**, 339–61.

Strathmann, R. R. and Vedder, K. 1977. Size and organic content of eggs of echinoderms and other invertebrates as related to developmental strategies and egg eating. *Marine Biology* **39**, 305–9.

Takahashi, M. M. and Okazaki, K. 1979. Total cell number and number of the primary mesenchyme cells in whole, 1/2 and 1/4 larvae of *Clypeaster japonicus*. *Development, Growth and Differentiation* **6**, 553–66.

Thorson, G. 1936. The larval development, growth and metabolism of Arctic marine bottom invertebrates, etc. *Meddeleser om Grønland* **100**, 1–155.

Thorson, G. 1946. Reproduction and larval development of Danish marine bottom invertebrates. *Meddeleser fra Kommissionen før Danmarks Fiskeri-og havundersøgelser. Serie Plankton* **4**, 1–523.

Thorson, G. 1950. Reproductive and larval ecology of marine bottom invertebrates. *Biological Reviews* **25**, 1–45.

Turner, R. L. and Lawrence, J. M. 1979. Volume and composition of echinoderm eggs: implications for the use of egg size in life-history models. In *Reproductive ecology of marine invertebrates* (ed. S. Stancyk), pp. 25–40. University of South Carolina Press, Columbia, SC.

Vance, R. R. 1973a. On reproductive strategies in marine benthic invertebrates. *American Naturalist* **107**, 339–52.

Vance, R. R. 1973b. More on reproductive strategies in marine benthic invertebrates. *American Naturalist* **107**, 353–61.

Vance, R. R. 1974. Reply to Underwood. *American Naturalist* **108**, 879–80.

PART V
Fossils and evolution

16

The phylogeny of the cystoids

C. R. C. PAUL

Department of Earth Sciences, Liverpool University, L69 3BX, England

INTRODUCTION

The purpose of this chapter is to present a summary of the relationships between the various groups of cystoids *s.l.* (= blastozoans of Sprinkle 1973a) based on a cladistic analysis of key taxa. This confirms Sprinkle's original assertion (1973a, p. 3) that the cystoids form a natural, monophyletic group distinct from the crinoids, and characterized by all three of the following: an aboral stalk, tesselate thecal plating, and brachioles. The aboral stalk alone characterizes the Pelmatozoa. Brachioles are also found in *Lepidocystis* and *Kinzercystis*, but they have primitive imbricate thecal plating. The relationships of the early Cambrian echinoderms were described by Paul and Smith (1984), but see also Smith (this volume, Fig. 7.1). Other characters which cystoids inherited from more primitive echinoderms include: five ambulacra with biserial flooring plates that form part of the thecal wall, and arranged in a 2–1–2 pattern, i.e. the unbranched A ambulacrum lies opposite the C–D interambulacrum which contains the anus, hydropore, and gonopore. Cladistic analysis has resulted in some considerable realignment of constituent groups as well as the realization that a few well established taxa are polyphyletic. I have also found that some characters traditionally used in classification, such as pore-structures and distal branches of the ambulacral system, are more homeoplasic than previously thought when compared with the morphology of the oral area and the arrangement of thecal plates. As far as space permits, justification of all such reinterpretations is presented below. However, there is insufficient room to illustrate the morphological features discussed, so although I have personally examined most of the taxa involved, I shall refer to the most up to date or most accurate published description of them. In particular, many details will be found in Moore (1968) and Sprinkle (1973a, 1982). Since there has been considerable realignment of taxa, it is not easy to discuss relationships between the currently accepted classes of cystoids. The best approach seems to be to present the phylogeny and then outline the reasons for it. Labelling of the ambulacra follows Carpenter's system.

THE PHYLOGENY OF THE CYSTOIDS

This is summarized in Figs 16.1–16.5. The genus *Gogia* lies at the centre of the radiation of the cystoids. *Gogia* first appeared in the latest Lower Cambrian, but flourished in the Middle Cambrian of North America where Sprinkle (1973a, p. 75) recognized twelve species. Where the oral surface is seen, it bears five ambulacra in the standard 2–1–2 arrangement. The ambulacra had biserial flooring plates and towards their tips gave rise alternately on either side to erect, biserial brachioles. *Gogia* also has an aboral stalk attached to the sea floor, but both the relative development of the stalk and its demarcation from the theca varies between species.

Gogia seems to have given rise to three principal lines of evolution (Fig. 16.1). The first line involved the loss of the stem and the development in a radial position of circumoral plates which bear thin epithecal food grooves. This line led to the Middle Cambrian genus *Lichenoides* and thence to directly attached diploporites of the superfamily Sphaeronitida. The second line retained an irregular polyplated stalk, but developed large erect unbranched biserial arms. It included the Upper Cambrian genus *Nolichuckia* and led to the fistuliporite rhombiferans. The latter retain a meric stem in which the skeletal elements instead

Echinoderm phylogeny and evolutionary biology (ed. C. R. C. Paul and A. B. Smith). Clarendon Press, Oxford, 1988.

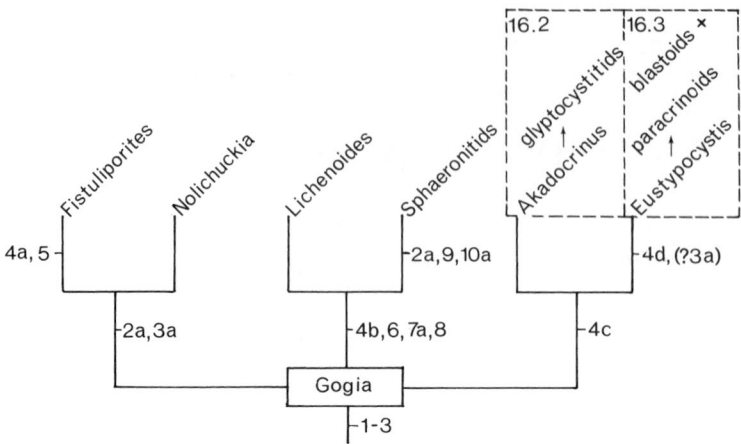

Fig. 16.1 A cladogram to illustrate the early radiation of the cystoids *sensu latu*. Areas in boxes are shown in Figs 16.2 and 16.3. Since the middle lineage lacks a stem it is uncertain to which of the other two lineages it is more closely related. In this and all other figures numbers refer to the characters listed in Table 16.1.

TABLE 16.1 Characters used in cladistic analysis

Plesiomorphic state	Derived state(s)
1 Imbricate thecal plating	Tesselate plating
2 Epispires	(a) Epispires lost
	(b) Epispires reappear
3 Brachioles	(a) Erect, unbranched, non-pinnate arms
	(b) Modified (biserial opposite) brachioles
	(c) Erect, biserial, pinnate arms with 1 brachiole per 3 arm plates
	(d) Erect, biserial, pinnate arms with 1 brachiole per arm plate
4 Polyplated stalk	(a) Meric stem
	(b) Stem lost
	(c) Holomeric stem with annular columnals
	(d) Holomeric stem with disc-shaped columnals
5	Humatirhombs
6 Interradial orals	Radial orals
7 Irregular thecal plating	(a) Reduced, regular plating
	(b) Radial columns of plates
	(c) Interradial columns of plates
8	Epithecal food grooves which lack cover or flooring plates
9	Diplopores
10 Basals undifferentiated	(a) 7 basals
	(b) 4 basals
	(c) 2 basals
	(d) 5 basals
	(e) 3 basals
11 Theca globular or cylindrical	(a) Flattened theca
	(b) Bud-shaped theca
	(c) Lenticular theca

TABLE 16. contd.

Plesiomorphic state	Derived state(s)
12 Five ambulacra	(a) 3 ambs
	(b) 4 ambs
	(c) 2 recumbent, asymmetrical, uniserial, 'hemipinnate' ambs
13 Normal column	(a) Reduced tapering column
	(b) Column of 1–2 columnals
14 Periproct on oral surface	(a) Periproct in marginal plates
	(b) Large, lateral periproct
15 Synostosial columnal articulations	Synarthrial columnal articulations proximally
16	5 infralaterals, 5 laterals
17	6 radials
18 All ambs the same	First two brachioles to left in ambs B and D only
19	Pectinirhombs
20 Three equal basals	Smallest basal opposite periproct
21 Stem lumen circular	Lumen triangular or oval
22 Hydropore horizontal or oblique	Hydropore vertical
23 Six periorals	4 periorals
24 Mouth at thecal summit	Mouth offset from summit
25 Pore-structures absent	(a) Pore-structures across all sutures
	(b) Pore-structures across R:D sutures only
26	Hydrospires
27 Erect ambulacra	Recumbent ambs
28	(a) Facetal plates
	(b) Single lancet plate per amb
29 Hydropore present	Hydropore lost
30 Pore-structures absent	(a) Specialized (un-named) pore-structures
	(b) 'Foerstepores'
	(c) Trans-sutural slits
31 Thecal plates irregular	(a) 5 columns of plates
	(b) >5 columns of plates
	(c) Single generation of plates
32 Erect brachioles	Recumbent brachioles
33 Convex or planar thecal plates	Concave plates
34	(a) Periproct to left of adjacent amb
	(b) Periproct to right of adjacent amb
35 Two recumbent, assymmetrical, 'hemipinnate' ambulacra	(a) 3 erect arms, 1 branched
	(b) 3 recumbent arms, 1 branched
	(c) Straight peripheral ambs
	(d) Curved ambs
	(e) 4 ambs curved counter clockwise
	(f) 2 ambs curved clockwise
36 Oval thecal outline	Circular thecal outline
37 Erect 'pseudopinnules'	Recumbent 'pseudopinnules'
38 Two unbranched ambulacra	2 branched ambs
39 Long ambulacra	Short ambs confined to oral area
40 One oral opening	Two openings

A blank in the column for plesiomorphic state indicates the character is absent. Characters 7 and 31 appear to be the same, but as paracrinoids have only two ambulacra it is not possible to define radial and interradial positions on the theca. Similarly, a wide variety of pore-structures develop among the cystoids. It is deemed simpler to use separate characters for the appearance of different types in different major groups. So characters 5, 9, 25, 26, and 30 refer to fistuliporites, diploporites, *Lysocystites* and coronates, eublastoids, and parablastoids, respectively.

of being disc-like or annular columnals are composed of several pieces or meres. The third lineage led to the development of a true column and was by far the most complex and important line. It can be split into at least two subdivisions one leading to the glyptocystitid rhombiferans and the other to the paracrinoids, coronates, and blastoids. This trichotomy leading from *Gogia* is based more on the characters of the stalk (when present) than on the ambulacra and it involves the evolution of erect arms independently in two lineages. The alternative argument that cystoids with erect arms are derived directly from *Lepidocystis*, is rejected here because within the Glyptocystitida alone unbranched, erect arms and recumbent, pinnate arms evolved independently, while erect, pinnate arms evolved in the hemicosmitids. My experience with more derived cystoids suggests that ambulacral structure is highly variable.

Each of the three important lineages will be discussed in the sections that follow, with a final section on the problem groups like the hemicosmitid rhombiferans and aristocytitid diploporites which have such distinctive oral surfaces that derivation from other groups proves very difficult. It should perhaps be emphasized at this stage that *Gogia* includes several species. Although it seems to have been at the centre of a major radiation, the first signs of the trends mentioned here can be seen in different species of *Gogia* itself.

The sphaeronitid diploporites (Fig. 16.1)

Lichenoides is a very distinctive 'eocrinoid' from the Middle Cambrian of Czechoslovakia and was described in detail by Ubaghs (1953). It appears to be relatively derived as it has no stem and a small number of thecal plates arranged in definite circlets. The plates are swollen, and bear on the external surface advanced epispires with elongate channels on either side of the sutural pore. The plates surrounding the oral opening are radial in position, with a sixth in the C-D interradius. The five radially-positioned circumorals bear narrow, shallow epithecal food grooves some of which may extend onto plates of the next lower circlet. The food grooves lack either flooring or cover plates and end in small facets less than 1 mm across. Radially-positioned circumorals and epithecal food grooves without cover plates developed directly on the surfaces of thecal plates are shared derived characters found only in *Lichenoides* and sphaeronitid diploporites, suggesting a close relationship between the two. The food gathering structures that arose from the ambulacral facets are unknown in both groups, but are usually presumed to have been brachioles. In *Lichenoides* the base of the theca, where a stem might have been attached, is filled with a variable number of small plates. *Lichenoides* seems to have been free rather than permanently attached. Previously (Paul 1977, p. 134), I suggested that the swollen plates were hollow, by analogy with the inflated hollow plates of *Brockocystis tecumseth* Billings, and that *Lichenoides* might have floated upside down. I now believe that the plates were solid and that this was an adaptation to keep the theca in position on the substratum. All known specimens of *Lichenoides* are natural moulds so this point cannot be settled by sectioning the plates. However, even damaged thecae never show sediment penetrating the swellings of the plates as might be expected if they had originally been hollow.

Sphaeronitids were directly attached to the substratum, but only loosely, possibly by the epidermis. Several genera have been described recently by Paul (1971, 1973), Bockelie (1984), and Paul and Bockelie (1983). Although their attachment areas directly mould the substratum, sphaeronitids are typically found free from their substrata with undamaged attachment areas, even when these were obviously hard as, for example, when they grew on trilobites or brachiopods (Paul 1971, p. 37; 1973, p. 14). Sphaeronitids are only found still attached to their substrata when the substratum was another echinoderm. All sphaeronitids have seven basals and seven circumorals, where the number is known. In *Haplosphaeronis* there are no other plates (Paul 1973, p. 27; Bockelie 1984, p. 33), but all other genera have a variable number of plates in between these circlets (up to several hundred in *Sphaeronites*; Paul and Bocklie 1983, p. 704). The mouth was covered by a unique palate composed of six plates (Paul 1971, p. 7). The hydropore lay to the left of the oro-anal line across plate suture C01:C06, the gonopore was associated with plate C07. The anus was covered by a simple pyramid of plates and was usually close to the mouth.

Lichenoides can be derived from species of *Gogia* such *G. granulosa* or *G. hobbsi*. The former has advanced epispires with elongate grooves externally, relatively few thecal plates and already shows a reduction in the stalk; the latter has very few thecal plates and a reduced stalk, but these features could simply reflect its small size. The transformation involves:

(1) complete loss of the stem;
(2) reduction in the total number, and thickening of thecal plates and their organization into definite circlets;
(3) modification of the oral area by loss of the original ambulacra and development of epithecal food grooves on radially-positioned circumoral plates.

The sphaeronitids can be derived from *Lichenoides* by direct attachment to the substratum, loss of epispires and development of true diplopores, and fixing the number of plates in the basal and circumoral circlets at seven. The positions of the gonopore, hydropore and anus remain unknown in *Lichenoides* so I cannot comment on changes in these orifices.

The fistuliporite rhombiferans (Fig. 16.1)

The genus *Nolichuckia* is not that well known. Sprinkle (1973a) gives all available information. It is, however, the only Upper Cambrian genus to retain a polyplated stalk. This stalk is thin and clearly differentiated from the oval to globular theca. At the base of the theca there appear to have been five or more plates forming the attachment of the stem, i.e. a crude basal circlet. The theca has irregular plating, no epispires and, apparently, a reduced oral area. From the oral area several (possibly five) erect, unbranched, biserial, non-pinnate arms arise directly. *Nolichuckia* is similar to species of *Gogia*, such as *G. kitchnerensis*, in which there are few epispires and a stem clearly separated from the theca. Some species of *Gogia* even seem to have had a primitive basal circlet of five or more plates. *Nolichuckia* differs only in the total lack of epispires and the modified ambulacral structure with erect arms rather than brachioles arising from recumbent ambulacra.

Fistuliporite rhombiferans include the genera *Arachnocystites*, *Echinosphaerites*, *Heliocrinites*, *Stichocystis*, and *Caryocystites*. Several have recently been thoroughly described by Bockelie (1981a,b, 1982). They are characterized by a meric stem in which the circumference of the stem is covered by a regular number of meres, usually 4–7, arranged in columns or circlets. In other words it is a polyplated stalk, but with the plates (meres) arranged in an organized fashion. In *Arachnocystites* the distal termination of the stem was attached to the substratum by a disc-like polyplated holdfast. Barrande (1887) figured and described this holdfast as '*Cystidea nugatula*'. The stem is clearly separated from the theca by a circlet of 4–7 basal plates which alternate with the meres of the stem. The theca is composed of a large number of plates (several tens to over 2000) all of which bear rhombic pore-structures, the humatirhombs (Paul 1972a). The rhombs are composed of canals, called fistulipores, that opened internally and through which body fluids flowed in life. The mouth is situated in a raised oral prominence off which 2–4 erect, biserial arms arose. The earliest species have only two arms (Bockelie 1981a). These arms may bifurcate immediately to give a maximum of eight branches, but are non-pinnate. They have a characteristic triangular cross-section with the food groove at the apex of the triangle (Bockelie 1981b, fig. 2B,C). Two series of cover plates are present in *Arachnocystites* and *Echinosphaerites*. No hydropore is known, but the gonopore lies to the right of the oro-anal line, and is covered by a small pyramid of (usually three) triangular plates. The anus is well separated from the mouth and covered by a simple pyramid of five anal plates. Fistuliporites can be derived from *Nolichuckia* by the development of a truly meric stem and humatirhombs, and the reduction of the number of arms to two.

Cystoids with true columns (Figs 16.2–16.5)

The third main evolutionary line leading from *Gogia* is by far the largest and possibly involves all cystoids with holomeric (single piece) columnals. The first Middle Cambrian cystoids with true columns, *Eustypocystis* and *Akadocrinus*, differ quite significantly, so it may be that there was not just a single line. For the present it is more parsimonious to assume that a true column was evolved only once within the cystoids. Early columns were frequently heteromorphic with larger and smaller columnals alternating along the stem. The first suggstions of such a pattern are already present in *Gogia prolifica* and *G. palmeri*, where the plates of the stalk are clearly of two sizes in rings around the stem. Provisionally, I am accepting *Akadocrinus*, with its annular columnals forming a stem with a large lumen, as the more primitive because it retains epispires. From *Akadocrinus* one line led to the glyptocystitid rhombiferans. *Eustypocystis* has disc-like columnals, not clearly alternating in size (height or diameter), lacks epispires and has modified ambulacra. It lies at the base of a second line leading to the paracrinoids, coronates and blastoids. Although the ambulacral structure of *Akadocrinus* is unknown, what I infer to be later descendants along the line to glyptocystitids (including the latter), retain the plesiomorphic 2–1–2 ambulacral structure with biserial flooring plates and alternating brachioles. Again this suggests that *Akadocrinus* is the less apomorphic of the two Middle Cambrian genera with holomeric stems.

The glyptocystitid rhombiferans (Fig. 16.2)

This line is relatively well documented, although the relationships of some possible offshoots remain in doubt. It commences with *Akadocrinus*, which is well described in Sprinkle (1973a), and has a long column composed of thin, annular columnals which surround a lumen as much as 80 per cent of the diameter of the stem. Previous suggestions of a fusellar structure to the columnals are rejected here as a misinterpretation of crushed columnals. Even so, I have never seen oblique cracks in columnals of *Akadocrinus*. The cup is composed of six or seven circlets of plates with small epispires along their sutures. The oral surface is flat and clusters of brachioles arise from its edges. Regrettably, the details of the oral surface are unknown, but it is definitely not modified to produce free arms nor is there evidence for a reduced number of ambulacra. *Akadocrinus* probably gave rise to *Cambrocrinus* simply by the loss of epispires and the development of the characteristic radiating ridges on the plates in the latter. The two genera are otherwise almost identical in their known characteristics. Since there is no evidence for a lateral periproct in either genus, I am assuming that the gonopore, hydropore and anus lay in the C–D interambulacrum on the oral surface, as in the ancestral *Gogia* and the inferred descendant, *Ridersia*. In *Cambrocrinus* the proximal stem tapers in diameter away from the theca and, although published details are unclear, there appears to be some sort of alternating articulation between columnals (Orlowski 1968, pl. 3, figs 5b, 6a, and 10).

The main line continues with *Ridersia*, recently described by Jell et al. (1985) from the Upper Cambrian of Australia. This genus has a tapering proximal stem, a cup which is composed of three definite circlets of plates, basals, infralaterals, and laterals. The oral surface has the standard five ambulacra with the periproct covered by a simple pyramid of plates in the C–D interambulacrum, where the gonopore and hydropore are also found. Brachioles arise from the tips of the ambulacra near the edge of the flattish oral surface. The only innovation from the pattern seen in *Gogia* is that the oral surface now has single large inter-radially

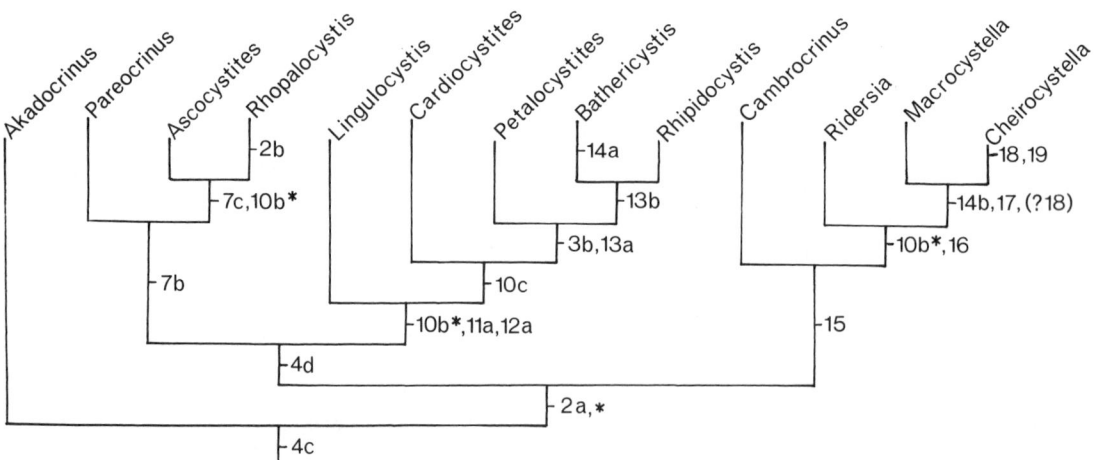

Fig. 16.2 A cladogram to illustrate the suggested phylogeny of the rhipidocystids and glyptocystitid rhombiferans. *The number of basals is unknown in *Akadocrinus*, *Pareocrinus*, and *Cambrocrinus*, but since four basals (character 10b) appear at these points in all three descendant lineages it seems plausible that the number was fixed in *Akadocrinus* or between *Akadocrinus* and its descendants.

positioned oral plates, six in all as there are two in the posterior (C–D) interambulacrum.

Macrocystella, the first of the glyptocystitid cystoids (Paul 1968a), has a stem with a tapering proximal portion and a long, thin, whip-like distal portion. The proximal portion is composed of annular columnals of two types with a very distinctive synarthrial articulation (Paul 1968a, 1984). The cup is composed of four basals, five infralaterals, five laterals, six radials and seven orals. It incorporates a large lateral periproct covered by a flexible, plated, periproctal membrane, enclosed within which is a small anal pyramid. The orals form a flat oral surface, lie between the five ambulacra, and the gonopore and hydropore lie across suture 01:06. The five ambulacra have biserial flooring plates which form part of the thecal wall and give rise to two to five brachioles near their tips. There is some evidence that *Macrocystella* had already evolved a unique arrangement of brachioles in the ambulacra whereby in all five ambulacra the first brachiole branched off to the left (as viewed looking from the mouth towards the tip), but in ambulacra B and D (I and IV) the second also branched to the left. Thereafter brachioles branched alternately in all ambulacra. The pattern is present in all Ordovician glyptocystitids with five ambulacra each bearing several brachioles. *Macrocystella* was, I believe, ancestral to the first glyptocystitid with pectinirhombs, *Cheirocystella* Paul, 1972b. The two genera differ solely in the absence of true pectinirhombs in *Macrocystella* and their presence in *Cheirocystella*.

Derivation of *Akadocrinus* from *Gogia* requires evolution of true columnals together with a reduction in the number of thecal plates and their organization into circlets. *Cambrocrinus* can be derived from *Akadocrinus* by the development of a tapering proximal stem with its distinct articulation, loss of epispires and formation of ridged thecal plates. Evolution of *Ridersia* from *Cambrocrinus* requires further reduction in the number of plate circlets with a fixed number of plates in each (perhaps already present in *Cambrocrinus*) and possibly the development of large interradial oral plates, although these too may have developed earlier (see below). Finally, *Macrocystella* can be derived from *Ridersia* by migration of the periproct to a lateral position, its enlargement and the development of a flexible plated cover. Thecal plating remained fixed but a fourth (radial) circlet of plates was added between the laterals and the orals, perhaps associated with the migration of the periproct from the oral surface. As mentioned above, *Macrocystella* possibly also developed the unique pattern of branching of brachioles found in early glyptocystitids, the first of which, *Cheirocystella*, differs only in the possession of true pectinitrhombs.

Pareocrinus, from the Middle and Upper Cambrian of Siberia, is not well known (see Ubaghs 1967), but may represent a side branch off this main lineage arising after the loss of epispires. *Pareocrinus* has a column, possibly with disc-like rather than annular columnals, and thecal plates arranged in five or six columns, rather than in circlets. Each column has five plates above the basal circlet. The oral surface is unknown. The arrangement of plates in columns suggests that *Pareocrinus* may have given rise to *Rhopalocystis* Ubaghs 1963, although this would have involved the reappearance of functional epispires in the adult reversing the trend to their loss. *Rhopalocystis* has a heteromorphic stem with alternating large and small disc-like columnals, a theca with a large conical base above which lie four basals and then radial columns of plates with more irregularly arranged interradials between them. The oral surface is identical to that seen in *Ridersia*, with five biserial ambulacra bearing regularly alternating brachioles, seven interradially positioned oral plates, three of which lie in the C–D interradius as do the gonopore, hydropore and periproct with its simple anal pyramid. The presence of an identical oral surface in these two branches strongly suggests that it is a synapomorphy of the group and that this type of oral surface was already developed in *Akadocrinus*. However, as mentioned above, no direct evidence of this is preserved.

Another genus to be considered in this side branch is *Ascocystites* from the Middle Ordovician of Bohemia, recent descriptions of which have been given by Sprinkle (1973a) and Ubaghs (1967). It has a heteromorphic column with alternating larger and smaller disc-like columnals, an elongate theca without epispires and with plates arranged in ridged columns. The oral surface is flat, bears five curved ambulacra with brachioles arising from the left side only and, apparently, at least five, large interradial orals. The positions of the gonopore, hydropore and anus are unknown. *Ascocystites* has unique features in the oral area, as well as being much younger than most of the other genera considered here. Its relationship to *Pareocrinus* is tenuous, to say the least, but no closer relationship can be demonstrated at present.

Finally, while considering the line to the glyptocystitid cystoids, I shall discuss the rhipidocystids, a group of four or five genera with flattened thecae which apparently lay down on the sea floor. Most genera along the main lineage to the glyptocystitids have four basals, and certainly this number characterizes the glyptocystitids. It may therefore be significant that the earliest known rhipidocystid, *Lingulocystis* from the Tremadoc of Bolivia and Lower Arenig of southern France, had four basals. Later genera seem to reduce the number to two. *Lingulocystis* and *Cardiocystites* have alternating thin and thick disc-like columnals. Plates of the marginal frame and some of the lateral plates of *Lingulocystis* are arranged in vertical columns, suggesting they may have been derived from *Pareocrinus*. *Lingulocystis* is distinctive in having a theca with a stout marginal frame within which were two thinly plated, apparently flexible, surfaces covered with numerous minute platelets. All other genera have relatively few thin plates, but retain a thickened periphery. This group also developed uniserial 'brachioles'. The transition from opposite biserial to alternating biserial plating is preserved within individual brachioles of *Petalocystites* (Sprinkle 1973a, fig. 31B). The uniserial brachioles of *Rhipidocystis* presumably arose by fusion of opposite biserial brachiolar plates. Again there is evidence for this in the brachioles of undescribed specimens in the National Museum of Natural History, Washington, DC. The best recent descriptions of these genera are in Sprinkle (1973a,b), Ubaghs (1960, 1968), and Bockelie (1981c).

Paracrinoids, coronates, and blastoids
(Figs 16.3–16.5)

The second main branch of the cystoids with true columns starts with *Eustypocystis* Sprinkle (1973a) from the Middle Cambrian of North America. This genus has a short holomeric column with round, disc-like columnals, a rather cylindrical theca of irregularly arranged plates that lack epispires and, apparently with a few short, erect biserial arm-like ambulacral appendages arising from a small oral prominence. Regrettably, no more details of the theca, oral surface, and ambulacral structure are known. However, it is reasonably clear that *Eustypocystis* does not have ambulacra which gave rise to several brachioles at their tips as virtually all the genera in the preceding lineage did.

Trachelocrinus, from the Upper Cambrian of North America, has a heteromorphic stem with alternating larger and smaller disc-like columnals, a pyriform theca with probably five basal plates, and a distinctly produced oral area. In between the basal and oral areas the plating is irregular and

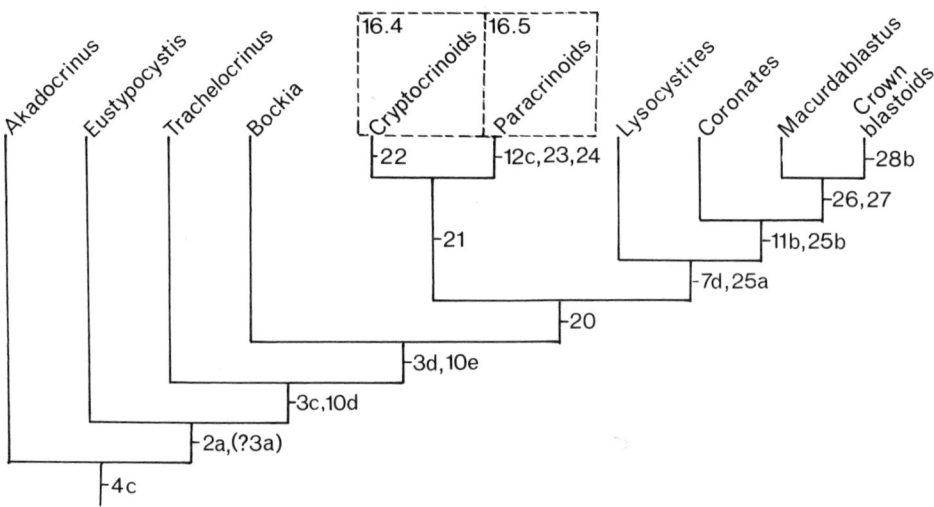

Fig. 16.3 A cladogram to illustrate the suggested phylogeny of the cryptocrinoids, paracrinoids, coronates, and blastoids. Areas in boxes are shown in Figs 16.4 and 16.5.

epispires are absent. The ambulacra consist of biserial erect, true arms with alternating, biserial, side branches (brachioles). Although the arms thus have a pinnate structure, brachioles arise from every third brachial plate on each side of the main brachial axis and the facets for side branches are shared between two brachial ossicles (Sprinkle 1973a, p. 125). This is in contrast to the arrangement in crinoids where lateral branches arise from facets on a single brachial plate, even in those crinoids which secondarily re-evolved biserial main arm branches.

Trachelocrinus may be the source of the parablastoids and some of the diploporites with stems, which have five basal plates and modified ambulacra, but this is one of the more doubtful associations. The next genus along the line towards the paracrinoids and blastoids is *Bockia* which is very similar to *Trachelocrinus*. Indeed Bockelie (1981c) assigned them both to the same family. *Bockia* differs from *Trachelocrinus* in having only three basals and erect, biserial, pinnate ambulacra in which every brachial plate gives rise to a brachiole which arises from a facet confined to a single brachial. In both genera the details of the plating in the oral area are uncertain (or unknown), but both have a distinct oral prominence. In *Bockia mirabilis* the three basals vary in size, while in *B. grava* the sutures between them are equally spaced at 120°, so that a definite smaller basal cannot be recognized. In at least one species of *Bockia* the lumen of the stem appears to have been triangular or oval (Bockelie 1981c, fig. 6d). *Bockia* had a lateral periproct, but the position of this orifice in *Trachelocrinus* is unknown.

The most important innovation of these two genera is the development of erect, pinnate, biserial ambulacra, even if the detailed structure differs. The varied ambulacra of the paracrinoids and cryptocrinoids on the one hand, and of the coronates and blastoids on the other, can be derived from these ambulacra. The term 'cryptocrinoids' is used here for a small group of genera which closely resemble paracrinoids, but do not preserve the key synapomorphy of paracrinoids, namely asymmetrical, uniserial ambulacra with side branches arising from one side only. Most of these genera have not been described recently, but Sprinkle (1973a, 1982, pp. 289–96) gives details of *Columbocystis* and *Bromidocystis*, respectively. Paracrinoids, cryptocrinoids, coronates, and blastoids share the commonly derived character of having three basals with the smaller consistently in the same position with respect to the anus. Surprisingly, it is not in a consistent position with respect to the hydropore in the former two groups. However, if the anus is regarded as being posterior, then the smallest basal usually lies just to the right of anterior, i.e. in the A–B interray of coronates and blastoids. Usually, this plate lies in the E–A interray with respect to the gonopore and hydropore in cryptocrinoids and paracrinoids, but ray designations are frequently uncertain in these forms which often have a reduced number of ambulacra and/or flattened thecae. Cryptocrinoids and paracrinoids also share the derived character that their circular stems have an oval or triangular lumen, at least proximally, although this character may have been present in *Bockia*.

Cryptocrinoids (Fig. 16.4)

These seem to be less derived than paracrinoids because most retain five ambulacra and, where ambulacral and brachiolar structures are known, both are biserial. As defined here, cryptocrinoids include eight Ordovician genera, the relationships of which are shown in Fig. 16.4. *Columbocystis* appears to be the least apomorphic, although the detailed structure of the ambulacra is unknown. The theca is globular with a flared oral platform and there are two generations of thecal plates. The mouth is surrounded by interradial perioral plates with three in the C–D interambulacrum which contains the gonopore and hydropore lying in a vertical orientation across a suture between two of the perioral plates. The anus is adoro-lateral in the B–C interray. It is possible that *Springerocystis* is very closely related to *Columbocystis* as it has a similar theca with two generations of plates, but it lacks the oral platform and nothing is known for certain about its ambulacra or oral plating.

All other cryptocrinoids have ambulacra associated with special facetal plates. These are present in *Cryptocrinites* and *Foerstecystis* inserted between the six interradial perioreals which directly surround the mouth. Both these genera are very similar to *Columbocystis*, but lack the oral platform. The only possible synapomorphies between these two genera are the apparent loss of the hydropore and the presence of very short food grooves which lack cover plates passing from the mouth to the ambulacral facets. The structures that arose from the facets are unknown. The

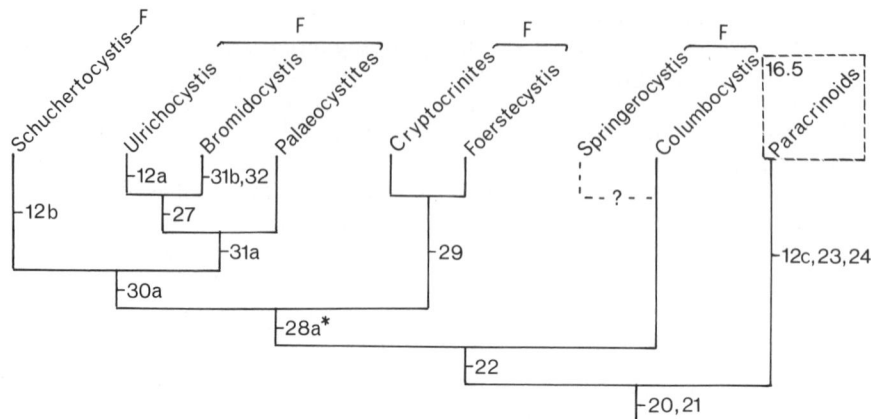

Fig. 16.4 A cladogram to illustrate the suggested phylogeny and classification of the cryptocrinoids. Area in box shown in Fig. 16.5. The four families, F, are the Columbocystidae, Cryptocrinitidae, Palaeocystitidae, and Schuchertocystidae. *Facetal plates (character 28a) may appear earlier because they cannot be recognized for certain in paracrinoids and cryptocrinoids with recumbent ambulacra, and because they are present in some cryptocrinoids and all blastoids *s.l.* I can find no significant character on which to distinguish *Cryptocrinites* and *Foerstecystis*. The latter may well be a junior synonym of the former.

remaining genera share the derived characters of possessing thecal pore-structures which, at least in *Bromidocystis* Sprinkle 1982 and *Ulrichocystis* are very similar to the humatirhombs of fistuliporite rhombiferans (see Paul 1972*a*, pl. 1, figs 3–6). All four genera have a vertical hydropore as in *Columbocystis*. *Palaeocystites*, *Bromidocystis*, and *Ulrichocystis* share thecae with many plates arranged in vertical columns, and *Palaeocrinus* retains less derived erect ambulacra (the structure of which is consequently unknown). *Bromidocystis* and *Ulrichocystis* both have recumbent, pinnate ambulacra with simple biserial flooring plates which do not form part of the thecal wall and with brachioles arising from a facet on a single ambulacral flooring plate, i.e. the ambulacral structure is identical to that of *Bockia* except that the main ambulacral axis is recumbent on the theca. *Bromidocystis* has the majority of its brachioles recumbent as well, while *Ulrichocystis* has reduced the number of ambulacra to three. Finally, *Schuchertocystis* has retained erect ambulacra, but reduced the number to four by losing the one in the A ray. Again their structure is unknown.

Paracrinoids (Fig. 16.5)

These form a similarly small group of Ordovician genera, 12 or 13 in all, with the synapomorphy of having two principal ambulacra which run towards and away from the periproct, and have uniserial flooring plates which give rise to uniserial 'pseudopinnules' on one side only. This unique ambulacral structure involves a reduction in the number of identifiable perioral plates to four, two on each side of the mouth, and argues strongly that they form a truly monophyletic group. They are closely related to the cryptocrinoids, sharing the derived characters of three basal plates with the smallest opposite the periproct and a stem with an oval or triangular lumen proximally. As with cryptocrinoids, some genera lack pore-structures while others developed them. The pore-structures are of two unique types not found in other cystoid groups. All genera of paracrinoids have recently been described by Parsley and Mintz (1975), Frest *et al.* (1979), Parsley, and Sprinkle (the last two in Sprinkle 1982). Their inferred relationships are shown in Fig. 16.5. It is instructive to compare this figure with the very similar cladogram derived independently by Smith (1984, fig. 9, p. 442). Paracrinoids appear to fall into five natural taxa (families), two of which share the derived character of possessing thecal pore-structures. On this interpretation the other three are more plesiomorphic. The genera *Platycystites*, *Globulocystis*, and *Arbucklecystis* are very similar. They have more or less lenticular thecae with two straight ambulacra passing down their edges, and the periproct lies to the left of the adjacent ambulacrum. They differ only in thecal outline (*Arbucklecystis* is pyriform,

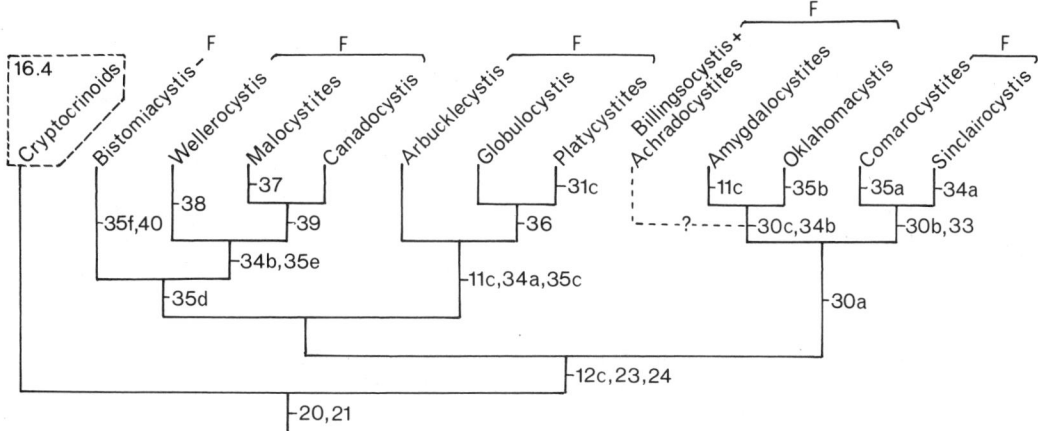

Fig. 16.5 A cladogram to illustrate the suggested phylogeny and classification of the paracrinoids. Area within box shown in Fig. 16.4. The five families recognized are the Comarocystitidae, Amygdalocystitidae, Platycystitidae, Malocystitidae, and the Bistomiacystitidae.

the other two circular) and the lack of intercalary plates in *Platycystites sensu stricto* (as defined by Frest *et al.* 1979).

Canadocystis, *Malocystites*, and *Wellerocystis* have more or less globular thecae with the periproct to the right of the adjacent ambulacrum. *Canadocystis* and *Malocystites* share short ambulacra. They differ principally in that both the ambulacra and the pseudopinnules are recumbent on the thecal surface in *Malocystites* whereas the inferred pseudopinnules were erect in *Canadocystis*. *Wellerocystis* has a more pyriform theca, longer ambulacra with two or three branches each and which gave rise to erect pseudopinnules presumably. As usual erect structures are rarely preserved. The final genus in this group is *Bistomiacystis* Sprinkle 1982, which is probably the most derived of all the paracrinoid genera. It has a spherical theca in which there are two oral openings, both of which gave rise to two curved ambulacra. These ambulacra are asymmetrical with facets on the right side only in contrast to all other paracrinoids which have facets on the left side only. Provisionally, I am associating *Bistomiacystis* with *Canadocystis*, *Malocystites*, and *Wellerocystis* because it has curved ambulacra, a globular theca and lacks pore-structures. I realize that, with the exception of curved ambulacra, these characters are plesiomorphic. Even the possession of curved ambulacra is a doubtful synapomorphy. Just as the ambulacral facets in *Bistomiacystis* are on the opposite side of the ambulacrum to all other paracrinoids, so too the direction of curvature (counterclockwise) is opposite to that in all the other genera with curved ambulacra.

The remaining two families share the synapomorphy of thecal pore-structures, although they are of two different types. In *Oklahomacystis* and *Amygdalocystites* there are covered trans-sutural slits which are sometimes associated with an ornament of triangular pyramids, and the periproct is on the right side of the adjacent ambulacrum. *Achradocystites*, and more doubtfully *Billingsocystis*, may have similar pore-structures and belong in the family Amygdalocystitidae, but too many morphological features are unknown to be certain of this or where precisely they fit in the scheme. The remaining genera, *Comarocystites* and *Sinclairocystis*, share unusual plates with concave outer surfaces and so-called foerstepores (Parsley and Mintz 1975). In *Sinclairocystis*, which has recumbent ambulacra, the periproct is to the left of the adjacent ambulacrum, while *Comarocystites* has erect ambulacra. In both genera and *Oklahomacystis* some species have the ambulacrum opposite the periproct branched to give a total of three.

Coronates, blastoids, and Lysosystites (Fig. 16.3)

These represent the most derived of the cystoid groups. They share a common plate arrangement of three basals, with the smallest, azygous basal in the A–B interray, five radials and more than five deltoids, the extra plates lying in the C–D interambulacrum which contains the periproct and the gonopore. Furthermore, in all of them the ambulacra are associated with a special plate, the lancet of eublastoids. Whether the possession of these ambulacral plates is a synapomorphy of this group is open to question as similar 'facetal' plates occur in cryptocrinoids such as *Palaeocystites*. It is possible that this innovation arose before the line to cryptocrinoids and paracrinoids separated from that to the coronates and blastoids.

The morphology of coronates is very conservative and has recently been described by Brett *et al.* (1983), while Donovan and Paul (1985) have discussed their origin and relationships. Coronates can be derived from *Bockia* by reduction in the size of the theca and number of thecal plates which become arranged regularly, by the development of internal canals across the radial:deltoid sutures, evolution of the lancet plate from which the erect, biserial, pinnate ambulacra arose, and apparently by the loss of the hydropore. *Lysocystites*, a peculiar Middle Silurian genus with a globular theca, has specialized pore structures in the basals, radials, and deltoids, and elongate 'lancet' plates between the deltoids, but from which plesiomorphic erect ambulacra (of unknown structure) arose. It appears to have been a blind offshoot from the main coronate-blastoid line. Donovan and Paul (1985) inferred that it branched off between the coronates and eublastoids since it has elongate 'lancet' plates.

To derive *Macurdablastus* Broadhead 1984, the first eublastoid, from coronates involves the loss of the coronal processes and their internal canals, and the development of true hydrospires.

PROBLEM GROUPS

Although the sections above give the impression that relationships can be determined between most early Palaeozoic cystoids, in fact, there remain several serious problems. In this section I wish to outline these briefly. Among major groups with well known morphology but obscure relationships are:

(1) all diploporites other than the sphaeronitids;
(2) the hemicosmitid rhombiferans;
(3) the parablastoids;
(4) the solitary edrioblastoid, *Astrocystites*.

The last is so unusual in its ambulacral structure that its relationships with other cystoids remain almost totally obscure and I shall not discuss it further.

Diploporites other than sphaeronitids include three major taxa (superfamilies), the protocrinitids, aristocystitids and asteroblastids. All three share diplopores and thecal orifices in the C–D interray. However, the latter is plesiomorphic and so does not help elucidate relationships, while the former can be shown to have been acquired independently at least twice. The argument is as follows. The stem is absent or drastically reduced in protocrinitids and aristocystitids, while asteroblastids retain it as adults. Aristocystitids have a major modification in the oral area with only two ambulacra forming a very deep and wide food groove within which a small oval mouth occurs and which is roofed over with complex cover plates. Hence, of the three superfamilies, protocrinitids seem most closely related to sphaeronitids. However, their five ambulacra bear cover plates and are composed of flooring plates which form part of the thecal wall, while the mouth is surrounded by interradial perioral plates, therefore, they could have diverged from the *Lichenoides*/sphaeronitid line after the loss of the stem, but before the radial orals with their epithecal food grooves which lack cover plates developed, i.e. before *Lichenoides* evolved. In this case diplopores evolved independently in the two most closely related diploporite superfamilies. There is, thus, not a single apomorphic character to unite the diploporite superfamilies. Aristocystitids have such a unique ambulacral structure that they could have arisen from almost anywhere, although provisionally the sphaeronitids seem the most likely sister group since both are cemented directly to the substratum.

The asteroblastids, to which I would add *Eumorphocystis* and *Regnellicystis* if they are not the same genus, have a functional heteromorphic stem as adults, four (or five?) basals and modified ambulacra in which at least some side branches are uniserial 'pseudopinnules'. They seem to form a natural group, but their origins are obscure. *Eumorphocystis* has the most unusual ambulacral structure of any cystoid. The proximal part of the five ambulacra is composed of biserial flooring plates which form part of the thecal wall and give rise to erect, biserial, brachioles alternately on either side. After about four brachioles the main ambulacra become free erect structures themselves, but with a triserial plate arrangement that gives rise to uniserial pseudopinnules in a pinnate arrangement. For a full description see Parsley (in Sprinkle 1982, pp. 280–8).

Hemicosmitids bear rhombs, called cryptorhombs, which are superficially very similar to the pectinirhombs of glyptocystitids (Paul 1968b). However, again it is possible to show that these two types of rhombs evolved independently. *Macrocystella* has all the morphological characters of the first glyptocystitid, *Cheirocystella*, except pectinirhombs. These characters include a unique stem divisible into proximal and distal parts with unique synarthrial articulations proximally; 27 thecal plates in exactly the same arrangement (four basal, five infralaterals, five laterals, six radials, and seven orals); a large lateral periproct surrounded by precisely the same thecal plates and covered with a flexible plated membrane enclosing the anal pyramid; a flat oral surface with five ambulacra which possibly already had the unique arrangement of brachioles in which ambulacra B and D had the first two brachioles off to the left and the others only the first. Hemicosmitids share none of these characters except four basals and similar, but not identical, rhombs. Even if the last character listed isn't present in *Macrocystella*, it is far more parsimonious to assume that cryptorhombs and pectinirhombs evolved independently than that *Macrocystella* independently evolved all the characters it shares with *Cheirocystella*. Thus, it seems that the slight difference between cryptorhombs, in which the canals (dichopores) always open in pores, and pectinirhombs where they invariably open in slits, is actually quite significant. Furthermore, the Dichoporita as defined by Paul (1968b) must be abandoned as a formal taxon.

If hemicosmitids are not closely related to glyptocystitids this does not help derive their relationships. The best recent account of their morphology is in Bockelie (1979). They are characterized by a simple heteromorphic stem, a pyriform to oval theca with a plate arrangement of four basals, six infralaterals, nine laterals, nine radials, and a variable number of oral cover plates; three ambulacra (A, C, and D) which are erect, biserial, and pinnate with biserial brachioles (see Sprinkle 1975); a lateral periproct covered by a simple anal pyramid and a hydropore, but no gonopore, in the C–D interambulacrum. Highly modified ambulacra are not found in any group associated with the lineage to the glyptocystitids. However, the ambulacral structure of hemicosmitids is not so dissimilar to those found in *Trachelocrinus* and *Bockia*, and basically the same as that found in coronates. It seems likely that hemicosmitids arose from this lineage, but at present I am uncertain precisely where.

Parablastoids include just three genera, two of which have recently been described by Sprinkle (1973a), and Paul and Cope (1982). They have a functional heteromorphic stem, a bud-shaped theca with variable plate arrangements, unique pore-structures called cataspires, five ambulacra with biserial flooring plates that form part of the thecal wall and give rise to biserial, erect brachioles alternately, and a central mouth surrounded by interradial periorals and surmounted by a unique crest plate. They are said to have five basals. Again, their morphology is so unique that relationships remain obscure.

CONCLUSIONS

The outline of the phylogeny of the cystoids given above is provisional and so I do not yet wish to introduce a new classification based upon it. Despite this, a few preliminary conclusions can be drawn. First, the Pelmatoza are a monophyletic group with *Kinzercystis* as the latest known common ancestor. They are characterized by the synapomorphies of an elongate aboral stalk and erect food gathering structures (brachioles) derived from aligned ambulacral cover plates. The Crinoidea also seem to be monophyletic, with *Lepidocystis* as their latest common ancestor. They are characterized by free arms derived from ambulacral flooring plates. The Cystoidea *s.l.* also form a monophyletic group with *Gogia* as their latest common ancestor. They are characterized by tesselate plating throughout the theca and stalk, thus enabling epispires to develop over all the theca. Within the cystoids the cryptocrinoids (not previously recognized as a separate taxon), paracrinoids, and blastoids (enlarged to include all genera with cups formed only of basals, radials, deltoids, and homologues of the lancet plate) all appear to be monophyletic and to belong in a larger monophyletic group that has *Bockia* as its latest common ancestor. Finally, although the Diploporita and Rhombifera are dismembered, their principal superfamilies remain monophyletic when allowance is made for some reassignments of individual genera, e.g. Paul (1984, p. 65) assigned *Glyptosphaerites* to the Sphaeronitida and renamed the superfamily containing the remaining genera the Protocrinitida, and as mentioned above, I would suggest that *Eumorphocystis* is more closely related to the asteroblastids than the protocrinitids.

REFERENCES

Barrande, J. 1887. *Systême Silurien du centre de la Bohême. Vol. 7. Classe des Echinodermes. Ordre des Cystidés.* R. Gerhard, Leipzig & Prague.

Bockelie, J. F. 1979. Taxonomy, functional morphology and palaeoecology of the Ordovician cystoid family Hemicosmitidae. *Palaeontology* **22**, 363–406.

Bockelie, J. F. 1981a. Functional morphology and evolution of the cystoid *Echinosphaerites*. *Lethaia* **14**, 189–202.

Bockelie, J. F. 1981b. A re-evaluation of the Ordovician cystoid *Stichocystis* Jaekel and the taxonomic implications. *Geologiska Föreningens i Stockholm Förhandlingar* **103**, 51–9.

Bockelie, J. F. 1981c. The Middle Ordovician of the Oslo Region, Norway. 30. The eocrinoid genera *Cryptocrinites, Rhipidocystis* and *Bockia*. *Norsk Geologisk Tidsskrift* **61**, 123–47.

Bockelie, J. F. 1982. Morphology, growth and taxonomy of the Ordovician rhombiferan *Caryocystites*. *Geologiska Föreningens i Stockholm Förhandlingar* **103**, 499–513.

Bockelie, J. F. 1984. The Diploporita of the Oslo Region, Norway. *Palaeontology* **27**, 1–68.

Brett, C. E., Frest, T. J., Sprinkle, J., and Clement, C. R. 1983. Coronoidea: a new class of blastozoan echinoderms based on taxonomic reevaluation of *Stephanocrinus*. *Journal of Paleontology* **57**, 627–51.

Broadhead, T. W. 1984. *Macurdablastus*, a Middle Ordovician blastoid from the southern Appalachians. *Paleontological Contributions of the University of Kansas, Paper* **110**, 1–9.

Donovan, S. K. and Paul, C. R. C. 1985. Coronate echinoderms from the Lower Palaeozoic of Britain. *Palaeontology* **28**, 527–43.

Frest, T. J., Strimple, H. L., and Coney, C. C. 1979. Paracrinoids (Platycystitidae) from the Benbolt Formation (Blackriveran) of Virginia. *Journal of Paleontology* **53**, 380–98.

Jell, P. A., Burrett, C. F., and Banks, M. R. 1985. Cambrian and Ordovician echinoderms from eastern Australia. *Alcheringa* **9**, 183–208.

Moore, R. C. (ed.) 1968. *Treatise on invertebrate paleontology. Part S. Echinodermata* 1 (2 vols). Geological Society of America and University of Kansas Press, Lawrence, Kansas.

Orlowski, S. 1968. Upper Cambrian fauna of the Holy Cross Mts. *Acta Geologica Polonica* **18**, 257–91.

Parsley, R. L. and Mintz, L. W. 1975. North American Paracrinoidea: (Ordovician: Paracrinozoa, new, Echinodermata). *Bulletins of American Paleontology* **68**, 1–115.

Paul, C. R. C. 1968a. *Macrocystella* Callaway, the earliest glyptocystitid cystoid. *Palaeontology* **11**, 580–600.

Paul, C. R. C. 1968b. The morphology and function of dichoporite pore-structures in cystoids. *Palaeontology* **11**, 697–730.

Paul, C. R. C. 1971. Revision of the *Holocystites* fauna (Diploporita) of North America. *Fieldiana: Geology* **24**, 1–166.

Paul, C. R. C. 1972a. Morphology and function of exothecal pore-structures in cystoids. *Palaeontology* **15**, 1–28.

Paul, C. R. C. 1972b. *Cheirocystella antiqua* gen. et sp. nov. from the Lower Ordovician of western Utah, and its bearing on the evolution of the Cheirocrinidae (Rhombifera: Glyptocystitida). *Geological Studies of Brigham Young University* **19**, 15–63.

Paul, C. R. C. 1973. British Ordovician cystoids. Part 1. *Palaeontographical Society Monographs*, 1–64, pls 1–11.

Paul, C. R. C. 1977. Evolution of primitive echinoderms. In *Patterns of evolution as illustrated by the fossil record* (ed. A. Hallam), pp. 123–58. Elsevier, Amsterdam.

Paul, C. R. C. 1984. British Ordovician cystoids. Part 2. *Palaeontographical Society Monographs*, 65–152, pls 12–26.

Paul, C. R. C. and Bockelie, J. F. 1983. Evolution and functional morphology of the cystoid *Sphaeronites* in Britain and Scandinavia. *Palaeontology* **26**, 687–734.

Paul, C. R. C. and Cope, J. C. W. 1982. A parablastoid from the Arenig of South Wales. *Palaeontology* **25**, 499–507.

Paul, C. R. C. and Smith, A. B. 1984. The early radiation and phylogeny of the echinoderms. *Biological Reviews* **59**, 443–81.

Smith, A. B. 1984. Classification of the Echinodermata. *Palaeontology* **27**, 431–59.

Sprinkle, J. 1973a. Morphology and evolution of blastozoan echinoderms. *Special Publication. The Museum of Comparative Zoology, Harvard University, Cambridge, Mass.*

Sprinkle, J. 1973b. New occurrence of the Ordovician eocrinoid *Lingulocystis* from Bolivia, South America. *Journal of Paleontology* **47**, 1113–6.

Sprinkle, J. 1975. The 'arms' of *Caryocrinites*, a rhombiferan cystoid convergent on crinoids. *Journal of Paleontology* **49**, 1062–73.

Sprinkle, J. (ed.) 1982. Echinoderm faunas from the Bromide Formation (Middle Ordovician) of Oklahoma. *Paleontological Contributions to the University of Kansas, Monograph* **1**, 1–369.

Ubaghs, G. 1953. Notes sur *Lichenoides priscus* Barrande, eocrinoide du Cambrien Moyen de la Tchecoslovaquie. *Institut royal des Sciences naturelles de Belgique Bulletin* **29** (34), 1–24.

Ubaghs, G. 1960. Le genre *Lingulocystis* Thoral (Echinodermata, Eocrinoidea) avec des remarques critiques sur la position systematique du genre *Rhipidocystis* Jaekel. *Annales de Paléontologie* **46**, 81–116.

Ubaghs, G. 1963. *Rhopalocystis destombesi* n.g. n. sp., Eocrinoïde de l'Ordovicien inférieur (Tremadocien supérieur) de Sud marocain. *Notes et Mémoires. Service des Mines et de la Carte Géologique du Maroc* **23**, 25–40, pls 1–3.

Ubaghs, G. 1967. Eocrinoidea. In *Treatise on invertebrate paleontology. Part S. Echinodermata 1* (ed. R. C. Moore), pp. S455–95. Geological Society of America and University of Kansas Press, Lawrence, Kansas.

17

The evolutionary palaeoecology of the Blastoidea

JOHNNY A. WATERS

Department of Geology, West Georgia College, Carrollton, GA 30118, USA

INTRODUCTION

Blastoids are an extinct class of blastozoan echinoderms ranging in age (Fig. 17.1) from Caradocian (Upper Ordovician) to Guadalupian (Upper Permian). Blastoids are divided into two orders, the Fissiculata and the Spiraculata, by the presence or absence of exposed hydrospire slits and spiracles. The fissiculates have been extensively restudied in recent years in monographs by Breimer and Macurda (1972), and Macurda (1983). These studies have placed the Fissiculata on a firm taxonomic footing and have increased our understanding of the details of their functional morphology, ontogeny, and phylogeny. In contrast, the Spiraculata have not received comparable treatment and the taxonomic foundation of this order is less firm. Indeed, Horowitz *et al.* (1986) have concluded that the Spiraculata are polyphyletic in origin, having originated from at least seven fissiculate ancestors by bridging open hydrospire slits with the side plates and forming spiracular pores. Although the Spiraculata represents a grade in evolution rather than a clade, the concept is retained herein for the sake of convenience until an alternative classification can be defined. The suprageneric classification of the Spiraculata largely dates from the blastoid volume of the *Treatise on invertebrate palaeontology* (Moore 1967). Treatise classification of blastoids into families has always been considered tentative (Moore 1967, p. S392). Consequently, diagnoses of spiraculate families in the Treatise are inadequate and in need of considerable revision. Horowitz *et al.* (1986) revised the Pentremitidae and distributed genera formerly assigned to it among the Troosticrinidae, Hyperoblastidae, Ambolostomatidae (a new family), and an additional undefined new family. This paper also gave revised familial diagnoses for the families listed above. Individual genera in the Spiraculata have been restudied recently. As an example, Waters *et al.* (1985) have restudied the ontogeny and phylogeny of *Pentremites* and have documented the role of heterochrony in the evolution of the genus. Taxonomic revision of *Pentremites* in progress will reduce the number of valid species from 67 recognized by Galloway and Kaska (1957) to 18. Taxonomic revisions of other genera will probably yield similar results.

BLASTOID DIVERSITY

Horowitz *et al.* (1985) studied the taxonomic survivorship of blastoids and recognized 90 genera. In addition to this list, I recognize the following genera: *Macurdablastus* (described by Broadhead 1984), *Ellipticoblastus* (considered by Macurda as a synonym of *Orbitremites*), *Artuschisma*, and two new genera from Ireland and Great Britain that I am currently describing with G. D. Sevastopulo. Ninety-five blastoid genera are currently recognized (Appendix 1) although not all have been formally described. Inevitably, additional genera will be found so this number is subject to change. The Upper Carboniferous would seem to be a promising interval for the discovery of new blastoids. Macurda and Mapes (1982) described a fauna of very small Pennsylvanian blastoids from the United States found by washing very large quantities of shale and picking the residue. The potential for similar faunas in other localities is quite good.

Figure 17.2 is a smoothed plot of blastoid distribution through time constructed by averaging the diversity of three adjacent intervals to obtain a value for each time interval. For example,

Echinoderm phylogeny and evolutionary biology (ed. C. R. C. Paul and A. B. Smith). Clarendon Press, Oxford, 1988.

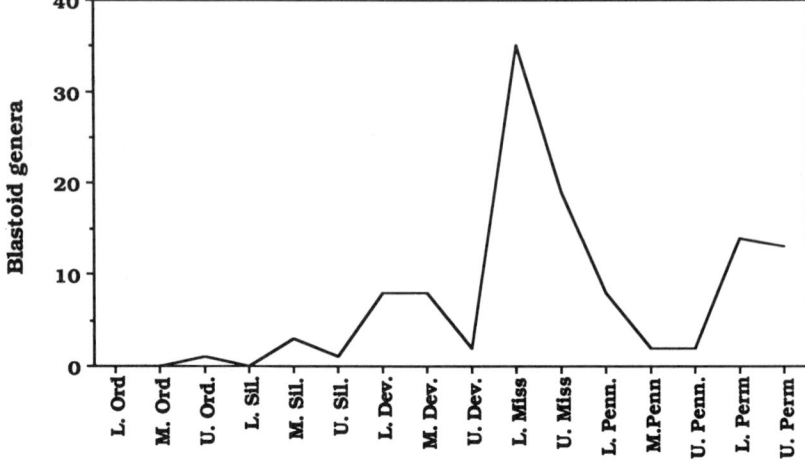

Fig. 17.1 Generic diversity of the Blastoidea. Figures 17.1–17.7 use blastoid data modified from Horowitz *et al.* (1985).

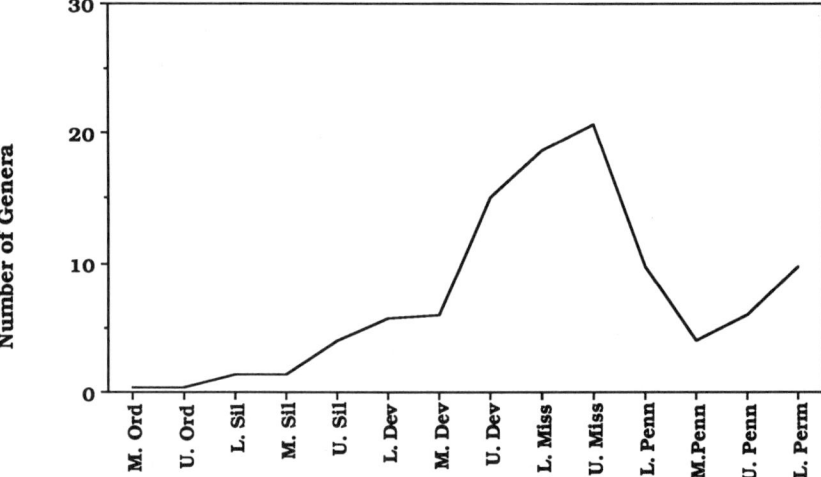

Fig. 17.2 Generic diversity of the Blastoidea smoothed by a three-term function.

the diversity of Lower, Middle, and Upper Silurian blastoids is averaged to yield the value for the Middle Silurian. Other values were constructed in a similar manner. A plot such as this is useful to filter out the effects of collecting bias, the variation in outcrop available for study, and smaller scale fluctuations in diversity. This graph shows very clearly the rise in blastoid diversity in Ordovician through Mississippian time and subsequent decline until the Permian. The Permian faunas, primarily from southeast Asia, represent a modest resurgence in blastoid diversity before their extinction in the Upper Permian.

Figure 17.3 is a plot of the generic diversity of fissiculate and spiraculate blastoids. The fissiculates show a steady increase in diversity to a maximum in the Lower Carboniferous. Although the generic diversity of fissiculates remained more or less constant in the Upper Carboniferous and Permian, the fissiculates dominate these blastoid faunas in terms of taxonomic diversity. The pattern of spiraculate diversity parallels that of the fissiculates through the Devonian. However, in the Lower Carboniferous, the spiraculates show a spectacular increase in diversity and dominate most blastoid communities in terms of abundance as well as taxonomic diversity. Upper Carboniferous and Permian spiraculates show a dramatic decline in diversity although individual spiraculate genera continue to dominate these faunas numerically. Examples include *Nodoblastus librovitchi* from the Lower Namurian of North Kazakhstan

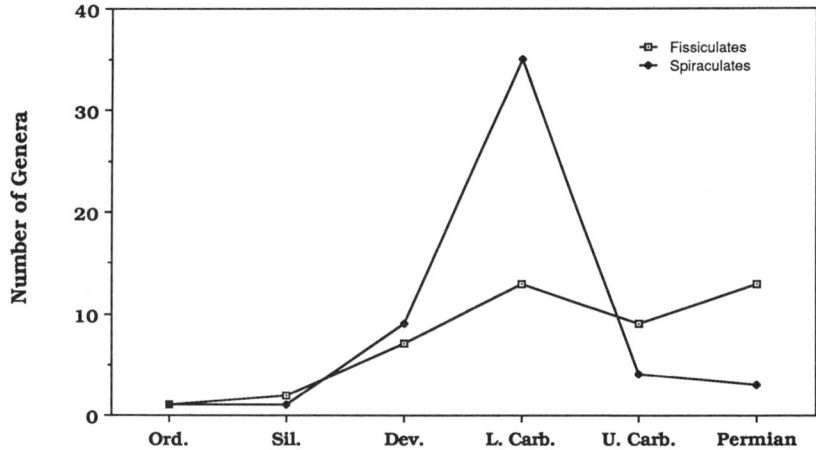

Fig. 17.3 Generic diversity of the fissiculate and spiraculate blastoids.

(Arendt et al. 1968) and *Deltoblastus* from the Permian of Timor.

Most blastoid genera were short-lived and ranged through only a single stage (Fig. 17.4). Horowitz et al. (1985) derived an average age duration of 9 Ma for blastoid genera and 5.1 Ma for species. These results were based on 10 Ma class intervals and one-half the value of the beginning and ending stage of the generic range. My work with the blastoid faunas in Ireland and Great Britain suggests that many generic ranges are much shorter than indicated by this study (see Waters and Sevastopulo, 1984, for a summary of the distribution of these blastoids). Most genera in the British and Irish faunas were confined to only one of the six local stages that define the Dinantian (Upper Tournaisian–Visean). Indeed, if the ranges are considered in even greater detail, very few genera actually range throughout any one of these six stages. Because this division of the Dinantian was based on sea level cycles, one could interpret this to suggest that most to these genera lived only through one transgressive-regressive cycle. Table 17.1 illustrates the different estimates for generic survivorship for the British and Irish fauna, and was constructed using those genera found only in this fauna. If each of the five stages in the British Visean were equal in duration and the blastoids ranged through the entire length of a single stage, then the average generic duration would be ~2.5 Ma for this fauna rather than a 6 Ma estimate obtained by the methodology of

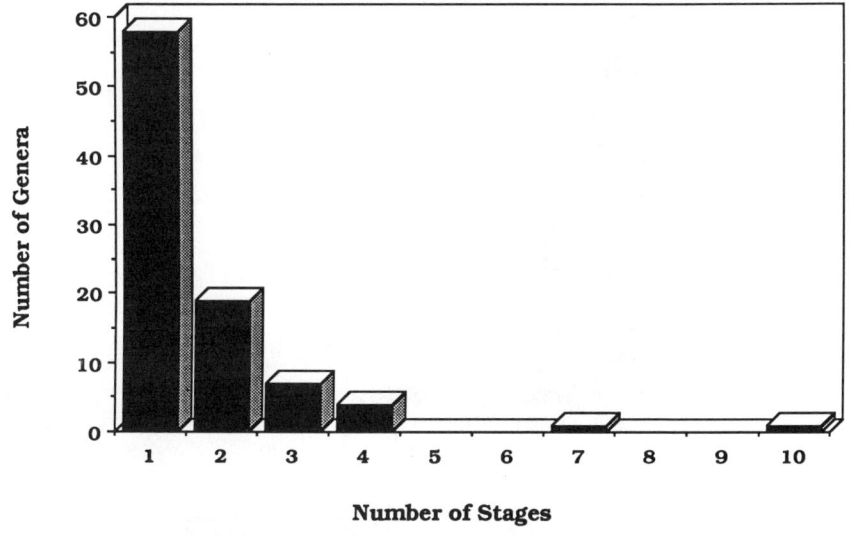

Fig. 17.4 Generic survivorship of blastoid genera. Most blastoids occurred only in a single stage.

TABLE 17.1 Survivorship of Irish and British Lower Carboniferous blastoids

	Horowitz et al. (1985)			Waters (this paper)	
		Stage (Ma)			
Genus	Range	Full	Half	Range	Ma (estimated)
Orbitremitid, n.g.	Tournaisian	8	4	U. Courceyan	<2
Ellipticoblastus	Visean	13	6.5	Chadian	<2.5
Acentrotremites	Visean	13	6.5	Chadian	<2.5
Monoschizoblastus	Visean	13	6.5	Asbian	<2.5
Astrocrinus	Visean	13	6.5	Brigantian	<2.5
Codaster	Visean	13	6.5	Brigantian	<2.5
Heteroblastus	Visean	13	6.5	Brigantian	<2.5
Orophocrinid, n.g.	Visean	13	6.5	Brigantian	<2.5

Horowitz et al. (1985). Although the British and Irish faunas are a small subset of the data used by Horowitz et al. (1985) for their analysis, the data suggest that their estimates may be off by 30–50 per cent if a more rigorous determination of blastoid ranges could be derived. These different figures for generic survivorship are potentially significant in the overall interpretation of blastoid palaeoecology. In previous papers, (Waters et al., 1981, 1985; Waters and Sevastopulo 1984), I have advanced the proposal that blastoids acted opportunistically in response to changes in crinoid community evolution. If the blastoids were in fact opportunistic, then they should have shorter ranges than other organisms. Although the detailed data on many generic ranges are not available to evaluate this thesis completely, I believe it remains a viable working hypothesis.

The blastoids show a tremendous range in the diversity and abundance within genera. Figure 17.5 shows a plot of the number of species per genus for the fissiculate blastoids. Of the 40 genera of fissiculate blastoids recognized by Breimer and Macurda (1972), and Macurda (1983), eighteen (45 per cent) are monospecific and thirteen (33 per cent) have only two species. Twenty-seven fissiculate species are known from fewer than six specimens. Similar results will probably be obtained for the spiraculates when they are monographically revised. Many blastoid

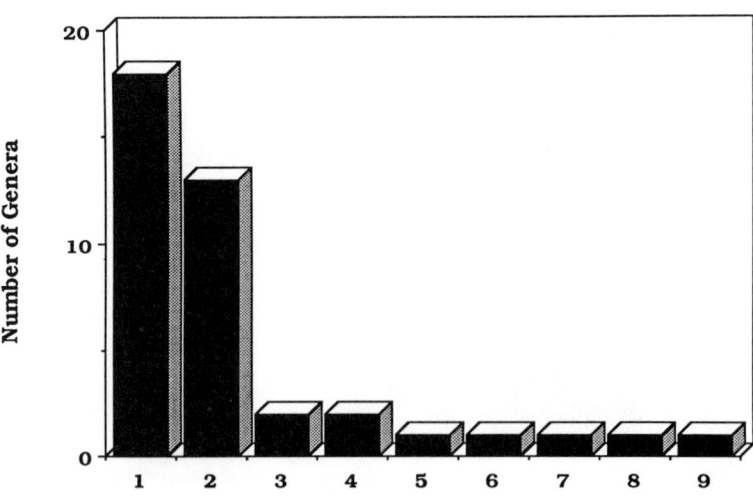

Fig. 17.5 Number of species in each genus of fissiculate blastoid.

genera are also geographically endemic, having been collected from a very few localities within a relatively small geographic region. To summarize, the typical blastoid genus is monospecific, relatively short lived (range limited to some part of a single stage), relatively rare in terms of abundance, and geographically restricted to one depositional basin. In contrast to this norm, however, some blastoid genera were very abundant (e.g. *Pentremites*, *Orbitremites*, *Deltoblastus*), geographically widespread (e.g. *Mesoblastus*, *Angioblastus*) and geologically longlived (*Orbitremites*). One goal of my blastoid studies is to explain why some genera were so diverse, widespread, or abundant while most were not. Were these genera functionally superior or just in the right place at the right time?

EVOLUTIONARY PALAEOECOLOGY OF BLASTOIDS

I believe the palaeoecological history of blastoids can best be interpreted within the context of crinoid community evolution in the Palaeozoic. Throughout most of their evolutionary history, blastoids were a component (sometimes minor, sometimes important) of echinoderm communities which were generally dominated by other groups (principally crinoids). Figures 17.6 and 17.7 show diversity plots of blastoids, monobathrid camerates, and cladid inadunate crinoids. The diversity of blastoids and monobathrids is highly correlated ($r = 0.65$; $P < 0.02$). Although the diversity of blastoids and cladids is not significantly

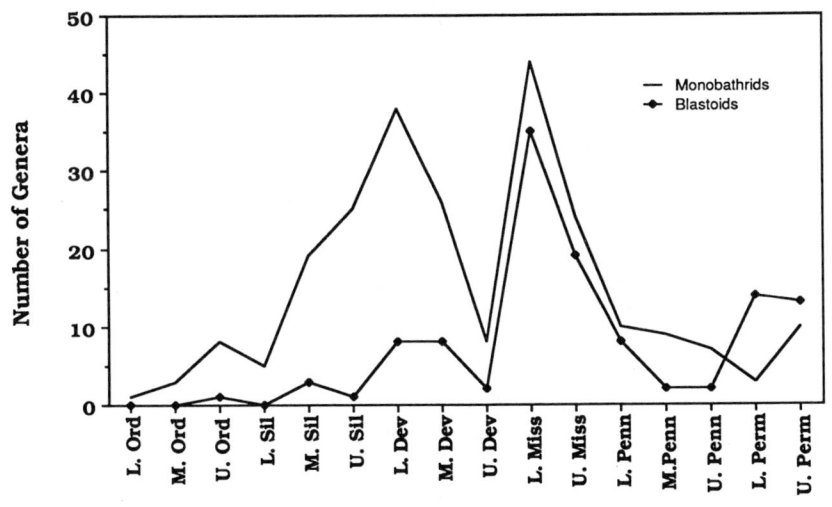

Fig. 17.6 Diversity of blastoids and monobathrid camerate crinoids. Crinoid data abstracted from the *Treatise on invertebrate paleontology*.

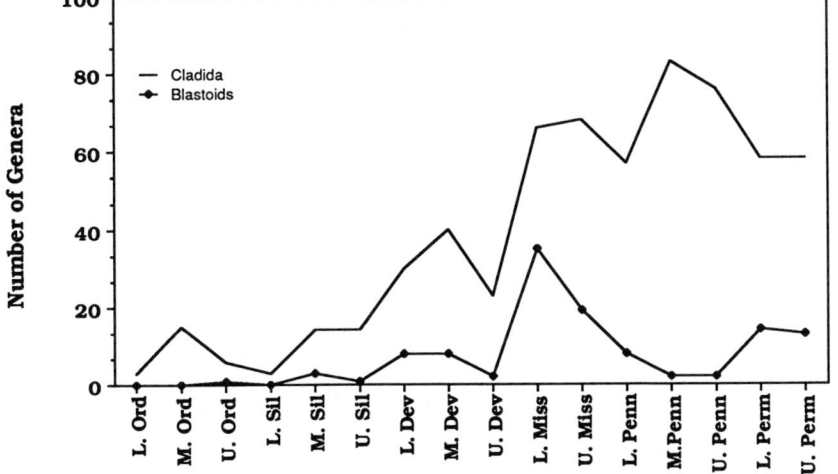

Fig. 17.7 Diversity of blastoids and cladid inadunate crinoids. Crinoid data abstracted from the *Treatise on invertebrate paleontology*.

correlated, diversity changes in the two groups were similar until the Upper Mississippian–Pennsylvanian (Visean–Namurian) when declines in monobathrids and blastoids were offset by an increase in the diversity of cladids.

Crinoid communities were stratified into tiers above the sea floor and food was apparently partitioned within each tier (Ausich 1980). Data to support these conclusions include stem length, width of the ambulacral grooves on the arms, and shape of the filtration fan for different crinoid species. Ausich documented that crinoids in the Borden delta (Lower Carboniferous of the eastern United States) occupied specific niches within this tiering/food selection continuum. Kammer (1985) studied crinoid communities in other parts of the Borden delta, and was able to document that different types of filtration fans were adapted to different current velocities and sediment regimes in the delta complex. As these examples illustrate, many advances have been made in interpreting the complex ecological structure of crinoids found in these communities. All these crinoid communities also contain blastoids. However, blastoids have not been successfully placed within this niche-partitioning model. In many cases the specific information (stem lengths and brachioles) needed to make the interpretation is missing. The details of blastoid functional morphology currently cannot be interpreted within the niche-partitioning model. Present information on stem lengths suggest that blastoids occupied the lower and lower intermediate tiers (<20 cm above the sea floor). Brachiolar widths from different blastoid genera are similar and do not show species-specific variability as do the crinoids. The brachioles of blastoids must have formed a simpler filtration fan than crinoids with their complex patterns of arm branching. Clearly, the detailed functional morphology of blastoid filtration fans is an area in need of additional study. At this point we can only conjecture whether they were competing for a specific sized food resource or used their generalized filtration fan to capture a wider size range of food than did the crinoids.

In most Palaeozoic echinoderm communities, blastoids occupied limited niches within the crinoid communities and as a result are generally rare when compared with the most abundant crinoid genera found in a given fauna. However, blastoids did become the numerically dominant echinoderm in several instances. In each case, blastoids become more diverse taxonomically and morphologically as well as the numerically dominant echinoderm after a major pertubation in the crinoid community. This pattern is well documented in the Upper Palaeozoic, but is less clear in the Lower Palaeozoic.

The history of blastoids can be viewed in three phases (Fig. 17.1). The initial radiation in the Middle Ordovician (Phase 1) was followed by an increase in diversity and geographic range in the Devonian and culminated in the Frasnian–Famennian extinction. Phase two included the widespread diversification of blastoids in the Lower Carboniferous. In some cases they became the dominant echinoderm in these communities. This phase ended in the sudden decline in blastoid diversity and abundance in the Upper Carboniferous. Phase three involves the re-radiation of blastoids in the Tethyan seaway during the Permian.

Phase 1

The oldest blastoid genus, *Macurdablastus*, was described by Broadhead (1984) from the Benbolt Formation in Tennessee which is Caradocian (Upper Ordovician) in age. The specimens are not well preserved and cannot be classified to order (fissiculate or spiraculate). Sprinkle (1973) described a single silicified deltoid plate from the Antelope Valley Formation (Middle Ordovician) in Nevada that he ascribed to the fissiculate blastoids. No additional material has been described from this locality, and further information is not currently available on this occurrence.

Sprinkle (1982) has provided a summary of echinoderm communities during the Middle Ordovician with his study of the Bromide Formation of Oklahoma. The echinoderm fauna from the Bromide contained over 11,000 complete specimens belonging to 61 genera and 13 classes. Three zones within the Bromide produced most of the echinoderm specimens. The Lower Echinoderm Zone yielded 6800 specimens in 31 genera. This zone was dominated by *Hybocrinus*, a hybocrinid inadunate (51 per cent), and *Platycystites* (23 per cent), a paracrinoid. The Upper Echinoderm Zone produced about 3700 echinoderms in 20 genera. This zone is dominated by the paracrinoid *Oklahomacystis* (93 per cent). The third zone in the Pooleville Member of the Bromide, has produced 650 specimens in 30 genera from a series of sections higher in the Bromide. This fauna is dominated by the camerate *Archaeocrinus* (29 per cent).

These three faunas are found in very different depositional environments. The Lower Echinoderm Zone was deposited on a gently sloping carbonate ramp during a marine transgression in water depths of 15–20 m in areas of the richest echinoderm faunas. Deposition occurred below normal wave base, but above storm wave base. The Upper Echinoderm Zone was deposited on a carbonate platform near the maximum Bromide transgression. Water depths varied, but were probably below normal wave base during deposition of this zone. The Pooleville Zones occurred in a sequence of biomicrites deposited in moderate to deep water conditions on a carbonate ramp near the end of the upper Bromide regression.

Bromide echinoderms lived at the time of maximum class diversity and very high generic diversity for echinoderm communities and occurred in a variety of depositional environments and blastoids are noticeably absent. The Bromide is slightly older than the Benbolt of Tennessee, so perhaps the blastoids simply had not evolved by Bromide time. Both the Bromide and Benbolt formations contain many blastozoan taxa or ecological correlatives such as cystoids, paracrinoids, parablastoids, and hybocrinid inadunate crinoids which occupied the lower tiers of the communities. Blastoids did not become diverse or abundant until many of these morphologically similar taxa declined in diversity and abundance.

Three genera of blastoids (*Decaschisma*, *Troosticrinus*, and *Polydeltoideus*) are known from the Silurian (Fig. 17.8). All these genera are well known in North America, but typically only a single genus is found in a given formation. The only exception is *Polydeltoideus* which is known from a single specimen found in Czechoslovakia. Population studies of all three genera (Macurda 1983; J. A. Waters unpublished observations) show that they are morphologically conservative and suggest that they occupied a rather narrow niche in Silurian echinoderm communities.

In the Devonian, however, blastoid distribution changed substantially. Devonian blastoids are well known from a variety of localities (Figs 17.9 and 17.10) and generally yield diverse, relatively abundant blastoids with several genera co-existing for the first time in the evolutionary history of blastoids.

Spanish Lower Devonian blastoids represent the best Lower Devonian fauna known. Summarized by Breimer and Dop (1974), this fauna consists of species of *Cryptoschisma*, *Pentremitidea*, *Pleuroschisma*, *Caryoblastus* (all fissiculates), *Hyperoblastus*, a spiraculate, and *Conuloblastus*, intermediate between the two orders. The Middle Devonian in the northeastern United States and southern Ontario has yielded an abundant fauna

Fig. 17.8 Palaeobiogeographic distribution of Wenlockian–Ludlovian (Silurian) blastoids. In this and following figures, fissiculates are represented by the narrower symbol, spiraculates by the more globose symbol.

Fig. 17.9 Palaeobiogeographic distribution of Siegenian–Emsian (Devonian) blastoids.

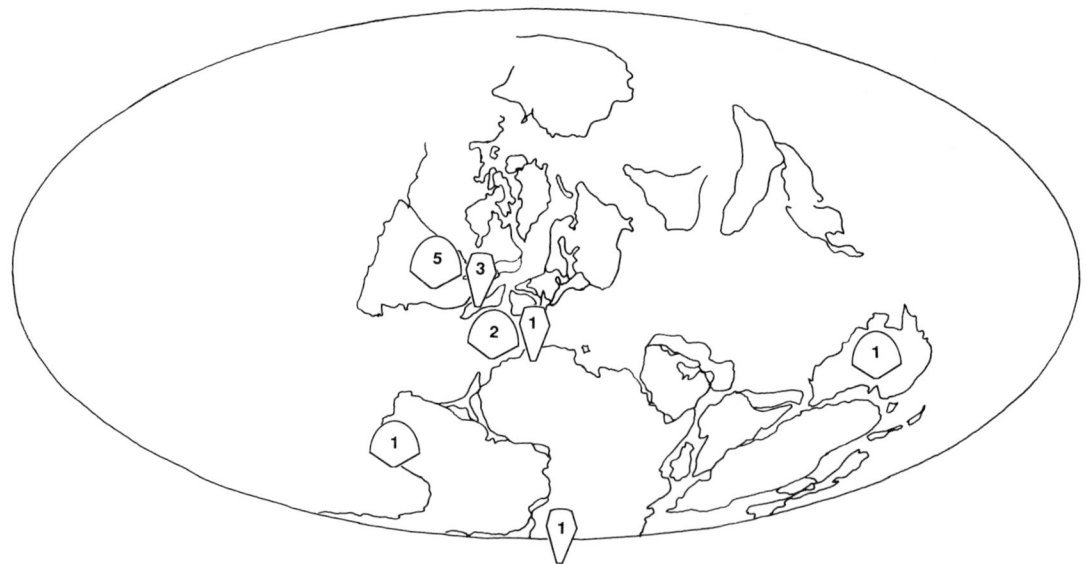

Fig. 17.10 Palaeobiogeographic distribution of Eifelian–Givetian (Devonian) blastoids.

of *Nucleocrinus*, *Hyperoblastus*, *Brachyschisma*, *Pleuroschisma*, *Devonoblastus*, *Heteroschisma*, and *Eleutherocrinus*. Middle Devonian blastoids show a very wide geographic distribution (Fig. 17.10) although maximum diversity remained around the margins of the Atlantic basin. Additional Devonian blastoids are known from Bolivia (Macurda 1979), South Africa (Macurda 1983), and Australia. The extinction event at the Frasnian–Famennian boundary severely affected the blastoids as well as a significant portion of the shallow marine fauna. Only *Hyperoblastus* survived into the Famennian. Among the echinoderms, blastoids suffer a 75 per cent reduction, monobathrid camerates an 80 per cent decline, but the cladid inadunates only a 50 per cent decline, in generic

Fig. 17.11 Palaeobiogeographic distribution of Tournaisian (Lower Carboniferous) blastoids.

Fig. 17.12 Palaeobiogeographic distribution of Visean (Lower Carboniferous) blastoids.

diversity. Although the Frasnian–Famennian extinction was severe, it did not affect the long term diversity increases of any of these groups (Figs 17.6 and 17.7).

Phase 2

Major changes in dominance and diversity of blastoids and crinoids accompanied fundamental shifts in the sedimentation patterns during the Lower Carboniferous particularly in the United States and Europe (Figs 17.11 and 17.12). Lane (1972) summarized the synecology of Mississippian crinoid communities in the United States and concluded that Lower Mississippian crinoid communities were dominated by monobathrid camerates while later faunas were dominated by cladid inadunate crinoids. The change in dominance occurred at or very near the Osage–Meramec boundary (early Visean). Diplobathrid camerates largely became extinct and only six families of monobathrids survived into the Meramec. Blastoids also suffered extinctions at this time. Only eight of 34 genera (24 per cent) present in the Tournaisian survived into the Visean.

During the Osage, monobathrid camerate crinoid communities reached a zenith on the broad carbonate platforms of the Burlington and Keokuk Limestones (and correlatives). Blastoids were an integral part of this community and 20 genera are known from the Burlington. This assemblage is probably the largest blastoid assemblage in terms of generic diversity. A majority of the genera were globose spiraculates. Some of the genera were abundant and diverse but most are relatively rare and geographically restricted. Ausich and Lane (1985) analysed 33 Osage–Meramec crinoid assemblages and concluded that communities dominated by monobathrid camerates were restricted to higher energy carbonate platforms such as the Burlington and Salem, while communities with diverse inadunate components were found in areas with significant clastic influx such as the Borden Delta. The blastoid genera found in these deltaic deposits are mostly fissiculates not associated with the platform carbonate communities. Meramecian carbonate platform communities were numerically dominated by batocrinid and platycrinid camerates; poteriocrinid inadunates were more diverse, but less numerous than camerates in these platform carbonate communities. Analysis of these echinoderm faunas from the Monteagle Limestone near Huntsville, Alabama, suggests that they persisted unchanged to the Meramec–Chester boundary (Horowitz and Waters 1972).

At the Meramec–Chester boundary (late Visean) the dominant Meramecian monobathrids (batocrinids, platycrinids) were again reduced in abundance and diversity. Monobathrids became minor elements of Chesterian crinoid communities that were dominated by poteriocrinine inadunates, although the monobathrid families Dichocrinidae and Acrocrinidae underwent a small scale diversification. Blastoids also underwent another extinction event at the Visean–Namurian boundary (Middle Chester). Of 22 genera of blastoids present in the Visean, 15 became extinct (68 per cent) and only eight (32 per cent) survived into the Namurian. Although crinoids appear not to have been so severely affected, among the monobathrids the extinction of *Globocrinus* and *Talarocrinus* were important events in eastern North America.

Near the Mississippian–Pennsylvanian (Namurian) boundary, poteriocrinine inadunates underwent tremendous radiation and many important Chesterian crinoid genera became extinct by the end of the period (Waters *et al.* 1981).

Pentremites Say is one of the most widely distributed blastoid genera occurring throughout the Mississippian in North America and extending into the Lower Pennsylvanian. It is known from the Appalachians west to Arizona as well as the Canadian Rockies, the Brooks Range of Alaska and Colombia in South America. Waters *et al.* (1981, 1985) have summarized the evolution of *Pentremites* and concluded that major speciation events and changes in diversity in this genus are related to changes in Mississippian crinoid communities, and were triggered by the ability of the genus to adapt to changing sedimentological regimes. *Pentremites* was a minor member of Osagean carbonate platform communities that survived and appeared to benefit from extinctions among the camerates and other blastoids. *Pentremites* became significantly more abundant in Meramecian carbonate platform communities, invaded deltaic communities for the first time, and successfully competed with Chesterian poteriocrine inadunates. With the further demise of monobathrids at the Meramec–Chester boundary, *Pentremites* again benefited from decreased competition and became one of the most abundant and diverse echinoderms of the Chesterian. The decline and extinction of *Pentremites* in the Lower Pennsylvanian was probably the result of the continuing expansion of poteriocrine inadunates and the resulting extinction of the majority of Chesterian echinoderms.

European Lower Carboniferous blastoids also exhibit changes in response to changing depositional environments and crinoid community evolution (Waters and Sevastopulo 1984). Late Tournaisian (Courceyan) blastoids are best known from Tournai, Belgium (Macurda 1967). The fauna consists of one spiraculate, *Mesoblastus crenulatus*, and five fissiculates, *Katoblastus konincki* Macurda, *K. puzos* (Münster), *Orophocrinus orbignyanus* (de Koninck), *Phaenoblastus caryophyllatus* (de Koninck and Le Hon) and

Xenoblastus sp. (Breimer and Macurda 1972). Additional faunas consisting of *Orophocrinus orbignyanus*, *O. pentangularis*, a new genus of orbitremitid and two fragmentary genera are known from southern Ireland and Britain (Waters and Sevastopulo 1984). These faunas are coeval and were all found in the Black Rock Limestone facies which was widespread at this time (Lees 1982). Although the blastoids are found in the same facies, they are endemic; only *Orophocrinus* is found both in Belgium and the British Isles. The section at Hook Head, County Wexford, Ireland, provides some insight into the distribution of these faunas. The blastoids at Hook Head are concentrated at the leading edge of the marine transgression at the base of the Hook Head Formation and the trailing edge of the regression associated with the Bullockpark Bay Dolomite (Waters and Sevastopulo 1984). Only a single specimen has been found in the deeper water facies exposed above the Bullockpark Bay Dolomite at Hook Head. The blastoids are associated with a variety of crinoids including *Dialutocrinus* and *Platycrinites* (both monobathrid camerates), but are stratigraphically much more restricted than are the crinoids.

During the latest Tournaisian (Upper Courceyan) and Lower Visean (Chadian), buildups of the Walsortian facies were widespread across Europe (Lees 1982). Blastoids were associated with the build-ups and are particularly well known from the Craven Basin in Britain and its extension, the Dublin Basin, in Ireland. The blastoids are commonly found in the flank facies again associated with diverse crinoid communities. Faunas associated with the build-ups include the fissiculates *Orophocrinus celticus*, *O. praelongus*, *O. verus*, *O. pentangularis*, *Phaenoschisma acutum*, and the more abundant spiraculates *Ellipticoblastus ellipticus*, *Mesoblastus elongatus*, and *M. angulatus*.

The factors affecting the distribution of blastoids among the various reefs in the Craven and Dublin basins are not known (probably temporal or ecological), but some general statements can be made. Blastoids are rare compared to crinoids and were a minor element of the crinoid communities. For example, five fissiculates species are known from the Feltrim Hill reef north of Dublin (Hudson et al. 1966), but collectively they constitute less than 2.5 per cent of the echinoderms collected (~500 specimens). Spiraculates such as *Ellipticoblastus ellipticus* are absent here, although they are found at the Carrickboy Quarry, County Longford, to the west of Dublin.

A more detailed stratigraphy of the reefs in the Craven Basin near Clitheroe has been worked out (Metcalfe 1980) and the reefs can be placed in stratigraphic order. Older reefs such as Coplow Knoll contain an abundant crinoid fauna characterized by actinocrinitid camerates. Blastoids are very rare at Coplow; the total reported faunas consists of a few specimens of *Orophocrinus pentangularis*. The geographically adjacent, but slightly younger reefs exposed at the Bellmanpark Quarry and Salthill Quarry contain different crinoid assemblages (Wright 1928) and substantially more abundant blastoid faunas. The actinocrinitids were replaced by *Amphorocrinus* and *Platycrinites* as the most abundant camerates in these younger reefs. Salt Hill contains a diverse (*Orophocrinus*, *Mesoblastus*, *Ellipticoblastus*, and *Phaenoschisma*), reasonably abundant (except *Phaenoschisma*) blastoid fauna in the crinoid-rich flank beds. At Bellmanpark and other coeval reefs in the area, blastoids are more closely associated with the core facies occurring just above or lateral to the reef core. Although the details are sketchy, Chadian crinoid communities in the Craven basin underwent a reorganization of some significance and blastoids benefited becoming significantly more abundant.

Mesoblastus is the most geographically widespread of the genera associated with the Waulsortian and is found in Waulsortian facies in France (Macurda and Racheboeuf 1975), New Mexico (an undescribed species) and unknown lithologies in China. Curiously, *Mesoblastus* has not been found in Ireland despite the widespread occurrence of the Waulsortian facies there.

Middle Visean (Arundian and Holkerian) blastoids are not known from Britain and are rare in Ireland. They are known only from fragments in the Bundoran Shale, a prodeltaic shale in western Ireland (Waters and Sevastopulo 1984). Macurda and Racheboeuf (1975) have described *Gongyloblastus armoricanus* (a spiraculate) from Laval, Brittany, France, which is the only European blastoid of this age.

Late Visean blastoids are well known from Ireland and Britain although none has been described from the European continent. These blastoids exhibit the extremes in environmental setting, range and abundance. *Monoschizoblastus rofei* is a very abundant spiraculate blastoid endemic to northwestern Ireland. It is primarily associated with shallow water reefs in the Dartry Limestone, but is also found in non-reef facies. In contrast, the only blastoid described from the Asbian of Britain is the rare fissiculate *Acentrotremites ellipticus* (Cumberland) which is known from three specimens collected in Somerset and Wales. On the other hand, the Brigantian of Britain contains an abundant and diverse blastoid fauna associated with a variety of environments. *Orbitremites derbiensis*, *Astrocrinus tetragonus* (a diminutive stemless fissiculate), and *Codaster acutus* are associated with Brigantian reefs in Yorkshire. They are best known from exposures at Grassington (Joysey 1955), where *O. derbiensis* is known from hundreds of specimens. These species are also known from reefal deposits in Ireland.

The Brigantian of Britain was a time when the sediment pattern shifted to a pattern of cyclic limestone-clastic deposition (Yoredale facies) similar to the Chesterian of the US. Blastoids are also found in these facies. The Scottish fauna consists of *Astrocrinus tetragonus* and *Hadroblastus benniei* (a rare fissiculate). Although very small, *Astrocrinus tetragonus* is known from hundreds of specimens picked from washings of the Lower Limestone Group of Fife by James Wright in the early 1900s.

Heteroblastus cumberlandi Etheridge and Carpenter is a spiraculate known from a single locality near Hexham, Northumberland. This species occurs in a shale with siderite nodules and is associated with a molluscan fauna. Two Irish faunas indicate that blastoids also occupied very shallow water environments at this time. The Meenymore Formation (a shallow water limestone with clay partings) at Gleniff, County Sligo, contains a new species of *Hadroblastus*, a new species of *Orbitremites*, as well as *Astrocrinus tetragonus*. All these blastoids are very small and are associated with a diverse echinoderm fauna including numerous inadunate crinoids, microcrinoids, asteroids, cyclocystoids and ophiuroids (Waters and Sevastopulo 1984). A second Meenymore locality at Cashel, County Fermanagh, contains *Orbitremites* sp, *Phaenoschisma* sp., and new orophocrinid genus in a fauna dominated by cephalopods and rugose corals. Collections from Lisdowney, County Kilkenny, provide the other shallow water fauna. Blastoids here include a new species of *Orbitremites* (very abundant), *Codaster acutus*, and *Astrocrinus tetragonus*, and occur with an atypical crinoid community consisting of microcrinoids, *Synbathocrinus*, *Cyathocrinites*, *Cydonocrinus*, and *Neolageniocrinus*. All these occurrences foreshadow the apparent environmental shift of Pennsylvanian blastoids.

Pennsylvanian communities (Fig. 17.13) continued to be dominated by poteriocrinine inadunates. The Pennsylvanian crinoid fauna included eight genera of camerates, seven genera of flexibles and 109 genera of poteriocrinine inadunates. Blastoids are extremely rare in Pennsylvanian rocks of North America; the few specimens described are fissiculates (see Breimer and Macurda 1972; Macurda and Mapes 1982). The best known Pennsylvanian blastoid is *Pentremites rusticus* Hambach (a spiraculate) from the Morrowan of Arkansas and Oklahoma. It is a vestige of the long successful lineage of *Pentremites* from the Mississippian. The successful European orbitremitid lineage is represented by *Orbitremites derbiensis moscovi* Arendt from the USSR. Arendt et al. (1968) described a fauna of *Nodoblastus librovitchi* (a spiraculate) and four fissiculates (*Dolichoblastus shimanski*, *Kazachstanoblastus carinatus*, *Mastoblastus ornatus*, and *Artuschisma rossica*) from the Lower Namurian of Kazachstan USSR. These blastoids are found in a cephalopod rich shale and limestone. *Nodoblastus* is very abundant and forms a major element of the echinoderm community in a fashion similar to *Pentremites* or *Orbitremites* from the Lower Carboniferous. These forms do not have any known successors nor does the orophocrinid *Pentablastus supracarbonicus* (Sieverts-Doreck 1951) from the Upper Namurian of Spain.

The most widespread late Palaeozoic fissiculate is *Angioblastus*. *Angioblastus ellesmerensis* Breimer and Macurda occurs in the Atokan of Ellesmere Island in the Canadian Arctic. *Angioblastus dotti* (Moore and Strimple) is found in the Missourian of Oklahoma and *A. caddense* (Strimple and Mapes) in Texas. *Angioblastus* and *Orbitremites* are the only known blastoids with both Pennsylvanian and Permian representatives. The only other described Pennsylvanian blastoid is *Malchiblastus australis* (Etheridge), a spiraculate from the Bashkirian of Queensland in Australia. Pennsylvanian blastoids are found in two environmental settings—reefal (e.g. *Angioblastus dotti* and *A. ellesmerensis*) or clastic which are dominated by molluscs (e.g. *Angioblastus caddense*).

Phase 3

Permian blastoids are widespread (Fig. 17.14), but the most diverse faunas are found in southeast Asia and Australia. The fissiculates are the most cosmopolitan with *Angioblastus* being known from Bolivia, the Urals, and Indonesia. It also occurred in the Pennsylvanian of North America. Poorly known blastoids have been found in the Canadian archipelago (spiraculate or fissiculate) and Sicily (spiraculate; Breimer and Macurda 1973). The fissiculates reach a second peak of diversity and some genera (e.g. *Neoschisma*) are

Fig. 17.13 Palaeobiogeographic distribution of Bashkirian–Moscovian (Upper Carboniferous) blastoids.

Fig. 17.14 Palaeobiogeographic distribution of Guadalupian (Permian) blastoids.

found in both Indonesia and Australia. The spiraculates are less diverse but one Mississippian genus (*Orbitremites*) found in the British Isles and USSR ranges into the Permian of Timor. The Timor fauna is the most diverse and abundant of the Permian faunas. Fifteen blastoid genera are known from Timor. Fissiculates are the most diverse (13 genera), but spiraculates dominate in terms of abundance. *Deltoblastus* is known from tens of thousands of specimens. Although this fauna is very well known from museum collections, the palaeoecology of the Timor fauna is very poorly understood. *Deltoblastus* is also known from Permian deposits in Kashmir (Gupta and Webster 1976).

SUMMARY

Blastoid genera in general are monospecific, relatively short lived, relatively rare, and geographically restricted to one depositional basin. In contrast, a few genera were either very abundant (*Pentremites*, *Deltoblastus*) very long lived (*Orbitremites*), geographically widespread (*Orbitremites*, *Angioblastus*), or some combination of the three. Blastoid and monobathrid camerate crinoid diversity are significantly correlated. Perturbations in the structure of crinoid communities preceded intervals of blastoid dominance in the echinoderm communities. In some instances, these changes can be related to changing sedimentological regimes (Mississippian–Pennsylvanian of the US), but in other instances this is not the case (Waulsortian of Great Britain and Ireland).

Although our understanding of the functional morphology of fissiculate and spiraculate blastoids is not sufficient to interpret their niche differences, the two orders have substantially different palaeoecological histories. Fissiculates and spiraculates first co-occur in the Devonian. Fissiculates are more diverse and geographically widespread, but spiraculates tend to dominate in terms of abundance. The Mississippian of the US was the zenith of blastoid diversity and abundance. Globose spiraculates living in platform carbonate communities account for the bulk of this diversity increase. Fissiculates lived in clastic-dominated communities and were less diverse and abundant. One spiraculate, *Pentremites*, successfully invaded clastic communities and became the most successful of all the blastoids. In the Lower Carboniferous of Europe, blastoids successfully adapted to a variety of environments—shallow marine shelf, Waulsortian facies (deep water buildups) shallow reefal facies and clastic dominated facies. Fissiculates are generally more diverse in a fauna, but spiraculates numerically dominate. The most successful group of European blastoids are the Orbitremitidae (Orbitremitid, n.g., *Ellipticoblastus*, *Monoschizoblastus*, and *Orbitremites*). The evolutionary history of this family is strikingly similar to that of the pentremitids in North America even though the two families were geographically endemic.

Blastoids suffered a substantial decrease in diversity in the Pennsylvanian (Upper Carboniferous) as advanced cladid inadunates increased their domination of echinoderm communities. The majority of known genera are fissiculates, but the only truly abundant blastoid of this interval is *Nodoblastus*, a spiraculate. Permian blastoids from the Tethyan seaway represent a resurgence of blastoid diversity. The best known faunas from Australia and Timor are numerically dominated by *Deltoblastus*, a spiraculate, although fissiculates account for the bulk of the generic diversity.

At this stage of our understanding of blastoids, whether a blastoid was successful or not seems to be related more to its ability to respond to locally occurring ecological events (crises?) rather than some sort of functional superiority. Clearly, much more work in all aspects of blastoid palaeobiology is needed to confirm or deny this assertion.

ACKNOWLEDGEMENTS

These blastoid studies have been made possible by grants from the National Science Foundation (DEB 77-23375) to Macurda, Horowitz, and Waters, the Petroleum Research Fund, the American Philosophical Society, the Eppley Foundation for Scientific Research and the West Georgia College Learning Resources Committee. The support of these organizations is gratefully acknowledged. The work has benefited from many discussions with A. S. Horowitz and D. B. Macurda, Jr.

REFERENCES

Arendt, Y. A., Breimer, A., and Macurda, D. B. 1968. A new blastoid fauna from the lower Namurian of North Kazachstan (U.S.S.R.). *Proceedings Koninklijke Nederlandse Akademie Van Wetenschappen* **71**, 159–74.

Ausich, W. I. 1980. A model for niche differentiation in Lower Mississippian crinoid communities. *Journal of Paleontology* **54**, 273–88.

Ausich, W. I. and Lane, N. G. 1985. Crinoid assemblages and geographic endemism in the Lower Mississippian (Carboniferous) of the United States continental interior. In *Comptes rendus, Neuvième Congrès International de Stratigraphie et de Geologie du Carbonifère 9* **5**, 216–24.

Breimer, A. and Dop, A. J. 1974. An anatomic and taxonomic study of some lower and middle Devonian blastoids from Europe and North America. *Proceedings. Koninklijke Nederlandse Akademie van Wetenschappen* **78**, 39–217.

Breimer, A. and Macurda, D. B., Jr. 1972. The phylogeny of the fissiculate blastoids. *Proceedings. Koninklijke Nederlandse Akademie van Wetenschappen* **26**, 1–390.

Breimer, A. and Macurda, D. B., Jr. 1973. Palaeozoic Blastoids. In *Atlas of palaeobiogeography* (ed. A. Hallam), pp. 207–12. Elsevier, Amsterdam.

Broadhead, T. W. 1984. *Macurdablastus*, a middle Ordovician blastoid from the southern Appalachians. *The University of Kansas, Paleontological Contributions, Paper* **110**, 1–10.

Galloway, J. J. and Kaska, H. V. 1957. Genus *Pentremites* and its species. *Memoirs of the Geological Society of America* **69**, 1–104.

Gupta, V. J. and Webster, G. D. 1976. *Deltoblastus batheri* from the Kashmir Himalaya. *Rivista Italiana Paleontologiese* **82**, 279–84.

Horowitz, A. S., Blakely, R. F., and Macurda, D. B., Jr. 1985. Taxonomic survivorship within the Blastoidea (Echinodermata). *Journal of Paleontology* **59**, 543–50.

Horowitz, A. S., Macurda, D. B., Jr., and Waters, J. A. 1986. Polyphyly in the Pentremitidae (Blastoidea, Echinodermata). *Bulletin of the Geological Society of America* **97**, 156–61.

Horowitz, A. S. and Waters, J. A. 1972. A Mississippian echinoderm site in Alabama. *Journal of Paleontology* **46**, 660–5.

Hudson, R. G. S., Clarke, M. J., and Sevastopulo, G. D. 1966. A detailed account of the fauna and age of the Waulsortian reef knoll limestone and associated shales, Feltrim, Co. Dublin. *Scientific Proceedings of the Royal Dublin Society, Series A* **2**, 273–86.

Joysey, K. A. 1955. On the geological distribution of Carboniferous blastoids in the Craven area, based on a study of their occurrence in the Yoredale Series of Grassington, Yorkshire. *Quarterly Journal of the Geological Society of London* **111**, 209–24.

Kammer, T. W. 1985. Aerosol filtration theory applied to Mississippian deltaic crinoids. *Journal of Paleontology* **59**, 551–60.

Lane, N. G. 1972. Synecology of Middle Mississippian (Carboniferous) crinoid communities in Indiana. *24th International Geological Congress, Section* **7**, 96–114.

Lees, A. 1982. The paleoenvironmental setting and distribution of the Waulsortian facies of Belgium and southern Britain. In *Symposium on the paleoenvironmental setting and distribution of the Waulsortian facies* (ed. K. Bolton, H. R. Lane, and D. V. Le Mone), pp. 1–16. El Paso Geological Society and University of Texas, El Paso.

Macurda, D. B., Jr. 1967. The Lower Carboniferous (Tournaisian) blastoids of Belgium. *Journal of Paleontology* **41**, 455–86.

Macurda, D. B., Jr. 1979. The Devonian blastoids of Bolivia. *Journal of Paleontology* **53**, 1361–73.

Macurda, D. B., Jr. 1983. Systematics of the fissiculate Blastoidea. *University of Michigan Museum of Paleontology Papers on Paleontology* **22**, 1–291.

Macurda, D. B., Jr., and Mapes, R. 1982. The enigma of Pennsylvanian Blastoids. *Proceedings of the Third North American Paleontological Convention* **2**, 343–5.

Macurda, D. B., Jr. and Racheboeuf, P. R. 1975. Devonian and Carboniferous spiraculate blastoids from Brittany (France). *Journal of Paleontology* **49**, 845–55.

Metcalfe, I. 1980. Conodont zonation and correlation of the Dinantian and early Namurian strata of the Craven lowlands of northern England. *Reports of the Institute of Geological Sciences* **80/10**, 1–70.

Moore, R. C. (ed.) 1967. *Treatise on invertebrate paleontology. Part S. Echinodermata 1* Geological Society of America and University of Kansas Press, Lawrence, Kansas.

Sieverts-Doreck, H. 1951. Echinodermen aus dem spanischen Ober-Karbon. *Paläontologische Zeitschrift* **24**, 104–19.

Sprinkle, J. 1973. Morphology and evolution of blastozoan echinoderms. *Special Publication. The Museum of Comparative Zoology, Harvard University*, 1–284.

Sprinkle, J. (ed.) 1982. Echinoderm faunas from the Bromide Formation (Middle Ordovician) of Oklahoma. *The University of Kansas. Paleontological Contributions, Monograph* **1**, 1–369.

Waters, J. A., Broadhead, T. W., and Horowitz, A. S. 1981. The evolution of *Pentremites* (Blastoidea) and Carboniferous crinoid community succession. In *Echinoderms. Proceedings of the international conference, Tampa* (ed. M. Lawrence), pp. 133–8. A. A. Balkema, Rotterdam.

Waters, J. A., Horowitz, A. S. and Macurda, D. B., Jr 1985. Ontogeny and Phylogeny of the Carboniferous blastoid *Pentremites*. *Journal of Paleontology* **59**, 701–12.

Waters, J. A. and Sevastopulo, G. D. 1984. The stratigraphical distribution and palaeoecology of Irish Lower Carboniferous blastoids. *Irish Journal of Earth Science* **6**, 137–54.

Wright, J. 1928. A rare *Euryocrinus* from the Carboniferous limestone of Coplow Knoll, Clitheroe. *Geological Magazine* **65**, 246–54.

NOTE ADDED IN PROOF

G. D. Webster (personal communication, 16 December, 1987) has recommended discounting the reports of echinoderms cited in papers he coauthored with Gupta because he has reason to doubt the validity of the Himalayan localities. Therefore, the record of *Deltoblastus* from the Himalayas in the text and on figure 17.14 (Guadalupian) should be ignored until the localities can be verified.

APPENDIX 1

Geological ranges and geographic distribution of blastoid genera. Data are modified from Horowitz et al. (1985).

Genus	Order	Range in stages	N. Am.	Europe	Africa	S. Am	Timor	Aust.	Asia
Macurdablastus	?	Caradocian	1						
Decaschisma	Fissiculate	Wenlockian	1						
Troosticrinus	Spiraculate	Wenlockian	1						
Polydeltoideus	Fissiculate	Ludlovian–Pridolian	1	1					
Leptoschisma	Fissiculate	Gedinnian	1	1					
Angulatoblastus	Spiraculate	Siegenian–Emsian		1					
Belocrinus	Spiraculate	Siegenian		1					
Caryoblastus	Fissiculate	Emsian		1					
Conuloblastus	Spiraculate	Emsian		1					
Cryptoschisma	Fissiculate	Emsian		1					
Hyperoblastus	Spiraculate	Emsian–Famennian	1	1	1				
Pachyblastus	Fissiculate	Emsian			1	1			
Pentremitidea	Fissiculate	Emsian		1	1				
Brachyschisma	Fissiculate	Emsian–Givetian	1						
Eleutherocrinus	Spiraculate	Eifelian–Givetian	1						
Heteroschisma	Fissiculate	Eifelian–Givetian	1						
Nucleocrinus	Spiraculate	Eifelian–Givetian	1	1					
Pleuroschisma	Fissiculate	Emsian–Givetian	1						
Devonoblastus	Spiraculate	Givetian	1	1					
Schizotremites	Spiraculate	Givetian	1	1				1	1
Petaloblastus	Spiraculate	Famennian	1						
Arcuoblastus	Spiraculate	Tournaisian	1						
Auloblastus	Spiraculate	Tournaisian	1						
Carpenteroblastus	Spiraculate	Tournaisian–Visean	1						
Costatoblastus	Spiraculate	Tournaisian	1						
Cryptoblastus	Spiraculate	Tournaisian	1						
Decemoblastus	Spiraculate	Tournaisian	1						
Dentiblastus	Spiraculate	Tournaisian	1	1					
Doryblastus	Spiraculate	Tournaisian		1					
Fissiculate genus 1	Fissiculate	Tournaisian–Namurian	1						
Globoblastus	Spiraculate	Tournaisian	1	1					
Hadroblastus	Fissiculate	Tournaisian–Visean	1	1					
Orbitremitid, n.g.	Spiraculate	Tournaisian		1					
Katoblastus	Fissiculate	Tournaisian	1						
Lophoblastus	Spiraculate	Tournaisian–Visean	1	1					
Mesoblastus	Spiraculate	Tournaisian–Visean	1	1					1
Metablastus	Spiraculate	Tournaisian–Visean	1						

Genus	Order	Range in stages	N. Am.	Europe	Africa	S. Am	Timor	Aust.	Asia
Monadoblastus	Spiraculate	Tournaisian	1						
Orbiblastus	Spiraculate	Tournaisian	1						
Orophocrinus	Fissiculate	Tournaisian–Visean	1	1					
Pentremites	Spiraculate	Tournaisian–Namurian	1			1			
Pentremoblastus	Fissiculate	Tournaisian	1						
Phaenoblastus	Fissiculate	Tournaisian	1	1					
Phaenoschisma	Fissiculate	Tournaisian–Namurian	1	1	1				
Ptychoblastus	Spiraculate	Tournaisian	1						
Pyramiblastus	Spiraculate	Tournaisian	1						
Schizoblastus	Spiraculate	Tournaisian	1						
Spiraculate genus 1	Spiraculate	Tournaisian	1						
Spiraculate genus 2	Spiraculate	Tournaisian	1						
Spiraculate genus 3	Spiraculate	Tournaisian	1						
Spiraculate genus 4	Spiraculate	Tournaisian	1						
Spiraculate genus 5	Spiraculate	Tournaisian	1						
Spiraculate genus 6	Spiraculate	Tournaisian	1						
Stronglyoblastus	Spiraculate	Tournaisian	1						
Xenoblastus	Fissiculate	Tournaisian	1						
Acentrotremites	Fissiculate	Visean		1					
Ambolostoma	Spiraculate	Visean–Namurian	1						
Astrocrinus	Fissiculate	Visean		1					
Codaster	Fissiculate	Visean		1					
Cribroblastus	Spiraculate	Visean	1						
Diploblastus	Spiraculate	Visean–Namurian	1						
Gongyloblastus	Spiraculate	Visean		1					
Granatocrinus	Spiraculate	Visean	1						1
Heteroblastus	Spiraculate	Visean		1					

Genus	Type	Range
Monoschizoblastus	Spiraculate	Visean
Nymphaeoblastus	Fissiculate	Visean–Namurian
Orbitremites	Spiraculate	Visean–Guadalupian
Orophocrinid n.g.	Fissiculate	Visean
Tricoelocrinus	Spiraculate	Visean
Artuschisma	Fissiculate	Namurian
Dolichoblastus	Fissiculate	Namurian
Fissiculate genus 2	Fissiculate	Namurian
Kazachstanoblastus	Fissiculate	Namurian
Mastoblastus	Fissiculate	Namurian
Nodoblastus	Fissiculate	Namurian
Pentablastus	Fissiculate	Namurian
Angioblastus	Fissiculate	Bashkirian–Guadalupian
Malchiblastus	Spiraculate	Bashkirian
Anthoblastus	Fissiculate	Sakmarian–Guadalupian
Nannoblastus	Fissiculate	Sakmarian–Guadalupian
Rhopaloblastus	Spiraculate	Sakmarian–Guadalupian
Sphaeroschisma	Fissiculate	Sakmarian–Asselian
Calycoblastus	Spiraculate	Asselian–Artinskian
Neoschisma	Fissiculate	Asselian–Guadalupian
Notoblastus	Fissiculate	Asselian–Artinskian
Thaumatoblastus	Fissiculate	Asselian–Guadalupian
Tympanoblastus	Fissiculate	Asselian–Artinskian
Australoblastus	Fissiculate	Artinskian
Pteroblastus	Fissiculate	Artinskian–Guadalupian
Timoroblastus	Fissiculate	Artinskian–Guadalupian
Ceratoblastus	Fissiculate	Guadalupian
Deltoblastus	Spiraculate	Guadalupian
Dipteroblastus	Fissiculate	Guadalupian
Indoblastus	Fissiculate	Guadalupian

18

The early evolution of the Crinoidea

STEPHEN K. DONOVAN

Department of Geology, University of the West Indies, Mona, Kingston 7, Jamaica

INTRODUCTION

The origin of the crinoids remains to the present day an unresolved problem, for the oldest representatives of the class, found in Lower Ordovician strata, are already diversified into groups, each of which displays its own distinctive essential features (Georges Ubaghs, in Moore *et al.* 1968, p. 1371).

The subject of this paper is the early evolution, not the origin, of the crinoids, but the two areas of investigation are obviously linked. In 1973 the first Cambrian crinoid, *Echmatocrinus brachiatus* Sprinkle, was described from the Middle Cambrian Burgess Shale of British Columbia. It is so unlike all later crinoids that only the uniserial arms and exceptional preservation of the tube feet enabled this identification to be made with certainty. *Echmatocrinus brachiatus* remains the only Cambrian crinoid known. Its relationship to other early echinoderms is now well understood (Paul and Smith 1984). However, what is still poorly known is its relationship to the next oldest crinoids, from the lower Ordovician, and the relationships of the various Tremadoc and Arenig crinoids, both to each other and to all later crinoids. Kelly (1986, in prep.) has proposed two alternative cladograms summarizing the relationships of the major groups of crinoids, based largely on the available evidence of crinoid ontogeny and crown morphology (Fig. 18.1). The point of variance between the models is whether the disparids (Fig. 18.1A) or the camerates (Fig. 18.1B) are more closely related to the cladids and flexibles. There is support for both hypotheses in features of the crown. On the basis of stem morphology, it is concluded that camerates did not give rise to disparids, cladids, or *Echmatocrinus*, and that disparids or cladids could not have been ancestral to *Echmatocrinus*. No camerate had a meric stem, whereas merism is found in the column of at least some early cladids, disparids and what may be the most primitive flexible (Lewis 1981).

Kelly has provided a theoretical framework of crinoid phylogenetic relationships based largely on ontogeny. In this paper I try to relate the evidence of the early radiation of the crinoids to this model, in an attempt both to test the model and to understand the relationships of the earliest crinoids. I have examined the variation within the crinoids shown by all species known from the Cambrian and lower Ordovician (Tremadoc and Arenig Series). By limiting myself to this relatively short stratigraphic interval I have chosen a manageable sample; Paul and Smith (1984) limited their analysis of the early echinoderm

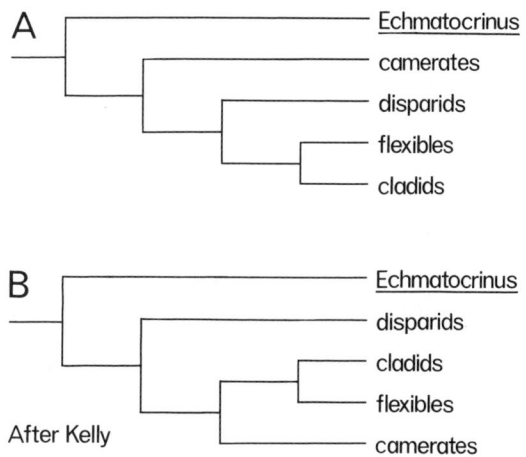

Fig. 18.1 Two cladistic models of the phylogenetic relationships of Palaeozoic crinoid groups (reproduced with the permission of S. M. Kelly; Kelly 1986, in prep.). A, disparids more closely related to flexibles and cladids than camerates. B, camerates more closely related to flexibles and cladids than disparids.

Echinoderm phylogeny and evolutionary biology (ed. C. R. C. Paul and A. B. Smith). Clarendon Press, Oxford, 1988.

radiation (mainly) to the Lower Cambrian for the same reason. Nevertheless, there are 21 crinoid species, more or less well preserved, known from this interval, although four still await description. Only four of these had been described when Ubaghs made his comments regarding the origin of the crinoids in 1968.

THE DIVERSITY OF EARLY CRINOIDS

There are 21 species of Cambrian and lower Ordovician crinoid known (Table 18.1), plus various columnal faunas that must include at least some crinoid material (Donovan 1986*b*). The morphology and affinities of the unique Cambrian species *Echmatocrinus brachiatus* Sprinkle have been discussed elsewhere (Sprinkle 1973; Sprinkle and Moore 1978; Paul and Smith 1984). It is sufficient to note here that it shows primitive features including a 'hohlwurzel' (multiplated holdfast), a broad, irregularly plated cup, perhaps eight, stout, unbranched, uniserial arms with tube feet preserved along them, and has no apparent regular symmetry elements. Brief discussion of the

TABLE 18.1 The early evolution of the Crinoidea

Class CRINOIDEA J. S. Miller,
  Subclass ECHMATOCRINEA Sprinkle and Moore
    Family ECHMATOCRINIDAE Sprinkle
      Echmatocrinus brachiatus Sprinkle
  Subclass CAMERATA Wachsmuth and Springer
    Order DIPLOBATHRIDA Moore and Laudon
      Family RHODOCRINITIDAE Roemer
        Proexenocrinus inyoensis Strimple and McGinnis
    ?Order MONOBATHRIDA Moore and Laudon
      Incertae familiae
        Camerate gen. et sp. nov.
  Subclass INADUNATA Wachsmuth and Springer
    Order DISPARIDA Moore and Laudon
      Family RAMSEYOCRINIDAE Donovan
        Ramseyocrinus cambriensis (Hicks)
        Ramseyocrinus vizcainoi Ubaghs
        Ramseyocrinus sp.
        Tetragonocrinus pygmaeus (Eichwald)
      Family PERITTOCRINIDAE Abel
        Perittocrinus radiatus (Beyrich)
        Tetracionocrinus transitor (Jaekel)
      Family EUSTENOCRINIDAE Ulrich
        Inyocrinus strimplei Ausich
        Pogonipocrinus antiquus (Kelly and Ausich)
      Incertae familiae
        '*Hybocrinus*' sp. A Lane
        Disparid sp. A nov.
        Disparid sp. B nov.
    Order CLADIDA Moore and Laudon
      Family AETHOCRINIDAE Ubaghs
        Aethocrinus moorei Ubaghs
        Aethocrinus murchisoni Donovan
      Family DENDROCRINIDAE Bather
        Compagicrinus fenestratus Jobson and Paul
      Incertae familiae
        Aethocrinid *s. l.* gen. et sp. nov.
      Incerti ordinis
        Incerti ordinis B Lane
        Incerti ordinis C Lane
        Incerti ordinis D Lane

20 lower Ordovician species is given below. The diversity of these early crinoids is not great and those specimens that are well known fall into perhaps six natural groups. Two crinoids that hitherto were regarded as early Ordovician are not included in this discussion. The diplobathrid camerate *Trichinocrinus terranovicus* Moore and Laudon 1943, was originally described as being of '... Early Ordovician (Canadian) age...' (Moore and Laudon 1943, p. 262), but it is most probably from the Lower Llanvirn (H. B. Whittington, written communication). Similarly, *Vosekocrinus granulatus* Jaekel 1918 (= *Pandoracrinus pinnulatus* Jaekel 1918; R. J. Prokop, written communication), also previously reported as being Lower Ordovician (Moore et al. 1978, p. T564), is from the Llanvirn Šárka Formation of Bohemia (R. J. Prokop written communication).

Subclass Camerata

Only two lower Ordovician camerates are known, both from the Middle Arenig. *Proexenocrinus inyoensis* Strimple and McGinnis 1972, was originally interpreted as a monobathrid xenocrinacean, either a xenocrinid (Strimple and McGinnis 1972, p. 72) or a tanaocrinid (Ubaghs 1978c, p. T441). A recent re-evaluation by Ausich (1986) has shown this species to be a diplobathrid rhodocrinitid. The second camerate, so far undescribed (Donovan and Cope, in press), is tentatively interpreted as a monobathrid xenocrinacean; however, it is possible that an infrabasal circlet may be concealed at the top of the column. Both genera show camerate apomorphies recognized by Kelly (1986); pinnulation, a rigid theca incorporating fixed brachials and fixed interbrachials, and a radial series bifurcating at the second primibrach and the second secundibrach. The stem of both species is holomeric, a feature common to all known camerates. Arms are uniserial.

Order Disparida, Family Ramseyocrinidae

The genus *Ramseyocrinus* Bates 1968, was formerly placed in the family Eustenocrinidae, but has been shown to differ radically from eustenocrinids (Donovan 1984). The type species, *R. cambriensis* (Hicks), is now recognized to have five arms (Cope 1988), as was suggested for *R. vizcainoi* Ubaghs 1983 (p. 49), rather than four arms and an anal tube (Bates 1968, p. 407; Donovan 1984, p. 624). The ramseyocrinids have a distinctive tetrameric stem, a feature otherwise only found in *Colpodecrinus* Sprinkle and Kolata 1982, and the perittocrinids. The dorsal cup of *Ramseyocrinus* is low, with four radials (discussion under relationships, below). Donovan (1986b, text-fig. 9) suggested a simple cladogram linking the four known ramseyocrinids, based on the features of the column. This cladogram may require revision following the recent description by Arendt (1985) of the crown of *Tetragonocrinus pygmaeus* (Eichwald), previously known only from dissociated stem material (Yeltysheva 1964). Arendt suggests that the crown consists of three arms only and an anal tube. However, details of the poorly preserved specimen are difficult to interpret from the published illustration. A fourth Arenig ramseyocrinid, *Ramseyocrinus* sp., has been collected from Morocco (Donovan and Savill 1988)

Order Disparida, Family Eustenocrinidae

The family Eustenocrinidae is probably an unnatural grouping, containing a variety of crinoids with a mixture of plating geometries, and requires revision. The genera *Ramseyocrinus* Bates (Donovan 1984) and *Ristnacrinus* Öpik (Donovan 1985) have recently been removed from the eustenocrinids. Two Lower Ordovician eustenocrinids have been described, *Pogonipocrinus antiquus* (Kelly and Ausich 1978, 1979) and *Inyocrinus strimplei* Ausich 1986, both having imperfectly known crowns. From the available information it is deduced that both species have five compound radials. The column is holomeric, composed of low columnals. The arms are non-pinnulate and uniserial. A complete understanding of *Pogonipocrinus* and *Inyocrinus* requires superior material. However, these two genera are almost certainly more closely related to each other than to any other Lower Ordovician crinoid.

Order Disparida, Family Perittocrinidae

The two members of this family, *Perittocrinus radiatus* (Beyrich) and *Tetracionocrinus transitor* (Jaekel), are of problematic stratigraphic position, reputedly from the Kunda Formation of Arenig to Llanvirn age (Ubaghs 1978b, p. T275). Both species were re-examined by Ubaghs (1971). All plates of the cup have radiating ribs which coalesce towards the centre. About 14 accessory plates are present between the angles of the basals, radials and the anal series (Ubaghs 1971, fig. 1). These accessory plates may be related to the development of respiratory structures (Ubaghs 1978a, p. T129). The perittocrinids are thus very different from the ramseyocrinids and the eustenocrinids, although probably more closely related to the former group, having a tetrameric stem.

Order Cladida, Families Aethocrinidae and Dendrocrinidae

Aethocrinus moorei Ubaghs 1969, is the oldest known cladid crinoid. Whether the dorsal cup contains fixed first primibrachials (Ubaghs 1969, 1972) or enlarged proximal pentameres (Philip and Strimple 1971; Jobson and Paul 1979) is undecided, although I favour the latter view. The discovery of a second species, *Aethocrinus murchisoni* Donovan 1986b, has not helped to clarify this situation because the new taxon is based on a pluricolumnal and a single brachial. The plate homologies with the closely related *Compagicrinus fenestratus* Jobson and Paul 1979, indicate that the plates identified by Ubaghs as infrabasals are swollen pentameres. A further aethocrinid-like taxon also seems to have lost the infrabasal circlet (G. C. McIntosh, written communication). The precise relationship of this new taxon to other members of this lineage has yet to be fully determined.

Subclass Inadunata, incerti ordinis

Lane (1970) described four species of early Ordovician crinoid, *incertae sedis* spp. B, C, and D, and *Hybocrinus* sp. A. Spp. B, C, and D are all based upon fragments of the arms only. They are non-pinnulate and, therefore, derived from inadunates. Thus, they are better regarded as being incerti ordinis. The identification of *Hybocrinus* sp. A must also be regarded as uncertain. The arms are poorly preserved, but are described as being '. . . distorted, broken and difficult to interpret . . .' (Lane 1970, p. 10). However, at least some of this 'breakage' appears to be branching (C. R. C. Paul, personal communication). If so, the specimen is not *Hybocrinus*, but an indeterminate disparid.

Two further species of inadunate from the lower Ordovician of Utah are to be described by Kelly (in prep.).

RELATIONSHIPS OF EARLY CRINOIDS

A speculative cladogram relating the known Cambrian and early Ordovician crinoids is presented in Fig. 18.2. This is derived from Fig. 18.1A. Although the stratigraphic data favours Fig. 18.1B (see Fig. 18.5), I have derived the figure supported by previous taxonomic groupings (disparids + cladids = inadunates), based on the evidence of crown and column. Six lower Ordovician taxa are omitted from Fig. 18.2: incerti ordinis B, C, D Lane, and '*Hybocrinus*' sp. A Lane because they are still poorly known; the two new species of disparid from Utah (Kelly in prep.) have yet to be fully evaluated.

Most Lower Ordovician crinoids are imperfectly known, so I have relied on features such as the stem more than would be usual. For example, of the known Lower Ordovician disparid families, the arrangement of dorsal cup plates is not fully known for both of the eustenocrinids, the column is poorly known for both and the arms of *I. strimplei* Ausich are poorly preserved; the arms of the perittocrinids are unknown and the only stem preserved is the proximal few columnals of the holotype of *T. transitor* (Jaekel) (Ubaghs 1971, fig. 2).

The ramseyocrinids and perittocrinids are linked by features of their stems and, perhaps, their dorsal cups. Apart from *Colpodecrinus* Sprinkle and Kolata, these are the only two crinoid genera with tetrameric stems (Donovan 1984; Moore *et al.* 1978, p. T563). The lowest circlet of cup plates in both families consists of four ossicles. These are certainly basals in the perittocrinids (Ubaghs 1971, fig. 1), but have been interpreted either as basals (Bates 1968, p. 407; Moore *et al.* 1978, p. T554; Ubaghs 1983, p. 49; Arendt 1985, fig. 1) or radials (Donovan 1984, pp. 623–4) in the ramseyocrinids. The circlet supported by these plates differs considerably between the perittocrinids (five radials, a radianal, and an anal X) and the ramseyocrinids (five first primibrachs only). Despite reports that *R. cambriensis* (Hicks) has split radials (Moore *et al.* 1978, p. T554), it appears that even the plates immediately above the lowest visible circlet are free and not fixed. It is probable that the perittocrinids and ramseyocrinids are more closely linked to each other than either is to the eustenocrinids.

Other relationships are either described elsewhere, obvious from the available information, or tentative. There are only two Lower Ordovician camerates, two eustenocrinids and two perittocrinids. The relationships of the ramseyocrinids are discussed in Donovan (1986b). The precise posi-

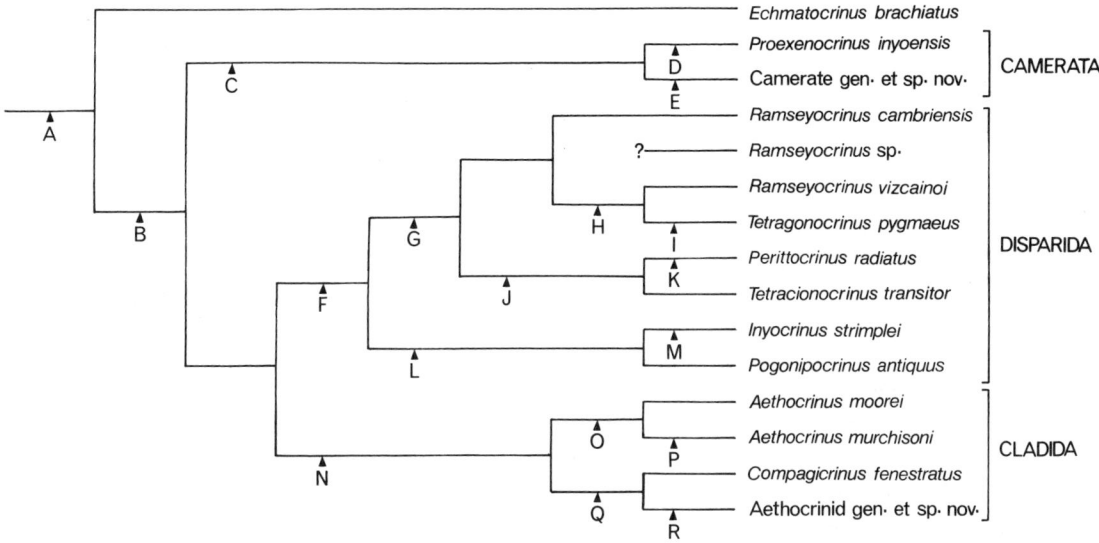

Fig. 18.2 Cladogram illustrating the phylogenetic relationships of Cambrian and Lower Ordovician crinoids. Characters (A–R) are as follows: A, loss of brachioles, gain of uniserial arm plates (Paul and Smith 1984); B, arm branching, development of stem, simplification of cup to regular circlets, simplification of crown to five rays; C, holomerism, pinnulation, rigid theca with fixed brachials and fixed interbrachials, radial series branching at $1Br_2$ and $2Br_2$ (Kelly 1986); D, cup dicyclic; E, cup monocyclic; F, cup monocyclic; G, stem tetrameric, tetralobate, basals reduced to four; H, stem regularly tetragonal; I, subsidiary canals developed corresponding to meric sutures, arms reduced to three (?), anal series developed; J, stem regularly tetragonal, accessory plates developed, anal series incorporated into cup; K, endothecal pore slits developed; L, holomeric stem, split radials, bowl-like crown; M, slender, upright crown; N, pentamerism, cup dicyclic, enlargement of most proximal meres, broad column, interradials developed, anal series incorporated into cup, Y plate in AB interray; O, column with spinose pentameres; P, rounded pentameres; Q, reduction of columnal height, loss of interradials, Y-plate lost; R, infrabasals lost (G. C. McIntosh written communication).

tion of *Ramseyocrinus* sp. from Morocco is problematic because the dorsal cup and stem are unknown. My interpretation of relationships within the cladids is tentative and relies heavily on the work of Jobson and Paul (1979) and G. C. McIntosh (written communication). Both morphology and stratigraphic position have influenced me in constructing this cladogram (Fortey and Jefferies 1982).

PALAEOBIOGEOGRAPHY AND BIOSTRATIGRAPHY

No consensus of opinion exists as to the true configuration of the present day continents in the early Ordovician. The reconstruction of Smith *et al.* (1981, maps 77, 78) produces a convincing distribution of early Ordovician crinoids (Fig. 18.3). Paul (1976) used an earlier restoration (Smith *et al.* 1973, pp. 34, 35) to analyse the distribution of all Ordovician echinoderm groups, but this produced some discontinuous distributions (for example, diploporites were concentrated between 30°N and 30°S, apart from a few occurrences close to the South Pole). The later model of Smith *et al.* (1981) results in a more continuous distribution of all Ordovician echinoderm groups.

All known Lower Ordovician crinoids are from North America, Greenland, Europe, and Morocco, and occur within about 20° of the palaeo-equator (*Echmatocrinus brachiatus* Sprinkle, from the Middle Cambrian, was also approximately equatorial; Smith *et al.* 1981 maps 81, 82). Other present day continents lay on the equator at this time but early Ordovician crinoids are unknown from Australia (P. A. Jell, written communication), Antarctica and Asia. In the Tremadoc and Arenig North America and Greenland were closely associated and were separated from Europe by the Iapetus Ocean, an early

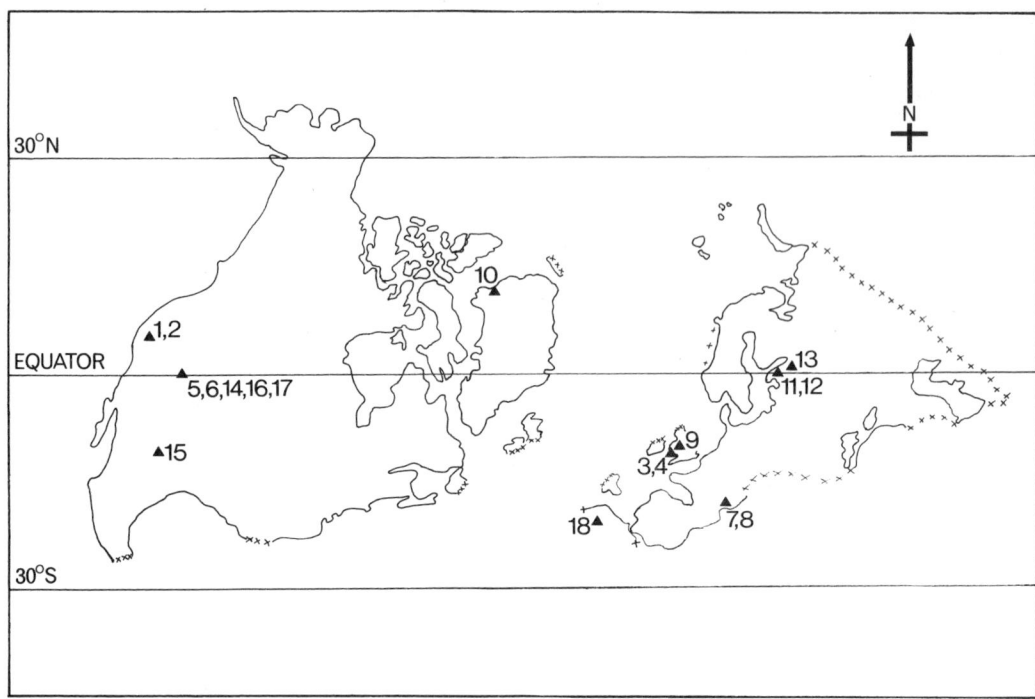

Fig. 18.3 Palaeogeographic restoration of North America and Europe during the Arenig (based on Smith *et al.* 1981, maps 77, 78). Early Ordovician crinoid localities marked by closed triangles. 1, *Proexenocrinus inyoensis* Strimple and McGinnis. 2, *Inyocrinus strimplei* Ausich. 3, monobathrid(?) gen. et sp. nov. 4, *Ramseyocrinus cambriensis* (Hicks). 5, '*Hybocrinus*' sp. A Lane. 6, incerti ordinis B, C, D Lane. 7, *Aethocrinus moorei* Ubaghs. 8, *Ramseyocrinus vizcainoi* Ubaghs. 9, *Aethocrinus murchisoni* Donovan. 10, *Compagicrinus fenestratus* Jobson and Paul. 11, *Perittocrinus radiatus* (Beyrich). 12, *Tetracionocrinus transitor* (Jaekel). 13, *Tetragonocrinus pygmaeus* (Eichwald). 14, *Pogonipocrinus antiquus* (Kelly and Ausich). 15, aethocrinid gen. et sp. nov. 16, disparid sp. nov. A. 17, disparid sp. nov. B. 18, *Ramseyocrinus* sp.

'Atlantic' which later closed. Evidence from the later Ordovician shows that crinoids were able to migrate across Iapetus, presumably as planktonic larvae, but only one Tremadoc–Arenig 'family' has representatives known from both sides of the ocean. Ramseyocrinids, perittocrinids, and a new monobathrid are known from Europe; eustenocrinids and *Proexenocrinus* are limited to North America. Only the aethocrinids and their descendants, the early dendrocrinids, have a truly broad range, being known from the Montagne Noire in France, Shropshire in England, Greenland, and the southern USA.

The distribution of early Ordovician crinoids is not equatorial on the reconstruction of Cocks and Fortey (1982, Figs 2–4). North America and Greenland are approximately equatorial, but eastern Europe and Scandinavia lie at about 40–60°S, separated from western Europe (south of 60°S) by Tornquist's Sea. The simple model of tropical distribution thus disappears. Cocks and Fortey ably showed that their model was consistent with known facies and much of the faunal data. The distribution of early Ordovician crinoids is such that no genus occurs on more than one continental block. It is possible to propose arguments in support of both the Smith *et al.* (1981) and the Cocks and Fortey (1982) models. It is premature to make any broad interpretation of the palaeobiogeography of the early crinoids.

The stratigraphic distribution of the early crinoids is easier to analyse. Even accepting the minimum duration permitted by the various published time scales, the gap between *Echmatocrinus brachiatus* Sprinkle and the next oldest (late Tremadoc) crinoid, *Pogonipocrinus antiquus* (Kelly and Ausich), is still over 40 Ma (Fig. 18.4), an extraordinarily long time, particularly for a

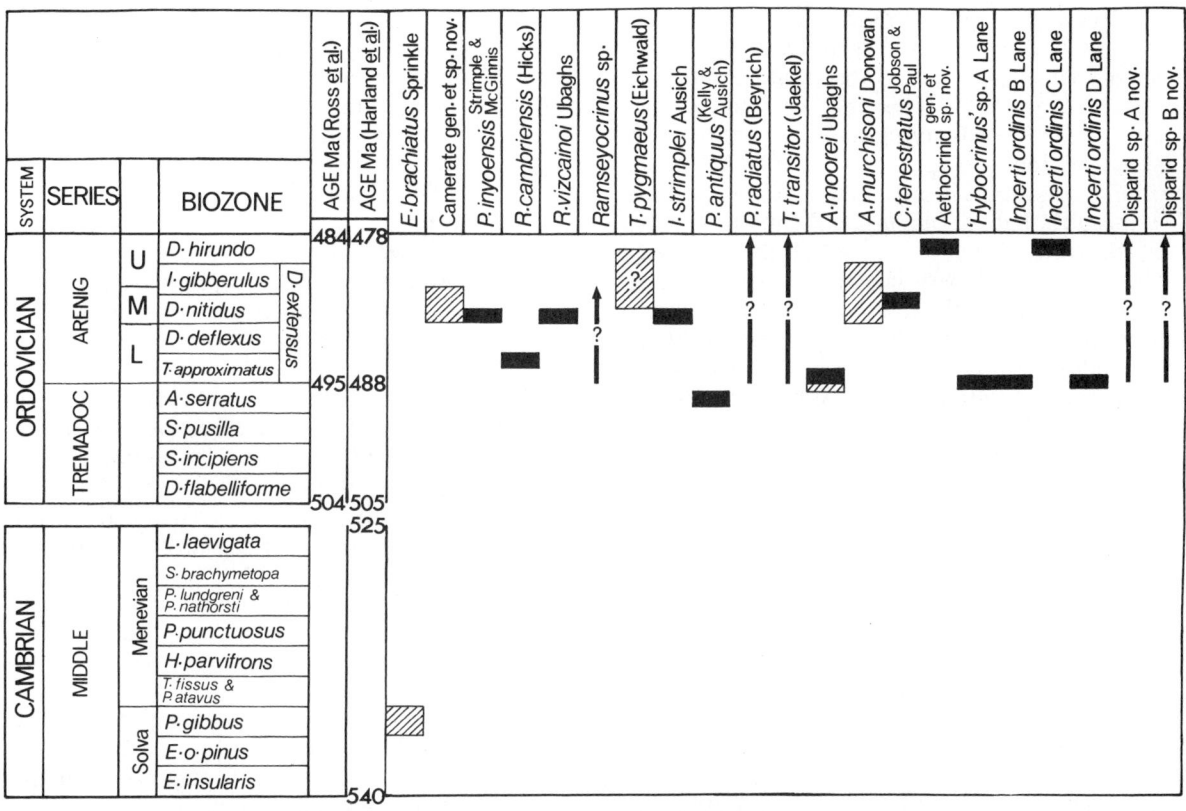

Fig. 18.4 Biostratigraphic distribution of early crinoids. Biozonation after Harland *et al.* (1982), Thomas *et al.* (1984), and Williams *et al.* (1972). Radiometric dates after Harland *et al.* (1982, p. 14) and Ross *et al.* (1982, sheet 1). Stratigraphic ranges indicated as closed boxes (tight stratigraphic control), hatched boxes (loose stratigraphic control) or arrows (stratigraphic position uncertain).

group with a skeleton that is easily preserved (either whole or fragmented) and whose fossils are so distinctive. However, this gap may be part of a more general phenomenon, as Upper Cambrian echinoderms are uncommon (C. R. C. Paul, personal communication). This gap includes the Cambro–Ordovician mass extinction event. However, the early Ordovician crinoid radiation postdates this event by eight to ten million years (Whittington 1966). It is, therefore, unlikely that this radiation is a direct consequence of the extinction, there being a considerable time interval between the two.

The stratigraphic distribution of early Ordovician crinoids indicates a hitherto unsuspected two-tier radiation, rather than the single tier radiation that was previously recognized (for example, Ubaghs 1978*b*, fig. 207). After the long time gap following *E. brachiatus* Sprinkle, six species 'suddenly' appear around the Tremadoc–Arenig boundary (*serratus-approximatus* biozones) in three regions: *Ramseyocrinus cambriensis* (Hicks) in south Wales, UK; *Aethocrinus moorei* Ubaghs in the Montagne Noire, France; *Pogonipocrinus antiquus* (Kelly and Ausich), '*Hybocrinus*' sp. A Lane, incerti ordinis B Lane, and D Lane in Utah (Fig. 18.3). No crinoids are known with certainty from the succeeding *deflexus* biozone (duration about 1.25 Ma; Ross *et al.* 1982, sheet 1), yet, in the following *nitidus-gibberulus* interval, perhaps seven new species appear: a new species of monobathrid(?) from south Wales, UK; *Aethocrinus murchisoni* Donovan from Shropshire, UK; *Proexenocrinus inyoensis* Strimple and McGinnis, and *Inyocrinus strimplei* Ausich from California; *Ramseyocrinus vizcainoi* Ubaghs from the Montagne Noire, France; *Tetragonocrinus pygmaeus* (Eichwald) from near Leningrad; and *Compagicrinus fenestratus* Jobson and Paul from northern Greenland. The only crinoids so far known from the end-Arenig *hirundo* biozone are a new aethocrinid gen. et sp. nov. (G. C. McIntosh, written communication) and incerti ordinis C Lane, although *A. murchisoni* Donovan may, in fact, be from this interval rather than slightly earlier (A. B. Smith, written communication). It is at this level that pelmatozoan columnal faunas first become common. At least some of these columnals must be derived from crinoids. Good Upper Arenig (*sensu* Fortey *in* Whittington *et al.* 1984, pp. 19–22) columnal faunas are known from Tourmakeady, Ireland (Donovan 1986a), Anglesey, north Wales (Donovan 1986b), and Österplana Quarry, Sweden (Paul and Bockelie 1983, pp. 690–1).

This stratigraphic distribution may have implications for crinoid phylogeny. The disparids and cladids appeared almost simultaneously in the fossil record, about 4 Ma before the first camerate (Ross *et al.* 1982, sheet 1). The stratigraphic evidence indicates the following order of appearance for the crinoids: *Echmatocrinus*; disparids and cladids approximately simultaneously; camerates; and flexibles. It is, therefore, possible that the camerates were derived either from the disparids or the cladids (Fig. 18.5), providing the distribution of lower Ordovician crinoids is not just an artefact of the fossil record or due to some collecting bias. If the camerates evolved directly from the cladids, we have a cladogram identical to Fig. 18.1B, that is, the stratigraphic distribution has support from the morphological evidence of the crown. However, morphology of the crinoid crown and stem was thought to favour Fig. 18.1A (above). This dilemma could be resolved if it could be shown that camerates evolved from a cladid group that had already evolved a holomeric column, but no such crinoid is known from the Lower Ordovician. Therefore, I suggest that any

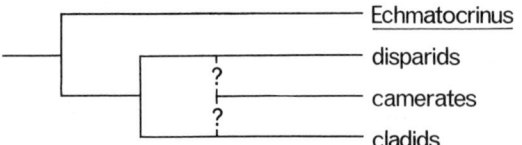

Fig. 18.5 Dendrogram showing inferred relationships of crinoids as indicated by the Cambrian and Lower Ordovician fossil record (cf. Figs 18.1B and 18.4).

choice between these two diagrams must, at present, be arbitrary. It demonstrates that we still need to discover more early crinoids to help complete the picture.

CONCLUSIONS

(1) Kelly's cladistic models of relationships within the Palaeozoic crinoids, based mainly on crown ontogeny and morphology, are supported both by biostratigraphy and stem morphology. Both models are equally probable.
(2) Early crinoids were limited to North America, Greenland, Morocco, and Europe. This distribution may either have been equatorial or equatorial to sub-polar.
(3) There is a 40 Ma gap between the only Cambrian crinoid and its early Ordovician descendants.
(4) The early Ordovician crinoid radiation seems to have been a two-tier event, in the top Tremadoc/lowest Arenig and in the Middle Arenig.
(5) The earliest known inadunate predates the first camerates by about 4 Ma. Early Ordovician inadunates are much commoner in the fossil record than camerates.

ACKNOWLEDGEMENTS

I thank Stuart Kelly and George McIntosh for allowing me to use their unpublished results. In particular, Stuart has most generously given me permission to incorporate his unpublished cladograms. I also thank all of those people who replied to my enquiries in mid-to late 1986; Russell D. White, Harry Whittington, Adrian Rushton, Peter Jell, Rudolf Prokop, Jeremy Savill, Georges Ubaghs, Andrew Smith, N. Gary Lane, Sam Morris, John Cope, A. Prieur, and Thomas E. Bolton. Andrew Smith gently goaded me into writing this paper and Chris Paul was always a happy source of information on early pelmatozoans. This paper was improved following constructive comments by an anonymous referee.

REFERENCES

Arendt, Yu. A. 1985. Ancient triradiate crinoid without pentameral symmetry. *Dokladȳ Akademia Nauk SSSR* **282**, 702–4 (in Russian).

Ausich, W. I. 1986. The crinoids of the Al Rose Formation (lower Ordovician, Inyo County, California, U.S.A.). *Alcheringa* **10**, 217–24.

Bates, D. E. B. 1968. On '*Dendrocrinus*' *cambriensis* Hicks, the earliest known crinoid. *Palaeontology* **11**, 406–9.

Cocks, L. R. M. and Fortey, R. A. 1982. Faunal evidence for oceanic separation in the Palaeozoic of Britain. *Journal of the Geological Society of London* **139**, 465–78.

Cope, J. C. W. 1988. A reinterpretation of the Arenig crinoid *Ramseyocrinus*. *Palaeontology* **31**, 229–35.

Donovan, S. K. 1984. *Ramseyocrinus* and *Ristnacrinus* from the Ordovician of Britain. *Palaeontology* **27**, 623–34.

Donovan, S. K. 1985. The Ordovician crinoid genus *Caleidocrinus* Waagen and Jahn, 1899. *Geological Journal* **20**, 109–21.

Donovan, S. K. 1986a. Pentameric pelmatozoan columnals from the Arenig Tourmakeady Limestone, Co. Mayo. *Irish Journal of Earth Sciences* **7**, 93–7.

Donovan, S. K. 1986b. Pelmatozoan columnals from the Ordovician of the British Isles. Part 1. *Palaeontographical Society Monographs* 1–68.

Donovan, S. K. and Cope, J. C. W. (in press). A new camerate crinoid from the Arenig of South Wales. *Palaeontology*.

Donovan, S. K. and Savill, J. J. 1988. *Ramseyocrinus* from the Arenig of Morocco. *Journal of Paleontology* **62**, 283–5.

Fortey, R. A. and Jefferies, R. P. S. 1982. Fossils and phylogeny—a compromise approach. In *Problems of phylogenetic reconstruction*, Systematics Association Special Publication **21**, (eds K. A. Joysey and A. E. Friday), pp. 197–234.

Harland, W. B., Cox, A. V., Llewellyn, P. G., Pickton, C. A. G., Smith, A. G., and Walters, R. 1982. *A geologic time scale*. Cambridge, University Press, Cambridge.

Jaekel, O. 1918. Phylogenie und System der Pelmatozoen. *Paläontologische Zeitschrift* **3**, 1–128.

Jobson, L. and Paul, C. R. C. 1979. *Compagicrinus fenestratus*, a new lower Ordovician inadunate crinoid from north Greenland. *Rapport Grønlands Geologiske Undersogelse* **91**, 71–81.

Kelly, S. M. 1986. Classification and evolution of Class Crinoidea. *Abstracts of the 4th North American Paleontological Convention*. A23.

Kelly, S. M. and Ausich, W. I. 1978. A new lower Ordovician (middle Canadian) disparid crinoid from Utah. *Journal of Paleontology* **52**, 916–20.

Kelly, S. M. and Ausich, W. I. 1979. A new name for the lower Ordovician crinoid *Pogocrinus* Kelly and Ausich. *Journal of Paleontology* **53**, 1433.

Lane, N. G. 1970. Lower and middle Ordovician crinoids from west-central Utah. *Geological Studies of Brigham Young University* **17**, 3–17.

Lewis, R. D. 1981. *Archaetaxocrinus*, new genus, the earliest known flexible crinoid (Whiterockian) and its phylogenetic implications. *Journal of Paleontology* **55**, 227–38.

Moore, R. C., et al. 1968. Developments, trends, and outlooks in paleontology. *Journal of Paleontology* **42**, 1327–77.

Moore, R. C., Lane, N. G., Strimple, H. L., and Sprinkle, J. 1978. Order Disparida Moore and Laudon 1943. In *Treatise on invertebrate paleontology, part T, Echinodermata* 2(2) (ed. R. C. Moore and C. Teichert), pp. T520–64, Geological Society of America and University of Kansas Press, Lawrence, Kansas.

Moore, R. C. and Laudon, L. R. 1943. *Trichinocrinus*, a new camerate crinoid from lower Ordovician (Canadian?) rocks of Newfoundland. *American Journal of Science* **241**, 262–8.

Paul, C. R. C. 1976. Palaeogeography of primitive echinoderms in the Ordovician. In *The Ordovician System: proceedings of a Palaeontological Association symposium, Birmingham, September 1974* (ed. M. G. Bassett), pp. 553–74. University of Wales Press and National Museum of Wales, Cardiff.

Paul, C. R. C and Bockelie, J. F. 1983. Evolution and functional morphology of the cystoid *Sphaeronites* in Britain and Scandinavia. *Palaeontology* **26**, 687–734.

Paul, C. R. C. and Smith, A. B. 1984. The early radiation and phylogeny of the echinoderms. *Biological Reviews* **59**, 443–81.

Philip, G. M. and Strimple, H. L. 1971. An interpretation of the crinoid *Aethocrinus moorei* Ubaghs. *Journal of Paleontology* **45**, 491–3.

Ross, R. J., et al. 1982. The Ordovician System in the United States. *Publication of the International Union of Geological Sciences* **12**, 1–73.

Smith, A. G., Briden, J. C., and Drewry, G. E. 1973. Phanerozoic world maps. *Special Papers in Palaeontology* **12**, 1–42.

Smith, A. G., Hurley, A. M., and Briden, J. C. 1981. *Phanerozoic paleocontinental world maps*. Cambridge University Press, Cambridge.

Sprinkle, J. 1973. Morphology and evolution of blastozoan echinoderms. *Special Publication. The Museum of Comparative Zoology, Harvard University*, 1–284.

Sprinkle, J. and Kolata, D. R. 1982. 'Rhomb-bearing' camerate. In Echinoderm faunas from the Bromide Formation (middle Ordovician) of Oklahoma. *The University of Kansas, Paleontological Contributions, Monograph* **1**, 206–11.

Sprinkle, J. and Moore, R. C. 1978. Echmatocrinea. In *Treatise on invertebrate paleontology, Part T, Echinodermata* 2(2) (ed. R. C. Moore and C. Teichert), pp. T405–7. Geological Society of America and University of Kansas Press, Lawrence, Kansas.

Strimple, H. L. and McGinnis, J. R. 1972. A new camerate crinoid from the Al Rose Formation, lower

Ordovician of California. *Journal of Paleontology* **46**, 72–4.

Thomas, A. T., Owens, R. M., and Rushton, A. W. A. 1984. Trilobites in British stratigraphy. *Special Reports of the Geological Society of London* **16**, 1–78.

Ubaghs, G. 1969. *Aethocrinus moorei* Ubaghs, n. gen., n. sp., le plus ancien crinoïde dicyclique connu. *The University of Kansas, Paleontological Contributions, Paper* **38**, 1–25.

Ubaghs, G. 1971. Un crinoïde énigmatique Ordovicien: *Perittocrinus* Jaekel. *Neues Jahrbuch für Geologie und Paläontologie Abhandlungen* **137**, 305–36.

Ubaghs, G. 1972. More about *Aethocrinus moorei* Ubaghs, the oldest known dicyclic crinoid. *Journal of Paleontology* **46**, 773–5.

Ubaghs, G. 1978a. Skeletal morphology of fossil crinoids. In *Treatise on invertebrate paleontology, part T, Echinodermata* 2(1) (ed. R. C. Moore and C. Teichert), pp. T58–216. Geological Society of America and University of Kansas Press, Lawrence, Kansas.

Ubaghs, G. 1978b. Origin of crinoids. In *Treatise on invertebrate paleontology, part T, Echinodermata* 2(1) (ed. R. C. Moore and C. Teichert), pp. T275–81. Geological Society of America and University of Kansas Press, Lawrence, Kansas.

Ubaghs, G. 1978c. Camerata. In *Treatise on invertebrate paleontology, part T, Echinodermata* 2(2) (ed. R. C. Moore and C. Teichert), pp. T408–519. Geological Society of America and University of Kansas Press, Lawrence, Kansas.

Ubaghs, G. 1983. Echinodermata. Notes sur les echinodermes de l'Ordovicien inférieur de la Montagne Noire (France). In *Calymenina, Echinodermata et Hyolitha de l'Ordovicien inférieur de la Montagne Noire (France, Méridionale)* (ed. R. Courtessole, L. Marek, J. Pillet, G. Ubaghs, and D. Vizcaino), *Mémoires de la Société d'Études Scientifiques de l'Aude, Carcassonne*, pp. 33–55.

Whittington, H. B. 1966. Phylogeny and distribution of Ordovician trilobites. *Journal of Paleontology* **40**, 696–737.

Whittington, H. B., Dean, W. T., Fortey, R. A., Rickards, R. B., Rushton, A. W. A., and Wright, A. D. 1984. Definition of the Tremadoc Series and the series of the Ordovician System in Britain. *Geological Magazine* **121**, 17–33.

Williams, A., Strachan, I., Bassett, D. A., Dean, W. T., Ingham, J. K., Wright, A. D., and Whittington, H. B. 1972. A correlation of Ordovician rocks in the British Isles. *Special Reports of the Geological Society of London* **3**, 1–74.

Yeltysheva, R. S. 1964. Stems of Ordovician sea lilies of the Baltic area (lower Ordovician). *Voprosȳ Paleontologii* **4**, 59–84 (in Russian).

19

Ontogeny and phylogeny of disparid crinoids

G. D. SEVASTOPULO AND N. G. LANE*

Department of Geology, Trinity College, Dublin, Ireland
*Department of Geology, Indiana University, Bloomington, Indiana, USA

INTRODUCTION

The *Treatise on invertebrate paleontology* (Part T) (Moore et al. 1978) contains descriptions of all the genera of crinoids proposed up to the late 1970s. The genera were assigned to higher level taxa, many of which formed components of earlier schemes of crinoid classification, but some of which were new (see Lane 1978 for a history of the classification of the Crinoidea). Since the publication of the *Treatise*, many new taxa of crinoids have been described and the classification has been adjusted to accommodate them. However, there has been little attempt to analyse the classification as a whole, and particularly to assess the extent to which it reflects the phylogenetic relationships of crinoid taxa. Such an analysis, to be complete, requires much more information than is available on the morphology of most crinoids, particularly with regard to the many features (such as the anatomy of the columnals and brachial plates of Palaeozoic crinoids), that traditionally have not been given much weight in taxonomy. In this paper we are concerned with the classification of a relatively small, but none the less important, group of Palaeozoic crinoids, the Disparida. The taxonomic position of this order within the *Treatise* classification is outlined below.

Moore et al. (1978) recognized five subclasses within the class Crinoidea: Echmatocrinea, Camerata, Inadunata, Flexibilia, and Articulata. They assigned the following orders to the Inadunata: Coronata, which have subsequently been shown to be blastozoans (Sprinkle 1980); Cladida, in which the aboral cup consists of three ranges of plates—radials, basals, and infrabasals; Disparida, in which the cup consists of radials and basals only, with infrabasals not developed (monocyclic condition); and Hybocrinida, a small group with a monocyclic cup containing a radianal which supported the anal series.

In what follows, we give a summary of the morphology of the disparid crinoids; describe their ontogeny, particularly those aspects that have a bearing on phylogenetic relationships; discuss their relationships to other crinoids; and finally, attempt to recognize relationships within the order. For details of taxa mentioned in the text, without a specific reference to the literature, the reader should consult the *Treatise on invertebrate paleontology* (Moore et al. 1978).

MORPHOLOGY OF THE DISPARIDA

The Disparida are small crinoids; many are referred to as microcrinoids because their thecae in maturity are less than 2 mm high. The aboral cup, which is generally an upright truncate cone, consists of a circlet of basal plates, surmounted by a circlet of radial plates. The basal circlet generally consists of five (the primitive condition), three, or one plate. A few taxa, such as some genera of calceocrinoids, have four basals. The radial circlet in advanced taxa consists of five radials with articular facets on their distal surfaces to which the arms were attached. In less advanced taxa, more than one plate occurs between the basal circlet and the top of the cup in one or more of the rays. The lowest of such plates has been termed an inferradial and the plate that follows it serially the superradial, the two together being referred to as a compound, split, or bi-radial. The ontogeny of *Homocrinus*, a disparid with compound radials, described below, throws into doubt the homologies of the two elements of the compound radial,

Echinoderm phylogeny and evolutionary biology (ed. C. R. C. Paul and A. B. Smith). Clarendon Press, Oxford, 1988.

but for the present we continue to use the traditional terminology. In a few primitive forms (for instance, *Peniculocrinus* Moore 1962) additional plates, identified as fixed brachials, are reported to occur between the top of the superradials and the top of the cup. While five-fold symmetry of the radial circlet is the norm, there are a few taxa in which there are four radials (for example, *Ramseyocrinus*) or three radials (*Holynocrinus*).

There is very little information regarding the tegmen in most disparids. In almost all taxa whose tegmen has been described (for instance, *Haplocrinites*, *Pisocrinus*, and allagecrinids, such as *Litocrinus*, Lane and Sevastopulo 1981), it is formed by five oral plates, arranged in a distinctive pattern with the largest plate, which commonly bears a hydropore, in the CD interray. Warn and Strimple (1977) have shown that *Cincinnaticrinus* (*Heterocrinus* of the *Treatise* and earlier literature), previously interpreted to have a many-plated tegmen, has, in fact, five orals, similar in arrangement to those described above. However, Warn (1982) has described *Doliocrinus* as having a multiplated tegmen. In some taxa, such as *Allocatillocrinus*, five orals are present in immature thecae, but are apparently resorbed in mature individuals (G. D. Sevastopulo, unpublished observations).

The arms of disparids are varied in structure. They may be unbranched or branch isotomously or heterotomously. It is possible to construct a morphological series ranging from heterotomous branching to the development of ramules; in a very few genera, such as *Doliocrinus* (Warn 1982), ramules occur on every brachial, and thus, are topologically pinnules. In general, the brachials are uniserial. In several taxa (for example, the Acolocrinidae Brett 1980; and the Allagecrinidae), the radials are axillary, each bearing a number of fine uniserial arms.

The anal series also varies widely. In most disparids, the anal sac consists of a uniserial, arm-like structure which articulates with the left side of the C ray radial or superradial, or is attached to the left side of a primibrachial in the C ray. In a few taxa (for example, *Eustenocrinus*) the most proximal anal plate articulates with the full width of the C radial and no arm is developed in the C ray. The structure of the sac is more elaborate in some taxa: in *Daedalocrinus* it is much expanded and in *Ohiocrinus* it is spirally coiled.

The columns of disparids are, in general, poorly known. The large majority are formed of holomeric columnals; *Ibexocrinus* and several other genera have pentameric columns; in perittocrinids, *Ramseyocrinus*, and related genera the column is tetrameric; while *Ectenocrinus* is trimeric proximally. Cirri are unknown except in the highly specialized myelodactylids.

ONTOGENY OF DISPARIDS

The early ontogeny of Palaeozoic crinoids is known for several disparids, but only for a few genera of cladids and a single possible camerate. This is because the thecal plates of most cladids, camerates, and flexibles were not firmly united in the early post-metamorphic stages of ontogeny and so were dispersed after the death of the crinoid. The thecal plates of juvenile disparids are relatively thick and the thecae were rigid. Post-larval ontogenies are now known for many species distributed among four of the eight superfamilies of Disparida recognized in the *Treatise*—Homocrinacea, Pisocrinacea, Allagecrinacea, and Belemnocrinacea. The lack of information about the ontogeny of several important disparid groups is a result of the small amount of research on the early growth of these crinoids. There is every likelihood that directed search will lead to the discovery of juvenile specimens in other taxa.

It is convenient to describe the ontogeny of disparids of three basic types: those with undivided non-axillary radials; those with undivided axillary radials; and those with compound radials.

As an example of the first type, we take the genus *Litocrinus*, a common Carboniferous microcrinoid. The mature theca in this genus is commonly about 1 mm high. It consists of a low upright basal circlet, formed from three basals or a single fused plate; five radials each with a single arm articular facet; and five oral plates. The ontogeny of a number of species of this genus (formerly assigned to *Kallimorphocrinus*, see Lane and Sevastopulo 1982) has been described by various authors (see for instance, Strimple and Koenig 1956), and is known for several additional species (G. D. Sevastopulo, unpublished observations).

The earliest growth stages preserved are commonly about 250 μm high and consist of a basal circlet, five radials without arm articular facets, and five oral plates. During growth, arm articular facets appear on the distal margins of the radials in a set pattern. The first to form is almost always in the C ray and the second in the E ray, although in a few cases this order is reversed (the rays are identified by reference to the large oral plate that lies in the CD interray). In collections containing large numbers of juvenile individuals, specimens with two arm facets are common, but those with a single facet are rare. We interpret this as showing that the arms in the C and E rays developed almost simultaneously. The next facet to develop was usually in the B ray. Strimple and Koenig (1956) recorded that the D ray facet developed before that of the B ray in *Litocrinus angulatus*. This order of development is less common, but does occur in other species; even within a single species, juveniles with three facets are found with both the normal arrangement (C, E, and B ray facets) and more rarely with the third facet in the D ray position. Except for these cases, the fourth facet developed in the D ray; the A ray facet was always the last to form.

The standard pattern of development of radial facets in *Litocrinus*—C, E, B, D, A—is also found in *Synbathocrinus*, another genus with undivided non-axillary radials (Moore and Ewers 1942; Koenig 1965). Mature individuals of *Synbathocrinus* differ from *Litocrinus* in several respects: there is an arm-like anal sac whose lowest plate rests in a notch between the tops of the C and D radials; there are usually no oral plates preserved; and the radial facets extend adaxially and the full width of the radial. Armless juveniles of *Synbathocrinus* are almost identical with those of *Litocrinus*. During growth they acquired arm facets that were initially narrow but then widened and grew adaxially. After the radial facet in the A ray had formed the notch for the anal sac developed. The fate of the oral plates that are prominent in the theca of juveniles is not known, but it seems probable that in mature individuals they were resorbed.

There is no evidence in *Litocrinus* and *Synbathocrinus* of fusion of thecal plates during growth: the smallest specimen of any species has the same number of basal plates as a mature individual; and there are no traces of divided radials in the juveniles.

Growth histories are known for the following genera with axillary undivided radials (all members of the family Allagecrinidae): *Allagecrinus* (see, for instance, Lane and Sevastopulo 1985); *Allocatillocrinus* (G. D. Sevastopulo, unpublished observations); *Gongrocrinus* (Moore 1940; see also Burdick and Strimple 1982); *Kallimorphocrinus* (Lane and Sevastopulo 1982); and *Thaminocrinus* (Wright 1941). In all of these, the early growth stages are identical to those of *Litocrinus*. After the first five radial facets had formed, a notch where the anal series articulated developed on the left shoulder of the C radial. Thereafter, additional radial facets developed on particular radials in a fairly regular sequence. The C and E radials remained with a single facet (except in *Thaminocrinus*, where additional facets developed on the C radial); in general, arm facets were added in the sequence D, B, A; and then in some forms with numerous arms, more facets were added on the A and D radials than on the B radial. In these, the rank of each radial, based on the number of arms it bore, is commonly the reverse of that based on the order of development of the first five facets. New facets were added on the left hand side of the B and D radials, and alternately on the left and right hand side of the A radial. There is no doubt that the new facets were accommodated by simple lateral growth of the radials, rather than by the addition and fusion of small pararadials of the sort that occur in *Jaekelicrinus*. There is good evidence that in *Kallimorphocrinus* (Lane and Sevastopulo 1982) and *Allocatillocrinus* (G. D. Sevastopulo, unpublished observations) the orals were resorbed in larger individuals.

The ontogeny of disparids with compound radial plates is known in detail only in *Homocrinus* and, in outline, in *Pisocrinus*. A detailed description of the growth and development of the theca of *Homocrinus* will be published elsewhere. Only a summary is presented here (Fig. 19.1). The mature theca (Fig. 19.1e) is an upright narrow cone consisting of a basal circlet of five plates, and a radial circlet with undivided radials in the A and D rays, and compound radials in the B, C, and E rays. Arm articular facets extend across the full width of the distal margin of the A and D radials and the B, C, and E superradials. There is an articular facet on the left shoulder of the C superradial for the arm-like anal sac.

The youngest juveniles of *Homocrinus* that we have found are approximately 250 μm high and consist of basal, radial, and oral circlets (Fig. 19.1a). The basal circlet is composed of five plates similar to those of the mature theca. The oral circlet is also composed of five plates, one of

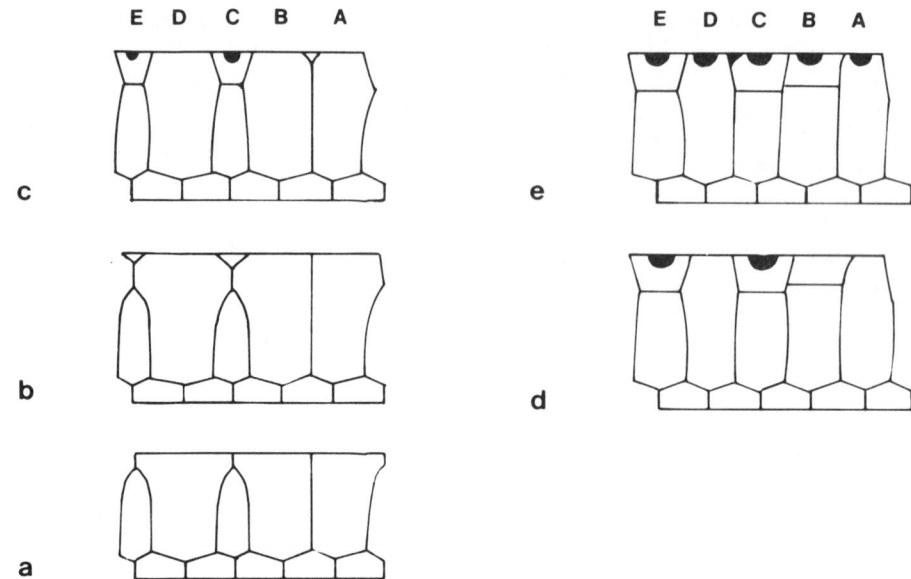

Fig. 19.1 Inception of radials and development of arm articular facets during the growth of *Homocrinus*. (a) Earliest growth stage found, with only inferradials in the E and C rays. (b) Inception of the C and E superradials. (c) Development of the C and E radial facets and inception of the B superradial. (d) Leftward shift of the B superradial to a radial position. (e) Mature specimen with five-arm articular facets and a facet for the anal series on the C radial. Plating diagrams are not drawn to the same scale and the oral plates have been omitted.

which is larger than the other four and is in contact with each of them, and is therefore identified as the CD oral. The radial circlet has an arrangement of plates that differs from that seen in disparids with undivided radials. It consists of three large plates that are identified, by reference to the position of the CD oral, as being in the A, B, and D rays, and two smaller plates in the C and E rays. The C and E ray radials are arch-shaped and their apices do not quite reach the top of the aboral cup whose perimeter is formed of the distal margins of the A, B, and D radials.

In slightly larger specimens, a small triangular plate developed at the top of the cup in the C ray position between the shoulders of the B and D radials. This is identified as the superradial of the mature theca. It was separated from the C ray inferradial by a considerable length of the suture between the B and D radials. The next plate to develop was the E ray superradial (Fig. 19.1b), which also first appeared as a small triangular plate at the top of the cup between the A and D radials, separated from the E inferradial by a length of the AD interradial suture. The C and E superradials grew to meet their corresponding inferradials, and arm articular facets developed on their distal margins. Both facets were well developed before the appearance of the B superradial (Fig. 19.1c). Unlike the C and E superradials, the B superradial was inserted in an interradial position in contact with its corresponding inferradial: it originated as a triangular plate directly above the AB interradial suture at the upper right corner of the B radial. During ontogeny, it expanded to the left, eventually to overlie the B inferradial with a nearly horizontal sutural contact, and thus moved into a radial position (Fig. 19.1d). After this, arm articular facets developed on the B superradial and on the A and D radials. The material on which we have worked does not allow us to determine whether the B or D ray facet developed first, but there is no doubt that the last to form was the A ray facet. The facet on the left shoulder of the C superradial for articulation of the anal series developed after the B and D ray arm articular facets, but we are unable to determine whether it succeeded the development of the A ray arm facet.

Thus, the order of inception of arms in *Homocrinus* is apparently the same as in the other disparids described above. The development of the superradials also follows the same sequence. (It should be noted that primary radial plates in *Homocrinus* are the inferradials and the undivided radials of current terminology). We intend to discuss the homologies and terminology of these plates and the superradials elsewhere; for the time being, we use the currently accepted terminology).

The ontogeny of *Pisocrinus* is under investigation at present by G.D.S. in collaboration with Dr Christina Franzén-Bengtson. Preliminary findings are mentioned below in the discussion on phylogeny. The pattern of inception of superradials and arms is similar to that in *Homocrinus*.

All of the disparid taxa for which the ontogeny is known either have undivided radials or compound radials in the B, C, and E rays. Other disparids with different plating arrangements, such as the Cincinnaticrinidae (formerly Heterocrinidae) with compound radials in the C and E rays only, may be found to have a different order of inception of arms and superradials, but until their ontogeny is established, we will retain the hypothesis that the ontogenetic pattern we have described is the normal one among disparids.

The early ontogeny of cladids is known only from the unusual superfamily Codiacrinacea, many of whose members are abrachiate even when mature. The ontogeny of one member of this family, *Cranocrinus*, which had five arm-bearing radials in maturity, is known in detail (Arendt 1970; G. D. Sevastopulo, unpublished observations). Arendt (1970, Fig. 5, p. 83) has shown that the theca of juveniles of a Permian species of *Cranocrinus* initially consisted of three circlets of plates: the lowest is formed of three infrabasals, above which there are five basals (in interradial positions); the roof of the theca is made up of five orals, one of which, regarded as the CD oral, is larger than the other four. The first structural change to the theca during growth was the development of a small gap, the anal vent, on top of the basal in the C ray position, immediately below the interoral suture. As the theca grew, the basals moved in a clockwise direction eventually attaining an interradial position. A minute radial plate developed in the C ray position on the right side of the distal margin of the CD basal just below the interoral suture. According to Arendt, similar small radials were then introduced in sequence in the D, B, E, and A rays. Arm articular facets then began to develop, first on the C radial, and then on the D, B, E, and A radials in sequence.

An undescribed species of *Cranocrinus* from the Carboniferous of northwest Ireland also developed small radials first in the C and D rays, but unlike the species described by Arendt, the arm articular facets were normally developed first on the D radial (although there is considerable variation in the order of development); the C radial facet was commonly the last or second last to develop; and the facet on the A radial was seldom the last—in marked contrast to the order in the disparids, where it is invariably the last.

The early pattern of development of other codiacrinacean cladids is similar to that of *Cranocrinus* in as much as the anal vent develops before the introduction of radials. Very rarely the anal vent is filled by a plate (as in *Streblocrinus* and some specimens of *Lampadosocrinus*), which is judged to be homologous with the radianal of other cladids. The subsequent development of the radials and the arms is poorly known. The development of radials seems to have been variable in *Embryocrinus* (Arendt 1970, fig. 39), an abrachiate form. In those genera, like *Monobrachiocrinus*, that had only one arm, the radial that bore it is always in the D ray

Because of the paucity of the data, we are forced to use the information gained from codiacrinaceans to make generalizations about the ontogeny of the cladids as a whole. We conclude that the earliest plates of the post-metamorphic juvenile cladid were infrabasals, basals, and orals (as in Recent comatulids); that the anal vent/radianal of cladids developed before the radials and arms; and that the order of development of radials and arms, which in disparids seems to be strictly regulated, is apparently less regular in cladids and does not follow the same pattern.

The early ontogeny of camerate crinoids is similarly poorly known. Although the juvenile specimens of different Ordovician camerate taxa described by Brower (1973) were very small (in several specimens, thecal heights were less than 2 mm), they all already had arms. Koenig (1965) has described the only growth series which includes minute armless thecae that possibly belong to a camerate crinoid. He showed that the thecae of juvenile specimens attributed to the Devonian platycrinitacean *Cyttarocrinus eriensis*, a monobathrid camerate, were initially formed of three circlets of plates—basals, radials (which lacked arm articular facets), and orals. During growth, radial facets appeared on the radials in the order C, E, B, D, A. No specimen showed a primanal plate, which therefore would have developed after the arms. The fact that this order of development of the arms is exactly the same as in disparids is most interesting. Unfortunately, Koenig could not be certain that the specimens he described were the juveniles of *Cyttarocrinus*. There is a substantial size gap between the largest specimen in the ontogenetic series he described and the holotype (and only known mature specimen) of *Cyttarocrinus eriensis*. Furthermore, although Carboniferous platycrinitacean crinoids are locally very common in rocks which have been the most fruitful sources of microcrinoids, their early ontogenetic stages have not been found. On the other hand, Koenig showed that both the holotype of *Cyttarocrinus eriensis* and the specimens making up the ontogenetic series have an elliptical stem facet (which is not a common feature), and a comparable calical shape. Although the evidence is almost evenly balanced, we follow Koenig in tentatively identifying the specimens in the growth series as juvenile *Cyttarocrinus*, rather than as a disparid similar to *Litocrinus* (but with subquadrate arm articular facets more like those of *Haplocrinites*).

To summarize, the thecae of early post-metamorphic juveniles of disparids, and possibly also of monobathrid camerates, consisted of basals, radials lacking arm articular facets, and orals. In contrast, the primary plates of the juvenile thecae of cladids were infrabasals, basals, and orals, The radianal/anal vent in cladids appeared before the growth of the radials; in contrast, in disparids (and possibly also in monobathrid camerates), the development of the anal series succeeded the initial development of the arms. During the growth of disparids (and possibly of monobathrid camerates), arms were added in a set order, usually first in the C ray; then in the E ray, the B ray, the D ray, and finally in the A ray. In contrast, radials and arm articular facets occurred in a less regulated manner in cladids and the order differed from that in disparids.

RELATIONSHIP OF DISPARIDS TO OTHER MAJOR GROUPS OF CRINOIDS

The classification adopted at subclass level in the *Treatise* groups together the orders Cladida, Disparida, and Hybocrinida in the subclass Inadunata. Thus, if it truly expresses phylogenetic relationships amongst the Crinoidea, the sister group of the Cladida should be either the Disparida or the Hybocrinida. In view of the synapomorphies shared by advanced cladids and the subclass Articulata, such as the well developed muscular articulation of the arms, this seems unlikely. Similarly, there is little doubt that the subclass Flexibilia is derived from cladid ancestors (Lewis 1981). It seems more likely that relationships between these groups are that the cladids are the stem group of the articulates, and that the flexibles are a sister group of the cladids plus articulates. The ontogenetic data discussed above are consistent with the hypothesis that articulates and cladids are more closely related than cladids and disparids.

The relationship of the disparids to the Hybocrinida, the cladid/articulate/flexible group, the camerates and *Echmatocrinus* is difficult to resolve and the ontogenetic data are not helpful. The Hybocrinida, which probably are a monophyletic group, and the Disparida share a monocyclic aboral cup, but the monocyclic condition probably arose more than once. The presence of a radianal in hybocrinids may indicate a closer relationship with cladids. It would be very interesting to establish the early ontogeny of hybocrinids. The apparent similarity of the order of development of the arms and anal series in the ontogeny of disparids and monobathrid camerates, if real at all, may reflect a primitive rather than a derived condition, and so may not indicate closeness of relationship. This topic is discussed further by Donovan (this volume).

RELATIONSHIPS WITHIN THE DISPARIDA

In the *Treatise*, Moore et al. (1978, T20 et seq.) recognized eight superfamilies within the Disparida. The morphological characters that may be used in classification of the group have been outlined above. Here, we briefly examine features of this classification and changes subsequently made to it. At present it is not possible to do more than speculate on the relationships of the major groups of disparids.

The superfamily Homocrinacea contains crinoids similar to *Homocrinus*, with divided radials in the B, C, and E rays. As recognized by Moore (1962) and other authors, the thecal structure of many groups of disparids can be derived plausibly from homocrinacean ancestors. The structure of the theca in the several genera assigned to the superfamily is broadly similar; taxonomic distinctions are based mostly on features of the arms and the stem. Several genera possess meric columnals, which are judged to be primitive: *Ibexocrinus*, *Apodasmocrinus* (Warn 1982), and *Daedalocrinus* (Warn and Strimple 1977) have pentameric stems; and *Ectenocrinus* has a trimeric stem, at least proximally.

The relationships of the superfamily Calceocrinacea to other disparids, and relationships within the superfamily itself, have been discussed in detail by several authors (see Ausich 1984; Brower 1982). There is little doubt that the superfamily is monophyletic and was derived from a homocrinacean ancestor. The autapomorphy which unites it is the possession of a specialized basal circlet which was hinged to the A, E, and D radial plates. In *Cremacrinus*, the most primitive genus, the basal circlet consists of four plates and there are four radials, with only an anal series in the C ray. In more advanced genera, a single basal replaces the two plates in the DE and EA interays, and the B ray superradial and arm did not develop. The E ray super- and inferradial, which are in contact in the more primitive genera, became separated by a considerable length of the AD interradial suture in a manner reminiscent of juvenile *Homocrinus*.

The superfamily Pisocrinacea contains a monophyletic group of genera in which the radial circlet consists of three large plates in the A and D rays, and the BC interray (the first two of which bear arms), and small arm-bearing superradials in the B, C and E rays (Fig. 19.2). Preliminary observations on ontogenetic series of *Pisocrinus* by G.D.S. and Dr Christina Franzén-Bengtson suggest that during growth the superradials and arms develop in the same manner as in *Homocrinus*. It seems almost certain that the Pisocrinacea were derived from homocrinacean ancestors. Moore (1962, p. 14) reviewed suggestions as to the origin of the large plate in the BC interray, which had previously been identified as a radianal and an inferradianal. He concluded that it was better classified as a unique sort of two-ray fused inferradial. The structure of the theca of *Pisocrinus* is topologically identical to that of a *Homocrinus* in an intermediate stage of growth, except that the C ray and E ray inferradials have failed to develop. If this analogy is correct, the large plate in the BC interray is likely to be the hypertrophied B inferradial, as was suggested by Moore *et al.* (1978, p. T534).

The Belemnocrinacea is an artifical assemblage. McIntosh (1979) has shown that two of the nominal genera it contains are based on abnormal specimens of cladid crinoids and has suggested that *Belemnocrinus*, itself, may be pseudomonocyclic. The relationships of several other genera assigned to the superfamily are not clear, but one of the constituent families, the Synbathocrinidae, is certainly related to the Homocrinacea. The autapomorphy that unites the Synbathocrinidae is the characteristic architecture of the arm articular facets. Primitive members of the family (*Abyssocrinus*, *Ramacrinus*, and *Theloreus*) have compound radials in the B, C, and E rays and five basal plates. More advanced genera have undivided radials and three basals. It seems likely that the earliest synbathocrinids evolved from a genus like *Homocrinus* by adaxial extension of the radial facets. Whether the change from compound to undivided radials later in their evolution was as a result of fusion (as is often stated) or of non-development of the superradial element cannot be answered from fossil evidence, because the changes that occurred affected the juvenile crinoids at a stage before they became rigidly plated.

The Allagecrinacea is also an artificial assemblage. It contains several genera with divided radials, such as *Anamesocrinus*, *Haplocrinites*, and *Tunguskocrinus*, which are derived from homo-

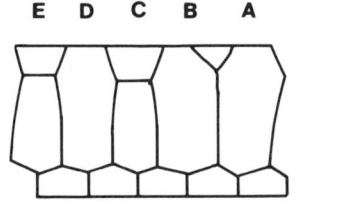
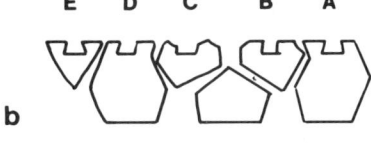

Fig. 19.2 Comparison of the cup plating of (a) immature *Homocrinus* and (b) mature *Pisocrinus*. Plating diagrams are not drawn to the same scale and the basal plates have been omitted from *Pisocrinus*.

crinaceans, but are not closely related to each other. The two large families, Catillocrinidae and Allagecrinidae, are each probably monophyletic (if the Acolocrinidae Brett 1980 are excluded from the Catillocrinidae, and *Allocatillocrinus* is transferred to the Allagecrinidae). However, their relationships to each other and to other disparids are not clear.

The Cincinnaticrinacea (Heterocrinacea of the *Treatise*) is characterized by compound radials in the C and E rays only. Whether this condition arose only once, or more than once, is not known. It is also not clear whether cincinnaticrinaceans were derived from homocrinacean ancestors by suppression or fusion of the B ray superradial, or directly by the disappearance of more than one superradial from other disparids in which there were five divided radials. Primitive cincinnaticrinaceans have brachials fixed into the aboral cup (*Ohiocrinus* and others), pentameric stems (*Cincinatticrinus* and others), and in *Othneiocrinus*, an anal series that is articulated with the first primibrach. The small superfamily Anomalocrinacea is like the Cincinnaticrinacea in possessing compound radials in the C and E rays only, but otherwise has few characters in common with it.

The superfamily Myelodactylacea contains several important groups of disparids some of which we judge are not closely related. One of the more interesting is the Eustenocrinidae, in which the radials are regarded as being divided in all five rays. In *Eustenocrinus* and *Peniculocrinus* primibrachs are fixed in the aboral cup. Primitive forms like these may have given rise to the homocrinaceans by loss of the superradials in the A and D rays. The Iocrinidae, with a cup in which there are no divided radials and with an anal series arising from the upper left hand margin of the first primibrach in the C ray, have few features in common with the Eustenocrinidae. The Myelodactylidae are judged to be a monophyletic group. The synapomorphy which unites them is their bilaterally symmetrical stem. They are similar to the Iocrinidae in having no divided radials and an anal series seated on the upper left shoulder of the first primibrach in the C ray. They may be the sister group of the Iocrinidae.

The relationships of the small, but morphologically unusual superfamily Perittocrinacea, are discussed by Donovan (this volume). In conclusion, it is evident that the classification in the *Treatise* does not clearly reveal those phylogenetic relationships which can be discerned amongst the Disparida. Studies of ontogeny of other groups of disparids may lead to insights into the ways in which they are related and thus eventually to a more natural classification.

ACKNOWLEDGEMENTS

Our research on *Homocrinus* has been supported by NSF grant No. BSR 83-14868. Critical review of the manuscript by Dr S. K. Donovan was most helpful.

REFERENCES

Arendt, Yu. A. 1970. Morski lilii gipokrinidy. *Trudȳ Paleontologicheskogo Instituta Akademia Nauk SSSR* **128**, 3–220. (Hypocrinid sea lilies.)

Ausich, W. I. 1984. Calceocrinids from the early Silurian (Llandoverian) Brassfield Formation of southwestern Ohio. *Journal of Paleontology* **58**, 1167–85.

Brett, C. E. 1980. *Paracolocrinus*, a new inadunate crinoid genus from the Rochester Shale (Silurian, Wenlockian) of New York. *Journal of Paleontology* **54**, 913–22.

Brower, J. C. 1973. Crinoids from the Girardeau Limestone (Ordovician). *Palaeontographica Americana* **7**, no. 46, 263–499.

Brower, J. C. 1982. Phylogeny of primitive calceocrinids. In Echinoderm fauna from the Bromide Formation (Middle Ordovician) of Oklahoma. *The University of Kansas Paleontological Contributions, Monograph* **1**, 90–110.

Burdick, D. W. and Strimple, H. L. 1982. Genevievian and Chesterian crinoids of Alabama. *Bulletin of the Geological Survey of Alabama* **121**, 1–274.

Koenig, J. W. 1965. Ontogeny of two Devonian crinoids. *Journal of Paleontology* **39**, 398–413.

Lane, N. G. 1978. Historical review of classification of Crinoidea. In *Treatise on invertebrate paleontology*, part T, Echinodermata 2(1), (ed. R. C. Moore and C. Teichert), pp. T348–59. Geological Society of America and University of Kansas Press, Lawrence, Kansas.

Lane, N. G. and Sevastopulo, G. D. 1981. Functional morphology of a microcrinoid: *Kallimorphocrinus punctatus* n. sp. *Journal of Paleontology* **55**, 13–28.

Lane, N. G. and Sevastopulo, G. D. 1982. Growth and systematic revision of *Kallimorphocrinus astrus*, a Pennsylvanian microcrinoid. *Journal of Paleontology* **5**, 244–59.

Lane, N. G. and Sevastopulo, G. D. 1985. Redescription of *Allagecrinus americanus* Rowley, 1895, a late Devonian microcrinoid. *Journal of Paleontology* **59**, 438–59.

Lewis, R. D. 1981. *Archaetaxocrinus*, new genus, the

earliest known flexible crinoid (Whiterockian) and its phylogenetic implications. *Journal of Paleontology* **55**, 227–38.

McIntosh, G. C. 1979. Abnormal specimens of the Middle Devonian crinoid *Bactrocrinites* and their effect on the taxonomy of the genus. *Journal of Paleontology* **53**, 18–28.

Moore, R. C. 1940. Relationships of the Family Allagecrinidae, with descriptions of new species from Pennsylvanian rocks of Oklahoma and Missouri. *Journal of the Scientific Laboratories of Denison University* **35**, 55–137.

Moore, R. C. 1962. Ray structure of some inadunate crinoids. Echinodermata. *The University of Kansas. Paleontological Contributions, Article* **5**, 1–47.

Moore, R. C. and Ewers, J. D. 1942. A new species of *Synbathocrinus* from Mississippian rocks of Texas, with description of ontogeny. *Journal of the Scientific Laboratories of Denison University* **37**, 92–106.

Moore, R. C., et al. 1978. Systematic descriptions. In *Treatise on invertebrate paleontology*, part *T*, Echinodermata *2(1)* and *2(2)*. (ed. R. C. Moore and C. Teichert), pp. T404–1027. Geological Society of America and University of Kansas Press, Lawrence, Kansas.

Sprinkle, J. 1980. Origin of blastoids: new look at an old problem. *Geological Society of America Abstracts with Programs* **12**, 528 (abs.).

Strimple, H. L. and Koenig, J. W. 1956. Mississippian microcrinoids from Oklahoma and New Mexico. *Journal of Paleontology* **30**, 1225–47.

Warn, J. M. 1982. Long-armed disparid inadunates. In Echinoderm faunas from the Bromide Formation (Middle Ordovician) of Oklahoma, *The University of Kansas. Paleontological Contributions, Monograph* **1**, 78–89.

Warn, J. and Strimple, H. L. 1977. The disparid inadunate superfamilies Homocrinacea and Cincinnaticrinacea (Echinodermata: Crinoidea), Ordovician-Silurian, North America. *Bulletins of American Paleontology* **72**, 1–138.

Wright, J. 1941. *Allagecrinus biplex* Wright—A revision of the species, with notes on other Scottish Allagecrinidae. *Geological Magazine* **78**, 293–305.

20

The evolution of feeding structures in Palaeozoic crinoids

THOMAS W. BROADHEAD

Department of Geological Sciences, University of Tennessee, Knoxville, TN 37996-1410, USA

INTRODUCTION

One of the most distinctive and varied features of members of the echinoderm class Crinoidea is the arrangement of both skeletal and soft tissue components of the feeding system. Although the basic organization of the soft tissues appears to be little changed throughout the history of the class, the skeletal supports for these soft parts have undergone many and often repeated changes in form. The analysis of changes, primarily in branching patterns, in the skeletal support system, when evaluated in terms of the known development of modern crinoids, allows both a determination of evolutionary patterns and processes in ancient crinoids, and a consequent evaluation of the taxonomic values of these features.

The primary purpose of this discussion is to review the major patterns of organization of the skeletal support for the feeding system in the major groups of Palaeozoic crinoids. The emphasis here is the recognition of primitive and derived features as they occur within these major groups. Additionally, evolutionary patterns and processes have been suggested for several innovations that collectively characterized the major evolutionary changes in crinoid arm structure during the Palaeozoic. Attention is not given to the actual methods of food capture hypothesized for fossil crinoids; several more detailed treatments (e.g. Lane and Breimer 1974; Lane and Macurda 1975) are important contributions to this area of investigation.

Echinoderm phylogeny and evolutionary biology (ed. C. R. C. Paul and A. B. Smith). Clarendon Press, Oxford, 1988.

ORGANIZATION AND ONTOGENY OF THE FEEDING SYSTEM IN MODERN CRINOIDS

The soft-tissue components of the feeding system of crinoids are dominantly the exposed parts of the water vascular system, a basic coelomic feature of the phylum. The water vascular system of most modern crinoids consists of a ring canal, from which branch (1) a number of stone canals emptying into the main coelomic cavity, (2) labial podia, and (3) radial canals. Notably, the stone canals of modern crinoids do not connect to the outside of the body by means of a madreporite or hydropore, although the presence of such structures has been suggested for some Palaeozoic cladids and flexibles (see review by Ubaghs 1978, pp. T195–7). The radial canals extend into the arms and undergo bifurcation in all but five-armed forms. The pattern of subordinate branching from the radial canal, either into pinnules or as podia, is typically alternate on either side of the canal.

During the ontogeny of a modern crinoid, the water vascular system is one of the last major features to form, and the skeletal support for the arms and pinnules lags slightly behind the formation and extension of the soft parts of the water vascular system. Generalizations concerning the larval ontogeny of modern crinoids (Hyman 1955, pp. 75–87) are based primarily upon the development of various species of *Antedon*. Rudiments of the radial plates are not present until the end of the cystidean stage, which is marked by rupture of the membrane over the summit, opening of the oral plates, and extension of the primary podia. There are three primary podia per ray, and a pair of secondary podia develops interradially during the pentacrinoid stage. Hyman noted (1955, p. 85) that 'the late appearance of the arms indicates that they [=radial plates] are not true thecal or calycinal plates but the first plates of the brachial series'. There is no reason to dispute this conclusion, at least for the Articulata; recent studies of microcrinoids (e.g. Lane and Sevastopulo 1982) suggest a similar ontogenetic relationship among some cladids.

The early development of the arms involves the loss of the two primary podia adjacent to the initial one in each ray. The initial podium elongates to become the radial canal of the arm, from which podia develop as lateral branches. For a short time only, the tip of the central podium functions as the terminal tube foot of the arm. Brachial plates develop nearly synchronously with the elongation of the radial canal.

HOMOLOGY OF ARMS, RAMULES, AND PINNULES

The homologous nature of the three basic 'levels' of skeletal feeding structures has been suggested previously and most recently by Paul and Smith (1984) in a study of the phylogenetic relationships among early echinoderms. Developmental patterns of modern crinoids also indicate that branching of the water vascular system, with accompanying skeletal support, is basically the same process for ramification of the arm as it is for pinnulation —an essentially dichotomous branching at the growing tip of the radial canal. Only the development of podia, as observed in modern crinoids, appears to differ from this pattern, but it is not possible to compare modern podial development with that of representatives of the extinct subclasses.

Thus, the terms 'arm' and 'ramule', and possibly also 'pinnule' represent arbitrary divisions of the skeletal supports for the branching radial canal and its terminal extensions, the podia. Additionally, the term 'ray trunk' only describes the highly modified principal arm structure observed in some camerate crinoids. The following is a brief reconsideration of the meaning of these terms.

An *arm* is the major central support of the feeding system in each ray. At the base, the first primibrachial plate of the arm rests upon the upper articular facet of a radial plate. The brachial plates comprising the arm are arranged in a uniseries or biseries. The arm is either unbranched (atomous or five-armed forms) or branched, with branching usually occurring initially near the base of the arm. Branching is dichotomous or heterotomous, and equal or unequal. Brachial plates may support one or more (hyperpinnulate) pinnules or none at all (apinnulate). 'Arm trunks', found in some camerates, are highly modified principal arms that usually support regularly spaced subordinate, biserial, pinnulate arm branches. All main arm branches of a pinnulate crinoid bear pinnules. All crinoids, with the exception of several small cladids that include progenetic paedomorphs and possibly also some larval forms, are known to possess arms with skeletal supports.

A *pinnule* is the small, ultimate part of the skeletal support of the feeding system and is always a subordinate, uniserial, unbranched, lateral branch from a single brachial plate of the arm. Lane and Breimer (1974) have suggested that the pinnules of Palaeozoic crinoids commonly had a combined muscular-ligamentary articulation to the host brachial in contrast to the largely ligamentary articulations among brachials and between 'ramules' and the principal arm. At the base, the first pinnular plate articulates with a brachial plate of the arm, and successive pinnules are located on alternate sides of the arm, usually one per brachial plate. Pinnules are common features of the poteriocrinine Cladida, Camerata, and Articulata, but are usually not thought to be present in the Flexibilia.

The term *ramule* is sometimes used to describe a subordinate arm (i.e. a small branch in a heterotomous branching pattern). In most instances, ramules are greatly abbreviated subordinate arm branches or branching systems, as in a few Flexibilia (e.g. *Onychocrinus*), several Camerata (e.g. *Melocrinites, Ctenocrinus, Trichotocrinus, Trybliocrinus, Rhipidocrinus*, some *Hexacrinites*), and some Cladida (e.g. *Barycrinus, Pellecrinus, Grenprisia, Bathericrinus*). In a smaller number of cases, structures referred to as 'ramules' are probably large pinnules (e.g. the camerate *Simplococrinus*, the cladid *Clathrocrinus*).

The recognition of a minor skeletal structure as a subordinate arm or a pinnule is important in the classification of Palaeozoic crinoids, but appears to have suffered inconsistency in the goal to achieve a convenient and workable classification. Of particular interest is the embryologic origin of exotomous and endotomous arm branching patterns. Broadhead (1987) has suggested that exotomous branching in camerate crinoids was the result of pinnule differentiation, a process observed in some modern crinoids by which a pinnule grows to become a pinnule-bearing arm. Most exotomous and endotomous branching observed in fossil crinoids follows the pattern where each ramification takes place on the second brachial following the previous ramification. This appears to be a reflection of the alternating nature of pinnulation by Bather's (1890) and Frest et al.'s (1979) definition. If this is true, then many structures referred to as armlets or ramules in many of the so-called 'apinnulate' cladids and disparids, are more properly termed pinnules.

MAJOR EVOLUTIONARY TRENDS

It is difficult to relate arm branching patterns to the phylogeny of the Crinoidea, because most patterns recur at different times in only distantly related groups. Nonetheless, these patterns can be examined in the temporal context of crinoid distribution, and their common recurrence in time can be evaluated in terms of their evolutionary origin. The taxonomic importance of these features is discussed later in this paper. The important crinoids and groups of crinoids discussed here are *Echmatocrinus*, the subclasses Camerata and Flexibilia, and the main inadunate 'orders' Cladida and Disparida. The cladids and disparids are treated at the same level as the flexibles and camerates based largely upon the conclusions reached by Kelly (1982) concerning plate homologies of cladid and disparid cups plus the subsequent comment by Smith (1984) that the Inadunata represented a paraphyletic group. In cladistic terms, the cladids also are a paraphyletic group because they gave rise to two crown groups: Flexibilia and Articulata. The other inadunate 'order', Hybocrinida, has been formally elevated to a subclass (Hybocrinea) by Rozhnov (1985). The hybocrinids were a short-lived group with simple, unbranched, apinnulate arms that varied both in number and in the erect or recumbent nature of the ambulacra (Sprinkle and Moore 1978; Rozhnov 1985). They are not treated further here because of their limited stratigraphic occurrence and minor role in the evolution of the class.

The 'base line' for a discussion of crinoid evolution is, of necessity, the earliest crinoid, *Echmatocrinus brachiatus* Sprinkle from the Burgess Shale (Middle Cambrian) (Fig. 20.1). The small number of details concerning the arms

Fig. 20.1 Feeding structures of *Echmatocrinus brachiatus* Sprinkle showing short, simple, uniserially plated arms with large podia. V-shaped impressions on brachial plates are interpreted by Sprinkle (1973) to have been sites of muscle insertion. Calyx plating not shown. (Modified from Sprinkle 1973.)

provide important insight into later evolution of feeding structures. First, there are approximately ten arms in *E. brachiatus*, the exact number not being known with confidence (Sprinkle 1973, p. 182). It is possible that, at this early stage in evolutionary development, crinoids did not possess a well-developed pentamerous symmetry and that the arm number was not a regular multiple of five. This, however, seems unlikely because the pentamerous grade was already well established in several more ancient representatives (Paul and Smith 1984). It is possible that the arms of *E. brachiatus* are dichotomously branched, as is common in most later species, but calyx plating is obscure on known specimens. Secondly, the plating of the arms is uniserial, establishing that characteristic (not surprisingly) as the primitive plating configuration of crinoid arms. This contrasts with the biserial plating of ambulacra and brachioles of many earlier echinoderms (see Sprinkle 1973; Paul and Smith 1984). Thirdly, the podia preserved in the holotype specimen are clearly large structures that extend alternately to the right and left side of each arm. This is the pattern observed in branches of the radial canal into the pinnules (if present), but not the podia of modern crinoids.

The crinoid stratigraphically closest to *Echmatocrinus* is the disparid *Pogonipocrinus* (stratigraphic interpretation of Donovan, this volume). It is essentially a five-armed crinoid with two or more dichotomous branches high on each arm. There are more than two primibrachial plates in each arm, all brachial plates are uniserial and equant, and pinnules are lacking. The long slender arms, in relation to cup height, contrast markedly with the short stout arms of the earlier *E. brachiatus*. Other early 'inadunates', such as the disparids *Ramseyocrinus cambriensis* (Hicks) (correctly reinterpreted by Donovan 1984), *Inyocrinus*, and *Vosekocrinus* or the cladid *Aethocrinus moorei* Ubaghs (I prefer the interpretation proposed by Philip and Strimple 1971, accepted also by Kelly 1982) appear to have been similar in general aspects of arm structure.

Innovations in arm structures during the Middle Ordovician include the development of different types of heterotomous branching in both cladids and disparids. True pinnulation also occurred at that time in the disparids (Frest *et al.* 1979). Other later modifications include atomy in some groups, and the development of multiple unbranched arms (sometimes regarded as pinnules) on individual radial plates in some disparids (e.g. *Acolocrinus*, Allagecrinidae). Modification of the symmetry of the feeding system, especially to a strongly bilateral pattern, characterizes many disparids, particularly the superfamily Calceocrinacea and a few later poteriocrinine cladids, such as *Anartiocrinus*. Reduction in number of arms (excluding the recumbent ambulacra exhibited by some hybocrinids) is relatively uncommon, except in disparids and cyathocrinine cladids. Abrachiate forms appear first in the early Mississippian (some members of cladid superfamily Codiacrinacea); their small size suggests that they are progenetic paedomorphs, although many may be the juveniles of brachiate forms.

The earliest camerate crinoids are known from the Lower Ordovician of North America and possess characteristics of the arms that, although representing primitive features in terms of camerate arms, suggest major innovation over the simple, uniserial, apinnulate, dichotomously branched arms of early disparids and cladids. The earliest camerate species, *Proexenocrinus inyoensis* Strimple and McGinnis is characterized by ten arms that branch above the radial plates only at the second primibrachial, uniserially arranged subequant cuneate secundibrachial plates, and widely spaced pinnules that extend from successive secundibrachials on alternating sides of the arms. With very few exceptions, this basic arm morphology continued throughout the history of the camerates (Broadhead 1987). Later modifications included development of the principal arm to form a 'ray trunk', additional dichotomies or heterotomies above the primibrachials commonly resulting from pinnule differentiation, shortening of brachial plates or development of biseriality, and closer spacing of pinnules.

The earliest known flexible crinoids, *Archaetaxocrinus burfordi* Lewis and *A. lanei* Lewis are from the lower Middle Ordovician of North America, and superficially resemble contemporary dendrocrinine cladids (e.g. species of *Cupulocrinus*) from which they apparently evolved. The arms of representatives of *Archaetaxocrinus* characteristically are uniserial, apinnulate, and dichotomously branched both at the first primibrach and usually also at higher levels. Although dichotomous arm branching was common throughout the history of flexibles, post-Ordovician modifications included the development of several different patterns of both equal and unequal heterotomous branching, and rare cases of atomy. Pinnulation is normally not considered to be a characteristic possessed by any flexible, but the distribution and size of extremely small subordinate arm branches in some flexibles with biendotomous branching, especially as shown by *Euonychocrinus simplex* Strimple and Moore, suggests that these may be pinnules.

INNOVATIONS IN ARM STRUCTURE

Evolutionarily derived features of the skeletal support for the feeding system are relatively easily determined by comparison with features exhibited by the earliest members of each subclass. In the case of arm structure, there appear to be few ambiguities concerning 'primitive' morphologies, i.e. there are no examples of extremely early forms that cannot be regarded as having evolved from a known earlier form.

Arm number

The number of arms possessed by a crinoid can be regarded as the number of primary branches proximal to the calyx. As with any definition, there is a certain arbitrary nature to this. Most proximal (beginning at the first or second primibrachial) dichotomies or heterotomies would be included, but higher and other subordinate branchings

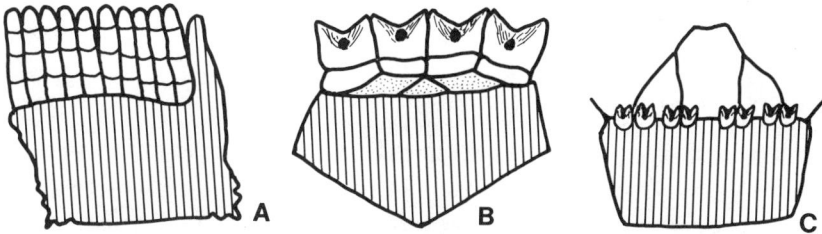

Fig. 20.3 Crinoid radial plates and arms with diminished proximal plating. (A) *Acolocrinus* superradial with numerous short arms arising directly from radial summit. (B) *Crotalocrinites* radial with primibrachial, secundibrachials, and outer tertibrachials resting upon radial summit. (C) *Parahexacrinus* radial and tegminal plates with openings for third order arms. Proximal brachials are either lacking or are greatly reduced and internal as in the case of *Paratalarocrinus* (Broadhead and Strimple 1979).

Isotomous arm branching (Fig. 20.4A) is the dominant pattern exhibited by known representatives of Lower Ordovician taxa. One or more dichotomies high above the calyx characterized both the early disparids and cladids; there was only one proximal dichotomy in earliest camerates, but by the Middle Ordovician common forms such as *Archaeocrinus* and *Abludoglyptocrinus* also exhibited higher dichotomies. Higher dichotomies are common features in many later camerates, although simple arms dominated. Flexible crinoids dominantly exhibit isotomous branching throughout their history, including the earliest forms, which characteristically exhibit both a proximal dichotomy and one or more higher dichotomies.

Heterotomous arm branching was a derivation from the primitive isotomous division of arms in cladids, disparids, and flexibles, but probably was the result of pinnule differentiation in camerates. In heterotomous branching, subordinate lateral arm branches are produced in a more or less regular pattern alongside the principal arm. The three chief types of heterotomy are (1) alternating heterotomy, (2) exotomy, and (3) endotomy.

Alternating heterotomy is characterized by the production of successive subordinate arm branches ('ramules') on alternating sides of the principal arm branch (Fig. 20.4 B,C). This appears to have been the most common form of heterotomy among the disparids, especially among calceocrinids; other early forms include *Cincinnaticrinus*, *Othneiocrinus*, and *Ohiocrinus*, all from the Middle Ordovician. Among cyathocrinine cladids, alternating heterotomy is uncommon, and is first represented in the Middle Ordovician by *Carabocrinus*. This is the most common form of heterotomy among dendrocrinines, beginning in the Middle Ordovician with *Grenprisia* and *Polycrinus*, and generally characterizes the Botryocrinidae. The unusual Lower Devonian genus *Eifelocrinus* appears to have been characterized by multiple alternating heterotomy (i.e. alternating second order branches, if not pinnules, extending from first order subordinate branches). The Upper Permian *Timorechinus* is possibly the only poteriocrinine cladid experiment with this branching pattern. In contrast to the disparids and cladids, camerate and flexible crinoids rarely exhibit alternating heterotomy; it is seen, however, in some species of *Periechocrinus* (Upper Silurian) and *Hexacrinites* (Upper Silurian–Upper Devonian) and several Actinocrinitidae (e.g. *Dialutocrinus*, *Strotocrinus*-Mississippian) in the camerates and in species of *Onychocrinus* (Mississippian) in the flexibles.

Endotomy is characterized by successive subordinate arm branches on the admedial side of each branch usually above the proximal dichotomy (Fig. 20.4D). This pattern is uncommon in early crinoids, but characterizes the Middle Ordovician disparids *Daedalocrinus*, *Tryssocrinus*, and *Tornatilicrinus*. Among the cladids, no representative of the Cyathocrinina is known with certainty to have endotomous arms, but specimens of *Palaeocrinus hudsoni* from the Middle Ordovician Bromide Formation appear to exhibit a poorly developed endotomy (Sprinkle 1982, pp. 156, 158). Endotomy is similarly rare in dendrocrinines, reported only from the Middle Ordovician *Metabolocrinus*. The earliest poteriocrinine cladid with endotomous arms is *Hallocrinus* (Lower Devonian), but endotomy was not common in that suborder until the Mississippian and Pennsylvanian. Endotomy

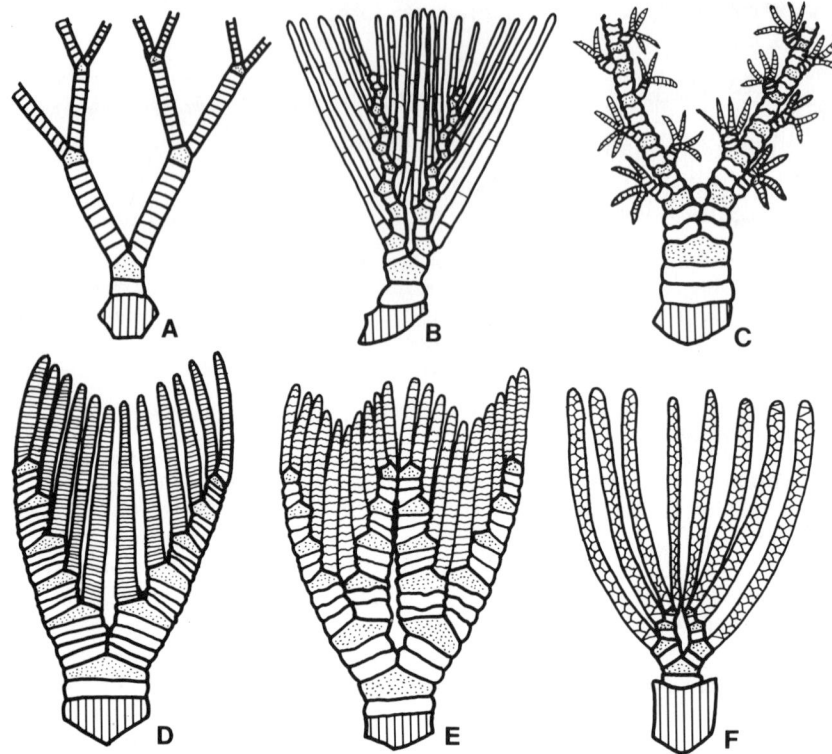

Fig. 20.4 Arm branching patterns in later Palaeozoic crinoids. Schematic representations of actual taxa; pinnules not shown for A, D, F. (A) Isotomous branching. (B) Alternating heterotomous branching. (C) Alternating heterotomous branching (simplified from *Onychocrinus*). (D) Endotomous branching. (E) Biendotomous branching. (F) Exotomous branching.

is exceedingly rare in camerates, being represented only by *Thamnocrinus*; this rarity is possibly related to the proposed method of branching by means of pinnule differentiation in many camerates (Broadhead 1987). Endotomy is not common in flexibles; it is first represented by *Homalocrinus* (Upper Silurian).

Two minor variants exist on the theme of endotomy: parendotomy, and biendotomy (Fig. 20.4E). The first, however, seems only to be a 'loosely organized' version of the second. In both instances, there are two dichotomous divisions, above which each pair of principal arm branches is internally endotomous. 'Parendotomy' occurs only in *Cydrocrinus* (Lower Mississipian) and in *Rhopocrinus* (Upper Mississippian), and biendotomy only in *Neozeacrinites* (Pennsylvanian–Permian) among the poteriocrinine cladids. It does not occur in camerates, but is relatively common among flexibles, particularly among the Sagenocrinitacea; the earliest biendotomous flexible is *Calpiocrinus* (Upper Silurian). Some Pennsylvanian genera, such as *Euonychocrinus* are characterized by a biendotomous pattern in which subordinate branches exhibit an almost consistent distribution on every other tertibrachial plate. This arrangement is consistent with pinnule distribution in pinnulate crinoids, except that it lacks small branches on the outer side of each arm. Although flexibles are usually regarded as having been apinnulate, such a regular distribution of subordinate branches simulating 'half' of a normal pinnule component suggests a possibility that these small branches may actually be pinnules.

Exotomous branching is characterized by subordinate arm branches only on the outer (abmedial or interradial) side of each arm above the initial dichotomy (Fig. 20.4F). It is absent in disparids and occurs only in a small number of late Palaeozoic cladids, the earliest of which is *Hyd*-

reionocrinus (Upper Mississipian). Exotomy is only slightly more common among camerates, the earliest being *Periechocrinus* (Middle Silurian), and the best developed being *Paradichocrinus* (Lower Mississippian). Exotomous branching is so far unknown among the Flexibilia.

Ray trunks

Ray trunks are highly modified principal arms that may have been limited to the Camerata. A ray trunk is commonly a uniserially plated arm supporting numerous biserial, pinnulate, subordinate arms in an alternating arrangement on either side of the ray trunk. Ray trunks evolved both by modification of an otherwise discrete principal arm (forms with 10 ray trunks) (Fig. 20.5A) and by lateral fusion of two adjacent principal arms along the medial plane of the ray (forms with five ray trunks) (Fig. 20.5B). The first type is represented by *Rhipidocrinus* (Middle Devonian), *Steganocrinus* (Lower Mississippian), and *Eucladocrinus* (Lower Mississippian). The second type is represented by Middle Silurian to Upper Devonian representatives of the Melocrinitidae, the earliest being *Promelocrinus*, and by *Manillocrinus* (Lower Carboniferous). Two main variants of these patterns are seen in:

(1) *Cytidocrinus* (Lower Mississippian), in which each of the five ray trunks represents one of the two branches from a dichotomy at primibrachial 2, whereas the other branch is only a smaller, simple, biserial, pinnulate arm (Fig. 20.5C);
(2) *Trybliocrinus* (Lower Devonian), in which the origin of each ray trunk in the cup appears to have been a biserial fixed arm that led to a multiserial trunk supporting biserial subordinate arms.

The distribution of the pinnulate, biserial arms along ray trunks suggests that these developed due to pinnule differentiation. Brower (1976) documented the ontogeny of *Promelocrinus anglicus* Jaekel and noted that all arm growth was terminal and not intercalary. He concluded that new biserial arms were probably produced throughout the life of the crinoid, a pattern also suggested by other crinoids possessing ray trunks.

Pinnulation

The earliest pinnulate crinoid is the camerate *Proexenocrinus inyoensis* from the Lower Ordovician. Throughout the history of camerates, pinnules were an important and virtually ubiquitous feature, most notably being absent in representatives of the aberrant family Reteocrinidae. Pinnulation appears to have developed independently in the cladids (*Eopinnacrinus*) and in at least two groups of disparids (Frest *et al.* 1979) and possibly also (if free structures actually are pinnules) in the disparids *Agostocrinus* and *Acolocrinus*, all during the Middle Ordovician. After that time, pinnulate disparids and cladids were extremely uncommon until the early Devonian, with the first appearance of poteriocrinine cladids. The evolutionary success of pinnulation in that group is obvious when considering the dominance, both in numbers of individuals plus specific diversity held by the Poteriocrinina from the Upper Mississippian

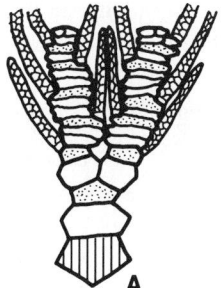 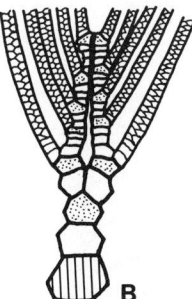 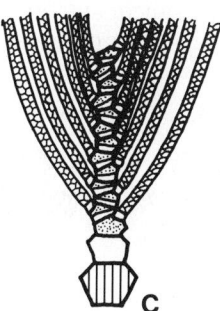

Fig. 20.5 Schematic representations of 'ray trunks'; pinnules not shown for A, B, C. (A) Ten-trunk form with hypertrophied principal arm of uniserial plates supporting biserial pinnulate subordinate branches (*Rhipidocrinus*). (B) Five-trunk form resulting from fusion of adjacent principal arms, each of which is uniserial and supports biserial pinnulate subordinate arms (*Promelocrinus*). (C) Five-trunk form resulting from heterotomous branching at secundibrach 2 to form a single uniserial trunk with biserial pinnulate subordinate arms (*Cytidocrinus*).

nearly to the end of the Palaeozoic. As mentioned previously, flexibles are commonly regarded as having been apinnulate. The most notable possible exception to this is *Euonychocrinus* from the Upper Pennsylvanian (Fig. 20.6E).

In addition to the presence or absence of pinnules, both pinnule form and distribution are evolutionary features deserving mention. Pinnules are usually composed of a relatively small number of plates (commonly 3–6). Actual dimensions of pinnulars are highly variable; lengths may exceed by several times the height of any brachial plate in the free part of the principal arm (e.g. many camerates) or may be nearly the same length or shorter than the height of the supporting brachial (e.g. most cladids, many camerates). Widely spaced (Fig. 20.6A,C) and narrowly spaced (Fig. 20.6B,D) pinnules are largely a function of brachial height; early camerates such as *Proexenocrinus inyoensis*, and *Trichinocrinus terranovicus* possess widely spaced pinnules as a primitive characteristic of camerates (Broadhead 1987). Closely spaced pinnules in both cladids and in camerates resulted mostly from either the reduction in height of pinnule-supporting brachials or in the development of biserial plating of the arms.

Hyperpinnulation is an additional modification of pinnule distribution in which more than one pinnule articulates with a single brachial plate. Hyperpinnulation evolved independently among different groups of camerates and poteriocrinine cladids as the result of crowding of pinnules or by fusion of adjacent brachial plates.

Arm plating

The primitive state in crinoid arm plating is the uniserial arm with subequant brachial plates (Fig. 20.6A). The two most notable modifications of this condition are a reduction in height/width ratio, and the development of a biserial arrangement of brachial plates. These two derived characteristics appear to be most closely, but not exclusively in the case of h/w ratios, related to decreasing space between adjacent pinnules that would result in a more dense array of podia. Increase in feeding capacity of apinnulate crinoids due to shortening of brachial plates is debatable. Reduction in brachial height probably increased flexibility of the arm, but podial spacing was probably regular, making the number of podia more dependent upon arm length than upon number of brachial plates per unit length in the arm.

There appears to be no particular pattern to the development of shortened brachials (Fig. 20.6B); this feature occurs throughout the camerates and poteriocrinine cladids at different times among different families. Most representatives evolved from forms with higher or nearly equant brachial plates (e.g. some species of the Mississippian camerate subfamily Dichocrininae; Broadhead 1981).

The biserial arrangement of arm plates (Fig. 20.6D) is exclusively associated with pinnulate arms and was most common throughout the evolutionary history of the camerates. This arrangement was most commonly derived from a uniserial arrangement characterized by moderately high, cuneate brachials (Fig. 20.6C). The Middle Ordovician genus *Hercocrinus* possibly includes the earliest reported biserial-armed camerates, although at least three species of biserial-armed camerates (*Bromidocrinus nodosus*, Camerate species A, *Diabolocrinus arbucklensis*) occur slightly later (Kolata 1982, p. 171). Cladids achieved the biserial grade of arm plating much later in the early Carboniferous. The oldest known

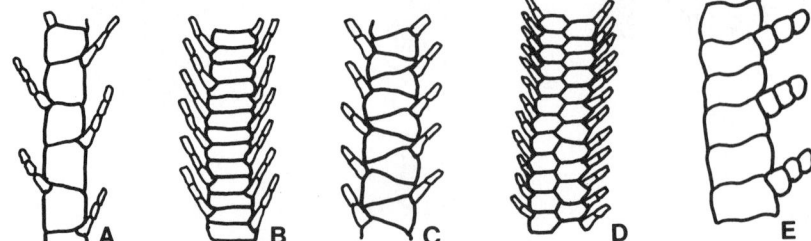

Fig. 20.6 Arm plating and pinnule distribution patterns. (A) Tall uniserial brachial plates with widely spaced pinnules. (B) Short uniserial brachial plates with resulting closely spaced pinnules. (C) Cuneate uniserial brachial plates with moderately widely spaced pinnules. (D) Biserial brachial plates with closely spaced pinnules. (E) Arm of *Euonychocrinus* showing arrangement of ?subordinate arms, or possibly pinnules.

cladids with biserial arms are poteriocrinines that belong either to *Anemetocrinus* or *Hydreionocrinus*. Biserial arm plating was more common in North American poteriocrinines ranging from the Upper Mississippian to Middle Pennsylvanian, but declined rapidly to the Triassic. At no time, however, was biserial plating as common as uniserial plating in cladid arms.

TAXONOMIC SIGNIFICANCE

When consideration is given to the biological importance of the arms and associated skeletal supports for the subvective system, it is somewhat surprising that such vital structures have such limited taxonomic value. During the Palaeozoic, the water vascular system of crinoids appears to have been capable of exhibiting almost unlimited variation, and commonly underwent iterative evolution in form of its ultimate parts, as inferred from the preserved skeletal supports. The lack of pinnules in crinoids is a primitive characteristic of the class, but pinnulation was a vicariously derived feature. Attempts partly to characterize major groups, such as the disparids, based upon presumed consistency in some aspect of arm structure (in that case, lack of pinnules) have proven unfeasible. The same may yet be true for flexibles (i.e. lacking pinnules) and camerates (i.e. always possessing pinnules).

Within many currently recognized ordinal groupings, branching patterns tend to be characteristic more of individual genera than of higher taxonomic groups, such as families. Within a genus there is usually a marked consistency in most characteristics of arms, such as branching pattern, plating, pinnulation, and sometimes number. Differences recognized as specific discriminators do exist within genera, such as uniserial or biserial plating in *Phanocrinus* or uniserial or biserial plating and ten or twenty arms in *Dichocrinus*. Intraspecific variation is usually not at a large scale, but in some monobathrid camerates that show considerable variability in proximal arm branching (Lane 1963), the mechanism for branching, in that case pinnule differentiation (see Broadhead 1987) provides the explanation.

EVOLUTIONARY PATTERNS

When viewed as the temporal distribution of primitive and derived characteristics among the four major groups of Palaeozoic crinoids (disparids, cladids, camerates, flexibles), it is apparent that the 'plastic' nature of variation within crinoid arms has produced few synapomorphies but many plesiomorphies (Table 20.1). The early divergence of these four groups helped to ensure a representation of many of the small number of primitive characteristics. The most definite of these ancestral traits also shared with *Echmatocrinus* is uniserial plating. The problem of arm number is one that is arguable in terms of the relationship of *Echmatocrinus* (>five arms) to the other groups (5 branched or 10 unbranched arms). If the five-armed form is primitive in post-Cambrian crinoids, then that arm number plus dichotomous branching unite the cladids and disparids. Later reduction in number of primibrachial plates led to the 10-armed structures of the camerates and flexibles that evolved from the cladids. Beyond that point, most other features mentioned in this paper represent iteratively derived features in many smaller clades within the four chief Palaeozoic groups.

The mechanisms for these changes mostly involved modifications of developmental patterns and frequently can be understood in terms of the known ontogenetic patterns of modern crinoids and of the pervasive processes of heterochrony. Consideration must be given to the bifurcations of the radial canal that produce isotomous, subordinate major, or pinnular branches of the water vascular system. The branching system for each crinoid is genetically programmed, and the patterns of branching may be viewed in terms of genetic regulation.

The single proximal isotomous dichotomy is the most common branching pattern among living and fossil crinoids. In modern comatulid crinoids, it develops during the stalked pentacrinoid larval stage. When compared to the earliest members of the stem group (Cladida), it is apparent that this branching has been brought into calical proximity by reduction in the number of primibrachs, but represents an increase from one to two primibrachs in articulates when compared to most late Palaeozoic cladids. In ten-armed crinoids, there are few exceptions (e.g. *Cytidocrinus*, Fig. 20.5C) to the rule of isotomous division at the first dichotomy.

Higher divisions of the arms almost always occur at approximately the same levels, leading to the possibility of an ecological influence on

TABLE 20.1 Times of first appearance of primitive (*) and derived characteristics of crinoid arm structure in the principal Palaeozoic groups. x—Not known to occur.

Character	Disparida	Cladida	Camerata	Flexibilia
5-Armed	L. Ord.*	L. Ord.*	x	U. Perm.
10-armed	M. Ord.	M. Ord.	L. Ord.*	M. Ord.*
multibrachiate	x	x	M. Ord.	x
abrachiate	x	L. Miss.	x	x
reduction of proximal plates	M. Ord.	U. Sil.	L. Dev.	x
Dichotomous	L. Ord.*	L. Ord.*	L. Ord.*	M. Ord.*
alternating heterotomous	M. Ord.	M. Ord.	U. Sil.	L. Miss.
arm trunks	x	x	U. Sil.	x
endotomous	M. Ord.	M. Ord.	M. Dev.	U. Sil.
biendotomous	x	L. Miss.	x	U. Sil.
exotomous	x	U. Miss.	M. Ord.	x
Appinulate	L. Ord.*	L. Ord.*	M. Ord.	M. Ord.*
pinnulate	M. Ord.	M. Ord.	L. Ord.*	?M. Penn.
Uniserial	L. Ord.*	L. Ord.*	L. Ord.*	M. Ord.*
biserial	x	L. Carb.	M. Ord.	x

branching and branching density. In addition to a genetic predisposition toward branching, it is possible that there is a chemical stimulus, related to proximity of podia from adjacent arms, that affects the timing of branching at higher levels in those crinoids in which such branching is common. Thus, the lack of contact by podia from adjacent arms may stimulate additional branching in order to maintain a suitable density of arms, etc., within the filtration system.

Patterns of arm branching are related to two dominant processes, (1) direct development, and (2) pinnule differentiation. A third mechanism, augmentative regeneration, has been uncommon generally, and was especially rare among Palaeozoic crinoids. Direct development is the process attributed to the proximal dichotomy common to most crinoids, but probably was the primary process for dichotomous branching in most disparids, flexibles, and probably the higher dichotomies in camerates and cladids. In larval stages of modern crinoids, this occurs where the growing tip of the radial canal divides isotomously at the axillary second secundibrach, leaving two divergent branches.

Pinnule differentiation involves the development of a pinnulate arm from a pinnule, and has been observed in modern crinoids (Mortensen 1920) and suggested as a primary mechanism for the second proximal dichotomy in multibrachiate camerates and for heterotomous branching in camerates (Broadhead 1987). The extent to which pinnule differentiation has occurred in cladids is uncertain. Evidence that this process took place comes from the occurrence of subordinate arm branches (usually exotomous or endotomous) only on alternating brachial plates of the principal arm—a distributional pattern which simulates that for pinnules on one side of an arm branch. Similar patterns of subordinate branches also occur among flexibles with heterotomously branched arms, but the characterization of flexibles as being apinnulate would preclude the suggestion of pinnule differentiation as a mechanism there. Ray trunks of camerates probably are extreme examples of pinnule differentiation accompanying hypertrophy of the principal arm. The usual maintenance of uniserial plating in the principal arm, which supports biserial pinnulate subordinate arms, reflects the continuance of the simplified plating of the most proximal parts of the arm until the last proximal production of a subordinate arm.

Pinnules are both distinctive and controversial structures in the function and interpretation of crinoid arms. Although only the camerates and cladids were at times characteristically pinnulate, rare examples of pinnulation occur among the disparids and possibly among the flexibles, as suggested here.

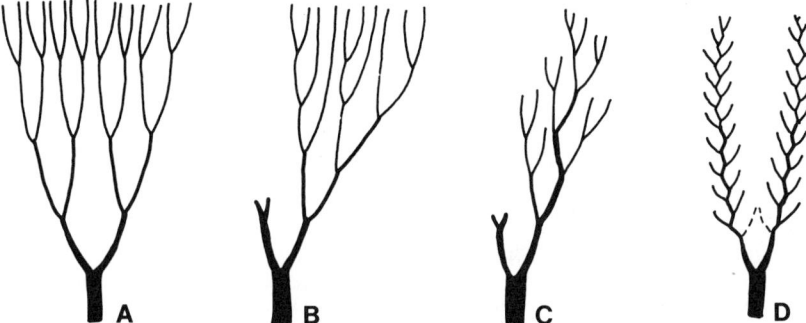

Fig. 20.7 The stages proposed by Bather (1900) in achieving 'holotomy' (pinnulation). (A) Isotomous. (B) Irregularly heterotomous. (C) Alternating heterotomous. (D) Holotomous (pinnulate). Although this evolutionary sequence may actually have occurred in different groups of cladids and disparids, the intermediate forms probably were unncesssary when pinnulation is viewed in terms of rapid evolution.

Recently, Frest *et al.* (1979) have reexamined the evolutionary origin of pinnulation among the disparids in terms of the earlier proposal by Bather (1890, 1900). The basic series of modifications (Fig. 20.7) from the primitive, dichotomously branched arm are (1) an irregular pattern of alternating heterotomy leading to (2) completely regular alternating heterotomy with a single unbranched, small subordinate division (i.e. a pinnule) on each brachial of the principal arm. Frest *et al.* proposed such a trend for the evolution of pinnules in the disparid superfamilies Cincinnaticrinacea and Homocrinacea; it is reasonable to suggest that the same occurred in the Calceocrinacea. Although it is not entirely certain, the origin of pinnules among the cladids may have been similar. That group had a considerably greater 'prepinnulate' history during which a large number of branching patterns appeared in different groups.

Ubaghs (1978) suggested that the sudden appearance of pinnules in camerate crinoids possibly represented a different mechanism for pinnule origins than may have characterized the disparids and cladids. As is the case for the origin of arm branching types, a long prepinnulate history with numerous intermediates (*sensu* Bather 1890, 1900) is unnecessary. Rather, a rapid evolutionary change (mutation of regulatory gene(s) controlling branching) could easily have resulted in an immediate pinnulate descendant from an apinnulate ancestor.

ACKNOWLEDGEMENTS

I thank C. R. Clement, C. R. C. Paul, and an anonymous reviewer for offering suggestions which have materially improved this paper, T. C. Labotka for computer assistance, and the University of Tennessee College of Liberal Arts Faculty Development Program, and the Department of Geological Sciences Discretionary Fund, which have permitted my participation in the conference.

REFERENCES

Bather, F. A. 1890. British fossil crinoids. II. The classification of the Inadunata Fistulata. *Annals and Magazine of Natural History*, 6th Series **5**, 373–88.
Bather, F. A. 1900. The Echinodermata. The Pelmatozoa. In *A treatise on zoology* Vol. 3 (ed. E. R. Lankester). pp. 38–204. Black, London.
Broadhead, T. W. 1981. Carboniferous camerate crinoid subfamily Dichocrininae. *Palaeontographica*, Abt. A, **176**, 81–157.
Broadhead, T. W. 1987. Heterochrony and the achievement of the multibrachiate grade in camerate crinoids. *Paleobiology* **13**, 177–86.
Broadhead, T. W. and Strimple, H. L. 1979. *Hyrtanecrinus*, a new Carboniferous camerate crinoid genus from eastern North America. *Journal of Paleontology* **54**, 35–44.
Brower, J. C. 1976. *Promelocrinus* from the Wenlock at Dudley, *Palaeontology* **19**, 651–80.
Donovan, S. K. 1984. *Ramseyocrinus* and *Ristnacrinus* from the Ordovician of Britain. *Palaeontology* **27**, 623–34.
Frest, T. J., Strimple, H. L., and McGinnis, M. R. 1979. Two new crinoids from the Ordovician of Virginia and Oklahoma, with notes on pinnulation in the Disparida. *Journal of Paleontology* **53**, 399–415.
Hyman, L. H. 1955. *The invertebrates: Echinodermata*. McGraw-Hill, New York.

Kelly, S. M. 1982. Origin of the crinoid orders Disparida and Cladida: possible inadunate cup plate homologies. *Proceedings of the Third North American Paleontological Convention* **1**, 285–90.

Kolata, D. 1982. Camerates. In Echinoderm faunas from the Bromide Formation (Middle Ordovician) of Oklahoma *The University of Kansas. Paleontological Contributions, Monograph* **1**, 170–205.

Lane, N. G. 1963. Meristic variation in the dorsal cup of monobathrid camerate crinoids. *Journal of Paleontology* **37**, 917–30.

Lane, N. G. and Breimer, A. 1974. Arm types and feeding habits of Paleozoic crinoids. *Proceedings. Koninklijke Nederlandse Akademie van Wetenschappen* B, **77**, 32–9.

Lane, N. G. and Macurda, D. B., Jr. 1975. New evidence for muscular articulations in Paleozoic crinoids. *Paleobiology* **1**, 59–62.

Lane, N. G. and Sevastopulo, G. D. 1982. Microcrinoids from the Middle Pennsylvanian of Indiana. *Journal of Paleontology* **56**, 103–15.

Mortensen, T. 1920. Notes on some Scandinavian echinoderms with descriptions of two new ophiuroids. *Videnskabelige Meddelelser fra Dansk Naturhistorisk Forening i Kjøbenhavn* **72**, 45–79.

Paul, C. R. C. and Smith, A. B. 1984. The early radiation and phylogeny of echinoderms. *Biological Reviews* **59**, 443–81.

Philip, G. M. and Strimple, H. L. 1971. An interpretation of the crinoid *Aethocrinus moorei* Ubaghs. *Journal of Paleontology* **48**, 491–3.

Rozhnov, S. V. 1985. Morfologiya, simmetriya i sistematicheskoye polozheniye morskikh liliy gibokrinid. *Paleontologicheskii Zhurnal* **1985**, 4–16. (In Russian.)

Smith, A. B. 1984. Classification of the Echinodermata. *Palaeontology* **27**, 431–60.

Sprinkle, J. 1973. Morphology and evolution of blastozoan echinoderms. *Special Publication. The Museum of Comparative Zoology, Harvard University*, Cambridge, Mass.

Sprinkle, J. 1982. Large calyx cladid inadunates. In Echinoderm Faunas from the Bromide Formation (Middle Ordovician) of Oklahoma *The University of Kansas Paleontological Contributions, Monograph* **1**, 145–69.

Sprinkle, J. and Moore, R. C. 1978. Hybocrinida, in *Treatise on invertebrate paleontology, part T. Echinodermata 2.* (ed. R. C. Moore and C. Teichert), pp. T564–74. Geological Society of America and University of Kansas Press. Lawrence, Kansas.

Ubaghs, G. 1978. Skeletal morphology of fossil crinoids. In *Treatise on invertebrate paleontology, part T. Echinodermata 2*, (ed. R. C. Moore and C. Teichert), pp. T58–216. Geological Society of America and University of Kansas Press. Lawrence, Kansas.

21

The phylogeny of post-Palaeozoic crinoids

M. J. SIMMS

Department of Earth Sciences, Liverpool University,
L69 3BX, UK

INTRODUCTION

Miller (1821) was the first to recognize that most post-Palaeozoic crinoids differ significantly in their morphology from their predecessors in the Palaeozoic, and so he assigned them to a distinct group, the Crinoidea Articulata. The most recent classification scheme, now generally adopted, is that of Rasmussen (1978) in which he divided the Articulata into seven orders; the Millericrinida, Cyrtocrinida, Bourgueticrinida, Isocrinida, Comatulida, Uintacrinida, and Roveacrinida. This differed from some previous schemes (Jaekel 1918; Matsumoto 1929) in that he excluded the Encrinidae, instead classifying them as poteriocrinine inadunates.

Previous opinions on the phylogeny of post-Palaeozoic crinoids, or selected groups within the Articulata, have also varied (Hagdorn 1983; Klikushin 1982; Moore, in Rhodes 1967; Pisera and Dzik 1979; Rasmussen 1978; Roux 1978, 1981; Ubaghs 1978; Taylor 1983). However, of these only Taylor (1983) has used cladistic analysis whilst none of the others appears to have been based upon any formalized analysis of character distribution. My interpretation of articulate relationships is based primarily upon a cladistic analysis (Fig. 21.1), as formulated by Hennig (1966), and further discussed by Wiley (1981) and Ridley (1986). I have taken account of stratigraphic information (Fig. 21.2) as well since, if the fossil record is reasonably representative of the history of the group, it should provide additional evidence in support of the cladogram.

Morphological terms used in this account are defined in Moore and Teichert (1978), and Donovan (1984). Brief definitions of new taxa are given in Appendix 2.

Echinoderm phylogeny and evolutionary biology (ed. C. R. C. Paul and A. B. Smith). Clarendon Press, Oxford, 1988.

MORPHOLOGY, STRATIGRAPHIC DISTRIBUTION, AND INFERRED RELATIONSHIPS OF POST-PALAEOZOIC CRINOIDS (FIGS 21.1 AND 21.2)

Definition of the Articulata

Although the subclass Articulata was first introduced by Miller as long ago as 1821 the concise and unambiguous definition of the group has always been problematical. Articulates have usually been recognized through the possession of one or more of a suite of characters, but I have found it impossible to identify any single character which is both unique to the group as well as being readily recognizable in fossil material. Characters used in previous diagnoses have included the following. 1. Cup simple dicyclic or cryptodicyclic. 2. Anal plate absent in adult. 3. Flexible tegmen with central peristome and exposed food grooves. 4. Axial nerves enclosed in canals which penetrate the basals, radials, and brachials. 5. Articulation between radial and arm, and between some or all brachials, muscular. 6. Arms uniserial. 7. Arms pinnulate.

Clearly, not all of these characters are diagnostic of the articulates alone. Cups with a dicyclic structure are encountered in the camerates, flexibles, and inadunates also, as are uniserial arms. Furthermore, if the Encrinidae are included within the Articulata on other grounds, as I have done here, then the possession of uniserial arms is no longer diagnostic of the articulates. Pinnulate arms are encountered in certain camerates and inadunates as well as in the articulates, although the possession of muscular articulations in the arms is confined to the articulates and certain of the late Palaeozoic cladid inadunates. The anal plate is absent from the cup in all adult articulates but this is also true of a number of cladids and at

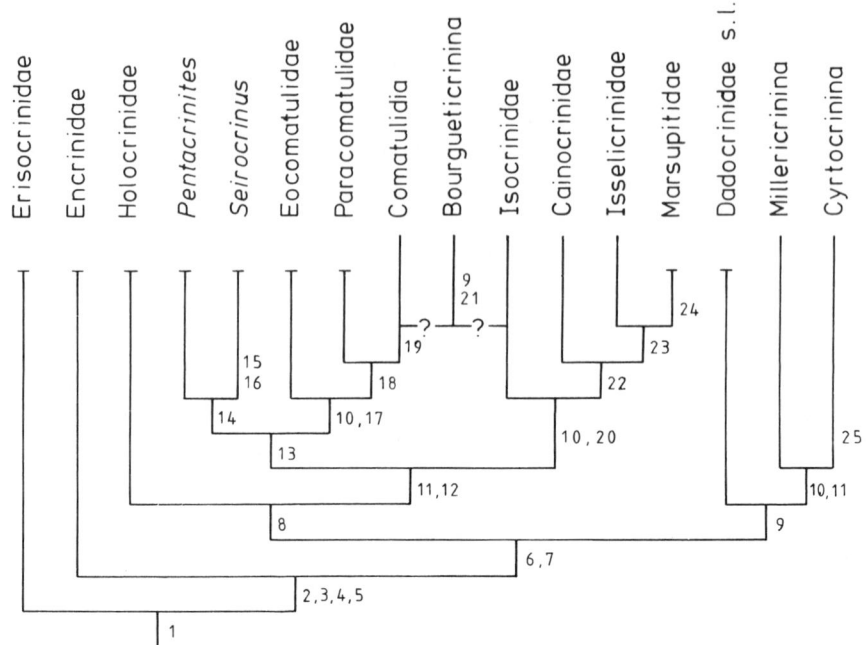

Fig. 21.1 Key to characters used in cladogram.
1. Muscular articulations on proximal brachials.
2. Axial nerves in cup and arms enclosed in canal.
3. Ambulacral grooves on tegmen open.
4. Anal plate absent in adult.
5. Development of 'synarthrial-type' cirri.
6. Uniserial arms in adult.
7. Synarthry at IBr1-2 and IIBr1-2.
8. Synarthrial stem articulations in larvae.
9. Loss of cirri.
10. Reduction of tegmen height.
11. Cryptic infrabasals.
12. Abandonment of holdfast in adult.
13. Endotomous arm branching.
14. Suppression of syzygies.
15. Proximal pinnules fixed in tegmen.
16. Separation of radials by basals.
17. Loss of internodals.
18. Reversion to simple arm branching.
19. Replacement of stem by centrodorsal (fusion of nodals).
20. Synostosial articulations beneath nodals.
21. Synarthrial stem articulations in adult.
22. Syzygy at IBr1-2.
23. Syzygy at II Br1-2.
24. Loss of stem.
25. Fusion of basals into ring.

least one flexible (Lane 1979). Enclosure of the axial nerves within canals penetrating the basals, radials, and brachials is also found in some inadunates as well as rarely in a few camerates and flexibles. The possession of a flexible tegmen with a centrally positioned peristome and exposed food grooves appears to be a character unique to the articulates, although these structures have such a low preservation potential that it is rarely possible to demonstrate this in fossil material.

The conclusion from this is that no single character can be used to define the articulates, with the possible exception of the tegminal structure. For fossil material, the three most diagnostic characters are the absence of an anal plate in the cup, the enclosure of the axial nerves in a canal penetrating the basals, radials, and brachials, and pinnulate arms. The presence of one or other of these in a particular crinoid is not necessarily diagnostic of the group, but a combination of the three would appear to be unique to the Articulata.

Erisocrinidae

The Erisocrinidae is a mid-Carboniferous to Permian group of poteriocrinine cladid inadunates of comparatively advanced evolutionary type, considered by some (Jaekel 1892; Hildebrand 1926) to be closely related to the Encrinidae. They are considered here for the purposes of outgroup comparison, but are not necessarily

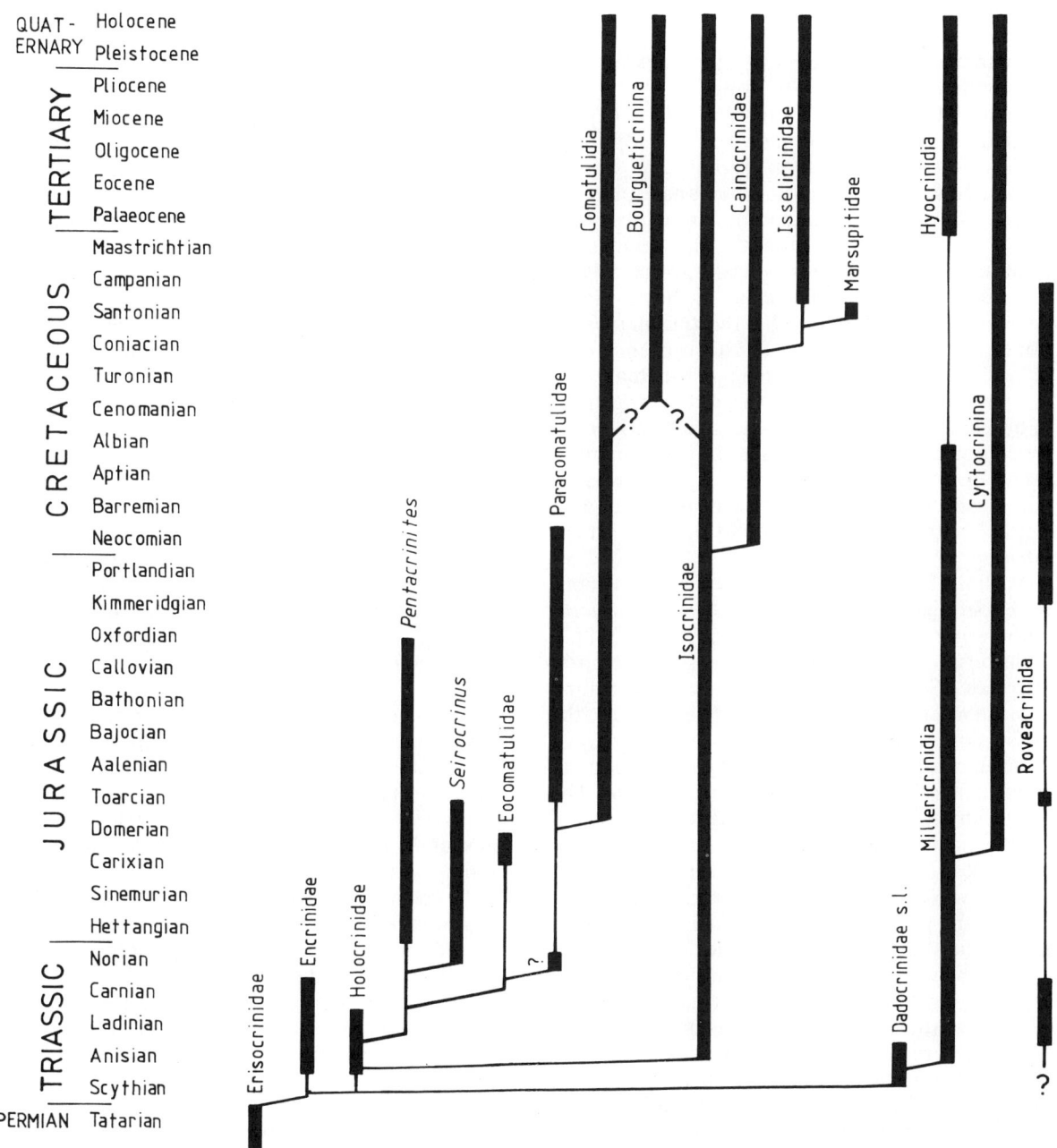

Fig. 21.2 Stratigraphic distribution and inferred phylogeny of post-Palaeozoic crinoids.

regarded as lying on the direct line of ancestry of the articulates since their precise phylogenetic position is unclear.

Erisocrinids have a circular stem lacking cirri and with columnals articulating by symplexies with radially arranged crenulae. The cup is moderately large and is dicyclic, with approximately equal-sized basals and radials, and smaller subhorizontal infrabasals. The anal plate is reduced to a vestigial remnant in a notch between the C and D radial articula. The tegmen of poteriocrinine inadunates typically forms a large ventral sac. The radial articula have highly developed muscular fossae, but the axial nerves are not enclosed within canals. The arms branch only once, have a biserial arrangement of brachials, and are pinnulate. Although the biserial arrangement of brachials in the Erisocrinidae is fairly advanced by comparison with the uniserial arrangement in many earlier Poteriocrinina, it is plesiomorphic with respect to the uniserial arrangement, with muscular articula distally, found in most articulates.

Willink (1979) has described two very advanced poteriocrinine inadunates, *Nowracrinus ornatus* (Etheridge) and *Tasmanocrinus mariensis* Willink, of uncertain systematic position, from the early Permian (Sakmarian) of south-east Tasmania. *N. ornatus* resembles many articulates in the possession of uniserial arms with well-developed muscular articulations, pierced by a pair of axial canals, alternating with syzygial articulations. The paired structure of the axial canals is unlike that of most articulates, where it is only a single canal, but resembles the arrangement found in the Encrinidae. However, this species differs from all articulates in the possession of an anal plate in the cup and in the structure of the pinnules which branch isotomously several times. The unique specimen of *T. mariensis* bears many similarities to certain Mesozoic Isocrinida in the possession of uniserial, pinnulate arms with oblique muscular articulations and isotomous branching on IBr2, reduced basals and infrabasals, and a pentastellate stem. However, a large anal plate is present in the radial circlet whilst the axial nerves are not enclosed in canals. These features demonstrate that, despite a number of similarities, neither of these species can be assigned to the Articulata and that they may be further removed from the ancestry of that group than are the comparatively primitive Erisocrinidae.

Encrinidae

The Encrinidae, although possessing biserial brachials, have several features not found in the Erisocrinidae, but characteristic of most articulates. In particular, no trace of the anal plate is seen in the cup which is perfectly symmetrical, whilst the basals, radials, and brachials are pierced by a pair of axial canals. The infrabasals are greatly reduced, by comparison with the Erisocrinidae, lying concealed between the basals and the top columnal, and the tegmen is also comparatively low. Although the arms only branch once and are predominantly biserial, the proximal few secundibrachs are typically uniserial. This demonstrates that, like the Encrinidae, all cladids with biserial arms in maturity almost certainly had uniserial arms as juveniles.

The stem is circular in the distal region, with simple radiating crenulae, and columnals have a fairly constant diameter and height. In the proximal region, however, columnals may have a pentagonal outline and the crenulae are arranged into five petaloid areas (Hagdorn 1982, Fig. 16) closely resembling the typical arrangement seen in isocrinids. Furthermore, pronounced alternation of columnals is found in this proximal region, with a hierarchical sequence of internodals intercalated between nodal columnals, as described for isocrinid stems (Donovan 1984). The nodals of most Encrinidae typically lack cirri but Hagdorn (1982) has described and figured specimens of *Chelocrinus schlotheimi* (Quenstedt) in which cirri are present on up to nine nodals below the cup. A very similar pattern of stem development is also found in the extant genus *Proisocrinus*. This has a very long stem which gradually tapers from the base towards the crown. The proximal region of the stem has pentaradiate columnals alternating in size and with cirri at intervals, the frequency of these cirrinodals decreasing rapidly away from the crown. However, in the median and distal part of the stem the columnals are cylindrical, do not alternate in size, and have a multiradiate arrangement of coarse marginal crenulae. Furthermore, cirri are entirely absent from this region of the stem. A similar suggestion of this trend towards the development of cirri in the proximal part of isocrinid stems only is also seen in the Jurassic genus *Pentacrinites* in which the cirri in the distal part of the stem are very much reduced in size and more widely spaced than in the proximal part of the stem (Simms 1986). This pattern of development, with cirri being developed only in younger columnals, is of considerable significance for the interpretation of the relationships of early articulates (see below).

The cirri of encrinids and other post-Palaeozoic crinoids articulate with the stem by a synarthrial-like articulation in which a transverse ridge on the cirral scar fits into a corresponding groove in the proximal articulum. This is quite different from the typical cirral articula of most Palaeozoic crinoids, in which there is a radial arrangement of crenulae surrounding the cirral lumen, but is identical to that found in isocrinids. This suggests that the cirri in articulate crinoids evolved independently of those in other crinoid groups and were not derived from a Palaeozoic ancestor. Brett (1981) has already suggested this on other criteria.

However, Moore and Jeffords (1968, p. 55, pl. 9, fig. 10) have described and figures the unique holotype of *Pentagonopterix insculptus*, from the Upper Carboniferous, which appears to have cirral scars identical to those of articulates. Even so, these cirral scars have pairs of fulcral ridges diverging laterally from the lumen, whereas articulate cirral scars have a single fulcral ridge. As the only specimen is a pluricolumnal, comprising a nodal and 14 attached internodals, it is impossible to ascertain the true significance of this particular specimen or even determine to which crinoid subclass it belongs.

The Encrinidae, however, although appearing to be the most plesiomorphic articulate group by retention of biserial arms, possess a number of apomorphies not found in the earliest articulates with uniserial arms. *Dadocrinus sensu lato* and the Holocrinidae differ from Encrinidae in the possession of dicyclic cups, with exposed infrabasals, and fairly high tegmina. This condition is not found even in juvenile Encrinidae (H. Hagdorn, personal communication) and so a direct derivation of articulates from this group, through paedomorphosis, seems improbable. Furthermore, in all articulates other than the Encrinidae the axial nerves are enclosed in a single rather than a paired canal. On the evidence available at present it seems best, therefore, to consider the Encrinidae as the sister group to all other articulates, sharing a common ancestor with them, but not on the direct line of descent. However, their cryptic infrabasals and reduced tegmen anticipate those of later articulates.

Uniserial arms

With the exception of the Encrinidae, discussed above, all articulates have a uniserial arrangement of brachials in the arms. This feature is encountered in juvenile Encrinidae and suggests that derivation of the arrangement in later articulates through neotenous retention of uniserial arms from an ancestor with biserial arms is as plausible as a direct descent from an ancestor with uniserial arms.

Relationship of Millericrinida and Isocrinida

The most important character by which the two main groups of uniserially-armed articulates, the Isocrinida and the Millericrinida, are distinguished is the presence of flexible cirri in the former and their absence in the latter. However, the relationship between the two groups is unclear, with a number of alternative hypotheses having been considered (Hildebrand 1926; Rasmussen 1978; Pisera and Dzik 1979; Taylor 1983). The synarthrial type of cirral articulation is shared by the Encrinidae and all Isocrinida, with the exception of the Bourgueticrinina and Marsupitidae which have secondarily lost them, and demonstrates the close relationship between these two groups. The Millericrinida entirely lack cirri yet the similarities between the cups of *Dadocrinus sensu lato* and the Holocrinidae, with exposed infrabasals, high tegmina and uniserial arms, suggest that they too are closely related. From this it can be inferred that the common ancestor of these two groups possessed cirri which were subsequently lost in the Millericrinida.

The stems of isocrinids and millericrinids are also characteristic and generally distinct from one another. Millericrinid columnals are typically circular with a simple radial arrangement of crenulae on articula whilst isocrinid columnals are frequently pentagonal or pentalobate with a pentaradiate arrangement of crenulae. These differences in stem morphology between the two groups invite comparison with the ontogenetic changes seen in the stem of encrinids, as discussed above. Some specimens of the encrinid *Chelocrinus* lack cirri entirely whilst others may have up to nine cirrinodals below the cup, although typically there are only two or three. By a simple change in the rate of development of the stem it would be possible to produce either the millericrinid or the isocrinid condition. A reduction in the rate of development of columnals as they move distally along the stem would result in the retention of cirrinodals and pentaradiate columnals for a much greater distance along the stem. In contrast, an accelerated rate of columnal development will lead ultimately to the suppression of cirri, and the development of circular columnals in both the proximal and distal parts of the stem.

Synarthrial stem articulations

A character shared by all Isocrinida so far examined is the presence of synarthrial columnal articulations in the pentacrinoid larval stage. Columnals with synarthrial articula are well documented in larval comatulids (Hyman 1955; Rasmussen 1978) and have also been found at comparable growth stages in early Jurassic Pentacrinitidae (Simms 1986) and Isocrinidae (Jäger 1985), and in recent Isocrinidae (Clark 1908a). They have not been seen in the Marsupitidae, for which the larval stages are unknown, nor have they been observed in the Holocrinidae, although this may well be attributable to collection failure since such columnals rarely exceed 0.5 mm and are easily overlooked. Millericrinid columnals as small as 0.4 mm diameter have been found in the early Jurassic, but show no trace of synarthrial articula. This character is therefore regarded here as a synapomorphy shared by all Isocrinida.

Holocrinidae

The Holocrinidae are the sister group of all other Isocrinida. They share several plesiomorphic characters with the Encrinidae, having the same type of cirral articulations, symplectial articulations throughout the stem and a fixo-sessile mode of life attached by a holdfast cemented to the substratum (Hagdorn 1983). They are apomorphic with respect to that group in possessing uniserial arms, and retaining pentaradiate columnals and cirrinodals throughout the stem, but are plesiomorphic with respect to them in the dicyclic arrangement of the cup and the height of the tegmen.

Pentacrinitidae

The Pentacrinitidae have high tegmina and symplexies throughout the stem, symplesiomorphies they share with the Holocrinidae. There are two genera, *Pentacrinites* and *Seirocrinus*, which share a number of apomorphic characters. They have a similar endotomous pattern of arm branching and syzygies are largely absent from the arms, although present at IIBr6-7 in *P. dichotomus* (M'Coy) and *P. dargniesi* Terquem and Jourdy. The muscular fossae of brachials are greatly reduced and the radials have a prominent aboral projection or spine. The cirri are characteristically lozenge-shaped in section and the stem, which shows indefinite intercalation of internodals, has numerous very small crenulae surrounding the long narrow areolae of symplectial articula. While some of these features may relate to a pseudopelagic mode of life (Simms 1986), there are too many other similarities for these all to be convergent.

Seirocrinus differs from *Pentacrinites* in having pinnules up to IVBr or VBr incorporated in the tegmen, in the separation of the radials by the basals to form a single ring in adults, and by the modification of the muscular articula in proximal brachials.

Eocomatulidae and Paracomatulidae

The Eocomatulidae fam. nov. are at present known only from a single monotypic genus, *Eocomatula* gen. nov., *Pentacrinus interbrachiatus* Blake is here designated the type. It shares the endotomous branching, long narrow symplectial areolae surrounded by numerous very small crenulae and lozenge shaped cirri found in *Pentacrinites*. It also resembles the Paracomatulidae (Hess 1951) in these latter two characters and in the presence of syzygies in the arms, the extremely short stem lacking internodals, the offsetting of cirral scars to either side of the radial midline and the inflation of the radial latera. The arrangement of offset cirral scars in eocomatulids and paracomatulids is very similar to that found in the proximal part of the stem in Pentacrinitidae whilst the inflation of the radial latera may perhaps represent a vestigial remnant of the aboral spine found in that group. The Paracomatulidae differ from the Eocomatulidae in having a simple ten-armed branching pattern and in the form of the stem. In paracomatulids this is very small and conical, like a segmented centrodorsal, with the last columnal rounded and the axial canal sealed. In eocomatulids the stem is larger, but, although very short, does not have a conical shape and the distal articulum retains the symplectial pattern, suggesting that abandonment of the stem occurred at a much later stage in ontogeny than was the case in paracomatulids. The Eocomatulidae appear to represent an almost perfect intermediate stage between the Pentacrinitidae and the Paracomatulidae.

Comatulidia

The Infraorder Comatulidia, as here defined, includes all Isocrinida which possess a cirriferous centrodorsal composed of a single ossicle and, with the exception of the Thiolliericrinidae, do not retain a stem beyond the pentacrinoid larval stage. It includes all the families formerly incorporated in the order Comatulida, as defined in the *Treatise* (Moore and Teichert 1978), with the exception of the Paracomatulidae which have a segmented centrodorsal. The Paracomatulidae, and their sister group the Eocomatulidae, are, however, united with the Comatulidia in the suborder Comatulidina since they represent morphological intermediates between the Pentacrinitidae and the Comatulidia, and their stems can effectively be regarded as segmented centrodorsals. The assignment of the Eocomatulidae to this group is somewhat arbitrary, however, since they might equally well be placed with the Pentacrinitidae on other characters.

The comatulids have undergone a very considerable diversification since the Mesozoic (Meyer and Macurda 1977), but the relationships of the different groups within this infraorder are uncertain. Their diversity is such that it is not possible to undertake any revision of their classification here.

Isocrinina

The suborder Isocrinina is here designated to include all Isocrinida formerly included within the family Isocrinidae as defined in the *Treatise* (Moore and Teichert 1978). They share the synapomorphy of synostosial articulations beneath the cirrinodals (Hagdorn 1983; Donovan 1984). Fourteen genera were recognized by Rasmussen (in Moore and Teichert 1978), but Klikushin (1982) has added more from the Triassic and early Cretaceous. The total is now probably more than twenty. Earlier attempts to subdivide this large group have largely failed to gain acceptance on account of the conflicting results obtained by using different characters. Most have been based either on the configuration or microstructure of columnal articulations (Rasmussen 1978; Roux 1970, 1974, 1977, 1981; Klikushin 1979, 1982) or on the type of articulation at IBr1-2 (Carpenter 1882, 1884; Rasmussen 1978; Oji 1985).

The use of columnal articulations to determine phylogenetic relationships above generic level seems questionable. Even within a species there may be considerable variation in this character between, or even within, individuals. Furthermore, morphological changes associated with heterochronous evolution may produce apparently quite distinct and morphologically dissimilar articulation patterns in two very closely related genera, as has been demonstrated for *Chladocrinus* and *Balanocrinus* (Simms 1985). By ignoring columnal articulations and considering only the proximal brachial articulations a much clearer pattern of relationships is evident within the Isocrinina. In all of those with a synarthry at IBr1-2 the articulation at IIBr1-2 is also a synarthry. Similarly, for most genera with cryptosyzygies at IBr1-2 there is also a cryptosyzygy at IIBr1-2. Both Carpenter (1879) and Clark (1908b) considered the articulations at IBr1-2 and IIBr1-2 to be homologous, as suggested by these two groups, but this is not supported by the situation in three other genera of Isocrinina, as well as in the comatulid family Zygometridae. In these forms the articulation at IBr1-2 is cryptosyzygial, but that at IIBr1-2 is synarthrial.

On the basis of the brachial articulations, therefore, the Isocrinina can be divided into three distinct groups. The first group retains the ancestral condition, with synarthries at both IBr1-2 and IIBr1-2, found in most Millericrinida, in the Pentacrinitidae, Holocrinidae, Bourgueticrinina and in most Comatulidina. It corresponds to the 'Old Group' of Oji (1985) and, since it includes the genus *Isocrinus*, constitutes the emended family Isocrinidae. All Triassic and Jurassic genera belong here, together with the extant genera *Hypalocrinus* and *Neocrinus*. The second group, here termed the Cainocrinidae fam. nov., have cryptosyzygies at IBr1-2, but retain synarthries at IIBr1-2. It comprises three genera; *Cainocrinus*, *Nielsenicrinus*, and *Teliocrinus*, the earliest of which, *Nielsenicrinus*, is found in the early Cretaceous (Neocomian). The third and final group, the Isselicrinidae (Klikushin 1979), comprises all Isocrinina in which both IBr1-2 and IIBr1-2 are cryptosyzgial. This includes the general *Annacrinus*, *Austinocrinus*, *Cenocrinus*, *Diplocrinus*, *Doreckicrinus*, *Endoxocrinus*, *Isselicrinus*, *Metacrinus*, and *Saracrinus*. The earliest representative is *Austinocrinus* from the late Cretaceous (Campanian).

Bourgueticrinina

The Bourgueticrinina comprise a group of small mid-Cretaceous to Recent crinoids which are characterized by a stem with synarthrial articulations at all growth stages, but without nodals or cirri. They have generally been classified with the Millericrinida on account of their lack of cirri and the presence, in *Bourgueticrinus*, of a supposed proximale (Gislén 1938). However, synarthrial articulations are unknown in the stem of millericrinids at any growth stage and although postulated in juvenile Apiocrinitidae (Gislén 1938) there is no evidence to support this.

Larval comatulids have long been known to have synarthrial columnals, these being retained in adult Thiolliericrinidae. Gislén (1924) noted the similarity between bourgueticrinids and the Thiolliericrinidae, and considered that early Cretaceous thiolliericrinids lay close to the ancestry of the bourgueticrinids. Thiolliericrinids typically differ from bourgueticrinids by the presence in the former group of a distinct centrodorsal which usually bears cirri. Cirri are absent in the early Cretaceous genus *Loriolocrinus*, however, and so this may represent an intermediate form between the two groups. The basals in bourgueticrinids are much larger than in thiolliericrinids, but this may easily be accounted for through heterochrony. A possible direct relationship between the two groups is further supported by their stratigraphic distribution, with bourgueticrinids appearing only a short time after the last known, and morphologically most similar, thiolliericrinid. Rasmussen (1978) also favoured the neotenous evolution of bourgueticrinids from a comatulid ancestor, but others (Pisera and Dzik 1979; Roux 1977, 1978) have opposed this idea and have retained them in the Millericrinida. I consider that the greater weight of evidence is in favour of their neotenous derivation from some group within the Isocrinida, either the Isocrinina or Comatulidia, and I classify them here accordingly as a suborder, the Bourgueticrinina, of the Isocrinida. However, more precise assignment within the Isocrinida is hampered by a lack of suitable characters by which to compare them with other groups. The main diagnostic characters of the Bourgueticrinina, in particular the synarthrial stem articulations, are juvenile features found in all Isocrinida and serve only to establish their affinity with this group.

Marsupites and Uintacrinus

The Marsupitidae, which is here taken to be equivalent to the Uintacrinida as defined in the *Treatise* (Moore and Teichert 1978), are a bizarre group of large stemless pelagic crinoids of worldwide distribution, but restricted to a narrow stratigraphic interval in the late Cretaceous (Santonian). There are only two genera, *Marsupites* and *Uintacrinus*, which do not resemble each other or any other articulate crinoids and consequently were placed in separate families within a distinct order, the Uintacrinida. Previous ideas concerning their origin have differed widely (Rasmussen 1978; Pisera and Dzik 1979).

Much of this confusion has arisen from the use of plesiomorphic characters in the cup or by merely comparing superficially similar characters in other groups. The stem, with its associated suite of characters, is unknown in this group and the structure of the cup has proved undiagnostic. Sufficient characters may be present in the arms, however. In both *Marsupites* and *Uintacrinus* the articulations at both IBr1-2 and IIBr1-2 are syzygial, strongly indicating that these two genera are, in fact, closely related. Furthermore, this is identical to the arrangement found in the Isselicrinidae and suggests that they are also closely related. I therefore regard the Marsupitidae as a highly specialized offshoot of the Isselicrinidae, although they are so highly modified that they do not appear to have retained any other diagnostic characters by which this hypothesis might be tested. However, the possibility that the syzygial articulations at IBr1-2 and IIBr1-2 evolved independently in this group and the Isselicrinidae cannot be discounted, particularly in the light of the convergence in this character shown by the Isselicrinidae and the comatulid families Zygometridae and Comosteridae. An alternative possibility is that the Marsupitidae arose from either of these comatulids, although they do not have any pre-Tertiary fossil record by which to judge this.

Only one other group of post-Palaeozoic crinoids are known to have adopted a pelagic lifestyle. These are the roveacrinids, a Triassic to Cretaceous group discussed in greater detail below. Morphologically, they resemble the marsupitids in having lost all trace of a stem, but otherwise they are quite dissimilar, being very much smaller and entirely lacking syzygial articulations, with those at IBr1-2 and IIBr1-2 being synarthrial or synostosial. Thus, although there is a stratigraphic overlap between the marsupitids and roveacrinids, and despite the similarity of lifestyle, derivation of the former from the latter seems unlikely.

Dadocrinidae *sensu lato*

The earliest representatives of the Millericrinida are found in the early Triassic (uppermost Scythian) (Hagdorn 1985) and have been assigned to the family Dadocrinidae, the type species of which is *Dadocrinus gracilis* (von Buch). This species resembles all subsequent millericrinids in possessing a cryptodicyclic cup, but is distinct from other species assigned to the genus *Dadocrinus*, such as *D. kunischi* Wachsmuth and Springer, which have exposed infrabasals (H. Hagdorn, personal communication). These dicyclic millericrinids should perhaps be assigned to a distinct family, whilst *D. gracilis* can be considered as the earliest representative of the family Millericrinidae. For the present, however, I have retained all species of *Dadocrinus* in the family Dadocrinidae *sensu lato*. All have comparatively high tegmina (Gislén 1924) and, with the exception of *D. gracilis*, a dicyclic cup, both characters shared with the Holocrinidae and the hypothetical common ancestor of millericrinids and isocrinids.

Millericrinina

The suborder Millericrinina differs from the Dadocrinidae *sensu lato* in the cryptodicyclic arrangement of the cup and the reduction in tegmen height. *Dadocrinus gracilis*, from the Middle Triassic (Anisian), may represent an intermediate form, but the group is otherwise rather poorly known before the Middle Jurassic. Furthermore, problems arise in their classification since this is based very largely on plesiomorphic characters. Two distinct families, the Millericrinidae and the Apiocrinitidae, are known from the Middle Jurassic to Lower Cretaceous. The latter group shows modifications in the proximal brachial articula and in the proximal part of the stem. Only one undescribed genus of Millericrinina is known from the early Jurassic (Simms, in press). Its precise systematic position is uncertain, although some brachial articula bear similarities to those in Apiocrinitidae. The history of Mesozoic Millericrinina is not well documented, although it is evident that there was a modest diversification in the early to mid Jurassic (Taylor 1983). Mesozoic Millericrinina can all be assigned to the infraorder Millericrinidia, but are very rare in the Cretaceous, although columnals from as late as the Cenomanian have been tentatively referred to this group (A. S. Gale, personal communication). However, a second infraorder, the Hyocrinidia, are still extant and have also been described on the basis of Palaeocene columnals. The stem is long and slender, with radiating crenulae on columnal articula, as in other millericrinids, and the crown also is relatively unspecialized. In some genera, however, the basals may be partly or completely fused, a characteristic of the Cyrtocrinina discussed below. The hyocrinids have been considered to be derived from them (Rasmussen 1978). Since other hyocrinids have distinct basals, I consider it more probable that the fusion of the basals in these two groups is a convergent character, and that they are more closely allied with the Millericrinidia despite the considerable gap in the fossil record between the last Millericrinidia and first Hyocrinidia. Whether the distinction between these two nominal groups can be retained remains to be resolved, however.

Cyrtocrinina

The Cyrtocrinina represent a distinct offshoot from the Millericrinina adapted for a hardground or reef-dwelling lifestyle. They are generally rather small and compact and the basals are fused into a continuous ring or are not distinguishable at all. The stem is quite short or may be entirely absent with the cup attached directly to the substratum. The earliest definite representatives are encountered in the early Jurassic (Hettangian), but Pisera and Dzik (1979) have suggested that a Middle Triassic species, *Calathocrinus digitatus* Meyer, might be ancestral to this group. They were a fairly diverse group during the Jurassic and Cretaceous (Pisera and Dzik 1979), and are still extant.

Roveacrinida

The Roveacrinida are a group of small, stemless, pelagic crinoids ranging in age from the Middle Triassic (Ladinian) to the Upper Cretaceous (Campanian). They possess few articulate apomorphies, other than possession of uniserial arms with muscular brachial articulations and absence of an anal plate, and so there have been many widely divergent interpretations of their origins.

At present none of these can be regarded as more than mere speculation.

They already form a distinct group at their earliest appearance in the mid-Triassic and their subsequent evolution is further obscured by their rather disjunct stratigraphical distribution, with virtually no record between the Middle Triassic and the Upper Jurassic. Such a spasmodic fossil record appears to be characteristic of pelagic and pseudopelagic crinoids, having been noted amongst the Pentacrinitidae and the Marsupitidae.

Their small size and slender morphology suggests derivation from some other articulate group through paedomorphosis, but gives no indication of the identity of that group. Furthermore, the possibility that the three families grouped together in this order, the Triassic Somphocrinidae, late Jurassic to late Cretaceous Saccocomidae, and Cretaceous Roveacrinidae, are less closely related to each other than to other articulate groups cannot be discounted.

THE EVOLUTION OF POST-PALEOZOIC CRINOIDS: A SCENARIO

Introduction

Cladistic analysis enables a tentative interpretation to be made of the relationship of members within a taxonomic group. This cladogram can be tested by considering stratigraphic information about the group and then, if the correlation between the two sources of information is sufficiently close, a scenario can be developed for the evolution of that group.

Scenario

The most recent common ancestor of the Articulata can be considered to have had a circular stem, lacking cirri, with simple radiating crenulae, and attached to the substratum by a cemented holdfast (Fig. 21.3). The cup was dicyclic, with infrabasals exposed on the side of the cup, there was a vestigial anal plate, a fairly high tegmen with closed ambulacral grooves and arms which were predominantly biserial, but had muscular articulations in the proximal few, uniserial, brachials. The axial nerves occupied a very deep groove in the adoral surface of radials and brachials. This ancestral form may have been closely related to the Erisocrinidae, but is at present unknown. It was presumably extant in the late Permian or early Triassic. Derivation of the first true articulate from this form was achieved by the complete loss of the anal plate, enclosure of the axial nerves in a canal, or pair of canals, penetrating the basals, radials and brachials, the development of open ambulacral grooves and the appearance, in the proximal region of the stem only, of a pentaradiate arrangement of crenulae and nodals bearing cirri with a synarthrial-like articulation. An offshoot from this 'proto-articulate' led to the Encrinidae in which the plates of the cup assumed a cryptodicyclic arrangement, by reduction in the size of the infrabasals, and the tegmen became lower and thinner. Both of these trends were later to be paralleled in other articulates.

The next evolutionary step was the neotenous retention in the adult of the uniserial brachial arrangement, found in the juveniles of the ancestor, and the development of muscular articulations throughout the arms. All subsequent Articulata can probably be derived from this hypothetical ancestor, which may have arisen in the early Triassic. A more precise understanding of the relationship of the Roveacrinida to other articulates is not possible at present and they will not be considered further.

From this ancestral uniserial form there arose two quite distinct lineages, the Millericrinida and the Isocrinida. Accelerated development of columnals in the former group led to the suppression of the cirri found on the proximal nodals of the ancestral form and the appearance of circular columnals, like those of the distal stem of the ancestor, throughout the stem. These early millericrinids, the Dadocrinidae *sensu lato*, are first known from the uppermost Lower Triassic (uppermost Scythian) (Hagdorn 1985). Reduction of the infrabasals, to produce a cryptodicyclic condition in the cup, led to the Millericrinidia, the earliest of which is *Dadocrinus gracilis* from the early Middle Triassic (Anisian). Subsequent evolution of the Millericrinidia led to a variety of distinct offshoots, such as the Apiocrinitidae and *Ailsacrinus* (Taylor 1983), in the mid and late Jurassic although the obvious limitations of their fixo-sessile mode of life (Fig. 21.3) may well account for the post-Jurassic decline of this group. There is a considerable gap in the fossil record between the Middle Triassic and Middle Jurassic forms with only a single genus known from the Lower Jurassic. This Lower Jurassic form may lie

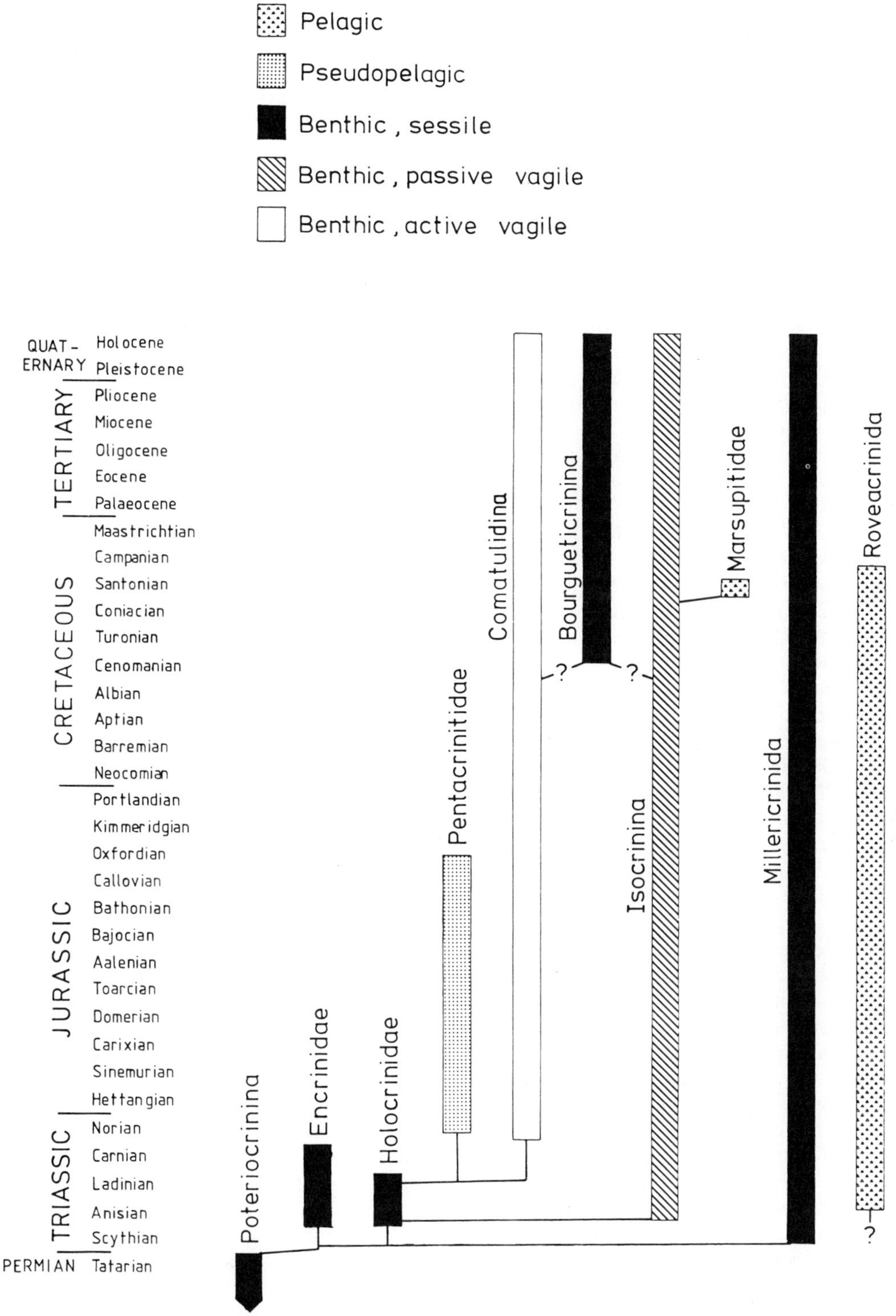

Fig. 21.3 Lifestyles and evolutionary pathways of post-Palaeozoic crinoids.

close to the common ancestry of the Millericrinidae and the Apiocrinitidae. The Millericrinidia appear to have undergone a decline in the late Jurassic and early Cretaceous and are last known from the mid-Cretaceous (Cenomanian). A second group, the Hyocrinidia, are known from the Palaeocene and Recent, and may have been derived from the Millericrinidia. One particular offshoot of the Millericrinidia became adapted for reef environments and are characterized by fusion of the basals into a continuous ring. This group, the Cyrtocrinina, are first known from the early Jurassic (Hettangian) and subsequently gave rise to a variety of small, hardground or reef-dwelling forms, some of which are still extant.

The Isocrinida diverged from the Millericrinida through the neotenous retention of cirri and pentaradiate columnals throughout the stem. Several species of Holocrinidae are known from the Middle Triassic and, like the contemporaneous Encrinidae and Dadocrinidae *sensu lato*, were still dependent upon hard substrata for attachment despite the development of cirri along the stem (Fig. 21.3). The Holocrinidae were probably ancestral to all subsequent Isocrinida which, from the Middle Triassic onwards, form two distinct lineages. These two lineages followed quite separate evolutionary pathways in moving away from the hard substrata upon which their ancestors had been dependent, using the cirri instead for attachment of the adult.

The larvae still required a hard substratum for initial settlement and the use of floating driftwood for this led to the adoption in the early history of the Pentacrinitidae of a pseudopelagic mode of life in the adults (Fig. 21.3). This group share a number of plesiomorphic characters with the Holocrinidae, but also developed several autapomorphies in association with their new lifestyle, such as endotomous branching and the suppression of syzygies (Simms 1986). Of the two genera, *Pentacrinites* and *Seirocrinus*, the latter genus shows even greater specialization for a pseudopelagic mode of life than does *Pentacrinites*. The morphology of *Seirocrinus* suggests it arose from *Pentacrinites* through peramorphosis. The earliest examples of *Seirocrinus* are known from the late Triassic (late Norian) (Klikushin 1982) but *Pentacrinites* is not known for certain before the early Jurassic (Hettangian). This apparent discrepancy between the known ranges of *Seirocrinus* and its presumed ancestor can almost certainly be attributed to the inadequacy of the fossil record, since even over the known range of this group in the late Triassic and Jurassic they show an extremely disjunct stratigraphic distribution.

The Pentacrinitidae survived only until the late Jurassic (Oxfordian), but an offshoot from them reverted to a benthic existence, adopting an active vagile mode of life (Fig. 21.3). This group, the Comatulidina, progressively lost the specializations found in the pentacrinitids and the stem underwent a rapid reduction in length. In the most primitive member of this lineage, the Eocomatulidae, the arms still retained the endotomous pattern although it was far less well developed than in the pentacrinitids. Syzygies reappeared in the arms and the tegmen became greatly reduced in size, whilst the stem, other than the few most proximal nodal columnals, was discarded in the adult. The Paracomatulidae represent an advance upon this, with a reversion to a simple arm-branching pattern and the loss of all but the proximal few nodals at a much earlier stage in ontogeny. Fusion of the columnals in this 'segmented centrodorsal' led to the Comatulidia, with a true centrodorsal. From the Middle Jurassic onwards the Comatulidia underwent a major diversification (Meyer and Macurda 1977) and are now the most diverse and widespread of extant crinoid groups.

The fossil record of the Comatulidina, particularly in the early part of their history, is rather poor. Suggestions of a comatulid trend are first encountered in *Pentacrinites dichotomus*, from the early Jurassic (late Carixian–Toarcian), which shows a reduction in stem length compared with other pentacrinitids and also redeveloped syzygies at IIBr6-7. The Eocomatulidae are known only from a single species from the early Jurassic (Carixian–Domerian) whilst the earliest true comatulid is also from the early Jurassic (lowermost Toarcian), Loriol 1884–89). Well authenticated paracomatulids are first recorded from the Toarcian but a rather poorly preserved specimen recently found in the late Triassic has also been tentatively interpreted as a paracomatulid (H. Hagdorn, personal communication). If this is actually the case, it is considerably earlier than the only known eocomatulid and also *P. dichotomus*, and causes a major discrepancy between the inferred relationships and the known fossil record. This may be attributable to collection failure or lack of recognition of fragmentary material since, even in the Jurassic, the known distribution of these groups is based almost entirely on intact material, the presence, or collection, of which at any particular horizon is largely fortuitous.

The major evolutionary innovation in the second of these isocrinid lineages was the development of synostosial autotomy planes beneath the cirrinodals. These enabled stalked benthic isocrinids to abandon the fixed mode of attachment found in their ancestors and move onto softer substrata where they used the cirri for anchorage, adopting a passive vagile lifestyle (Fig. 21.3). The position of the autotomy planes immediately beneath the cirrinodals ensured that the cirri were always in the optimum position at the distal extremity of the stem (Hagdorn 1983). Previously (Simms 1986), I assumed that the evolution of stem synostoses preceded the diversification of the Isocrinida, and that their absence in the Pentacrinitidae was a result of suppression of this character in association with a pseudopelagic mode of life. A more parsimonious assumption, however, is that the pentacrinitids diverged from the main isocrinid line prior to the development of this character.

The earliest representatives of the suborder Isocrinina are found in the early Middle Triassic (early Anisian), only slightly later than the first known Holocrinidae, but they are very poorly known (Hagdorn 1985). All pre-Cretaceous Isocrinina can be assigned to the family Isocrinidae which retain the primitive condition of synarthrial articulations at IBr1-2 and IIBr1-2. The Isocrinidae experienced a significant diversification in the mid to late Triassic although this has yet to be systematically documented. Only one species is found in the earliest Jurassic (Hettangian), however, suggesting that there was a major end-Triassic extinction of benthic isocrinids, comparable with that documented for bivalves by Hallam (1981), although the pseudopelagic Pentacrinitidae do not appear to have been affected. There was a further major phase of diversification in the early Jurassic which produced most of the later Jurassic and early Cretaceous lineages. Replacement of the synarthry at IBr1-2 by a cryptosyzygy gave rise to the Cainocrinidae, first known from the early Cretaceous (Neocomian). A similar transformation at IIBr1-2 then produced the Isselicrinidae which are first known from the late Cretaceous (Campanian). The replacement of the synarthries by cryptosyzygies in these two groups, which are now the dominant stalked isocrinids, has been attributed to the diversification of predatory teleost fish at about this time (Oji 1985).

The Marsupitidae may represent an offshoot of the Isselicrinidae since the two groups share syzygial articulations at IBr1-2 and IIBr1-2. They are highly specialized for a pelagic mode of life (Fig. 21.3) and it is unclear how their morphology might be derived from that of the Isselicrinidae. They have an extremely limited stratigraphical distribution in the late Cretaceous, being restricted to the upper part of the Santonian, but have a cosmopolitan distribution at this time. Although this is slightly earlier than the first known isselicrinid, from the overlying Campanian, the poor fossil record of Cretaceous Isocrinina may have prevented the recognition of any earlier representatives of the Isselicrinidae.

The suborder Bourgueticrinina are first encountered in the late Cretaceous (Turonian). They clearly arose from one of the contemporaneous isocrinid groups, probably either the Comatulidia or the Isocrinidae, although their extreme neotenous morphology masks their precise relationship to these groups. The earliest bourgueticrinids were apparently reef-dwellers (Klikushin 1975) and so they may have arisen initially through paedomorphosis as an adaptation to a turbulent environment, subsequently to diversify onto both hard and soft substrata. Their reversion to a fixo-sessile mode of life is almost certainly due to neotenous retention of the larval form of attachment.

SUMMARY

It is here considered that the crinoid subclass Articulata arose from an encrinid-like ancestor at some point in the late Permian or early Triassic. Considerable diversification along at least two distinct evolutionary pathways (Fig. 21.3) in the mid- to late Triassic was abruptly curtailed by an end-Triassic extinction which was survived by only a few representatives from each major group. A second major radiation took place in the early Jurassic, but thereafter the appearance of other new groups has been only spasmodic and largely restricted to the Cretaceous. Throughout their history heterochrony has been the dominant mechanism in the evolution of articulates at all taxonomic levels. Of the two main lineages, the Millericrinida have remained the more conservative and have not diversified to the extent seen in the Isocrinida. The Isocrinida are the dominant group today and, of these, the Comatulidia are by far the most successful and diverse.

The reinterpretation of the phylogeny of post-Palaeozoic crinoids, as outlined above, has necessitated a revision of the classification of articulate crinoids (see appendix). The classification scheme outlined by Rasmussen (in Moore and Teichert 1978) does not reflect phylogenetic relationships within the Articulata and so a system based upon cladistics has been employed for this new classification.

AKNOWLEDGEMENTS

I thank Dr. A. B. Smith and Professor A. Hallam for their supervision of this work, which was undertaken during the tenure of a N.E.R.C. CASE Studentship at Birmingham University and the British Museum (Natural History).

REFERENCES

Arendt, Y. A. 1974. Morskie lilii.Tsirtokrinidy [Sea lilies. Cyrtocrinids]. *Trudÿ Paleontologicheskogo Instituta. Akademia Nauk SSSR* **144**, 1–251.

Brett, C. E. 1981. Terminology and functional morphology of attachment structures in pelmatozoan echinoderms. *Lethaia* **14**, 343–70.

Carpenter, P. H. 1879. On the genus *Actinometra*. *Transactions of the Linnean Society (Zoology)* Series 2, **2**, 1–122.

Carpenter, P. H. 1882. Report on the results of dredging under the supervision of Alexander Agassiz, in the Gulf of Mexico (1877–78) and the Caribbean Sea (1878–79) by the U.S. Coast Survey steamer 'Blake', Lieut.-Commander C. D. Sigsbee, U.S.N., and Commander J. R. Bartlett, U.S.N., commanding. XVIII: The stalked crinoids of the Caribbean Sea. *Bulletin of the Museum of Comparative Zoology, Harvard* **10**, 165–81.

Carpenter, P. H. 1884. Report upon the Crinoidea collected during the Voyage of H.M.S. Challenger during the years 1873–76. Part I: General morphology, with descriptions of the stalked crinoids. *Reports of Scientific Results of the Exploration Voyage of H.M.S. Challenger, Zoology* **11**, 1–442.

Clark, A. H. 1908a. The axial canals of the recent Pentacrinitidae. *Proceedings of the United States National Museum* **35**, 87–91.

Clark, A. H. 1908b. The homologies of the arm joints and arm divisions in the recent crinoids of the families Comatulidae and the Pentacrinitidae. *Proceedings of the United States National Museum* **35**, 113–31.

Clark, A. H. 1908c. New genera of unstalked crinoids. *Proceedings of the Biological Society of Washington* **21**, 125–36.

Donovan, S. K. 1984. Stem morphology of the Recent crinoid *Chladocrinus (Neocrinus) decorus*. *Palaeontology* **27**, 825–41.

Dujardin, F. and Hupé, L.-H. 1862. Histoire Naturelle des Zoophytes échinodermes. *Librarie Encyclopédique de Roret*, 8 vol, 628 pp., 10 pl. Roret Paris.

Forbes, E. 1852. Monograph of the Echinodermata of the British Tertiaries. *Palaeontographical Society Monographs* 36 pp., 4 pl.

Gislén, T. 1924. Echinoderm Studies. *Zoologiska Bidrag från Uppsala* **9**, 330 pp.

Gislén, T. 1938. A revision of the recent Bathycrinidae. *Acta Universitatis Lundensis* n. ser. Avd. 2, **34**, 1–30.

Hagdorn, H. 1982. *Chelocrinus schlotheimi* (Quenstedt 1835) aus dem Oberen Muschelkalk (mol. Anisium) von Nordwestdeutschland. *Veröffentlichungen aus dem Naturkunde Museum Bielefeld* **4**, 1–34.

Hagdorn, H. 1983. *Holocrinus doreckae* n. sp. aus dem Oberen Muschelkalk und die Entwicklung von Sollbruchstellen im Stiel der Isocrinida. *Neues Jahrbuch für Geologie und Paläontologie, Monatshefte* **6**, 345–68.

Hagdorn, H. 1985. Immigrations of crinoids into the German Muschelkalk basin. In *Sedimentary and evolutionary cycles* (ed. U. Bayer and A. Seilacher), pp. 237–54. Springer-Verlag Berlin.

Hallam, A. 1981. The end-Triassic bivalve extinction event. *Palaeogeography, Palaeoclimatology and Palaeoecology* **35**, 1–44.

Hennig, W. 1966. *Phylogenetic systematics*. Illinois University Press, Urbana, Illinois.

Hess, H. 1951. Ein neuer Crinoide aus dem mittleren Dogger der Nordschweiz (*Paracomatula helvetica* n. gen. n. sp.). *Eclogae Geologicae Helvetiae* **43** (1950), 208–16.

Hildebrand, E. 1926. *Moenocrinus deecki*, eine neue Crinoidengattung aus dem fränkischen Wellenkalk und ihre systematische Stellung. *Neues Jahrbuch für Mineralogie, Geologie und Paläontologie, Beilagebände* **54**, 259–88.

Hyman, L. H. 1955. *The invertebrates*. Vol. 4: Echinodermata. McGraw-Hill, New York.

Jaekel, O. 1892. Uber Plicatocriniden. *Hyocrinus* und *Saccocoma*. *Zeitschrift der Deutschen Geologischen Gesellschaft* **44**, 619–96.

Jaekel, O. 1918. Phylogénie und System der Pelmatozöen. *Paläontologische Zeitschrift* **3**, 1–128.

Jäger, M. 1985. Die Crinoiden aus dem Pliensbachium (mittlerer Lias) von Rottorf am Klei unde Empelde (Süd-Niedersachsen). *Berichte der Naturhistorische Gesellschaft Hannover* **128**, 71–151.

Klikushin, V. G. 1975. Mechanics of the column in the Bourgueticrinidae. *Paleontologicheskiy Zhurnal* **9**, 121–4.

Klikushin, V. G. 1979. Microstructural features of

isocrinid stems. *Paleontologicheskiy Zhurnal* **13**, 88–96.

Klikushin, V. G. 1982. Taxonomic survey of fossil isocrinids with a list of the species found in the USSR. *Geobios* **15**, 299–325.

Lane, N. G. 1979. Upper Permian crinoids from Djebal Tebaga, Tunisia. *Journal of Paleontology* **53**, 121–32.

Loriol, P., de 1884–9. *Paléontologie francaise, ou description des fossiles de la France*, Sr. 1. *Animaux invertebrés. Terrain jurassique*: 11, *Crinoides*, pt 2, (ed. G. Masson), 580 pp. G. Masson, Paris.

Lowenstam, H. A. 1942. Mid-Triassic crinoid *Dadocrinus*. *Bulletin of the Geological Society of America, Series B*, **53**, 1832.

Matsumoto, H. 1929. Outline of a classification of Echinodermata. *Science Reports. Tohoku University* Series 2, **13**, 27–33.

Meyer, D. L. and Macurda, D. B. 1977. Adaptive radiation of the comatulid crinoids. *Paleobiology* **3**, 74–82.

Miller, J. S. 1821. *A natural history of the Crinoidea or lily-shaped animals, with observations on the genera Asteria, Euryale, Comatula and Marsupites*. Bryan, Bristol, 150 pp.

Moore, R. C. and Jeffords, R. 1968. Classification and nomenclature of fossil crinoids based on studies of dissociated parts of their columns. *The University of Kansas, Paleontological Contributions, Article* **46**, 1–86.

Moore, R. C. and Teichert, C. (ed.) 1978. *Treatise on invertebrate paleontology. part T. Echinodermata 2*, 3 vols. Geological Society of America and University of Kansas Press. Lawrence, Kansas.

Oji, T. 1985. Early Cretaceous *Isocrinus* from Northeast Japan. *Palaeontology* **28**, 629–42.

Orbigny, A. D., d'. 1852. *Prodrome du paléontologie stratigraphique universelle des animaux mollusques et rayonnés faisant suite au cours élémentaire de paléontologie et de géologie stratigraphique* **3**. Masson, Paris.

Pisera, A. and Dzik, J. 1979. Tithonian crinoids from Rogoznik (Pieniny Klippen Belt, Poland) and their evolutionary relationships. *Eclogae Geologicae Helvetiae* **72**, 805–49.

Rasmussen, H. W. 1978. Evolution of articulate crinoids. In *Treatise on invertebrate paleontology. part T. Echinodermata 2* (ed. R. C. Moore and C. Teichert), T302–16. Geological Society of America and University of Kansas Press, Lawrence, Kansas.

Rhodes, F. H. T. 1967. Permo-Triassic extinction. In *The fossil record*, Special Publication of the Geological Society of London 57–76.

Ridley, M. 1986. *Evolution and classification. The reformation of cladism*. Longman, London.

Roux, M. 1970. Introduction a l'étude des microstructures des tiges de crinoïdes. *Geobios* **3**, 79–98.

Roux, M. 1974. Observations au microscope électronique à balayage de quelque articulations entre les ossicules du squelette des Crinoïdes pédonculés actuels (Bathycrinidae et Isocrinidae). *Travaux du Laboratoire de Paléontologie, University de Paris, Faculté des Sciences, d'Orsay*. 9 pp.

Roux, M. 1977. The Stalk-joints of Recent Isocrinidae (Crinoidea). *Bulletin of the British Museum, Natural History (Zoology)* **32**, 45–64.

Roux, M. 1978. Ontogenèse, variabilité et évolution morphofonctionelle du pédoncule et du calice chez les Millericrinda (Echinodermes, Crinoïdes). *Geobios* **11**, 213–41.

Roux, M. 1981. Echinodermes: Crinoïdes Isocrinidae. *Memoires ORSTOM* **91**, 477–543.

Sieverts-Doreck, H. 1952. In *Invertebrate fossils* (ed. R. C. Moore, C. G. Lalicker, and A. G. Fischer). McGraw-Hill, New York, 766 pp.

Sieverts-Doreck, H. 1953. In Ubaghs, G. Classe des Crinoïdes. (ed. J. Piveteau) *Traité de paléontologie* **3**, 658–773. Figs 1–166. Masson, Paris.

Simms, M. J. 1985. The origin and early evolution of *Balanocrinus* (Crinoidea: Articulata). In *Echinodermata. Proceedings of the 5th International Echinoderm Conference, Galway* (ed. B. F. Keegan and B. D. S. O'Connor), p. 169. Balkema, Rotterdam.

Simms, M. J. 1986. Contrasting lifestyles in Lower Jurassic Crinoids: A comparison of benthic and pseudopelagic Isocrinida. *Palaeontology* **29**, 475–93.

Simms, M. J. (in press). British Lower Jurassic Crinoids. *Palaeontographical Society Monographs*.

Tate, R. and Blake, J. F. 1876. *The Yorkshire Lias*. J. van Voorst, London.

Taylor, P. D. 1983. *Ailsacrinus* gen. nov., an aberrant millericrinid from the Middle Jurassic of Britain. *Bulletin of the British Museum, Natural History (Geology)* **37**, 37–77.

Ubaghs, G. 1978. Origin of Crinoids. In *Treatise on invertebrate paleontology. part T. Echinodermata 2* (ed. R. C. Moore and C. Teichert), pp. T275–81. Geological Society of America and University of Kansas Press, Lawrence, Kansas.

Wiley, E. O. 1981. *Phylogenetics: the theory and practice of phylogenetic systematics*. Wiley, New York.

Willink, R. J. 1979. Some conservative and some highly-evolved inadunate crinoids from the Permian of eastern Australia. *Alcheringa* **3**, 117–34.

Zittel, K. A. von 1876–80. *Handbuch der Paläontologie, Band 1, Paläozoologie* **1**. Oldenbourg, Munich and Leipzig.

APPENDIX 1

A REVISED CLASSIFICATION OF THE ARTICULATA
Class Crinoidea Miller, 1821
Subclass Articulata Zittel, 1879
  emend Simms, herein, to include Encrinidae

Plesion (Family) Encrinidae Dujardin and Hupé 1862
Order Isocrinida Sieverts-Doreck 1952
   Plesion (Family) Holocrinidae Jaekel 1918
   Plesion (Family) Pentacrinitidae Gray 1842
   Suborder Comatulidina A. H. Clark 1908c (*nom. transl.*)
     Plesion (Superfamily) Paracomatulacea Hess 1951
       emend Simms, herein, to exclude Atelicrinidae
     Plesion (Family) Eocomatulidae (nov.)
     Plesion (Family) Paracomatulidae Hess 1951
   Infraorder Comatulidia A. H. Clark 1908c. (*nom. transl.*)
     emend Simms, herein, to exclude Paracomatulidae
   Suborder Isocrinina Gislén 1924 (*nom. transl.*)
   (=Isocrinidae Gislén 1924)
     Family Isocrinidae Gislèn 1924
     Family Cainocrinidae (nov.)
     Family Isselicrinidae Klikushin 1979
     Family Marsupitidae d'Orbigny 1852
       emend Simms, herein, to include Uintacrinidae
   Suborder Bourgueticrinina Sieverts-Doreck 1953 (*nom. transl.*)
   (=Bourgueticrinida Sieverts-Doreck 1953)
Order Millericrinida Sieverts-Doreck 1952
   emend Simms, herein, to include Cyrtocrinida
   Plesion (Family) Dadocrinidae s.l. Lowenstam 1942
   Suborder Millericrinina Sieverts-Doreck 1952
     emend Simms, herein, to exclude Dadocrinidae s.l. with dicyclic cup
   Infraorder Millericrinidia Sieverts-Doreck 1952 (*nom. transl.*)
     (=Millericrinina Sieverts-Doreck 1952)
   Infraorder Hyocrinidia Rasmussen 1978 (*nom. trans.*)
     (=Hyocrinina Rasmussen 1978)
   Suborder Cyrtocrinina Sieverts-Doreck 1952
     (=Cyrtocrinida Sieverts-Doreck 1952)
   Infraorder Cyrtocrinidia Sieverts-Doreck 1952 (*nom. transl.*)
     (=Cyrtocrinina Sieverts-Doreck 1952)
   Infraorder Holopodinidia Arendt 1974 (*nom. trans.*)
     (=Holopodina Arendt 1974)
Incertae sedis (Order) Roveacrinida Sieverts-Doreck 1952

Appendix 2

Definitions of new taxa

Plesion ((Family) Eocomatulidae—Paracomatulacea with short untapered stem retaining distinct distal articulum on terminal columnal. Arm branching feebly endotomous beyond IIIBr.

Type genus—*Eocomatula*, gen. nov. for *Pentacrinus interbrachiatus* (Blake, in Tate and Blake 1876).

Family Cainocrinidae—Isocrinina with cryptosyzygial articula at IBr1-2 and synarthrial articula at IIIBr1-2.

Type Genus—*Cainocrinus* Forbes 1852.

PART VI
Evolutionary biology

22

The ultrastructure of tube foot epidermal cells and secretions: their relationship to the duo-glandular hypothesis and the phylogeny of the echinoderm classes

J. DOUGLAS McKENZIE*

Department of Physiology, The Worsley Medical and Dental Building, The University, Leeds LS2 9NQ, UK

INTRODUCTION

From their presumed origins as simple respiratory evaginations of the water-vascular system (Nichols 1966b), tube feet have diversified into the wide range of specialized and sometimes bizarre structures found in modern echinoderms. This morphological diversity of form reflects the variety of functions that tube feet perform, particularly their involvement in feeding and locomotion. The importance of tube feet in these two fundamental aspects of echinoderm biology has prompted many investigations of their structure (see Nichols 1966a,b; Binyon 1972). Attempts have been made to find indications of the phylogenetic relationships of the echinoderm classes from studies of the gross structure of tube feet (e.g. Woodley 1967; Smith 1979), but their fine structure has promoted much less interest of this kind. This is despite the relatively extensive literature on the ultrastructure of the tube feet and the importance of these electron microscopical (EM) studies in elucidating the adhesive mechanisms of tube feet. As this paper will attempt to show, comparative EM studies have a role in phylogenetics as well as in the analysis of functional morphology.

While purely mechanical adhesive processes are evident in suckered tube feet, many rely entirely on chemical adhesion. Even in suckered tube feet, chemical adhesive processes occur as an adjunct to mechanical ones (Paine 1926). Chemical adhesion is the result of secretory activity by the tube feet. These secretions have been the subject of a large number of investigations using both light and electron microscopy, but their composition and mode of operation are still poorly understood. Adhesion has long been considered the result of 'mucus' secretions (Smith 1937). This opinion has been followed without criticism by most authors (e.g. Defretin 1952; Nichols 1959a,b, 1966a,b; Chaet and Philpott 1960; Buchanan 1962; Fontaine 1964; Souza Santos 1966a,b; Engster and Brown 1972; Binyon 1972; Hammond 1982). The theory of mucus induced adhesion has been profoundly influential in the interpretation of tube foot functional anatomy. Although it has never been blandly stated, there is an undercurrent of thought behind this mucus theory that envisages modern echinoderms to have a common, homologous mechanism of chemical adhesion. This has arisen, however, more because of a general acceptance of the stale tautology 'tube feet adhesives are mucus because mucus is the adhesive in tube feet', rather than from any rigorous interpretation of the data. Few workers have challenged this tautology, preferring instead to interpret anomalous results in one of two ways. Either tube foot secretions that are likely to be adhesives are termed mucus regardless of the histochemical accuracy of the description (e.g. Smith 1937) or by ascribing functions other than adhesion to secretions that

Echinoderm phylogeny and evolutionary biology (ed. C. R. C. Paul and A. B. Smith). Clarendon Press, Oxford, 1988.

*This work was carried out while the author was at the Department of Zoology, Queen's University of Belfast, Belfast, Northern Ireland.

have no histochemical basis for being described as mucus (e.g. Buchanan 1962). This process has been aided by the increasing vagueness and redundancy of mucus as a descriptive term; a problem by no means solely confined to echinoderms. Richards (1984) notes the wide use of mucus in describing invertebrate secretions and points to its lack of being anything other than: 'very broad, ambiguous and ultimately unhelpful definition'.

Mucus as the ubiquitous source of chemical adhesion in tube feet was obliquely questioned by Nichols (1966a) who noted the lack of mucus glands in holothurian tentacles, which are modified tube feet. Similar observations by Fish (1967) led him to propose the 'pharyngeal-mucus' model as an explanation of adhesion in dendochirote holothurian tentacles. In this model, mucus from the pharynx was smeared onto the tentacles every time they were placed into the mouth and provided the 'stickiness' of dendrochirote tentacles. However, Brumbaugh (1965) argued that the tentacles did possess secretions, albeit very small ones and it was these that made the tentacle sticky. This was supported by Fankboner (1978) whose EM results demonstrated the presence of secretions with electron-dense cores in the specialized papillae that Brumbaugh (1965) had claimed were the only adhesive parts of the tentacles. Neither of these authors, however, questioned that these secretions were some form of mucus. A more direct challenge came from Austin (1966, cited in Warner and Woodley 1975) who considered proteinaceous secretions, not mucus, to be the source of adhesion in *Ophiothrix* tube feet. This was expanded upon by Warner and Woodley (1975) who suggested that the mucus secretions present in *Ophiothrix* tube feet were important in the transport of particles after they had been captured by the proteinaceous secretions.

In general the term 'mucus' has meant 'acid mucopolysaccharide' to investigators of echinoderm tube feet, although some authors have used it in a wider context (e.g. Woodley 1967). The concept of acid mucopolysaccharides as adhesives was completely overturned by Hermans (1983) in an important and largely theoretical paper on duo-glandular adhesive systems. In it he argues that many adhesive systems consist of two secretions: a proteinaceous adhesive and an acid mucopolysaccharide de-adhesive. This hypothesis is discussed in detail below. While it is highly relevant to the study of tube foot functional anatomy, the duo-glandular theory also has a phylogenetic importance. Hermans (1983) places emphasis on the ancestral nature of chemical adhesion and suggests that nearly all tube feet share a homologous, evolutionarily conservative, duo-glandular adhesive system. This is in contrast to the adhesive 'suckers' found on some tube feet which are thought to have evolved at least twice (Nichols 1966b). The far-reaching implications of the duo-glandular theory demanded further scrutiny of the available information and the acquisition of new comparative data on tube foot secretions. This was one of the reasons why I extended my EM studies of dendrochirote tentacles to other tube feet; the other being a need for a comparative study of echinoderm surface coats. Including my work on dendrochirote tentacles; tube feet from four classes and twenty-three species were examined. It was not possible to work on crinoid tube feet which was unfortunate as very little information on the secretions of their tube feet exists. The methods employed and the full results can be found in McKenzie (1985, 1987). Shortage of space prevents any detailed rendition and discussion of the results of this study. Instead, only the salient points as they relate to the duo-glandular hypothesis and the phylogenetic relationships of the four classes studied are discussed. It is hoped that the results will show the potential value of comparative EM studies to evolutionary studies, as well as highlighting their more immediate interest.

THE DUO-GLANDULAR HYPOTHESIS AND TUBE FEET

Discussion of adhesive interactions is too often only concerned with their adhesive aspects. While this is justified for interactions that produce permanent adhesion, many biological adhesive interactions are temporary. The de-adhesive aspects of such temporary interactions are obviously as important and as necessary as the adhesive ones. This dichotomy of adhesion and de-adhesion is rarely acknowledged, but it is fundamental to the understanding of adhesion in many biological systems. Where adhesion is the result of a secretion, might its converse, de-adhesion, be produced by a second secretion? Such speculation originally arose to explain the rapid attachment and detachment of gastrotriches to sand grains

(Boaden 1968). It was soon applied to turbellarians, archiannelids and nematodes, with EM results adding morphological evidence to the theory that adhesion in these groups relied upon duo-glandular adhesive systems (see Hermans 1983 for review). The contribution made by Hermans (1983) to this hypothesis was to provide it with a plausible physico-chemical framework and to suggest that other adhesive structures, including echinoderm tube feet, possessed duo-glandular systems.

Hermans' model of the duo-glandular adhesive system as it applies to echinoderm tube feet can be briefly summarized as follows.

(1) Adhesion of tube feet is initiated by release of basic, proteinaceous secretions that bind the negatively charged tube foot surface to the interacting surface or particle.
(2) De-adhesion of the tube feet is invoked by carbohydrate dominated (=acid mucopolysaccharide=glycosaminoglycans) secretions outcompeting the adhesives for bonds on the tube foot surface.
(3) Proteinaceous adhesive secretions and carbohydrate dominated, de-adhesive secretions are associated with each other in the tube feet to act as duo-glandular adhesive systems.

This is summarized in Fig. 22.1, though 'GAG's decline' is my invention. This is a considerably more sophisticated model of tube foot adhesion than any previous explanation. It seeks to reverse completely the commonly held supposition that the acid mucopolysaccharide secretions in tube feet are adhesive. Detailed theoretical arguments for the model's physico-chemical rigour were given in Hermans (1983), and Thomas and Hermans (1985). These will not be elaborated here except to say that adhesives in other organisms are often proteinaceous, e.g. the pili of bacteria (Costerton *et al.* 1981); the eosinophilic granules in the colloblasts of ctenophores (Hernandez-Nicaise 1984) and the 'stickiness' of some spiders' webs (Foelix 1982). As most surfaces in aquatic systems carry a negative charge (see Thomas and Hermans 1985), negatively charged secretions such as acid mucopolysaccharides will tend to repel, not bind such surfaces. While the background to the theory is sound, the factual evidence that tube feet utilize duo-glandular adhesive systems is scanty. Perhaps the best evidence is that tube feet either contain, or can be argued to contain, more than one secretion (Hermans 1983). However, the occurrence of two or more secretions in a tube foot need not indicate that they are part of a duo-glandular adhesive system or even that their functions are related. This is discussed in detail below. A more positive form of evidence would be if tube feet were found to contain a proteinaceous and an acid mucopolysaccharide secretion. Unfortunately most histochemical investigations of tube feet have been rather perfunctory in looking for proteinaceous

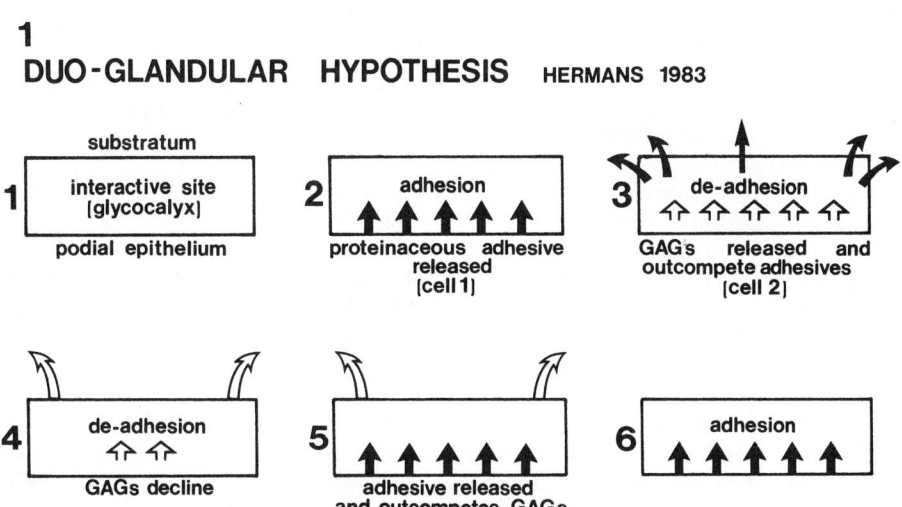

Fig. 22.1 The duo-glandular model of adhesion in tube feet, showing a cycle of adhesion, de-adhesion, and adhesion. GAGs, glycosaminoglycans.

secretions. Austin (1966, quoted in Warner and Woodley 1975) reported proteinaceous secretions from ophiuroid tube feet; while the histochemical nature of many of the secretions described as neutral or basic mucopolysaccharides is equivocal. A complication is that many tube foot secretions are heterogeneous and may contain both proteinaceous and mucopolysaccharide components. Good evidence that tube foot adhesives are proteinaceous comes from the tentacles of dendrochirote holothurians. The Type 1 cell secretions appear to contain basic proteins (McKenzie 1987) and, as these tentacles are sticky (Brumbaugh 1965; Fankboner 1978) and the Type 1 cells are the only ones to liberate their secretions onto the tube foot surface, these secretions are almost certainly the source of the tentacle's stickiness (McKenzie 1987).

Hermans (1983) maintained the dichotomy of adhesive and de-adhesive, proteinaceous and acid mucopolysaccharide, secretions for holothurian tentacles by suggesting that the highly acidic pharyngeal mucus secretions were acting as de-adhesives each time a tentacle was placed into the mouth. Holothurian tentacles cannot then be thought of as being strictly duo-glandular, but they can be considered to fit the spirit, if not the letter, of the model. EM studies of dendrochirote holothurians by McKenzie (1987) produced two relevant surprises. The first was that the entire tentacular surface, including the adhesive papillae, is covered with a glycocalyx, underlain by a substantial cuticle. The Type 1 cell secretions are completely obscured by the glycocalyx, with the only cell membranes exposed being the tips of cilia. The Type 1 cell secretions cannot be initiating adhesion by direct contact with their vesicles as Fankboner (1978) had reported. The secretions are instead released as tiny 'quanta' that reach the surface via the long, attenuated filaments of the glycocalyx (McKenzie 1987). The second finding was that the papillae contain elements of not one, but two secretory cells as well as support cell elements (Fig. 22.2). Does this mean that the papillae of dendrochirote tentacles are duo-glandular? They are certainly very similar to the adhesive organs in turbellarians that Tyler (1976) considered to be duo-glandular. Neither of the papillar secretions are, however, anything like the large, complex secretions that Hermans (1983) proposed as the de-adhesives in his model. More conclusively, the relationship of the surface coats

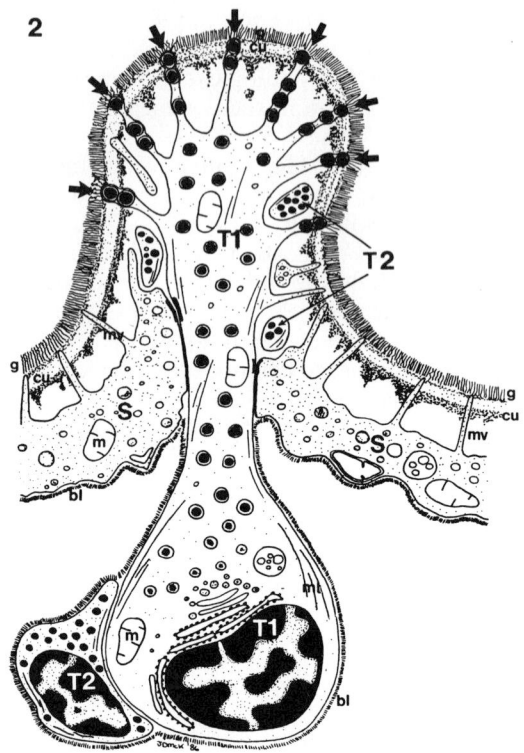

Fig. 22.2 Apical tuft cell. Diagrammatic Type 1 cell (T1) from a dendrochirote tentacle with associated Type 2 cell (T2) and support cells (s). The apex of the Type 1 cell protrudes from the tentacular surface to form a papilla. Secretory vesicles are transported within the microvilli to their release sites (heavy arrows) below the glycocalyx. Note the Type 2 cell elements within the papilla. Abbreviations: bl basal lamina; cu cuticle; g glycocalyx; m mitochondrium; mt microtubules; mv microvilli; s support cells; T1 Type 1 cell (apical tuft cell); T2 Type 2 cell.

to the two cell types (Type 1 and Type 2) reveal that the Type 2 cell secretions are not released onto the surface of the papilla and thus are not directly involved in the adhesion of the papilla. The ultrastructural evidence points to the Type 2 cells being neural or neurosecretory (McKenzie 1987).

The tentacular secretions of aspidochirote holothurians are essentially the same as dendrochirotes, but the tentacles of an apodous holothurian, *Leptosynapta*, possess goblet cells that contain a flocculent mucus as well as Type 1 and Type 2 cells (McKenzie 1985). As with the other holothurians, the tentacles of *Leptosynapta* are placed into a pharynx lined with mucus glands. Unless the *Leptosynapta* tentacles have an unlikely 'belt and braces' approach to de-adhesion then the tentacular mucus secretions must have some function other than as part of a duo-glandular adhesive system. Their probable function is as protective lubricants to prevent abrasion of the tentacles by the coarse sands through which *Leptosynapta* burrows. Interestingly, there are mucus cells over much of the body, but these secrete an acid mucus while those of the tentacles are neutral (Gotto 1984). This may be to lessen interference with the adhesive interactions between the Type 1 cell secretions and captured particles.

These results from holothurian tentacles do not prejudice the duo-glandular model since Hermans (1983) had already conceded them as exceptional. They do, however, indicate that considerable caution should be exercised when interpreting tube foot secretions in relation to the model. Before a secretion can be considered to have a role in adhesive interactions, there must be evidence that it is actually released onto the tube foot surface and that more likely explanations of its function do not exist.

Burke (1980) described two secretory cells in the tube feet of *Strongylocentrotus* and two species of *Lytechinus*. This led Hermans (1983) to suggest that echinoid tube feet were duo-glandular. However, only one type of secretory cell was found in the tube feet of *Diadema* (Coleman 1969), *Echinus* and *Psammechinus* (McKenzie 1985). The mucus cell illustrated in Coleman (1969) was a morula cell (see Byrne 1986 for review) and not involved in surface secretion. While there is a possibility that the species studied by Burke (1980) contain duo-glandular systems, and those studied by Coleman (1969) and McKenzie (1985) do not, this seems unlikely. All the secretory cells described by all three authors are apical tuft cells, similar to the Type 1 cells in holothurian tentacles. The latter are all thought to secrete adhesives (McKenzie 1985), as must at least one of the echinoid apical tuft cells. This would leave the remaining echinoid apical tuft cell as the only de-adhesive exception. It may transpire that the differences between the two cells described by Burke (1980) are more apparent than real, but further research on this is necessary. What is relevant to this present review is that because the tube feet studied by Coleman (1969) and McKenzie (1985) only contain one secretion these cannot be considered duo-glandular.

Both ophiuroid and asteroid tube feet have been the subject of a comparatively large number of histochemical and EM studies. Despite, or perhaps because of this, there remains confusion as to the number, nature and function of the tube foot secretions in these classes. A common feature is that all the surface secretory cells are goblet cells which open onto the surface via long ducts (Fig. 22.3). No apical tuft cells, such as are found in holothurian and echinoid tube feet, have been recorded. The significance of this is discussed below.

Only the tube foot secretions of two of the better studied ophiuroids (*Ophiothrix* and *Ophiocomina*) are considered here. Martinez (1977) described no fewer than six secretory cells (A–F) from the tube feet of *Ophiothrix*, some of which appear to be synonymous. McKenzie (1985) recognized three secretory cells, equivalent to the Type A, B and D cells of Martinez (1977), as well as a presumptive neuronal or neurosecretory cell (Type 2). Similar cells were found in *Ophiocomina* (McKenzie 1985). The Type B cells contain acid mucopolysaccharide secretions (Martinez 1977) and are equivalent to the cells containing acid mucopolysaccharides described by Buchanan (1962) and Fontaine (1964). The Type A and D cells both appear to be the cell containing proteinaceous secretions that was revealed by light microscopy (Buchanan 1962; Austin 1966, quoted in Warner and Woodley 1975). If we ignore the confusing problem of why ophiuroid tube feet should apparently have two adhesive secretions, then the morphological and histochemical evidence supports ophiuroids possessing duo-glandular adhesive systems.

There is, however, an alternative and, in my opinion, more plausible explanation of the function of the secretions in these tube feet. This rests upon the well founded observation of some ophiuroid species using mucus secretions as nets to catch suspended particles (Warner 1982). This initially seems difficult to reconcile with the concept of mucus secretions being de-adhesives but this can be done. The physico-chemical properties of negative charge, hydrophilicity and low interfacial surface energy, will confer de-adhesive properties to acid mucopolysaccharides but they will also be very cohesive, elastic and strong (Hunt 1970). Nets of mucus probably exploit these latter properties of mucopolysaccharides. Particles impinging onto mucus sheets will tend to deform, but not break, the sheet.

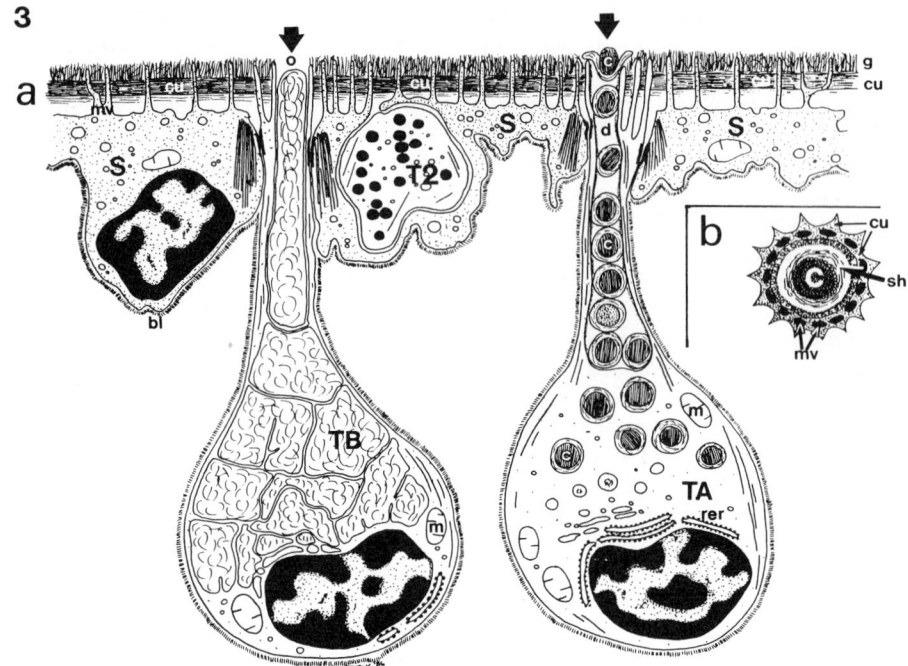

Fig. 22.3 Two goblet cells. Diagrammatic Type A and Type B cells from the sucker epithelium of an asteroid tube foot, with support cells (s) and presumptive Type 2 cell (T2). Black arrows show secretory orifices. Abbreviations: bl basal lamina; c core of Type A cell secretion; cu cuticle; d duct; g glycocalyx; m mitochondrium; mv microvilli; rer rough endoplasmic reticulum; s support cell; sh shell of Type A cell secretion: T2 Type 2 cell; TA type A cell; TB type B cell.

Where impingement is sufficiently strong, the sheet will flow round the particle and cohere to other parts of itself. No adhesive interaction is occurring but the particle becomes entrapped by this wrapping ability of the mucus sheet. Production of these sheets is an obvious possible function of the Type B cell secretions in species, such as *Ophiocomina*, which use these nets (Warner 1982), but this wrapping ability may also explain the function of these secretions in *Ophiothrix* tube feet. *Ophiothrix* captures particles which directly impinge onto the tube feet and it has been suggested that the mucus secretions are used for the subsequent transport of the particles (Warner and Woodley 1975). The mechanism proposed for this is that the impinging particles are captured by the proteinaceous adhesive secretions then wrapped in mucus released from the Type B cells. The mucus prevents further adhesion to the proteinaceous secretions and the resulting bolus of mucus and wrapped particles can be transported to the mouth by the mechanical activity of the tube feet (Warner and Woodley 1975). If so, the mucus secretions do have a de-adhesive role, but this is secondary to their wrapping and transport functions.

Clearly, the capture of food particles is a different situation from locomotory adhesive interactions though both involve adhesion and de-adhesion. In locomotion the loss of contact between the tube foot and the substrate after de-adhesion is desirable, but loss of contact with captured food particles would negate their capture. Thus, not only is it important to have an adhesive system that will allow the tube foot to adhere and de-adhere to captured particles, but there must also be the ability to retain these particles in a non-adhesive interaction so that they can be transported and ingested. The use of proteinaceous adhesives and mucus wrapping/de-adhesive secretions allows all three processes. A correlation between mucus secretions and microphagous feeding might then be expected in other tube feet. Such tube feet may well be found to contain secretory systems very similar to duo-glandular systems but their functions and mode of operation are subtly different. Aspidochirote and dendrochirote tentacles do not require wrapping secretions because the tentacles are placed directly into the mouth.

Asteroid tube foot secretions are the bulwark of the duo-glandular model as it applies to tube feet (Hermans 1983; Thomas and Hermans 1985). Most EM studies have been on suckered tube feet, but there is also a single study on an unsuckered one (Engster and Brown 1972). All the tube feet have a Type A secretory cell. This is always the most common type though its secretions show a degree of interspecific variation. Light microscopy suggested that the secretions were acid mucopolysaccharides (Chaet and Philpott 1964; Souza-Santos and Sasso 1968), but Harrison and Philpott (1966), using 'Thorostrast' staining, showed that the mucus component was restricted to the secretion's periphery. They suggested that the filamentous rods found in the core of the secretion were proteinaceous.

Despite this, Hermans (1983), and Thomas and Hermans (1985) claimed that Type A cell secretions are totally acid mucopolysaccharides and that the failure of thorium to stain the secretion's core (Harrison and Philpott 1966) was an artefact. Thomas and Hermans (1985) also claimed that there is a second cell type present in asteroid tube feet and this contains an electron dense secretion that is the proteinaceous component of a duo-glandular adhesive system. Unfortunately, they do not indicate in their 1985 paper which secretion this is supposed to be, making it difficult to comment upon. Hermans (1983) did, however, suggest that the small, electron-dense secretions found by Engster and Brown (1972) were adhesives. Engster and Brown themselves suggested that this cell was neurosecretory and so similar to the Type 2 cells in holothurians and ophiuroids (see above). McKenzie (1985) found similar cells in *Crossaster* and *Henricia*, though their secretions were larger (~ 200 nm) than those of other Type 2 cells (60–100 nm), including those found by Engster and Brown (1972). There was no evidence that these secretions were released onto the tube foot surface. A third cell type (Type B) has been reported from the adhesive areas of the tube foot and contains a flocculent mucus secretion (Souza Santos 1966b: McKenzie 1985). Type B cells also release their secretions onto parts of the tube foot other than the adhesive areas and so do not seem to be involved in adhesive interactions (McKenzie 1985). Interestingly, Type B cell secretions are much more variable and common in *Henricia*, which is a filter feeder (Jangoux 1982) than species which feed on larger prey (McKenzie 1985). This is tentative support for the argument that mucus secretions will correlate with microphagous feeding. The tube feet of the four species studied by McKenzie (1985) all appear to have only one secretion involved in adhesion and the results from other published studies seem to confirm this. If so, then there seems to be little evidence for considering asteroid tube feet as duo-glandular. No decision on this, however, can be made until the ultrastructural evidence of a second cell type, alluded to by Thomas and Hermans (1985), has been published.

The most interesting result in Thomas and Hermans (1985) is their photograph of the secretory print left by a starfish tube foot. This shows a clear pattern of anionic and cationic domains, and Thomas and Hermans (1985) suggest that this reflects the release of two secretions, one a basic protein or glycoprotein from their unnamed cells and the other an acid mucopolysaccharide coming from the Type A cells. The pattern of cationic domains shown on the tube foot seems, however, to be the mirror image of the Type A cell secretory ducts, suggesting that these released the proteinaceous secretions. It is unnecessary to have two cell types to produce footprints of the type demonstrated by Thomas and Hermans (1985). As mentioned above, the Type A cells secretions contain two components, an acid mucopolysaccharide shell and a core that may be proteinaceous (Harrison and Philpott 1966). Presumably the shell lubricates the passage of the secretion through the duct and prevents the core adhering to the duct. Similar shells are obvious around the cores of all the apical tuft cell secretions in echinoids and holothurians and thin shells are present around the cores of Type A and D cell secretions in ophiuroids. On the secretion reaching the surface the mucus shell must peel aside to reveal the adhesive core. This would result in a secretory pattern similar to an archery target, with a proteinaceous 'bull' surrounded by a ring of mucus. Where the outer rings of the 'targets' meet, the mucus secretions will tend to coalesce and the result will resemble a footprint similar to that illustrated in Thomas and Hermans (1985). To explain de-adhesion it would only be necessary to propose that the cores are ejected, probably by contractile activity of the duct or of filamentous elements through which the ducts pass.

The locomotory tube feet of holothurians have been left till last as they resemble duo-glandular systems more than any other type of tube foot does. They possess two secretory cells plus a Type 2 (neurosecretory?) cell. The first cell type is an apical tuft cell that is clearly homologous with the Type 1 cell in holothurian tentacles. As the latter type is almost certainly adhesive in function (see above), there is no reason to suppose an homologous cell type in the locomotory tube feet has a different function. The second cell (Type 3) is a goblet cell with large, complex secretions that show a high degree of interspecific variation (Harrison 1968; McKenzie 1985). Here at last is a tube foot whose secretions unambiguously fits the criteria for being considered duo-glandular. It possesses two (and only two) secretory cells which do definitely release their secretions onto (and only onto) the adhesive surface; one cell (Type 1) probably contains an adhesive proteinaceous secretion and one (Type 3) definitely contains an acid mucopolysaccharide component (Harrison 1968) that may be de-adhesive. Nor are there complications of interpretation such as involvement in microphagous feeding. The only piece of information that disturbs this duo-glandular harmony is that the Type 3 cell secretions contain, in addition to their acid mucopolysaccharide component, an electron dense component which may be proteinaceous (Harrison 1968). If it is proteinaceous, then it might be expected to be adhesive. The presence of two adhesive cell types in one tube foot is a challenge to explain, though it should be remembered that ophiuroid tube feet also seem to possess two adhesive cells. A possibility is that one of the adhesive cells is used in locomotion and the other is only released when the holothurian is stationary and requires a stronger adhesive interaction than when moving, perhaps to avoid losing its foothold during strong water movements.

What this brief review of tube foot secretions reveals is, firstly, a need for more thorough histochemical investigations, preferably at the EM level, of almost all the tube feet. More experimental investigations such as that of Thomas and Hermans (1985) would also be useful. It also shows that there is presently little evidence to support the hypothesis that duo-glandular adhesive systems are common in echinoderm tube feet. Some are clearly not duo-glandular as they only contain one secretion. Where two or more secretions are present there are often more plausible explanations of their functions than as being part of a duo-glandular system. Duo-glandular adhesive systems may occur: despite my suggested alternative, holothurian locomotory tube feet may use such a system and the model of adhesive interactions on ophiuroid tube feet (see above) differs from the duo-glandular model only in its emphasis. But the phylogenetic concept of the duo-glandular model as a common, evolutionary conservative mechanism in echinoderm tube feet is not upheld by the available data. Indeed, close scrutiny of the data reveals that there may be at least two quite different adhesive systems being employed in echinoderm tube feet. The basis of this lies not with the secretions, but with the types of secretory cells.

CELL TYPES AND CLASS RELATIONSHIPS

While tube foot secretions show considerable inter- and intra-class variation, there are only two basic types of cells which appear to be involved in surface secretory activity. These are the goblet cells and the apical tuft cells. Goblet cells have not yet been demonstrated in echinoid tube feet by EM but the light microscopical results of Nichols (1959b) suggest that they will be found in the tube feet of some species. They are found elsewhere in echinoids, notably the gut (de Ridder and Jangoux 1982). Goblet cells are generally large, flask-shaped cells with a long neck and a basally located nucleus (Fig. 22.3). Their characteristic feature is the single, microvillus duct which opens as a pore onto the secretory surface (Fig. 22.3). Secretions are transported to the surface via this duct and pass through the ring of supporting microvilli. Microtubules are often associated with the duct and may be responsible for the locomotion of the secretions (Engster and Brown 1972). The orifices of the ducts are not overlain by the glycocalyx, instead the glycocalyx appears to line the walls of the ducts, at least at their anterior. Some of the goblet cells which possess filamentous mucus secretions have an apical cilium (Martinez 1977; McKenzie 1985). Their secretions are thought to have assorted functions, including adhesion, particle wrapping, and possibly de-adhesion.

Apical tuft cells are also flask-shaped, have a basal nucleus, and may have a long, cytoplasmic 'neck'. Here the resemblance to goblet cells ends. Instead of the apex being drawn into a single duct, it splays into an array of microvilli (Fig. 22.2). This may form raised papilla, as in dendrochirote tentacles or a more flattened surface. It is this microvillar tuft that suggested the name for this cell type. Secretions are passed to the surface within single microvilli and are released below the glycocalyx. All the apical tuft cells appear to release adhesives. The principal differences between apical tuft cells and goblet cells are summarized in Fig. 22.4.

Apical tuft cells	Goblet cells
Only found in holothurians and echinoids	Found in asteroids, ophiuroids, holothurians and echinoids
All thought to secrete adhesives	Thought to secrete adhesives 'wrapping' secretions, lubricants etc.
Secretions released at microvilli tips	Secretions released through ducts
Multi-focal release of secretion	Unifocal release of secretion
Secretions released within surface coats	Secretions released outwith surface coats

Fig. 22.4 Summary of the major differences between goblet and apical tuft cells.

If, for simplicity, we consider only those goblet and apical tuft cells that are thought to secrete adhesives, it is evident that two very different forms of secretion are occurring. Both cell types will probably be releasing proteinaceous adhesives, but those of apical tuft cells interact with the glycocalyx while the goblet cell secretions do not. The latter may flow over the glycocalyx but apical tuft cell secretions must pass through it. [Scanning electron microscopy (SEM) suggests that the proteinaceous cores exposed at the orifices of goblet cells do not spread out much, though the mucus shells might.] The glycocalyx is such a fundamental surface feature and so likely to be important in determining the adhesive qualities of the surface (see McKenzie 1987) that the differences in release of secretion between goblet and apical tuft cells may reflect completely different types of adhesive interaction. This raises many questions about functional anatomy but the striking differences between the cell types also reveal evolutionary questions.

Since goblet cells are found in all four classes this implies that both the cell type and all four classes share a common origin, but this is neither surprising nor controversial. The phylogenetic interest derives largely from the restricted distribution of the apical tuft cells. This divides the classes into two groups: a holothurian\echinoid line which possesses goblet cells and apical tuft cells and an ophiuroid/asteroid line which has only goblet cells. This has a number of implications and a variety of possible interpretations. It is possible that the holothurian\echinoid line is an artefact of our limited data, with apical tuft cells being actually present, but currently undiscovered, in the other two classes. Only further research will demonstrate if this is the case, but this division is already based on a moderately large number of species and seems likely to be correct. All the remaining possibilities affect our understanding of the relationships of the four classes.

The simplest possibility is that the adhesive abilities of tube feet arose after partial radiation of the classes from a common ancestor. The ophiuroids and asteroids may have developed the use of goblet cells for tube foot adhesion independently or during a period of common ancestry. Goblet cells are such common structures that they are likely to have been present even in proto-echinoderms. Not only are they found in all four classes studied here, but very similar cells are found in many other phyla (see reviews in Berieter-Hahn *et al.* 1984, for examples) though this need not indicate homology. We can only guess when the ophiuroids and asteroids started to use their goblet cells for adhesive purposes and to what extent their mutual lack of apical tuft cells indicates a relationship between them. The holothurians and echinoids did not follow this pattern of development, but instead a new cell type, the apical tuft cell, arose in a common ancestor before these two classes diverged but after they had separated from the ophiuroids and asteroids. Of course, it is possible that apical tuft cells arose twice and separately, so that their distribution only indicates convergent evolution between the echinoids and holothurians rather than a phylogenetic relationship. This seems implausible. The apical tuft cell is very unusual, possibly even unique. Examples of goblet cells are easy to find in virtually all phyla but it is difficult to think of a cell type that is directly analogous to the apical tuft cell. It is unlikely that such an unusual cell type has arisen twice within a phylum. It could be argued, however, that if such a cell does evolve twice then it is more likely to do so within a phylum, as the species will share a more similar genetic potential. Such arguments are inevitably circular but it seems unlikely that the holothurian\echinoid line is an artefact of convergent evolution.

A second possibility is that the adhesive systems in modern tube feet are obscuring a common, ancestral one. An ancestral system could have been based on either apical tuft cells or goblet cells, but the latter are more promising material as they are found in all four classes. If an ancestral adhesive system was based on apical tuft cells then ophiuroids and asteroids must have lost this cell line in favour of exploiting their goblet cells as sources of adhesive secretions. This seems less likely than the converse hypothesis, that the goblet cells formed the ancestral adhesive cell type and the apical tuft cells arose and replaced the adhesive functions of the goblet cells in the holothurians and echinoids. This latter hypothesis would still imply a closer relationship between these two classes than with either the ophiuroids or asteroids. The Type 3 cell in the locomotory tube feet of holothurians may represent a relic of such an ancestral adhesive system if its function is adhesion. The evolution of apical tuft cells and their supplanting of a primitive adhesive system based on goblet cells would require the apical tuft cell to have some intrinsic superiority over goblet cells for adhesion. If this were not the case, then it would be difficult to suggest why the apical tuft cells arose and why they were not suppressed in favour of the more effective adhesive system based on goblet cells. This is another reason why it is unlikely that an ancestral adhesive system would have been based on apical tuft cells. Why should a presumably more effective system of goblet cell mediated adhesion only appear in the ophiuroids and asteroids when holothurians and echinoids also possessed goblet cells that could have been adapted to this superior adhesive system? It thus appears then, that if an ancestral adhesive system existed in the stem group of the modern echinoderms it is likely to have used goblet cells. Apical tuft cells probably arose later in a common holothurian\echinoid line and represent a superior adhesive system to the ancestral one. Alternatively, the adhesive functions of tube feet arose after the partial radiation of the modern classes and arose at least twice.

CONCLUSIONS

The duo-glandular hypothesis represents an attempt to provide a global framework, both for the mechanism and evolutionary history of tube foot secretions. Sadly, the current evidence does not support the duo-glandular model as a common, primitive adhesive system in tube feet, despite the aesthetic attractions of such a unifying hypothesis. The first conclusion is thus rather negative, but the implications of the differences in the secretory cells are much more positive and demonstrate the utility of EM in evolutionary studies. Of course, there are problems inherent in such speculation, as there are in all attempts to relate the morphology of extant species to phylogeny. These are, however, increased in EM studies because of the uncertainty that individual species have been examined in sufficient detail and that a wide enough range of species has been examined to allow meaningful consensus statements to be made. If apical tuft cells should be found in ophiuroids or asteroids, then the interpretation of the results would be immediately and drastically altered. Further EM studies might help to resolve which of the interpretations of the distribution of apical tuft cells is correct. Crinoid tube feet are an obvious place to look for indications of an ancestral adhesive system. As microphagus feeders it would be expected that they will possess an adhesive cell type and a second cell type that produces a mucus wrapping secretion. The nature of the adhesive cell will be critical to the interpretations of the tube foot cell types in the other classes.

The advantage of EM lies in its resolving abilities. This allows statements about comparative morphology to be more accurate and detailed. The use of SEM in taxonomic work is well founded, but few workers have exploited the potential of transmission electron microscopes (TEM) in such studies. One exception is Tyler (1976), who compared the ultrastructure of turbellarian adhesive organs. He was able to ascertain phylogenetic relationships between turbellarian families based on the ultrastructural differences in their adhesive organs. It is probable that the differences between apical tuft and goblet cells would never have been fully revealed without EM. Demonstrating these differences allows their evolutionary implications to surface and may stimulate further lines of research. For instance, do echinoids and holothurians share other cell types that ophiuroids and asteroids lack?

ACKNOWLEDGEMENTS

My work on tube feet was carried out while at the Department of Zoology, Queen's University, Belfast, and was funded from a SERC Postgraduate Studentship.

REFERENCES

Bereiter-Hahn, J., Matoltsy, A. G., and Richards, K. S. (eds) 1984. *Biology of the Integument, Vol. 1 Invertebrates*. Springer, Berlin, Heidelberg.

Binyon, J. 1972. *Physiology of Echinoderms*. Pergamon, Oxford.

Boaden, J. P. S. 1968. Water movement—a dominant factor in interstitial ecology. *Sarsia* **34**, 124–36.

Brumbaugh, J. H. 1965. The anatomy, diet and tentacular feeding mechanism of the dendrochirote holothurian *Cucumaria curata* Cowles 1907. Unpublished PhD Thesis, Stanford University.

Buchanan, J. B. 1962. A re-examination of the glandular elements in the tube feet of some common British ophiuroids. *Proceedings of the Zoological Society of London* **138**, 645–50.

Burke, R. D. 1980. Podial sensory receptors and the induction of metamorphosis in echinoids. *Journal of Experimental Marine Biology and Ecology* **47**, 223–34.

Byrne, M. 1986. The ultrastructure of the morula cells of *Eupentacta quinquesemita* (Echinodermata: Holothuroidea) and their role in the maintenance of the extracellular matrix. *Journal of Morphology* **188**, 179–89.

Chaet, A. B. and Philpott, D. E. 1960. Secretory structures in the tube feet of starfish. *Biological Bulletin of the Marine Biology Laboratory, Woods Hole* **119**, 308–9.

Chaet, A. B. and Philpott, D. E. 1964. A new subcellular particle secreted by the starfish. *Journal of Ultrastructural Research* **11**, 354–62.

Coleman, R. 1969. Ultrastructure of the tube foot sucker of a regular echinoid, *Diadema antillarum* Philippi, with special reference to the secretory cells. *Zeitschrift für Zellforschung und Mikroskopische Anatomie* **96**, 151–61.

Costerton, J. W., Irvin, R. T. and Cheng, K.-J. 1981. The role of bacterial surface structures in pathogenesis. *CRC Critical Reviews in Microbiology* **8**, 303–38.

Defretin, R. 1952. Sur les mucocytes des podia de quelques echinodermes. Comparison de leur secretion avec d'autres mucoprotides. *Comptes Rendus de l'Academie des Sciences, Paris* **234**, 1806–8.

Engster, M. S. and Brown, S. C. 1972. Histology and ultrastructure of the tube foot epithelium in the phanerozonian starfish *Astropecten*. *Tissue and Cell*, **4**, 503–18.

Fankboner, P. V. 1978. Suspension feeding mechanisms of the armoured sea cucumber *Psolus chitinoides* Clark. *Journal of Experimental Marine Biology and Ecology* **31**, 11–25.

Fish, J. D. 1967. The biology of *Cucumaria elongata* (Echinodermata: Holothuroidea). *Journal of the Marine Biological Association of the United Kingdom* **47**, 129–43.

Foelix, R. F. 1982. *Biology of spiders*. Harvard, London.

Fontaine, A. R. 1964. The integumentary mucous secretions of the ophiuroid *Ophiocomina nigra*. *Journal of the Marine Biological Association of the United Kingdom* **44**, 145–62.

Gotto, R. V. 1984. Observations on *Snaptiphilus tridens* (T. and A. Scott) an ectoparasite of holothurians. *Crustaceana*, Supplement 7, 214–16.

Hammond, L. S. 1982. Analysis of grain-size selection by deposit-feeding holothurians and echinoids (Echinodermata) from a shallow reef lagoon, Discovery Bay, Jamaica. *Marine Ecology Progress Series* **8**, 25–36.

Harrison, G. 1968. Subcellular particles in echinoderm tube feet II. Class Holothuroidea. *Journal of Ultrastructural Research* **23**, 124–33.

Harrison, G. and Philpott, D. 1966. Subcellular particles in echinoderm tube feet I. Class Asteroidea. *Journal of Ultrastructure Research* **16**, 537–47.

Hermans, C. O. 1983. The duo-gland adhesive system. *Oceanography and Marine Biology Annual Reviews* **21**, 281–339.

Hernandez-Nicaise, M. L. 1984. Ctenophora. In *Biology of the integument, Vol. I Invertebrates* (ed. J. Bereiter-Hahn, A. G. Matoltsy, and K. S. Richards), pp. 97–111. Springer-Verlag, Berlin.

Hunt, S. 1970. *Polysaccharide-protein complexes in invertebrates*. Academic Press, London.

Jangoux, M. 1982. Food and feeding mechanisms: Asteroidea. In *Echinoderm nutrition* (ed. M. Jangoux and J. M. Lawrence), pp. 117–60. Balkema, Rotterdam.

Martinez, J. L. 1977. Estructura y ultraestructura del epitelio de los podios de *Ophiothrix fragilis* (Echinodermata, Ophiuroidea). *Boletin Real Sociedad Española de Historia Natural (Biol.)* **75**, 275–301.

McKenzie, J. D. 1985. A comparative study of dendrochirote holothurians with special reference to the tentacular functional anatomy. Unpublished PhD Thesis, Queen's University of Belfast.

McKenzie, J. D. 1987. The ultrastructure of the tentacles of eleven species of dendrochirote holothurians studied with special reference to the surface coats and papillae. *Cell and Tissue Research* **248**, 187–99.

Nichols, D. 1959a. The histology of the tube-feet and clavulae of *Echinocardium cordatum*. *Quarterly Journal of Microscopical Science* **100**, 73–87.

Nichols, D. 1959b. The histology and activities of the tube feet of *Echinocyamus pusillus*. *Quarterly Journal of Microscopical Science* **100**, 539–55.

Nichols, D. 1966a. Functional morphology of the water-vascular system. In *Physiology of Echinodermata* (ed. R. Boolootian), pp. 219–44. Wiley, New York.

Nichols, D. 1966b. *Echinoderms*. Hutchinson, London.

Paine, V. L. 1926. Adhesion of the tube feet in starfishes. *Journal of Experimental Zoology* **45**, 361–6.

Richards, K. S. 1984. Introduction. In *Biology of the integument, Vol I Invertebrates* (ed. J. Bereiter-Hahn, A. G. Matoltsy, and K. S. Richards), pp. 1–4. Springer-Verlag, Berlin, Heidelberg.

Ridder, C. de and Jangoux, M. 1982. Digestive systems: Echinoidea. In *Echinoderm nutrition* (ed. M. Jangoux and J. M. Lawrence), pp. 57–116. Balkema, Rotterdam.

Smith, A. B. 1979. Peristomial tube feet and plates of regular echinoids. *Zoomorphologie* **94**, 67–80.

Smith, J. E. 1937. The structure and function of the tube feet in certain echinoderms. *Journal of the Marine Biological Association of the United Kingdom* **22**, 345–57.

Souza-Santos, H. 1966a. The ultrastructure of the mucous granules from starfish tube feet. *Journal of Ultrastructural Research* **16**, 259–68.

Souza-Santos, H. 1966b. Ultrastructure of a mucous gland cell found in the tube feet of the starfish *Asterina stellifera*. *Experientia* **22/12**, 812–13.

Souza-Santos, H. and Sasso, S. W. 1968. Morphological and histochemical studies on the secretory glands of starfish tube feet. *Acta Anatomica* **69**, 41–51.

Thomas, L. A. and Hermans, C. 1985. Adhesive interactions between the tube feet of a starfish *Leptasterias hexactis* and substrata. *Biological Bulletin of the Marine Biology Laboratories, Woods Hole* **169**, 675–88.

Tyler, S. 1976. Comparative ultrastructure of adhesive systems in the turbellaria. *Zoomorphology* **84** (1), 1–76.

Warner, G. F. 1982. Food and feeding mechanisms: Ophiuroidea. In *Echinoderm nutrition* (ed. M. Jangoux and J. M. Lawrence), pp. 161–84. Balkema, Rotterdam.

Warner, G. F. and Woodley, J. D. 1975. Suspension-feeding in the brittle-star *Ophiothrix fragilis*. *Journal of the Marine Biological Association of the United Kingdom* **55**, 199–210.

Woodley, J. D. 1967. Problems in the ophiuroid water-vascular system. *Symposia of the Zoological Society of London* **20**, 75–104.

23

Crystallographic axes of echinoid genital plates reflect larval form: some phylogenetic implications

RICHARD B. EMLET*

Department of Biology, University of California, Berkeley, California 94720, USA

INTRODUCTION

Larval forms are now well known for many species of echinoids. Approximately 50 species have been reared through metamorphosis and there is some information on the echinopluteus larva for at least 85 species (Emlet et al. 1987). As early as 1921, Mortensen recognized that larval shapes and internal skeletons of regular echinoids afforded useful systematic characters at the family level. Mortensen's (1921) conclusions were extended to include cidaroids (Mortensen 1937), and they remain valid today. Notable exceptions are species which produce large yolky eggs that undergo an abbreviated, non-feeding larval development (e.g. Mortensen 1921; Williams and Anderson 1975; Amemiya and Tsuchiya 1979).

In an extensive survey of the orientations of crystallographic axes (C-axes) of adult apical plates, Raup (1965) found evidence for family level patterns. In his search for the basis of this pattern he suggested that it was fixed during the development of the echinoid rudiment and he noted that several exceptions to the familial patterns were species with large yolky eggs (Raup 1965). Emlet (1985) showed that parts of the calcareous skeleton in feeding larvae determined the orientations of the C-axes of the four apical plates that grew from these larval rods. Because non-feeding larvae apparently lack an organized larval skeleton, these apical plates form *de novo* and all have similar crystallographic orientations. Emlet's study demonstrated that crystallographic patterns of adult apical plates could be used to determine trophic mode of larval echinoids.

The purpose of this study is to show that part of the variation in the orientations of C-axes of apical plates in species with feeding larvae can be linked directly to distinctive characteristics of family-specific larval forms. Once the basis of variation of crystal patterns is understood, these patterns can be used to infer structure of larval forms and provide most useful information for phylogenetic analyses. The orientations of C-axes of apical plates of species from the superorder (cohort) Diadematacea (*sensu* Smith 1984: including Diadematoida, Micropygoida, and Pedinoida; excluding Echinothurioida) are analysed to study the distribution of the unusual two-armed larvae, *Echinopluteus transversus*, presently known only in the Diadematidae. The results suggest a separation of the Aspidodiadematidae from the diadematoid lineage prior to separation of the Micropygidae from the remaining Diadematoida.

MATERIALS AND METHODS

Information on larval forms was collected from the literature and from species raised through metamorphosis by the author. One specific aspect of larval form was analysed; namely the spreading of postoral and posterodorsal arms outward and anteriorly from the anterio-posterior axis of the larval body. The angle between the anterio-posterior axis and a line coincident with the postoral and posterodorsal arms or skeletal rods

---

Echinoderm phylogeny and evolutionary biology (ed. C. R. C. Paul and A. B. Smith). Clarendon Press, Oxford, 1988.

*This work was carried out while the author was at the Department of Invertebrate Zoology, National Museum of Natural History, Smithsonian Institution, Washington, DC 20560, USA

was used to estimate the spread of larval arms. Angles were measured from photographs or drawings of larvae viewed from a dorsal or ventral aspect and thus were planar projections. Because the spread of larval arms changes with the stage of larval development, data were collected on later larval stages (six- and eight-armed echinoplutei) which are closer to the morphology displayed when the adult rudiment ('echinus rudiment') forms. Later stages of some species can move postoral and posterodorsal arms with muscles attached to the bases of skeletal rods and undergo shape changes when they are disturbed. I chose drawings of larvae that were in a natural posture, as judged from comparison with other figures of the same species and from much personal experience working with echinoplutei. A list of species and the sources of data on larval form can be obtained from the author.

Each echinoid skeletal plate behaves optically as a single crystal of calcite and the orientation of the crystal relative to the plate surface is easily described. The optical axis of symmetry through the crystal (C-axis) can be determined and its orientation relative to the plane of the plate can be measured. Orientations of the C-axes of genital plates were determined from petrographic sections made through the plane of the apical system. Sections were analysed on a petrographic microscope fitted with a universal stage (Emmons 1943). See Raup (1965) and Emlet (1985) for further details on methods of determination of C-axes of echinoid plates. Two measurements were taken on each of five genital plates to characterize the orientation of the C-axis: the angle of plunge and the azimuth of the C-axis. The angle of plunge is the angle of the C-axis of a given genital plate with the plane through the apical system. Planes of individual genital plates can deviate from the plane of the apical system and this deviation contributes to variation in orientations of C-axes. For the purposes of this study, the planes of the genital plates will be considered coincident with the plane through the apical system. The azimuth of a C-axis is its direction relative to the centre of the apical system. The C-axis is a line. In this study the line is polarized such that the leading end passes down through the apical system toward the oral side of the echinoid.

The relationship between family-specific larval forms and the orientations of C-axes of genital apical plates was analysed by comparing the spread of posterodorsal and postoral larval arms with the angles of plunge of C-axes of genital plates 3 (G3) and 5 (G5), respectively. G3 forms at the base of the right posterodorsal arm rod, and G5 forms at the base of the right postoral arm rod. Because the C-axes of the posterodorsal and postoral rods are parallel to the length of the rods (Okazaki and Inoue 1976; Emlet 1985), and because G3 and G5 form on proximal parts of these rods, widely spread arms should result in high angles of plunge for C-axes of the respective genital plates. Conversely, a larval form with arms that are more anteriorly directed (little spread) would be expected to result in C-axes of G3 and G5 which are nearly parallel to the plate surfaces.

Feeding larvae that lack a full larval skeleton prior to metamorphosis are also predicted to have a specific crystallographic pattern in their apical plates. Members of the genus *Diadema* are known to have an unusual larva with a single pair of very long postoral arms (Mortensen 1921, 1931, 1937). Because they lack posterodorsal arms and skeletal rods, genital plate 3 (G3) forms *de novo* at metamorphosis and should have a C-axis that is nearly perpendicular to the plate surface. This prediction is based on the observation that genital plate 1 (G1) in species with feeding larvae, and all genital plates in species with nonfeeding larvae (that apparently lack larval skeletons) form *de novo* at metamorphosis and have C-axes that are nearly perpendicular to plate surfaces.

To test this prediction, larvae of *Diadema mexicanum* were raised through metamorphosis and the formation of genital plates was observed to verify the origin of G3. Orientations of C-axes of apical plates in adult *D. mexicanum* were studied in petrographic thin sections. These crystallographic patterns were compared with those species known to have larvae with both posterodorsal and postoral arms and have C-axes for G3 and G5 that both depart from perpendicular to the plate surfaces. Larvae of *D. antillarum* were also reared to a late larval stage to verify that this species also has a two-armed larval form.

The distribution of the two-armed larval form was inferred by observation of the orientations of the C-axes of apical plates in 21 species from five families among three orders of the superorder Diadematacea (see Table 23.1). Data for seven of these species were originally collected by Raup (1965) in his study of the variation of apical systems (see Table 23.1).

Phylogenetic implications of echinoid genital plates 301

TABLE 23.1 Species of Diadematacea for which orientations of the C-axes of apical plates were determined. The species, the source of the specimen that was studied, and the inferred larval form are given. The classification used is that of Smith (1984). Data on species that are followed by (R) were originally collected by Raup (1965). Sources of material were the National Museum of Natural History, Washington, DC (NMNH), and the Museum of Comparative Zoology, Harvard University (MCZ). Larval forms are inferred to have a pair of postoral arms (po only) or a pair of postoral and a pair posterodorsal arms (po and pd).

Taxon	Source	Inferred Larval Form
Order Diadematoida Duncan 1889		
Family Diadematidae Gray 1855		
Astropyga pulvinata (R)	NMNH 17447, Gulf of California, 31 m, Albatross Stat. #3026	po only
Centrostephanus coronatus (R)	NMNH 27552, Gulf of California, 17 m, Albatross Stat. #2826	po only
Chaetodiadema granulatum	MCZ 8639, near Nosy Bè, Madagascar, 23–27 m	po only
Chaetodiadema pallidum	NMNH E9112, Hawaii, 'Albatross'	po only
Diadema antillarum (R)	NMNH 21211, off Puerto Rico	po only
Diadema mexicanum	all 3 specimens were collected at Isla Urava, Bay of Panama, 5 m	po only
Diadema savignyi (R)	Eniwetok (lagoon), less than 2 m	po only
Echinothrix calamaris (R)	Eniwetok (lagoon), less than 2 m	po only
Echinothrix diadema (R)	NMNH 27475, Papeete, Tahiti	po only
Family Lissodiadematidae Fell 1966		
Leptodiadema purpureum	NMNH, exchange from Bishop Museum, Hawaii, off Mahukona, Hawaii, 30 m	po only
Family Aspidodiadematidae Duncan 1889		
Aspidodiadema arcitum	NMNH 27542, off Molokai, Hawaii, 790–818 m, Albatross Stat. #4112	po and pd
Aspidodiadema hawaiiense	NMNH 27534, off Mokuhooniki Islet, Hawaii, 523 m, Albatross Stat. #4097	po and pd
Aspidodiadema jacobyi	NMNH E20361, Gulf of Mexico, east of Yucatán, Mexico, 40–165 m	po and pd
Aspidodiadema tonsum	NMNH E9032, Palawan Passage, Philippines	po and pd
Plesiodiadema antillarum	NMNH E15575, Gulf of Mexico, east of Brownsville, Texas	po and pd
Plesiodiadema horridum	NMNH 21060, off Galapagos Islands, 2489 m	po and pd
Plesiodiadema indicum	NMNH E9051, Kayoa Island, Philippines, 412 m	po and pd
Order Micropygoida Jensen 1981		
Family Micropygidae Jensen 1981		
Micropyga tuberculata (R)	NMNH E8641, Philippines, 445 m, Albatross Stat. #5565	po only
Micropyga violacea	NMNH E5973, Verde Island Passage, Philippines	po only
Order Pedinoida Mortensen 1939		
Family Pedinidae Pomel 1883		
Caenopedina cubensis	NMNH E12314, Caribbean, east of Belize, 296–329 m	po and pd
Caenopedina mirabilis	NMNH 31182, off Shikoku, Japan, 350 m	po and pd

RESULTS

Relationship between larval form and C-axis orientation in genital plates

The angles of plunge of the C-axes through apical plates G3 and G5 increase as a function of increasing spread of larval arms out from the larval body (Fig. 23.1; Spearman rank correlations: $r_s = 0.77$, $n = 16$, $P < 0.001$ for postoral arms and G5; $r_s = 0.68$, $n = 12$, $P < 0.02$ for posterodorsal arms and G3). The angle of plunge of C-axes of G3

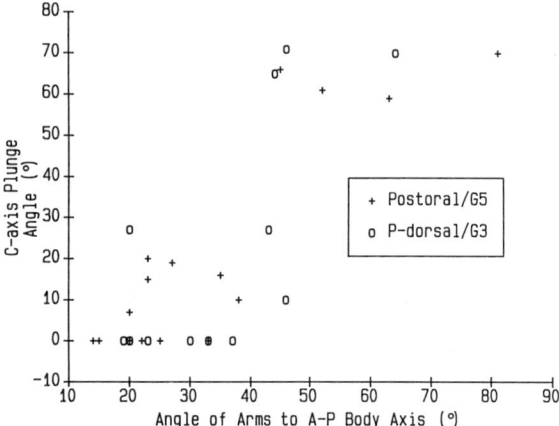

Fig. 23.1 Relation between spread of larval arms and angle of plunge of the C-axis through the plane of the genital plate. The angles of spread of posterodorsal and postoral arms are compared with C-axis elevation in G3 and G5 respectively. N = 16 for postoral/G5 and N = 12 for posterodorsal/G3 comparisons.

and G5 are grouped at or near 0° elevation or above 50° elevation. The reason why the plunge of the C-axis does not increase steadily with the spread of larval arms is not known. Several species have C-axes for G3 and G5 plunging in the intermediate range, but their larval forms are undescribed or known for only the earliest stages.

Part of the trend in Fig. 23.1 may be related to factors that influence the actual sites where the genital plates grow at the bases of the larval rods. The base of the postoral rods usually has three smaller rods that branch off at varying angles. These smaller processes, the lateral transverse rod, the ventral transverse rod, and the body rod, have the same crystallographic orientation as the main rod of the larval arm and are sites of muscle attachment. In some species the ends of these smaller processes are flattened and perforate in the region of muscle attachment. In cidarids and diadematids G5 tends to form in association with these flattened processes, while in strongylocentrotids the plate forms in association with the body rod process that is usually parallel to the main axis of the arm rod. How the angle of the larval arm determines the elevation of the C-axis is not understood, nevertheless the correlation allows a prediction of approximate larval form from the angle of plunge of the C-axis in G5. With the exception of certain diadematoids (see below), the angle of plunge of the C-axis of G3 also allows a prediction of approximate larval form.

The larvae of *D. mexicanum* and *D. antillarum* have postoral arms but lack posterodorsal arms prior to metamorphosis. Late in larval development of *D. mexicanum*, a pair of spicules was observed in the larval body at the location where the posterodorsal arms form in other species. These spicules appeared to be vestigial, having a planar branched form and lacking an arm rod for a main axis. At metamorphosis, G3 formed from one of these spicules in the plane of its growth, and the second spicule of the pair could not be followed. The C-axes of both the spicule and G3 that grew from it were roughly perpendicular to the growth plane. Due to a shortage of larval stages, I was unable to resolve whether these spicules were vestiges of the posterodorsal rods or the beginnings of G3 and ocular plate 4, respectively. Despite this uncertainty, it is clear that the usual posterodorsal arm rod does not form and does not direct orientation of the C-axis of the larval spicule.

Three species of *Diadema* are now known to have this unusual larval form. *Echinopluteus transversus* was originally recognized as a larval species from the Caribbean and Pacific (Mortensen 1921). Studies of other echinoid groups have led to the assignment of this unusual larval form to the diadematids (Mortensen 1937). In addition to *D. mexicanum* and *D. antillarum* raised for the present study, Mortensen (1921, 1931, 1937) has described the two-armed larval form of *Diadema setosum*. In his study of several species of *Echinopluteus transversus*, Mortensen (1921) identified posterodorsal spicules, but these never produced larval arms.

The crystallographic pattern of genital plates of *Diadema mexicanum* was compared with those of *Eucidaris thouarsi* and *Arbacia stellata*, both of which have feeding larval forms with both postero-dorsal and postoral arm pairs (Figs 23.2 and 23.3). Results are presented as stereographic projections of C-axis orientation relative to the apical centre and the plane of the apical system. The angle of plunge of the C-axis above the plate is represented by its distance from the centre of the plot. A point falling at the centre of the plot indicates a 90° plunge, i.e., a C-axis orientation perpendicular to a plane through the apical system or the genital plate. The outer circle represents a zero angle of plunge, i.e., a C-axis parallel to the plane of the plate. The inner circle represents an angle of plunge of 45°. In the same projection, azimuth of the C-axis orientation is plotted as a function of the quadrants. Azimuths that are toward the apical centre are plotted in the lower half of the circle, and azimuths that are away from the apical centre are plotted in the upper half of the circle. A C-axis falling on the vertical line, is orientated in a radial direction relative to the apical centre. Thus, a C-axis parallel to the plane of the plate and radially orientated would fall at the lower edge of the outer circle. The horizontal axis represents an orientation that is normal to the radial plane and thus is a plane tangent to the apical system. See Raup (1965) or a geology text for further explanation of how to interpret stereographic plots.

Two separate aspects of C-axis orientations characterize a feeding larval form. First, the C-axes of G3 and G5 should plunge at lower angles than those of other genital plates, especially G1. On a stereographic projection this pattern is represented by points for G3 and G5 that are further from the centre of the plot than points for other genital plates. Secondly, the azimuths of C-axes for G3 and G5 should have nearly opposite tangential directions relative to the apical centre. By convention, a polarity is assigned to the C-axis such that the direction down through the genital plate is the leading end of the C-axis. With this convention, points for G3 and G5 will show up on a stereographic projection to the left and the right, respectively, of the vertical line (radial plane) and often will be positioned closer to the horizontal line (transverse plane) than to the vertical line. (On the aboral surface of the echinoid, if the C-axes of G3 and G5 are projected onto the plane of the apical system, they will point toward interambulacrum 4.) The stereographic projections for C-axes of genital plates of *Eucidaris thouarsi* and *Arbacia stellata* illustrate these aspects (Fig. 23.2). Both species have feeding larvae that differ in the spread of larval arms. In comparison to *A. stellata*, the arms of *E. thouarsi* are more widely spread and the stereographic projection shows that the C-axes of G3 and G5 plunge at higher angles (closer to the centre of the plot) for this species.

The stereographic plots of *Diadema mexicanum* (Fig. 23.3, upper 3 plots) show a high angle of plunge of C-axes of genital plates. G5 is displaced to the right side of the plot and G3 is located closer to the centre than similar plots for *Eucidaris thouarsi* and *Arbacia stellata* (Fig. 23.2). The

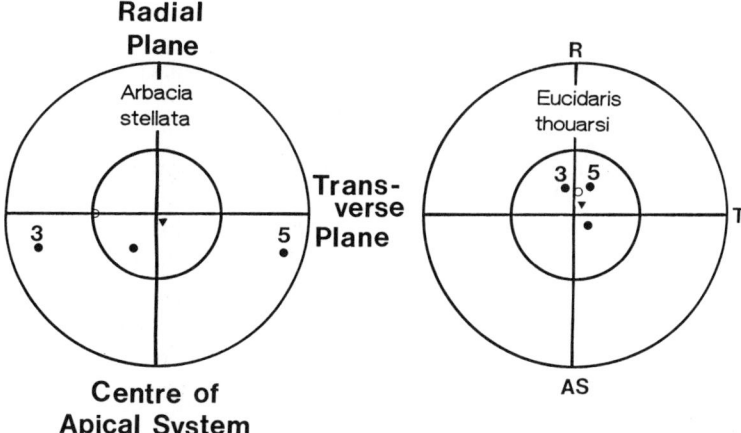

Fig. 23.2 Stereographic projections of C-axes of genital plates (G1–G5) for *Arbacia stellata* and *Eucidaris thouarsi*. Both species have feeding larvae with a full complement of larval arms. See text for explanation of figure. Symbols: ▼, G1; ●, G2; ●3, G3; ○, G4; ●5, G5.

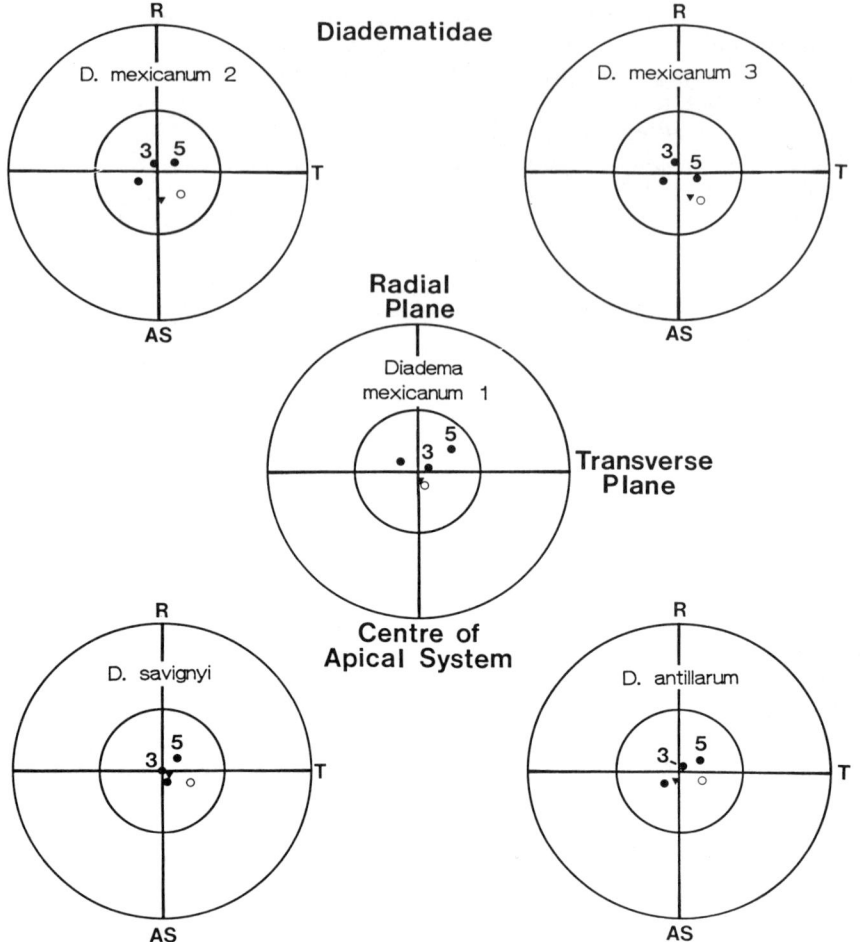

Fig. 23.3 Stereographic projections of C-axes of genital plates for three species of *Diadema*. The upper three plots are different specimens of *D. mexicanum* and the lower two plots are other species of *Diadema*. The two-armed larval form is known for *D. mexicanum* and *D. antillarum*.
Symbols: ▼, G1; ●, G2; ●3, G3; ○, G4; ●5, G5.

asymmetry of the points for C-axes of G3 and G5 on the stereographic projections are an indication of the unusual larval form of *D. mexicanum*. Because these larvae lack posterodorsal arms, the C-axis of G3 is not influenced by the larval rod, and thus the C-axis of G3 approaches perpendicular to the plate surface. As with other species, there is some intraspecific variation in crystal orientations found in genital plates of *D. mexicanum* (compare the upper 3 plots of Fig. 23.3). There are numerous sources of this variation including both natural and methodological ones, but the stereographic plots consistently show the asymmetrical positions of G3 and G5, and thus are indicative of a two-armed larval form. The two other species of *Diadema*, for which data on orientation of the C-axes have been collected, *D. antillarum* and *D. savignyi* (Table 23.1), show stereographic projections similar to *D. mexicanum* (Fig. 23.3 lower plots).

Crystallographic survey of larval form in the Diadematacea (including Pedinoida)

The survey of crystal patterns of genital plates of 21 species from the Diadematacea show an unambiguous distribution of inferred larval forms (Table 23.1, Figs 23.3–6). All species show patterns that indicate a feeding larval stage. Species of

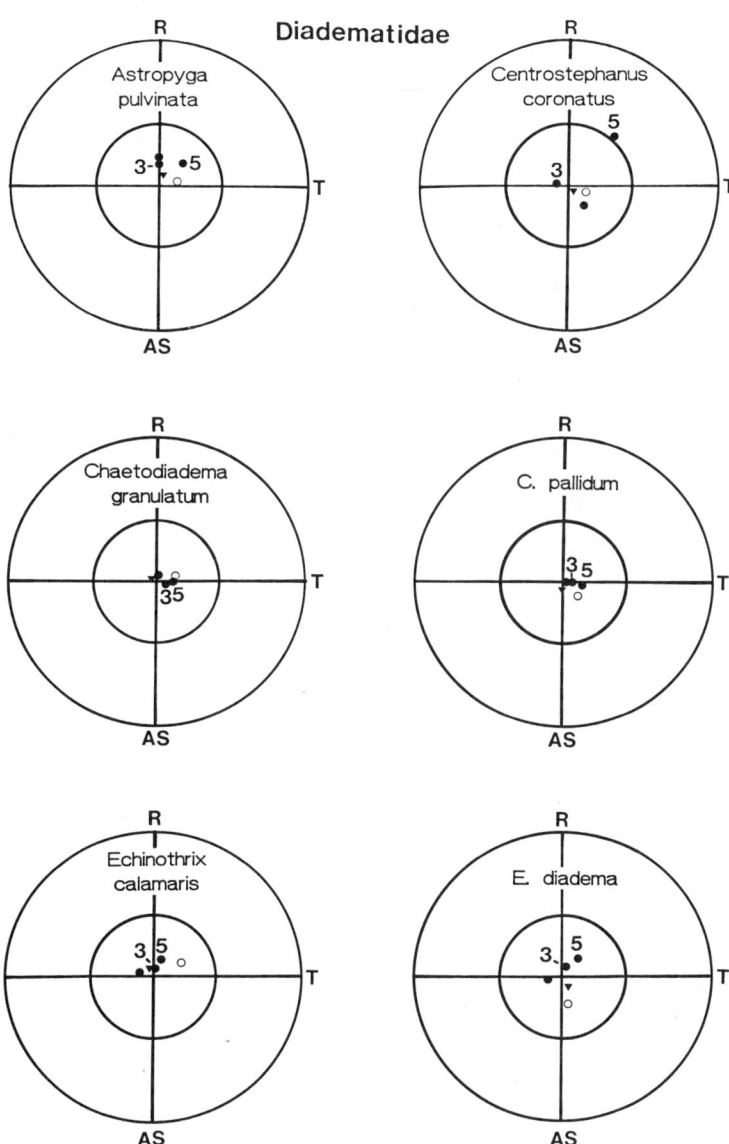

Fig. 23.4 Stereographic projections of C-axes of genital plates for six additional species of Diadematidae. Same symbols as Fig. 23.3.

the Pedinidae and Aspidodiadematidae are inferred to have larvae with a full complement of larval arms or at least with both posterodorsal and postoral arms (Figs 23.5 and 23.6). Species of the Diadematidae, Lissodiadematidae, and Micropygidae are inferred to have a two-armed larval form, that lack posterodorsal arms (Figs 23.3–5). The differences between species in the plunge of the C-axis of G5 suggest that the larval arms of species in the Diadematidae (except *Centrostephanus coronatus*) and Lissodiadematidae are more widely spread than those in the other families.

Several species in the Diadematidae (e.g. *Chaetodiadema* spp.) and the lissodiadematid, *Leptodiadema purpureum*, show a pattern of crystallographic orientation in which the C-axes are nearly perpendicular to the plate surfaces for all genital plates. This pattern is expected for species with non-feeding larvae that apparently lack an organized larval skeleton. The characteristics of the crystallographic patterns of the species studied here that distinguish them from species with non-feeding larvae are stereographic projections that show G5 consistently to the right of the vertical

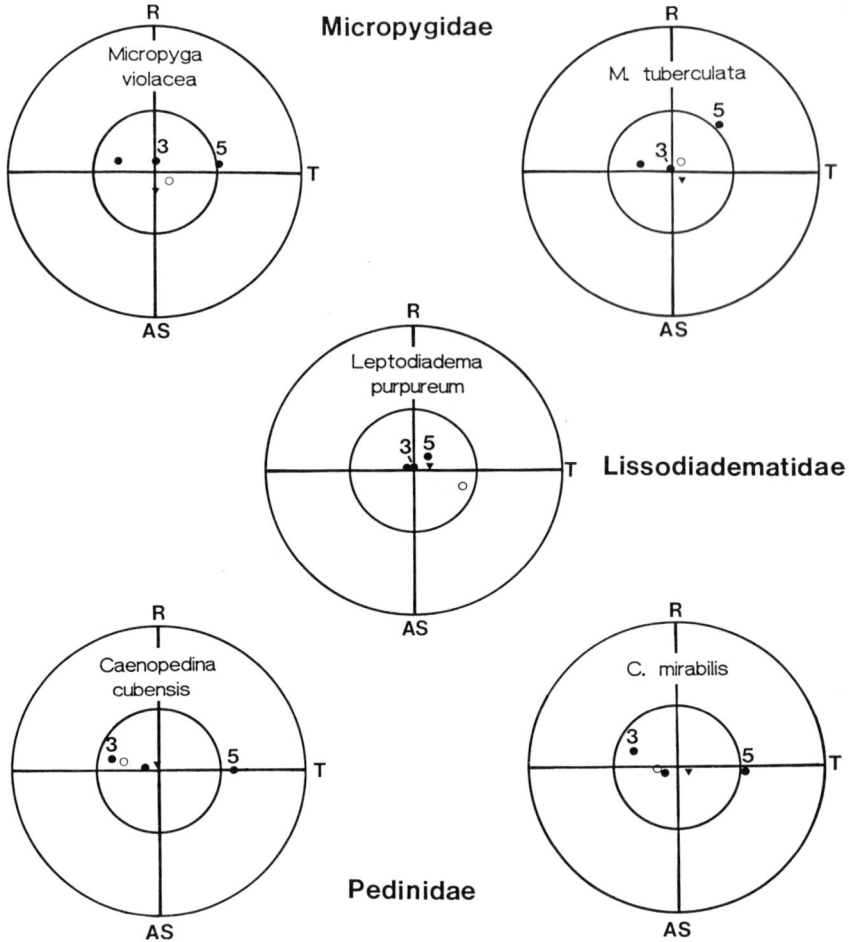

Fig. 23.5 Stereographic projections of C-axes of genital plates for two species of Micropygidae, one species of Lissodiadematidae, and two species of Pedinidae. Same symbols as Fig. 23.3.

line, and the higher plunges for G1, G2, and G3 relative to G4 and G5. Species with non-feeding larvae have genital plates with high C-axis plunge angles but without a specific relationship between points on the stereographic projection. Replicate specimens of species with non-feeding larval development show variable distribution of points concentrated near the centre of the stereographic projection (Emlet unpublished data).

Raup (1965, Table 2) reported that *Aspidodiadema nicobaricum* differed from other species of *Aspidodiadema* which he examined in having high C-axis orientations of G3 and G5. His specimens of *A. nicobaricum* and *A. meijerei* came from Hawaiian 'Albatross' collections that were identified by Mortensen (1940, pp. 46, 55) as *A. arcitum* and *A. hawaiiense*, respectively. My own specimen of *A. arcitum*, from the same lot as Raup's '*A. nicobaricum*', showed lower plunges for G3 and G5 and an unambiguous pattern for feeding larval development (Fig. 23.6). The discrepancy between these two specimens is apparently due to natural and methodological sources of variation.

DISCUSSION

The variation in crystallographic orientations of genital plates of species that have feeding larvae is at least partially explained by variation in larval form. The spread of larval arms out from the larval body is positively correlated with the angle of

Phylogenetic implications of echinoid genital plates 307

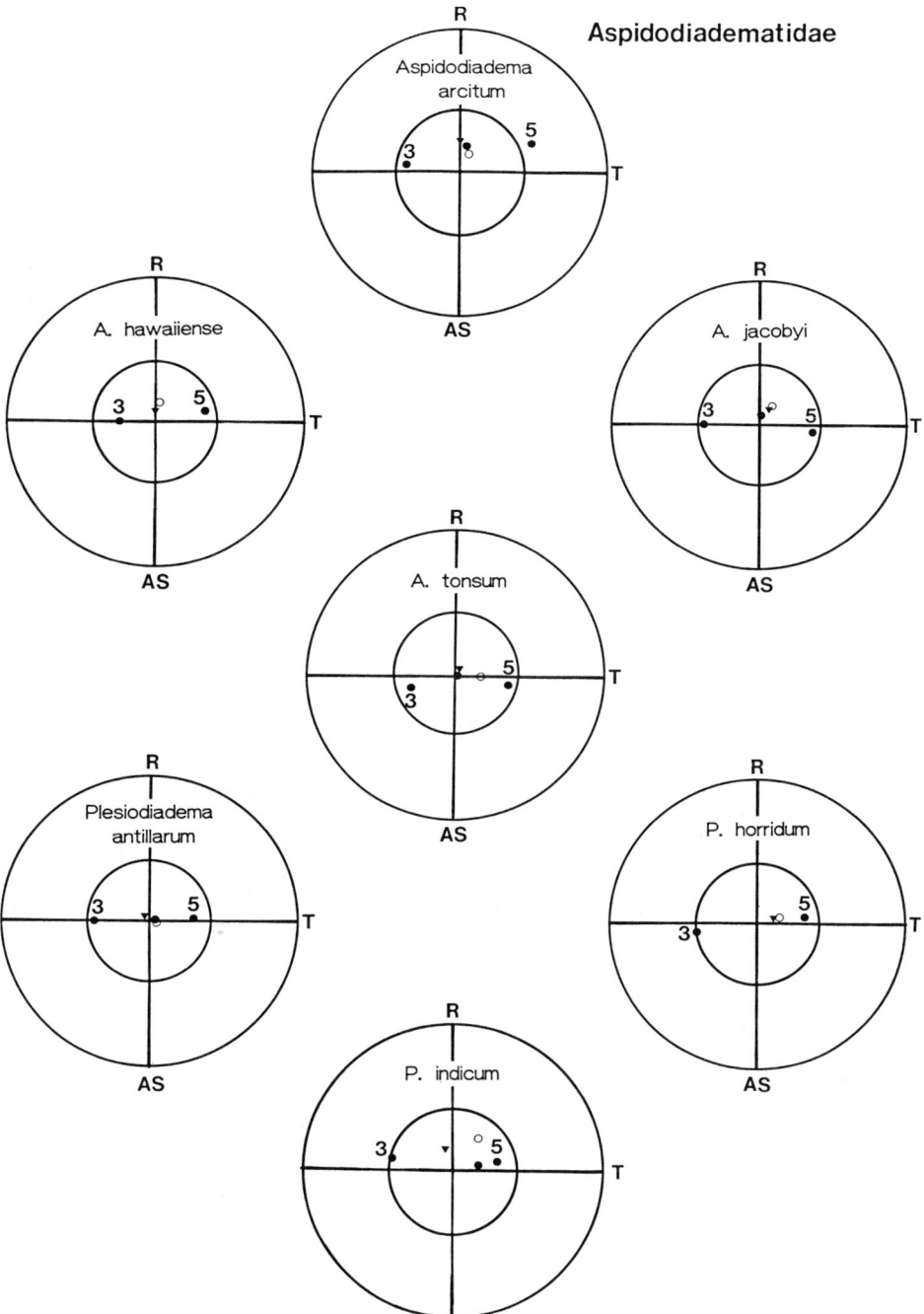

Fig. 23.6 Stereographic projections of C-axes of genital plates for seven species of Aspidodiadematidae. G2 is missing from sections of *Aspidodiadema hawaiiense* and *Plesiodiadema horridum*. Same symbols as Fig. 23.3.

plunge of the C-axis in genital plates 3 and 5. Furthermore, the presence or absence of the posterodorsal arms can be recognized in the crystal orientation patterns. Thus, the variation in apical plate crystallography, that was thoroughly described by Raup (1965) and found to produce family level 'signatures', appears to have as its basis family-specific larval forms. The connection between larval form and genital plate crystallography will allow a fuller interpretation of crystallographic patterns for Recent and fossil species. In addition to allowing the inference of trophic mode in development (Emlet 1985), crystallographic patterns can now be used to infer larval form.

The inference of larval form in the Diadematacea adds an important new character to a phylogenetic analysis at higher levels of classification of this superorder (cohort). Because feeding larval forms are recognizable at the family level (Mortensen 1921) and developmental patterns generally thought to be conservative, it seems reasonable to assume that a two-armed larval form like the *Echinopluteus transversus* of *Diadema* spp. has arisen only once.

Presently, the characters used to define the superorder Diadematacea (*sensu* Smith 1984) are all adult characters: the diadematoid ambulacral plates, thorns on the shafts of spines, the aulodont lantern morphology, grooved teeth, and CLNP (central lamellae-needles-prisms) microstructure of tooth plates (Jensen 1981; Smith 1981). On the basis of morphology of ambulacral plating, Jensen (1981) separated the Pedinoida from the Diadematacea, erecting a new superorder Pedinacea. Smith did not recognize this change and inferred that diadematoids arose from early pedinoids (Smith 1984, Fig. 1.5). Also based on derived characters of ambulacral plates, and tooth microstructure, Jensen (1981) separated the Micropygidae from the Diadematoida and erected a new order Micropygoida. According to Jensen's (1981) classification, the Diadematoida consists of the Diadematidae, Aspidodiadematidae, and Lissodiadematidae. Figure 23.7A shows the phylogenetic tree for the diademataceans (with pedinoids included) presented by Jensen (1981, Figs 38 and 40). Though no complete cladistic analysis has been published for Diademataceans, the following list contains characters that have been recognized as diagnostic for various groups by previous authors (Mortensen 1940; Fell 1966; Jensen 1981; Smith 1981).

(1) Hollow spines (plesiomorphy present in diadematoids and micropygoids).
(2) Solid spines (apomorphy of pedinoids).
(3) Hollow septate spines (aspidodiadematids).
(4) Reduction of gill slits (pedinoids).
(5) Non-crenulate tubercles (pedinoids, micropygoids, lissodiadematids).
(6) Apical system dicyclic (pedinoids).
(7) Diadematoid ambulacral plating (diadematoids).
(8) Plesiechinid ambulacral plating (pedinoids).
(9) Micropygoid ambulacral plating (micropygoids).
(10) Tooth microstructure of the *Diadema* type (diadematoids).
(11) Tooth microstructure of the micropygoid type (micropygoids).

Given the distribution of larval forms inferred from the crystallographic patterns of adult apical systems, the present classification is ambiguous on the origin of the two-armed larval form. Either the two-armed form evolved twice, once in the micropygid lineage and again in the diadematoid lineage, or the two-armed form evolved once in the ancestor of both lineages and reversed in the aspidodiadematid lineage. In order to retain parsimony of the larval character, its seems reasonable to suggest the two-armed larval form evolved only once in the ancestor of the diadematids and micropygids and after the separation from the aspidodiadematids (Fig. 23.7B). This phylogenetic tree retains all the distinctions recognized in adult characters listed above.

In order to avoid paraphyletic organization of the Diadematacea, either the aspidodiadematids should form a new order (i.e., Aspidodiadematoida) or the Micropygoida Jensen 1981, should be demoted to a family within the Diadematoida. Accepting the latter alternative, the Diadematoida would contain four families, Aspidodiadematidae, Diadematidae, Lissodiadematidae, and Micropygidae, with the aspidodiadematids the sister group of the other three families. Using characters of spines and lanterns, Mortensen (1940) recognized the suborders Aspidodiademina, Diademina, and Pedinina in the Aulodonta Jackson. His classification included the Diadematidae and Micropygidae in the Diademina because they have similar hollow spines. As mentioned above, Jensen (1981) emphasized differences in ambulacral plating structure and tooth microstructure to justify

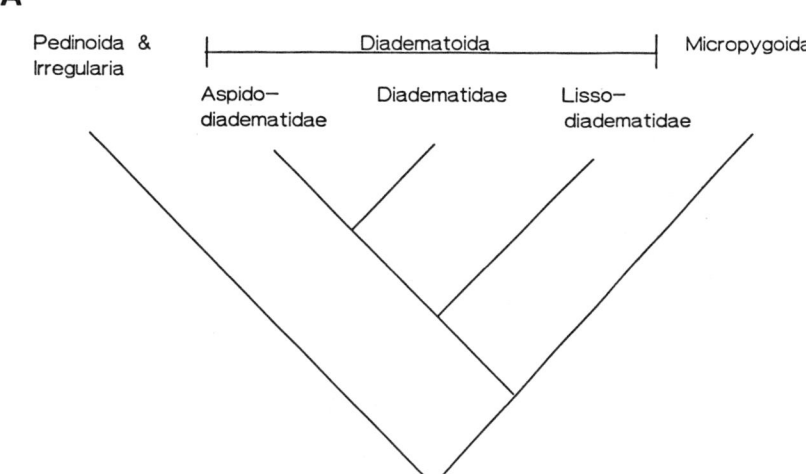

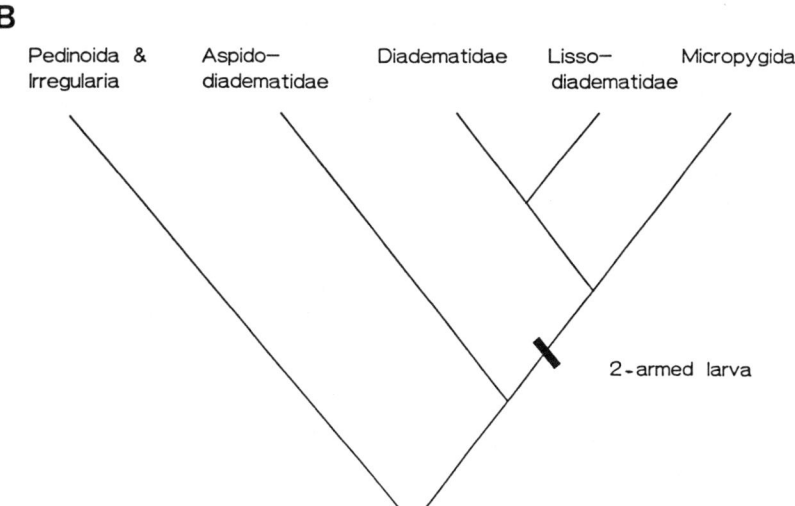

Fig. 23.7 A. Phylogenetic tree based on adult characters and presented by Jensen (1981). B. Phylogenetic tree from A. with rearrangements required to preserve parsimony of the two-armed larval form.

raising the micropygids to the level of an order, the Micropygoida. While there is little doubt that micropygids have distinct characters, both the ambulacral plating and tooth microstructure appear variable between and within the aspidodiadematids and diadematids. In fact, Jensen (1981) used the adjective 'diadematoid' to describe several different patterns of ambulacral plating and different patterns of tooth microstructure. In lissodiadematids, the pattern of ambulacral plating has been observed in only a few specimens (see Mortensen 1940), and their tooth microstructure has not yet been described. In the absence of clear evidence for assigning polarity to the ambulacral plate structure and tooth microstructure, it seems reasonable to be conservative and retain the Micropygidae as one of four families of the Diadematoida.

If the variation of crystallographic patterns within species of Diadematacea reflects the degree of spread of larval arms in addition to the presence or absence of a two-armed larval form, further resolution of the relationship between the families may be possible. Pedinids, aspidodiadematids, and micropygids, all show apical crystal patterns which indicate that their larvae may have similarly and less widely spread arms compared with larvae of lissodiadematids and diadematids, with the sole exception of *Centrostephanus coronatus* (compare Figs 23.3–6). If the pedinids are considered an outgroup for comparison with the remaining families, having less widely spread larval arms becomes a plesiomorphic character shared with aspidodiadematids and micropygids, and having more widely spread larval arms becomes an apomorphic character restricted to diadematids and lissodiadematids. (The pattern shown by *Centrostephanus coronatus* may indicate retention of the plesiomorphic character or may be due to natural or methodological variation in orientation of C-axes). These inferred differences in spread of larval arms imply that the diadematids and lissodiadematids are the most derived families and together form a sister group of the micropygids.

In conclusion, there is a developmental basis to some of the variation in the crystallographic patterns of the apical system that is related to larval form in species with feeding larvae. This relationship between crystal patterns of apical plates allows predictions to be made about larval forms. Because larval forms are known to reflect familial groups among the regular echinoids, larval form should allow resolution and reassessment of some higher level phylogenetic relationships in Recent and fossil echinoid lineages. Data presented here show that a re-evaluation of the Diadematacea is warranted.

ACKNOWLEDGEMENTS

This research was supported by a Smithsonian Institution Postdoctoral Fellowship. Studies on adult material were conducted in the Department of Invertebrate Zoology, National Museum of Natural History (NMNH), Washington DC. Echinoid larvae were raised at Naos Marine Laboratory, Smithsonian Tropical Research Institute [STRI], Republic of Panama. I owe special thanks to D. L. Pawson for encouragement and open access to echinoid collections of the NMNH and to D. Raup for allowing use of some of his original data on apical crystallographic patterns. I am grateful to J. Post (NMNH) for providing access to a universal stage and petrographic microscope, and to H. Lessios and J. Jackson (STRI) for providing supplies for raising larvae. J. Miller and R. A. Cameron kindly supplied algal cultures to feed larvae. I thank D. L. Pawson and P. M. Emlet for helpful comments on the manuscript.

REFERENCES

Amemiya, S. and Tsuchiya, T. 1979. Development of the echinothurid sea urchin *Asthenosoma ijimai*. *Marine Biology* **52**, 93–6.

Emlet, R. B. 1985. Crystal axes in Recent and fossil adult echinoids indicate trophic mode in larval development. *Science, New York* **230**, 937–40.

Emlet, R. B., McEdward, L. R., and Strathmann, R. R. 1987. Echinoderm larval ecology viewed from the egg. In *Echinoderm studies, Vol. 2* (ed. M. Jangoux and J. M. Lawrence), pp. 55–136. Balkema, Rotterdam.

Emmons, R. C. 1943. The universal stage (with five axes of rotation). *Memoirs of the Geological Society of America* **8**, pp. 1–205.

Fell, H. B. 1966. Diadematacea. In *Treatise on invertebrate paleontology, part U Echinodermata 3 (1)* (ed. R. C. Moore), pp. U340–65. Geological Society of America and University of Kansas Press, Lawrence, Kansas.

Jensen, M. 1981. Morphology and classification of the Euechinoidea Bronn, 1860—a cladistic analysis. *Videnskabelige Meddelelser fra Dansk Naturhistorisk Forening i Kjobenhavn* **143**, 7–99.

Mortensen, T. 1921. *Studies of the development and larval forms of echinoderms*. G. E. C. Gad, Copenhagen.

Mortensen, T. 1931. Contributions to the study of the development and larval forms of echinoderms, I–II. *Kongelige Danske Videnskabernes Selskabs Skrifter Naturiidenskalselig og Mathematisk Afdeling 9* **4** (2), 1–39.

Mortensen, T. 1937. Contributions to the study of the development and larval forms of echinoderms, III. *Kongelige Danske Videnskabernes Selskabs Skrifter Naturiidenskalselig og Mathematisk Afdeling 9*, **7** (1), 1–65.

Mortensen, T. 1940. *A Monograph of the Echinoidea. III.1. Aulodonta*. C. A. Reitzel, Copenhagen.

Okazaki, K. and Inoue, S. 1976. Crystal property of the larval sea urchin spicule. *Development, Growth and Differentiation* **18**, 413–34.

Raup, D. M. 1965. Crystal orientations in the echinoid apical system. *Journal of Paleontology* **39**, 934–51.

Smith, A. B. 1981. Implications of lantern morphology for the phylogeny of post-Palaeozoic echinoids. *Palaeontology* **24**, 779–801.

Smith, A. B. 1984. *Echinoid palaeobiology*. Allen and Unwin, London.

Williams, D. H. C. and Anderson, D. T. 1975. The reproductive system, embryonic development, larval development, and metamorphosis of the sea urchin *Heliocidaris erythrogramma* (Val.) (Echinoidea, Echinometridae). *Australian Journal of Zoology* **23**, 371–403.

24

Mutable collagenous tissues and their significance for echinoderm palaeontology and phylogeny

I. C. WILKIE AND R. H. EMSON*

Department of Biological Sciences, Glasgow College of Technology, Cowcaddens Road, Glasgow G4 0BA, UK
**Department of Biology, King's College, Campden Hill Road, Kensington, London W8 7AH, UK*

INTRODUCTION

Echinoderms possess the most unusual collagenous structures yet discovered in the Animal Kingdom. Although in terms of their ultrastructure and what is known of their biochemistry they strongly resemble the collagens that form the principal structural materials of most metazoans, they demonstrate unique properties: they can, within a timescale of less than a second to a few minutes, alter profoundly their tensile state, switching, for example, between stiff and compliant conditions, or even disintegrating irreversibly to permit the complete detachment of body parts during autotomy.

The extraordinary nature of some echinoderm collagenous structures has been recognized since the nineteenth century (see, for example, Hamann 1887; Perrier 1889; Cuénot 1891; von Uexküll 1900), yet it is only within the last 20 years or so that attention has become increasingly focused on their organization, biomechanics, and physiology. This revival of interest, sparked off by the papers of Serra-von Buddenbrock (1963) and Takahashi (1967a,b) has revealed the phylum-wide occurrence of the phenomenon and identified it as being of likely importance for many aspects of echinoderm biology.

Appreciation of these points has not been restricted to neontologists. Following the observations by Meyer (1971) on crinoid ligaments, echinoderm palaeontologists began to take account of these peculiar connective tissues in their interpretation of the functional morphology of extinct forms, particularly Palaeozoic pelmatozoans. This has also led to a certain degree of overspeculation, perhaps an indication of the extent to which the phenomenon has 'captured the imagination' of echinodermologists, but which has been compounded by widespread misunderstanding of the nature and properties of echinoderm collagenous tissues in general. Evidence for this is present in widely read textbooks and treatises. One of our aims in this chapter is, therefore, to provide a firm basis for future consideration of the role of connective tissues in the biology of extinct echinoderms.

The mutable collagenous tissues of echinoderms have been fully discussed in a number of recent reviews (Emson and Wilkie 1980; Motokawa 1984a; Wilkie 1984). Rather than go over the same ground, we intend firstly to assess current knowledge of these tissues, wherever possible by reference to work conducted and/or published since the previous reviews. We shall then argue that mutable collagenous tissues are likely to have been a significant component of the effector systems of extinct forms and an important enabling factor in the adaptive radiation of the Palaeozoic echinoderm fauna.

TERMINOLOGY

We define mutable collagenous tissues (MCTs) as collagenous tissues that evince variable tensility, i.e. the capacity to undergo rapid changes in mechanical properties. 'Variable tensility' was first used in this context by Biglow (1981), and the expression 'mutable collagenous tissue' was suggested originally by Eylers (*in litt.*). Although the

Echinoderm phylogeny and evolutionary biology (ed. C. R. C. Paul and A. B. Smith). Clarendon Press, Oxford, 1988.

latter term is somewhat vague, we deem it to be more appropriate than 'catch connective tissue' which has been advocated by Motokawa (1984a, 1985) following Rüegg's (1971) reference to 'connective tissue catch' in a review of smooth muscle tone. This describes adequately those echinoderm collagenous structures, such as the echinoid spinal ligament and holothurian dermis, that show reversible changes in tensile state and appear to subserve the function of catch muscles, but it is not applicable to other structures involved in autotomy processes, such as ophiuroid tendons and crinoid syzygial ligaments, that show irreversible disintegration and are unlikely to demonstrate reversible alterations in mechanical properties. 'Mutable collagenous tissues' encompasses both types of structure.

RECENT STUDIES ON MUTABLE COLLAGENOUS TISSUES

Morphology

Like the collagenous tissues of other phyla, MCTs consist mainly of extracellular materials associated with sparse cellular elements. The ultrastructure of the majority of MCTs is unexceptional in that, like vertebrate interstitial connective tissues, they are dominated by striated collagen fibrils with axial periodicity around 60 nm, interconnected by unbanded microfibrils of variable diameter (5–40 nm) and disposition. Attention has already been drawn to the microarchitectural diversity of these components across the phylum (Wilkie 1984), but recent investigations have shown that this diversity extends to collagenous structures within individual organs. The introvert dermis of dendrochirote holothurians is composed of three distinct layers: a superficial layer containing ossicles, collagen, and interfibrillar ground substance; a middle layer of many laminae of densely arranged parallel arrays of collagen fibrils (diameter 20–160 nm), attached 10–14 nm microfibrils, and bundles of muscle cells; and an inner layer dominated by ground substance with occasional thin collagen fibrils (diameter 40–60 nm) and more abundant 7–12 nm unstriated microfibrils (Byrne 1985a). Dendrochirotes are capable of auto-evisceration, a process which is facilitated by the loss of tensile strength and rupture of the introvert dermis. This dermis must also accommodate a considerable amount of flexion during routine protraction and retraction of the introvert for feeding purposes. How this might be facilitated by the dermal morphology is not clear, although it is interesting that the thin collagen fibrils of the inner layer resemble type III collagen which is associated with extensible mammalian connective tissues (Gay and Miller 1983).

Some crinoid stalks have the ability to autotomize (Emson and Wilkie 1980). As well as participating in the autotomy mechanism, stalk ligaments are likely to possess reversible catch properties, both on functional grounds and because the cirral ligaments of comatulid crinoids show such properties (Wilkie 1983). Grimmer and his co-workers (1984a,b, 1985) have examined the ultrastructure of the deciduous stalk of a larval comatulid and the permanent stalk of adult representatives of the orders Bourgueticrinida and Isocrinida. Their careful biometrical approach has revealed morphological differences between the various ligaments present in the adult stalks which are bound to reflect as yet unexplained functional differentiation (Table 24.1). Grimmer et al. (1984b) discussed the possible functional significance of bourgueticrinid stem ligaments, and we have attempted the same for isocrinid stem ligaments (Fig. 24.1).

Variable tensility is not restricted to interstitial collagenous elements. It has been shown that during autotomy of ophiuroid arms separation of the intervertebral muscles from the vertebral ossicles is achieved through an increase in the extensibility of tendinous fibres that normally link the muscles to the skeletal stereom. However, the tendons represent extensions of the basal lamina of the muscle cells, lack striated fibrils, and are likely to contain type IV collagen (Wilkie and Emson 1987).

Consistently present in or adjacent to all MCTs are granule-containing cell processes of neurosecretory-like cytology and uncertain function. In the ophiuroid arm the perikarya of these processes are located in a system of ganglion-like clusters associated with each ligament and even with the tendons of the intervertebral muscles (Wilkie 1979). Such a degree of organization has not been found in other echinoderm classes, but in all classes there is evidence that at least two types of juxtaligamental cells may be present in any one MCT, which are distinguishable by the size and sometimes the shape of their electron-dense granules (see Wilkie 1984, Table II). Table 24.2

TABLE 24.1 Ultrastructural features of the extracellular components of crinoid stalk ligaments. Data from Grimmer et al. (1984a,b, 1985)

Stalk	Ligament	Collagen fibril diameter (nm) (range, mean ±SD)	Collagen fibril periodicity (nm) (range, mean ±SD)	Interfibrillar material
Comatulid larval stalk	Intercolumnar ligaments	20–100	60–70	?
	Central through-going ligaments	20–80	60–70	?
Bourgueticrinid stalk	Intercolumnal ligaments	40–100 (67.7±14.3) Unimodal frequency distribution	57–63 (59.6±2.3)	Dense globules c. 5 nm; groups of filaments 12 nm in diameter
	Peripheral through-going ligaments	40–180 (means 60 and 120) Bimodal frequency distribution	59–72 (64.6±10.8)	More strand-like than globular; no 12 nm filaments present
	Central through-going ligaments	30–280 (means 70 and 130) Bimodal frequency distribution	57–62 (59.3±1.3)	Inconspicuous strands or globules; no 12 nm filaments present
Isocrinid stalk	Intercolumnal ligaments at synostosial articulations	40–100	60–70	No 12 nm filaments
	Intercolumnal ligaments at symplectial articulations	40–140	60–70	Groups of 12 nm filaments
	Peripheral through-going ligaments	40–200	60–70	No 12 nm filaments

TABLE 24.2 Evidence for presence of two types of granule-containing cells in echinoderm collagenous tissues, distinguished by the shape and/or size of their granules

Class and species	Structure	Granule type 1 (spherical) diameter (nm)	Granule type 2 (drug capsule-shaped) size (nm)	Reference
Asteroidea				
Acanthaster planci	Spine ligament	200 ± 40 (SD)	up to 400 long	Motokawa 1986
Crinoidea				
Comanthus japonica	Larval stalk ligament	100–150	220 × 500	Grimmer et al. 1984a
Democrinus conifer	Stalk ligaments	200	400	Grimmer et al. 1984b
Metacrinus rotundus	Stalk ligaments	150–200	600	Grimmer et al. 1985
Holothuroidea				
Eupentacta quinquesemita	Introvert dermis	150–300	180–550 × 125–220	Byrne 1985a

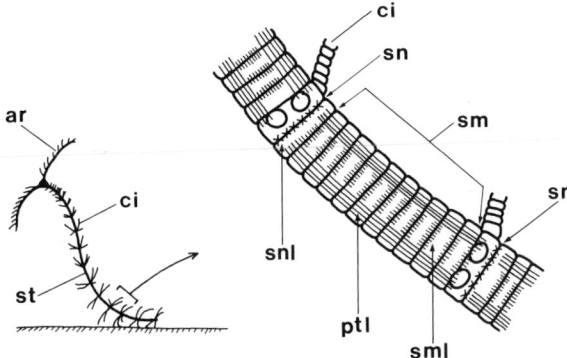

Fig. 24.1 Location and possible functions of isocrinid stalk ligaments: diagrammatic lateral view of an isocrinid crinoid and part of its stalk. In isocrinids, the capacity for stalk autotomy is confined to synostosial junctions (*sn*) between the columnal ossicles and does not involve the symplexies (*sm*). It is likely that synostosial ligaments (*snl*) can undergo irreversible disintegration at autotomy; symplectial ligaments (*sml*) may show reversible changes in tensility and have a role in postural control; and the peripheral through-going ligaments (*ptl*) may lack variable tensility, limit the degree of flexion permissible in unstiffened stalks, and provide elastic recoil. *ar*, arm; *ci*, cirrus; *st*, stalk (Redrawn and adapted from Grimmer *et al.* 1985; with permission of Springer-Verlag).

presents more data supporting this contention. Their distribution and their relationship with the motor nervous system (see below) suggest that the juxtaligamental cells control the tensile properties of MCTs, and it has been surmised that the two types represent separate 'stiffening' and 'relaxing' effectors (Motokawa 1982). Both of these suppositions are supported by our observations on ophiuroid tendons, since juxtaligamental processes are absent from tendons that do not rupture during autotomy and those processes associated with the 'autotomy tendons', which are likely to demonstrate only irreversible plasticization, contain only one type of electron-dense granule (Wilkie and Emson 1987).

Mechanical properties and the influence of various agents

Two aspects of the *in vitro* behaviour of MCTs imply the occurrence of variable tensility:

(1) the rapid change in their mechanical properties which can be invoked by certain agents (see below);

(2) the extraordinary variability in mechanical properties demonstrated by samples of the same structure from different animals (or even from one individual) subjected to identical testing regimes.

Such variability has been observed in the creep behaviour of ophiuroid disc integument (Wilkie *et al.* 1984), in the tensile stiffness of the sea-urchin catch apparatus (Diab and Gilly 1984), and in the stress relaxation and creep rates of holothurian dermis (Motokawa 1984*b,c*; Hayashi and Motokawa 1986). Holothurian dermis provides an extreme example of this: samples of dermis from *Stichopus japonicus* showed a wide range of viscosities, the maximum (35 MPa.s) being 450 times greater than the minimum (0.076 MPa.s: Motokawa 1984*b*). It has been shown independently by Motokawa (1984*b,c*), and Greenberg and Eylers (1984) that, despite this individual variability, the stress relaxation behaviour of holothurian dermis can be represented by a model consisting of two Maxwell elements in parallel (Fig. 24.2). In view of the complex organization of holothurian dermis, Motokawa (1984*c*) has been perhaps justifiably circumspect as to the precise significance of this model, although Eylers (1985) has ventured to correlate each of its components with specific processes in the tissues, and to claim for it predictive value with regard to the physicochemical basis of variable tensility.

Compared with former investigations, some recent work on MCT biomechanics is characterized by a more holistic approach, in the sense that greater account has been taken of the functioning of the tissues in the living animal. Motokawa (1984*c*) has attempted to relate marked differences in the tensile properties of the dermis of two holothurian species to their differing lifestyles: the dermis of *Actinopyga echinites* has a greater elastic stiffness and a five- to nine-fold longer relaxation time than that of *Holothuria leucospilota*. *A. echinites* shows little change in body shape and inhabits a high energy shore environment, whereas *H. leucospilota* is very mobile, able to change shape rapidly, and is found in calm lagoons. This recalls the 'ecomechanical' approach of Koehl (1977) to sea-anemone mesogloea and would be applicable to other echinoderm structures, such as the arms of asteroids and ophiuroids. Another sign of a 'return to reality' in MCT studies is the employment of testing regimes that mimic the

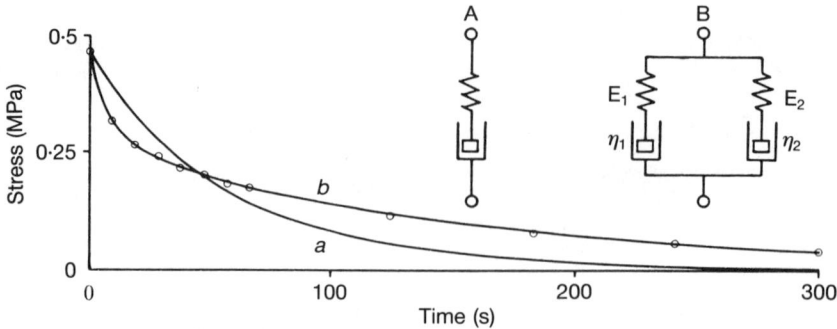

Fig. 24.2 Mechanical behaviour of holothurian dermis: stress-relaxation curve of a two-element model and that of a four-element model at 0.5 strain. Twelve points (hollow circles) were selected from the stress relaxation curve of a sample of dermis from *Actinopyga echinites* in artificial sea-water. The parameters of the mechanical models were selected so as to minimize the sum of the squares of the difference in stresses between the dermis and the model at these points. A simple Maxwell model shown in inset A generated the best fit stress-relaxation curve *a*. A four-element model shown in inset B generated the best fit curve *b*. The stress-relaxation curve of the four-element model agrees well with that of the dermis (Reprinted from Motokawa 1984c; with permission of The Company of Biologists Ltd.).

pattern of force experienced by tissues *in vivo*. Thus Diab and Gilly (1984) have measured the resistance of an echinoid spinal ligament ('catch apparatus') to imposed deflection of the spine about small angles, having discovered that deflections greater than 10° resulted in irreversible plastic flow. Ligaments exhibiting little initial resistance could be induced to adopt a highly resistive catch state by mild mechanical stimulation, this state representing a tensile stiffness close to that of vertebrate tendon at a similar strain (Fig. 24.3).

In general quantitative analyses of the mechanical properties of MCTs have indicated that in their extreme 'catch' phase some of these tissues can exhibit values for tensile strength, elastic stiffness, and viscosity approaching those of analogous mammalian collagenous elements such as tendon and skin. The important question is: 'What is the mechanism at the molecular level that permits these tissues to alter drastically their properties?'. One clue is provided by the ultrastructural appearance of MCTs fixed in their low viscosity condition or while undergoing autotomy: their collagenous fibrils are intact but disaggregated indicating that there has been a loss of interfibrillar cohesion (see, for example, Byrne 1985a). What, then, are the factors responsible for interfibrillar cohesion and how do they change to permit variable tensility? Possible answers to these questions have been inferred from the responses of MCTs to various externally applied agents that have known physico-chemical effects on connective tissue macromolecules and their interactions.

Manipulation of the external ionic environment can affect profoundly the mechanical properties of

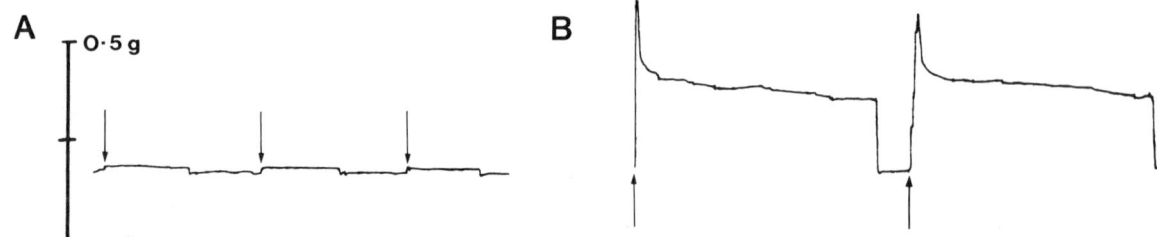

Fig. 24.3 Responses of a spinal ligament of *Strongylocentrotus franciscanus* to imposed 10° deflection (arrows). (A) The spine was deflected three times while the ligament was out of catch (i.e. relaxed). (B) The same spine was deflected twice after catch was induced by mechanical stimulation (Reprinted from Diab and Gilly 1984; with permission of The Company of Biologists Ltd.).

MCTs. The most comprehensive quantitative account of such effects is that of Hayashi and Motokawa (1986) on the dermis of *Holothuria leucospilota* (Fig. 24.4). Some of their findings are summarized in Table 24.3 and compared with data on the introvert dermis of *Eupentacta quinquesemita* from Byrne (1985b). Two general conclusions from these and other studies (e.g. Greenberg and Eylers 1984; Motokawa 1984b,c; Wilkie et al. 1984) are that:

(1) calcium ions are involved in interfibrillar cohesion, perhaps acting as electrostatic cross-bridges between ground substance macromolecules or between them and the collagen fibrils;
(2) the sensitivity of MCTs to sodium ion depletion and pH denotes the importance of ionic interactions for their mechanical stability.

It is perhaps the degree of their dependence on such interactions rather than on, for example, covalent linkages that may explain the unique properties of these tissues (Wilkie 1984), but the significance of the above investigations is difficult to assess when there is no comparable information on:

(1) echinoderm collagenous tissues that do not evince variable tensility;
(2) the collagenous tissues of most other invertebrate phyla.

Also, interpretation of such results continues to be handicapped by the paucity of data on the biochemistry of MCTs (see review by Bailey 1985).

An interesting alternative explanation for the role of calcium ions has been proposed by Diab and Gilly (1984) who showed that absence of calcium ions, as expected, prevents stiffening of the sea-urchin spinal ligament. However, they suggested that this might indicate the involvement of an extracellular, calcium-dependent transglutaminase (or similar enzyme) such as is responsible for

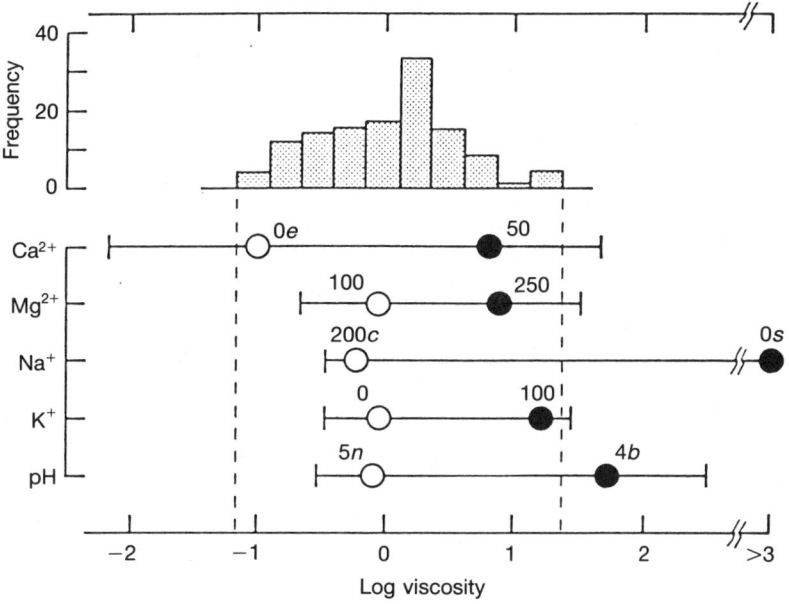

Fig. 24.4 Summary of the effects of ionic manipulation on the viscosity of the dermis of *Holothuria leucospilota*. The upper histogram shows the frequency distribution of viscosity of 123 samples of dermis in artificial sea-water; the lower figure shows the viscosity ranges which were observed by the manipulation of cation concentrations, the values being expressed relative to that observed in artificial sea-water (log value 0). In the lower figure, the maximum viscosity value and the minimum value which were observed for a particular ion species are connected by a horizontal line. The filled circle gives the mean value in the medium in which the maximum value was observed. The empty circle gives the mean value in the medium in which the minimum value was observed. The numerals beside the circles refer to the concentration (in mM) of the medium. *e*, with EGTA; *c*, choline substituted for Na^+; *s*, sucrose substituted for Na^+; *n*, no buffer; *b*, with buffer (Reprinted from Hayashi and Motokawa 1986; with permission of The Company of Biologists Ltd.).

TABLE 24.3 Summary of the effects of ionic manipulation on the viscosity of the body wall dermis of *Holothuria leucospilota* (Hayashi and Motokawa 1986) and the introvert dermis of *Eupentacta quinquesemita* (Byrne 1985b). ↑, increase; ↓, decrease; 0, no change; —, no data

Treatment	Effect on viscosity	
	Holothuria leucospilota	*Eupentacta quinquesemita*
Calcium-free sea-water	↓	↓
Sea-water containing 100 mM Ca^{2+}	↑	↑
Magnesium-free sea-water	↑	↓
Sea-water containing 250 mM Mg^{2+}	↑ (after transient ↓)	—
Sodium-free sea-water	↑	—
Potassium-free sea-water	0	—
Sea-water containing 100 mM K^+	↑	↓
pH less than 5 or greater than 8	↑	↑

the stabilization of fibrin and the coagulation of semen in mammals (Folk 1980), because stiffening was also blocked by low concentrations of the transglutaminase inhibitors putrescine and cadaverine. Such a mechanism would not preclude the direct involvement of calcium ions in macromolecular interactions, and leaves unexplained the ability of MCTs to relax or soften rapidly.

A review of all published ionic manipulation work on MCTs shows that, while their responses may conform to broad trends (those mentioned above being the most consistent), there may be considerable variability with regard to the direction of their responses (i.e. whether a particular treatment causes stiffening or relaxation), their magnitude, and their time-course (i.e. reaction time and recovery time). This is illustrated by the contradictory effects of magnesium and potassium ion manipulation quoted in Table 24.3, and may be due partly to differences in the biochemical organization of the tissues and partly to differences in the ionic dependence and activities of cellular elements, particularly neurons, controlling variable tensility.

Control of variable tensility

There is growing morphological and experimental evidence for the involvement of the nervous system in the control of variable tensility. Although their mode of operation is unknown, the neurosecretory neuron-like juxtaligamental cells are the only likely cellular elements that could be regulating the interfibrillar cohesion of MCTs. This is confirmed by the presence of unequivocal synapses between hyponeural axons and juxtaligamental cells serving oral arm plate ligaments of the ophiuroid *Ophiura ophiura* (Cobb 1985a). Similar varicose axons containing small agranular vesicles have been observed amongst the juxtaligamental perikarya of the sea-urchin spinal ligament by Peters (1985), who has gone so far as to include excitatory and inhibitory pathways to the ligament in a hypothetical consideration of the control of spine responses in *Echinus esculentus*. Close associations between conventional axons and juxtaligamental cells have also been observed in an asteroid spinal ligament (Motokawa 1986) and in the larval stem of a feather-star (Grimmer et al. 1984a).

Autotomy of the disc as practised by amphiurid brittlestars depends on the disruption of ligaments connecting the genital bars of the disc skeleton to the arms (Dobson and Turner 1983). Neural mediation of this process is implied by the depressant action of anaesthetics on disc autotomy in *Ophiophragmus filograneus* and by the ability of 5-hydroxytryptamine, noradrenaline, and γ-aminobutyric acid to shorten the response time to autotomy-inducing mechanical stimuli (Dobson 1985). Since they were applied to intact animals (at concentrations of up to 10^{-2} M), the sites of action of these agents are indeterminable and could be at the sensory or motor sides of a neural pathway; this experimental approach also does not preclude the participation of other factors such as circulating chemical substances. Chemical factors that can affect the mechanical properties of isolated MCT preparations appear to be ubiquitous in echinoderm coelomic fluids (Motokawa

1984a; Wilkie 1984) and have been detected most recently in the dendrochirote holothurian *Eupentacta quinquesemita* (Byrne 1986). Coelomic fluid from autotomizing *E. quinquesemita* causes evisceration if injected into intact animals. Although evisceration behaviour in this species was largely unaffected by acetylcholine and cholinergic antagonists, the stimulatory action of potassium ions and electrical current, which could be blocked by anaesthetics, suggested neural mediation. Byrne postulated that evisceration is invoked by a neuroendocrine pathway including the neurally stimulated release into the coelomic fluid of a chemical factor that directly or indirectly plasticizes the introvert dermis and other MCTs.

The stiffening response of isolated strips of holothurian body wall to photic stimulation (delivered by photographic flash) which is abolished by either removal of the epidermis or prior anaesthetization (Motokawa 1984d) is a remarkable indication of local neural control, and this has been confirmed by a thorough pharmacological investigation of the same preparation (Motokawa 1987). In strips of dermis subjected to constant load, acetylcholine (10^{-6}–10^{-3} M) causes a biphasic response consisting of an initial increase in viscosity followed by a drop in viscosity (Fig. 24.5A). Employing a range of cholinergic agonists and antagonists, Motokawa has shown that both muscarinic and nicotinic receptors are likely to be present in the dermis, the viscosity increase being mediated by nicotinic receptors, and the viscosity decrease by both nicotinic and muscarinic types (Fig. 24.5B,C). Stimulation with a medium containing 20–100 mM K^+ ions induces a similar biphasic response (stiffening then relaxation) in ophiuroid arm plate ligaments undergoing cyclical flexion (Wilkie in prep.). Prior

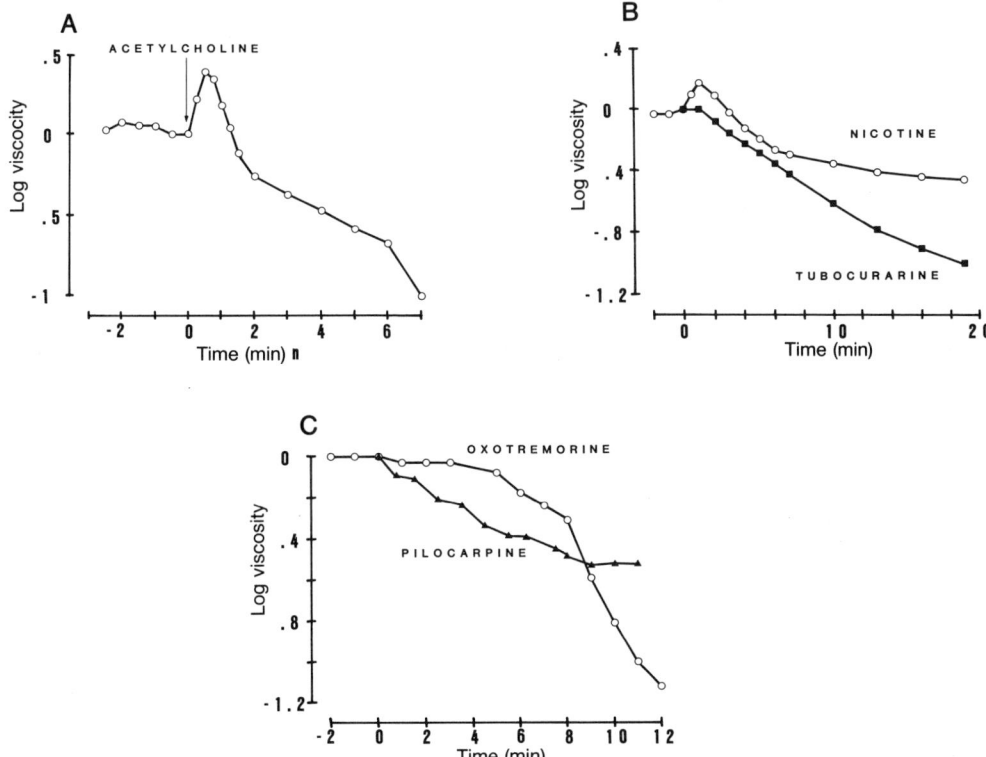

Fig. 24.5 Effects of pharmacological agents on the viscosity of the dermis of *Holothuria leucospilota* in creep tests. Values for viscosity are expressed relative to that observed when each test solution was applied. (A) Effect of 10^{-3} M acetylcholine. (B) Effects of cholinergic antagonists applied at time 0: 10^{-5} M nicotine and 10^{-4} M tubocurarine. (C) Effects of muscarinic agonists applied at time 0: 10^{-7} M oxotremorine and 10^{-5} M pilocarpine (Redrawn from Motokawa 1987; with permission of Pergamon Books Ltd.).

treatment with anaesthetics such as propylene phenoxetol, menthol, or chloroform reversibly abolishes relaxation and, unexpectedly, enhances stiffening. These biphasic responses and Motokawa's pharmacological findings provide evidence for dual innervation of MCTs. However, we are at present no nearer knowing whether there is any connection between dual innervation, the apparent occurrence of two types of juxtaligamental cells (see above, Table 24.2), and the presence of separate stiffening and relaxing factors in the coelomic fluid of some echinoderms than when these phenomena were previously reviewed (Motokawa 1984a; Wilkie 1984).

The uniqueness of mutable collagenous tissues

How unique are MCTs? We shall attempt to answer this question firstly by summarizing their main characteristics.

(1) The mechanical behaviour of isolated MCT preparations is often quantitatively unpredictable: the viscosity of fresh specimens of the same tissue can vary by more than two orders of magnitude (Motokawa 1984b).
(2) MCTs demonstrate rapid and profound changes in tensile properties: for example, during autotomy the tensile strength of the brittlestar intervertebral ligament can plunge to less than 0.1 per cent of its intact value in less than a second (Wilkie 1976).
(3) Although most MCTs consist largely of typical, cross-banded collagen fibrils, their variable tensility depends on changes in the forces responsible for interfibrillar cohesion rather than on alterations in the collagen fibrils themselves.
(4) Alteration of the ionic composition of the surrounding medium can produce tensile changes as large and as rapidly occurring as those seen in vivo (Hayashi and Motokawa 1986).
(5) It is likely that the variable tensility of most, if not all, MCTs is under neural control, the link between the nervous system and collagenous tissue being through the juxtaligamental cells.

With regard to the collagenous tissues of other phyla, the variability of the different parameters by which their mechanical behaviour can be described has rarely been commented upon. However, from the extensive literature on mammalian connective tissues the general impression is obtained that for a given species at a known stage in its life cycle and reproductive cycle, the mechanical properties of its collagenous structures when subjected to well-defined testing regimes are quantitatively consistent.

On the other hand, mammalian connective tissues are by no means immutable. Firstly, they show well-known changes in mechanical properties during maturation and senescence: the elastic modulus and tensile strength of rat skin and tail tendon increase two to four times during maturation and decline slowly thereafter (Vogel 1978); Fry et al. (1964) reported a remarkable 80-fold reduction in the 'viscous extensibility' of rat skin between the ages of 30 and 600 days. Secondly, more dramatic changes in tensile properties are exhibited by certain collagenous structures in the female body which are affected by the ovarian cycle or pregnancy, the most notable being the uterine cervix. This is perhaps the nearest functional equivalent to an MCT outside the Echinodermata, a point which has been reiterated frequently since Takahashi (1967b) first discussed the cervix in connection with the echinoid catch apparatus. During most of gestation the cervix acts as a strong, inextensible barrier preventing premature expulsion of the fetus. Then hormonal factors at or near term cause the cervix to switch to a low viscosity state that facilitates its dilatation and the movement of the fetus through the birth canal, and it returns to its previous inextensible state 24 hours post-partum (Harkness and Harkness 1959).

The timescale of these age- and reproduction-related changes is at the least two to three orders of magnitude greater than that for MCTs. Ageing changes in collagenous tissues have been correlated with increased intermolecular crosslinking of collagen (Vogel 1978), and relaxation of the cervix involves fibrillar dispersion which seems to be partly a result of changes in the composition of the ground substance and partly due to some other mechanism, perhaps enzymolysis or changes in the bonds linking ground substance components (see various reviews in Naftolin and Stubblefield 1980). All of these processes are believed to result from the modulation of fibroblastic activity. Like MCTs, mammalian collagenous tissues are sensitive to their ionic environment (calcium ions: Minns et al. 1973; Minns and Steven 1977; pH: Harkness and Harkness 1965; Viidik 1980), but generally to a much lesser extent. For example, the treatment of human tendon for 24 hours with the calcium chelator EDTA reduces stress relaxation (after 100 sec at 1.5 per cent strain), a measure of viscosity, by up to 29 per cent, whereas treatment for about 20 min with calcium-free sea-water reduces the viscosity of holothurian dermis by more than 80 per cent (Hayashi and Motokawa 1986). Unfortunately, there is no information on the ionic dependency of the uterine cervix.

This brief comparison has revealed that, as far as variable tensility and ionic sensitivity are concerned, MCTs differ from other investigated collagenous tissues only quantitatively, i.e. in the magnitude and rapidity of their tensile changes and reactions to ionic manipulation. What appear to be unique about MCTs are:

(1) the physicochemical basis of their variable tensility, which is likely to involve labile ionic bonds connecting ground substance macromolecules;
(2) the nervous control of their variable tensility, involving the unprecedented link between collagenous elements and motor nervous system through the juxtaligamental cells.

As observed by Eylers (cited by Wainwright 1980), each MCT displays in microcosm the spectrum of mechanical behaviour exhibited by different collagenous tissues across the other phyla. This functional flexibility—having 'conventional' mechanical properties which can be abandoned when appropriate in favour of variable tensility—may have been exploited in a number of ways.

(1) Echinoderms possess a wide repertoire of autotomy mechanisms, all investigated examples of which depend on the plasticization of MCTs (Emson and Wilkie 1980; Dobson 1985; Byrne 1985a).
(2) Observation of intact animals and *in vitro* experiments have shown that stiffening of MCTs is employed to maintain posture without the need for muscular activity in holothurian body wall and echinoid spinal joints. Although it is plausible that such a role is widespread throughout the phylum, this needs to be verified by studies on echinoderm energetics, *in vivo* monitoring of mechanical properties, and neurophysiological investigations of complementary muscles and of the nerves thought to be involved in MCT and muscle control.
(3) MCTs may be important for certain echinoderm reproductive mechanisms. A study of the fissiparous ophiuroid *Ophiocomella ophiactoides* has indicated that division may be accomplished by the plasticization of collagenous structures subtending the fission plane (Wilkie *et al.* 1984).
(4) Dafni (1986) has speculated that sutural ligaments in the echinoid test may demonstrate variable tensility, relaxing in rapidly growing sutures and stiffening when sutures stop growing. This is an interesting hypothesis which would be amenable to experimental investigation.

THE SIGNIFICANCE OF MCTS FOR ECHINODERM PALAEONTOLOGY

There is reliable evidence for the occurrence of MCTs in the five extant echinoderm classes (Wilkie 1984; and additional cases discussed above). Taking into account current views on phylogenetic relationships within the Echinodermata (e.g. Paul and Smith 1984; Smith 1984; Fig. 24.6), which claim that the latest common ancestor of living forms was a *Camptostroma*-like animal from which separate pelmatozoan and eleutherozoan lines arose, the possession of MCTs appears to be a synapomorphy (shared derived characteristic) of the crown group echinoderms (the latest common ancestor of living forms and all its descendants extinct and extant). In other words, it is highly unlikely that MCTs with their associated juxtaligamental elements evolved independently in articulate crinoids and the surviving eleutherozoan classes.

It follows from this that MCTs must have occurred in other, possibly all, Palaeozoic crown group echinoderms. We shall confine our discussion mainly to these echinoderms, mentioning helicoplacoids only briefly and carpoids not at all. In what situations might MCTs have been deployed in Palaeozoic echinoderms? Autotomy is practised by members of all five surviving echinoderm classes, and we noted above that all echinoderm autotomy mechanisms involve the irreversible disintegration of MCTs. There is evidence for autotomy in extinct echinoderms, although it is understandably sparse and often very circumstantial. It is implied by such phenomena as the frequent absence of the crown in myelodactylid crinoids (Donovan and Sevastopulo in prep.), the

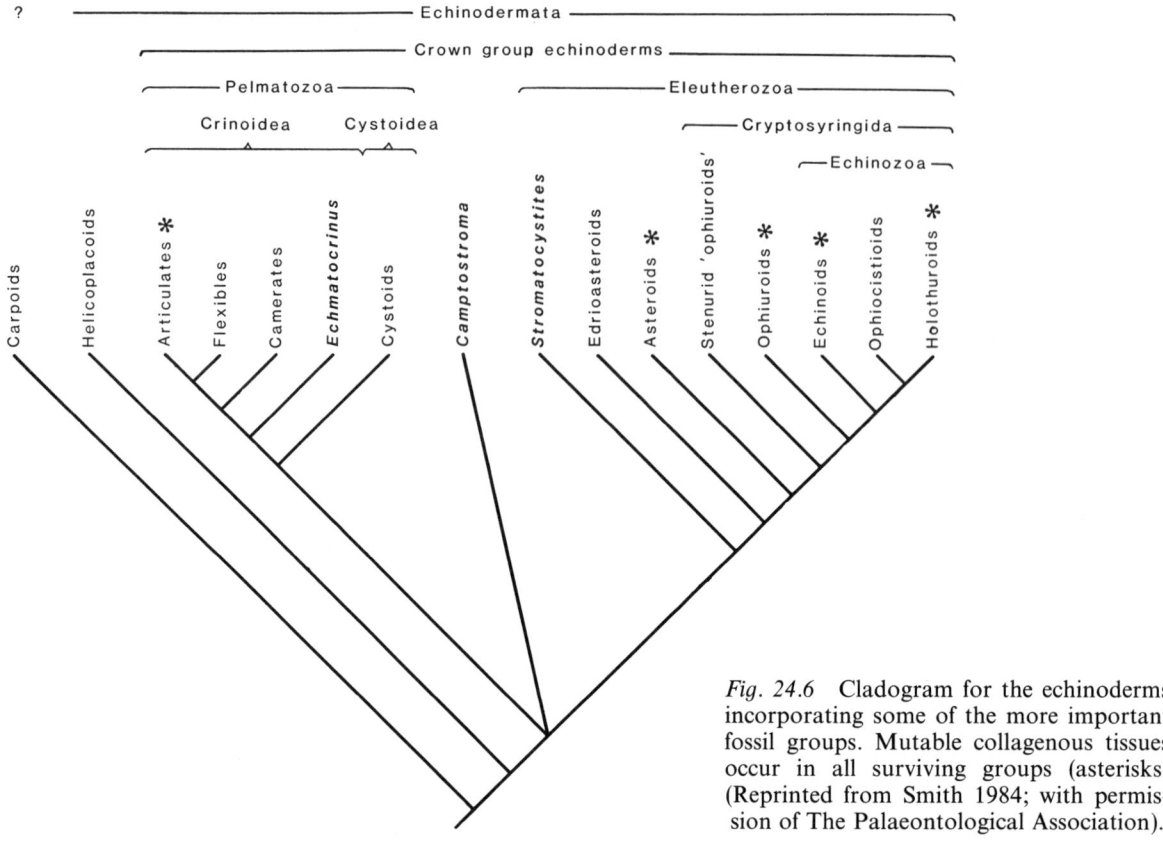

Fig. 24.6 Cladogram for the echinoderms incorporating some of the more important fossil groups. Mutable collagenous tissues occur in all surviving groups (asterisks) (Reprinted from Smith 1984; with permission of The Palaeontological Association).

apparently deciduous stems of certain diploporite cystoids (Kesling 1967) and inadunate and flexible crinoids (Ubaghs 1978a), and by the occurrence of regenerating arms in camerate crinoids (Ubaghs 1978a; for a comparison with modern articulate crinoids see Oji 1986) and fossil ophiuroids (Tasch 1980). It is possible that Palaeozoic echinoderms had as great a capacity for autotomy as modern forms, although in view of the lower predation pressure in the Palaeozoic (Ausich 1980; Sprinkle 1983; Aronson 1987) it may have served mainly to jettison diseased or damaged appendages, or to avoid adverse environmental conditions such as high sedimentation rates, rather than as an antipredator escape reaction.

We also observed in the previous section that, through reversible changes in tensility, MCTs are responsible for maintaining the posture of the body wall and appendages in some living echinoderms. Reversible changes in tensility have been demonstrated in two kinds of anatomical location (for references consult Wilkie 1984, and above):

(1) *the integumental dermis* of aspidochirote and dendrochirote holothurians (which is largely uncalcified), and of the ophiuroid arm and disc (which incorporates imbricated skeletal plates);
(2) *ligaments interconnecting the articulating ossicles* of crinoid intercirral joints, echinoid spinal joints, and asteroid spinal joints.

It is a reasonable assumption that integumental and articular MCTs have played an important role in the biology of surviving eleutherozoan classes since their first appearance in the Palaeozoic and neither these nor the extinct eleutherozoan classes will receive further consideration in this paper. We shall concentrate on the likely significance of integumental and articular MCTs for Palaeozoic pelmatozoans.

Paul and Smith (1984) have provided a clear account of the morphology and likely phyloge-

Phylogenetic significance of mutable collagenous tissues 323

tic relationships of the early Cambrian echinoderms (Fig. 24.7). All the earliest crown group echinoderms had an aboral region covered by small, usually imbricated plates, either in the form of a conical structure (*Camptostroma*) or a more differentiated stem as in lepidocystoids, the eocrinoid *Gogia*, and the primitive crinoid *Echmatocrinus*. The turret-shaped pyrgocystid edrioasteroids independently evolved a column of imbricating plates (Smith 1986). Paul and Smith believe that all these aboral structures were flexible and that at least those of *Camptostroma*, the lepidocystoids, and the pyrgocystids contained musculature which controlled their shape. A close modern equivalent is the disc integument of certain ophiuroids. We envisage that, as in the

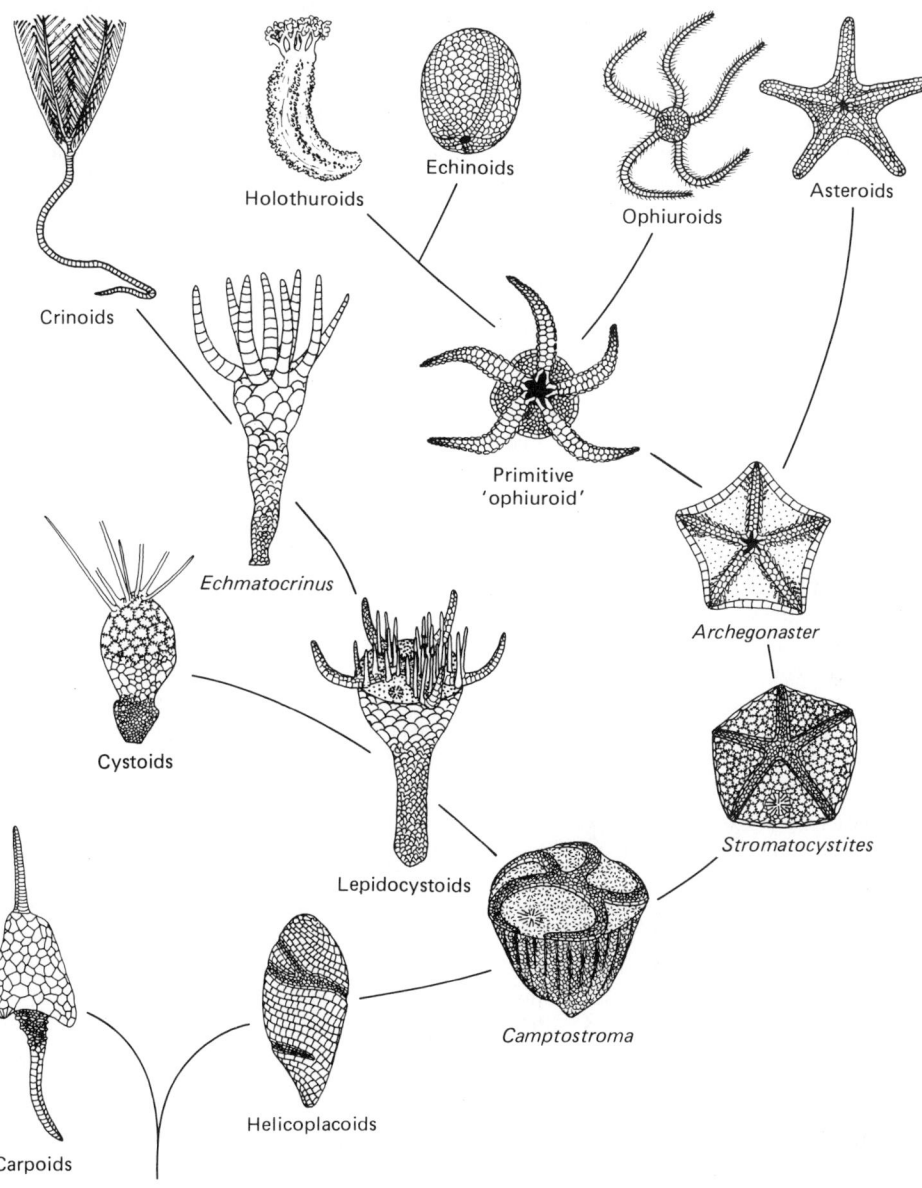

Fig. 24.7 Evolutionary tree for the early radiation and living classes of echinoderms (Reprinted from Paul and Smith 1984; with permission of Cambridge University Press).

ophiuroid integument, the imbricating plates of the Palaeozoic forms were embedded in the outer region of a collagenous dermis underlain by a layer of muscle fibres (Wilkie et al. 1984). The rigidity required of a functional stem would have depended not on constant muscular tone but on the collagenous dermis. However, the adaptive significance of an imbricated skeleton is that it facilitates distortion, and, by analogy with the ophiuroid arm and disc integument, this would have depended on the ability of the collagenous dermis to plasticize. The same arguments could be applied to the helicoplacoids at least some of whose plating is thought to have been imbricate (Paul and Smith 1984). Derstler (1982) has commented on the general functional advantages and reiterative evolution of imbricate plating in echinoderms in the context of 'variable ligaments'.

There are two likely anatomical locations for articular MCTs in Palaeozoic pelmatozoans: in the intercolumnal joints of their stems and in the joints of their exothecal appendages (brachia and brachioles). What is known about the soft tissue components at these sites? Speculation on the nature and disposition of the soft tissues of fossil skeletal articulations was formerly based only on the topology of the joint surfaces, but in recent years an additional source of information has been generated by SEM studies of fossil stereom microstructure interpreted against the correlation between soft tissue features, topology, and stereom microstructure in the joints of modern echinoderms (principal sources: Macurda 1976; Macurda and Meyer 1975, 1976; Macurda et al. 1978; Roux 1974; Smith 1980).

The stems of all but the most primitive crinoids and cystoids (in the broad sense, i.e. rhombiferans, diploporites, blastoids, eocrinoids etc.) consist of a stack of units called columnals, each of which is either a single washer-shaped ossicle (holomeric condition) or, more rarely, a ring of ossicles (meric condition). It is now known that in modern crinoids neither the various types of stem articulations nor the cirral articulations contain muscular elements (Holland and Grimmer 1981; Grimmer et al. 1984a,b, 1985), and this is also believed to have been the case in all extinct crinoids (Ubaghs 1978a). This has been deduced from general joint morphology, but is confirmed by descriptions of stereom structure in fossilized material (camerate Platycrinites: Sevastopulo and Keegan 1980; inadunate Pisocrinus: Ausich 1977; inadunate Myelodactylus: Donovan and Franzén-Bengston in prep.). The limited information available on the stems of blastoids and eocrinoids indicates that their intercolumnal articulations were symplexies or synostoses (Beaver 1967; Ubaghs 1967; Berg-Madsen 1986) which would have allowed limited flexibility and certainly lacked muscular elements. Controversial claims have been made that certain rhombiferan cystoids, notably members of the Glyptocystitida, possessed muscular stems which could control the attitude of the theca (e.g. Lepadocystis: Kesling 1967) or even act as propulsive organs (Pleurocystites: Paul 1967). These beliefs are founded solely on the stems having wide lumina which are supposed to have housed the musculature responsible for active flexion. That the stem was flexible is clear from the fulcral ridges of the columnals, whose axis of rotation varied between successive pairs of columnals (Paul 1967). However, in keeping with segmented appendages in other echinoderms, i.e. the arms of crinoids, asteroids, and ophiuroids, any stem musculature would have exhibited a segmental arrangement and have left distinctive insertion facets on the luminal sides of the columnals. No such facets have been reported. We think it more likely that the wide lumen of the glyptocystitid stem accommodated expanded coelomic spaces and that this condition evolved initially in the 'upright' members of the superfamily as a means of broadening the stem without increasing its weight. It must therefore be concluded that there is no evidence for pelmatozoan stems (other than the primitive types with imbricate plating) having had anything other than purely ligamentous stem articulations or having contained muscle at any other location.

The majority of articulations in Palaeozoic pelmatozoan exothecal appendages appear also to have been purely ligamentous. There is evidence, based on stereom microstructure and/or topology, for muscular brachial or 'brachio-radial' articulations in certain inadunate crinoids (cladid Poteriocrinina: Lane and Macurda 1975; Lane and Burke 1976; disparid Pisocrinus: Ausich 1977; unidentified: Sevastopulo and Keegan 1980) and in one camerate crinoid (Planacrocrinus: Ubaghs 1978b). All other camerates, two thirds of the inadunates (on the reckoning of Van Sant and Lane 1964), and all the flexible crinoids had only ligamentous brachial or brachio-radial articulations, judging from the trifascial, synostosial, or symplectial nature of these junctions (Ubaghs 1978a) and the limited information available on the stereom microstructure of the joint facets (unidentified flexible: Sevastopulo and Keegan 1980; inadunate *Barycrinus*: Ausich 1983). The primitive crinoid *Echmatocrinus* has V-shaped areas on the distal aboral edge of its lower brachials which Sprinkle and Moore (1978) suggested could represent muscle insertion sites. Since muscles in the brachial articulations of later crinoids are always located on the oral side of the fulcral ridge, it is much more likely that these scars are aboral ligament attachment areas. Joint topology indicates that the pinnules of camerate and inadunate crinoids lacked muscles with the possible exception of the mobile junction between the first pinnular and the arm (Ubaghs 1978a). There is, equally, no convincing evidence for muscular articulations within the brachioles or between the brachioles and the theca of any cystoid (see relevant sections of *Treatise on invertebrate paleontology S, Echinodermata 2*, Vols. 1 and 2). Macurda's (1973) observations on stereom microstructure confirm this for blastoids and there is an obvious need for comparable data on other cystoids.

From this survey it appears that muscle was a rarity in Palaeozoic pelmatozoans, at least muscle associated with skeletal articulations (which we shall refer to as 'skeletal musculature'), and that the vast majority of their joints, whether rigid or mobile, were subtended by ligaments alone. Did these animals therefore spend the whole of their lives in the same state of rigidity or unvarying flexibility? We believe that this is highly unlikely and that they must have possessed the wherewithal to cope with varying external factors such as changing current direction and velocity, changing sedimentation rates, or collision with other sessile or vagile animals. Since they possessed little, or no, skeletal musculature, such adaptability could have been provided only by MCTs. Stiffened MCTs could have served to maintain favourable feeding or defensive postures, and repositioning of the stem or exothecal appendages could have been achieved through a combination of MCT relaxation, elastic recoil of previously stretched (non-mutable?) ligaments, and external forces due to gravity, horizontal water currents, and water turbulence around the theca. Such mechanisms can account for the grasping function of the purely ligamentous cirri of present-day comatulid crinoids (Wilkie 1983); unfortunately, the involvement of MCTs in the postural control of the brachia and stem of the only extant pelmatozoans, i.e. the stalked crinoids, is still a matter for speculation (Breimer and Webster 1975; Breimer 1978; Grimmer et al. 1984a). Experimental investigation of the effector systems of these animals would test the feasibility of our contention that MCTs were an important factor in the biology of Palaeozoic pelmatozoans and that this should be taken into account in any attempt to interpret the functional morphology and behavioural possibilities of these animals.

THE SIGNIFICANCE OF MCTS FOR ECHINODERM PHYLOGENY

If the fossil record is truly representative, the echinoderm endoskeleton first appeared as small, loosely linked plates embedded in a collagenous dermis (cf. helicoplacoids, edrioasteroids, lepidocystoids, and *Echmatocrinus*: Paul and Smith 1984). In our opinion, the property of variable tensility arose initially in this dermis either before or at the same time as the stereomic skeleton, enabling the dermis to function as a tonic musculature which maintained body wall shape arrived at through the action of subdermal muscle layers. This function has persisted to the present day in the dermis of holothurians and probably in that of ophiuroids and asteroids. In view of their presence in the five surviving echinoderm classes, juxtaligamental elements must have occurred in these early dermal MCTs. We would speculate that evolving with them would have been a hyponeural nervous system which only secondarily became involved in the control of skeletal muscles when these began to appear in the later Palaeozoic and Mesozoic. Cobb (1985b) has recently suggested that the hyponeural system is mesodermal in embryonic origin. Should this be verified, then a key event in the history of the Echinodermata may have been the evolution of a unique *mesodermal neuroeffector system* which comprised initially:

(1) the mutable dermis and the articular ligaments that evolved from it;
(2) the endoskeleton;
(3) the juxtaligamental cells;
(4) the hyponeural nervous system;
(5) and (at a later stage) the skeletal musculature.

In the pelmatozoan lineage from *Camptostroma*,

it is likely that with the development of holomeric stems and heavily calcified thecae, subdermal muscle layers became redundant and were lost altogether. Although the lack of muscles in both the stems and exothecal appendages of perhaps the majority of Palaeozoic pelmatozoans may have been partly compensated for by the presence of MCTs, such a lack would still seem to be a disadvantageous feature. Clearly, however, it was not: echinoderms displayed an explosive adaptive radiation in the early Palaeozoic with, on Sprinkle's (1983) reckoning, 20 classes appearing between the earliest Cambrian and Middle Ordovician. Far from being a disadvantage, the possession of an effector system consisting of MCTs rather than muscles may have been one of the factors contributing to the success of the echinoderms at this time.

The early echinoderm fauna was dominated by sessile passive suspension feeders which during the course of the Palaeozoic exhibited a striking diversification (Ausich and Bottjer 1985). For example, Ausich (1980) observed that during the radiation of Palaeozoic crinoids maximum stem height increased resulting in vertical tiering, while at each vertical level different types of filtration fan evolved to maximize utilization of food resources and minimize competition. Obviously these echinoderms were competing against other suspension feeders such as corals, bryozoans, and brachiopods. Of necessity, suspension feeders have to maintain extended feeding postures for long periods of time. If maintenance of the feeding posture depended on the tonic contraction of conventional muscles, it would be energetically expensive and perhaps compromise the cost-effectiveness of suspension feeding as a nutritional strategy. Suspension feeders therefore show features that can be interpreted as adaptations for maintaining the posture of the body or feeding apparatus at low energy cost. Such features can be identified in coelenterates, bryozoans, and brachiopods (Table 24.4), and are to be expected in suspension-feeding echinoderms. One such feature may be the expansion of coelomic spaces and of energetically cheap endoskeleton as a strategy for increasing adult size and post-metamorphic growth rate (Emson 1985). However, perhaps the single most important adaptation of Palaeozoic echinoderms for minimizing energy expenditure during suspension feeding was the absence of metabolically demanding skeletal musculature from the body and their dependence for postural control on metabolically undemanding MCTs. This must have been especially important in animals having such an extensive endoskeleton, the active movement of which would have required a considerable bulk of muscle. That these animals had low metabolic requirements is probable, since modern echinoderms, even though they possess well-developed muscles, have very low respiratory rates (e.g. see Johnson 1973; LaBarbera 1982; Bianconcini et al. 1985). We would, therefore, suggest that the presence of MCTs was a pre-adaptation without which the diversification of Palaeozoic pelmatozoans could not have occurred.

From the Middle Ordovician to the end of the

TABLE 24.4 Features of coelenterates, bryozoans, and brachiopods that can be interpreted as energy-sparing adaptations for maintaining the posture of the body and feeding apparatus. The high paramyosin content of bryozoan and brachiopod muscle is included because paramyosin-rich catch muscles in bivalve molluscs can maintain tonic contraction with minimal expenditure of energy

Phylum	Energy-sparing adaptation(s)	References
Coelenterata	Body and tentacle extension maintained by passive properties of the mesogloea	Alexander 1962; Chapman 1970; Koehl 1977
Bryozoa	Eversion of lophophore maintained by paramyosin-rich muscles; extension of tentacles maintained by passive properties of collagenous 'basal lamina'	Gordon 1974
Brachiopoda	Body muscles have very high paramyosin content; lophophore extension may rely on elastic recoil of relaxing anterior adductor muscles acting as compression springs; extension of lophophore tentacles maintained by passive properties of collagenous 'cylinders'	Rudwick 1970; Winkelman 1976; Reed and Cloney 1977

Palaeozoic echinoderm class diversity declined from 17 to 6 (Sprinkle 1983). One factor in the elimination of the majority of these classes may have been their over-dependence on MCTs in environments where, due to increasing competition and predation pressure, the capacity for active movement was becoming a desirable asset. We argued above that cystoids entirely lacked skeletal muscles: no cystoids survived beyond the Palaeozoic. Flexible crinoids, which were non-muscular, and camerate crinoids, very few of which had skeletal muscles, failed to cross the Permo–Triassic boundary; of the inadunate crinoids, only the highly successful Poteriocrinina, which had muscular brachio-radial articulations, survived into the Triassic. It is believed that the Poteriocrinina gave rise in the Triassic to the exclusively post-Palaeozoic articulates with their extensively muscularized arms (Rasmussen 1978; Simms, this volume). A fascinating parallel is seen in the history of the eleutherozoan Asteroidea: joint topology and stereom microstructure indicate that most Palaeozoic asteroids lacked muscular ambulacral articulations, but that the evolution of the post-Palaeozoic Neoasteroidea was marked by a progressive muscularization of the ambulacral region (Gale 1987). However, if the absence of muscle and its consequent energetic advantages was an element in the earlier success of Palaeozoic pelmatozoans, the increased metabolic requirements accompanying the acquisition of an extensive musculature by articulate crinoids must have been offset by compensating factors such as the exploitation of new habitats offering more abundant food supplies or an improvement in the efficiency with which food was collected or assimilated.

CONCLUSIONS

Extant echinoderms possess highly unusual collagenous tissues which can undergo rapid, nervously mediated changes in tensile properties. We hope that our review has demonstrated that the investigation of these tissues has become a distinct and active area of current echinoderm studies which receives input from, and has implications for, many aspects of echinoderm research, e.g. ecology, behaviour, functional morphology, functional biomechanics, neurophysiology, energetics, and biochemistry. Although the evidence is as yet incomplete, we consider it likely that mutable collagenous tissues play an important role in the biology of all five surviving echinoderm classes. For this reason, and because palaeontology has provided remarkably little evidence for the presence of muscles associated with the skeleton of Palaeozoic echinoderms, especially the Pelmatozoa, we have argued that MCTs were crucial in the lives of extinct forms, and indeed that they, as part of a unique mesodermal neuro-effector system, were both an enabling factor in the early adaptive radiation of the echinoderms and in part responsible for the later Palaeozoic decline in class diversity.

We hope that the apparent extravagance of these views will encourage further work on the functional morphology and ecomechanics of living echinoderms, particularly stalked crinoids, and further microstructural investigations of fossil material. With regard to the latter approach, it is likely that the extension to other groups of work emulating Macurda's (1973) on blastoids, Ausich's (1977, 1983) on crinoids, and Gale's (1987) on asteroids would do much to solve several outstanding problems in echinoderm palaeontology and phylogeny, not the least of which is the carpoid controversy.

REFERENCES

Alexander, R. McN. 1962. Visco-elastic properties of the body-wall of sea anemones. *Journal of Experimental Biology* **39**, 373–86.

Aronson, R. B. 1987. Predation on fossil and Recent ophiuroids. *Paleobiology* **13**, 187–92.

Ausich, W. I. 1977. The functional morphology and evolution of *Pisocrinus* (Crinoidea: Silurian). *Journal of Paleontology* **51**, 672–86.

Ausich, W. I. 1980. A model for niche differentiation in Lower Mississippian crinoid communities. *Journal of Paleontology* **54**, 273–88.

Ausich, W. I. 1983. Functional morphology and feeding dynamics of the early Mississippian crinoid *Barycrinus asteriscus*. *Journal of Paleontology* **57**, 31–41.

Ausich, W. I. and Bottjer, D. J. 1985. Echinoderm role in the history of Phanerozoic tiering in suspension-feeding communities. In *Echinodermata. Proceedings of the fifth International Conference* (ed. B. F. Keegan and B. D. S. O'Connor), pp. 3–11. Balkema, Rotterdam.

Bailey, A. J. 1985. The collagen of the Echinodermata. In *Biology of invertebrate and lower vertebrate collagens* (ed. A. Bairati and R. Garrone), pp. 369–88. Plenum, New York.

Beaver, H. H. 1967. Morphology of blastoids. In

Treatise on invertebrate paleontology, part S, Echinodermata 1, Vol. 2 (ed. R. C. Moore), pp. 300–44. Geological Society of America and University of Kansas Press, Lawrence, Kansas.

Berg-Madsen, V. 1986. Middle Cambrian cystoid (*sensu lato*) stem columnals from Bornholm, Denmark. *Lethaia* **19**, 67–80.

Bianconcini, M. S., Mendes, E. G., and Valente, D. 1985. The respiratory metabolism of the lantern muscles of the sea urchin *Echinometra lucunter* L.-I. The respiratory intensity. *Comparative Biochemistry and Physiology* **80A**, 1–4.

Biglow, C. E. 1981. Investigation of variable tensility in echinoderm connective tissue. BSc thesis, University of Victoria.

Breimer, A. 1978. Paleoecology: autecology. In *Treatise on invertebrate paleontology*, part T, Echinodermata 2, Vol. 1 (ed. R. C. Moore and C. Teichert), pp. 331–43. Geological Society of America and University of Kansas Press, Lawrence, Kansas.

Breimer, A. and Webster, G. D. 1975. A further contribution to the paleoecology of fossil stalked crinoids. *Proceedings. Koninklijke Nederlandse Akademie van Wetenschappen B* **78**, 149–67.

Byrne, M. 1985a. Ultrastructural changes in the autotomy tissues of *Eupentacta quinquesemita* (Selenka) (Echinodermata: Holothuroidea) during evisceration. In *Echinodermata. Proceedings of the fifth International Conference* (ed. B. F. Keegan and B. D. S. O'Connor), pp. 413–20. Balkema, Rotterdam.

Byrne, M. 1985b. The mechanical properties of the autotomy tissues of the holothurian *Eupentacta quinquesemita* and the effects of certain physicochemical agents. *Journal of Experimental Biology* **117**, 69–86.

Byrne, M. 1986. Induction of evisceration in the holothurian *Eupentacta quinquesemita* and evidence for the existence of an endogenous evisceration factor. *Journal of Experimental Biology* **120**, 25–40.

Chapman, D. M. 1970. Re-extension mechanism of a scyphistoma's tentacle. *Canadian Journal of Zoology* **48**, 931–43.

Cobb, J. L. S. 1985a. The motor innervation of the oral plate ligament in the brittlestar *Ophiura ophiura* (L.). *Cell and Tissue Research* **242**, 686–8.

Cobb, J. L. S. 1985b. Intracellular studies on the nervous system of an echinoderm. In *Echinodermata. Proceedings of the fifth International Conference* (ed. B. F. Keegan and B. D. S. O'Connor), pp. 617–21. Balkema, Rotterdam.

Cuénot, L. 1891. Études morphologiques sur les echinodermes. *Archives de Biologie, Paris* **11**, 313–680.

Dafni, J. 1986. A biomechanical model for the morphogenesis of regular echinoid tests. *Paleobiology* **12**, 143–60.

Derstler, K. 1982. Variable ligaments and imbricate plates in echinoderm skeletons. In *Echinoderms: Proceedings of the International Conference, Tampa Bay* (ed. J. M. Lawrence), p. 77. Balkema, Rotterdam.

Diab, M. and Gilly, W. F. 1984. Mechanical properties and control of non-muscular catch in spine ligaments of the sea urchin, *Strongylocentrotus franciscanus*. *Journal of Experimental Biology* **111**, 155–70.

Dobson, W. E. 1985. A pharmacological study of neural mediation of disc autotomy in *Ophiophragmus filograneus* (Lyman) (Echinodermata: Ophiuroidea). *Journal of Experimental Biology* **94**, 223–32.

Dobson, W. E. and Turner, R. L. 1983. Morphological and histological basis of disc autotomy in a brittle star. *American Zoologist* **23**, 1026.

Emson, R. H. 1985. Bone idle—a recipe for success? In *Echinodermata. Proceedings of the fifth International Conference* (ed. B. F. Keegan and B. D. S. O'Connor), pp. 25–30. Balkema, Rotterdam.

Emson, R. H. and Wilkie, I. C. 1980. Fission and autotomy in echinoderms. *Oceanography and Marine Biology Annual Review* **18**, 155–250.

Eylers, J. P. 1985. Dynamic connective tissues: the echinoderm paradigm. In *Echinodermata. Proceedings of the fifth International Conference* (ed. B. F. Keegan and B. D. S. O'Connor), p. 153. Balkema, Rotterdam.

Folk, J. E. 1980. Transglutaminases. *Annual Review of Biochemistry* **49**, 517–31.

Fry, P., Harkness, M. L. R., and Harkness, R. D. 1964. Mechanical properties of the collagenous framework of skin in rats of different ages. *American Journal of Physiology* **206**, 1425.

Gale, A. S. 1987. Phylogeny and classification of the Asteroidea (Echinodermata). *Zoological Journal of the Linnean Society* **89**, 107–32.

Gay, S. and Miller, E. J. 1983. What is collagen, what is not? *Ultrastructural Pathology* **4**, 365–77.

Gordon, D. P. 1974. Microarchitecture and function of the lophophore in the bryozoan *Cryptosula pallasiana*. *Marine Biology* **27**, 147–63.

Greenberg, A. R. and Eylers, J. P. 1984. Influence of ionic environment on the stress relaxation behavior of an invertebrate connective tissue. *Journal of Biomechanics* **17**, 161–6.

Grimmer, J. C., Holland, N. D., and Kubota, H. 1984a. Fine structure of the stalk of the pentacrinoid larva of a feather star, *Comanthus japonica* (Echinodermata: Crinoidea). *Acta Zoologica, Stockholm* **65**, 41–58.

Grimmer, J. C., Holland, N. D., and Messing, C. G. 1984b. Fine structure of the stalk of the bourgueticrinid sea lily *Democrinus conifer* (Echinodermata: Crinoidea). *Marine Biology* **81**, 163–76.

Grimmer, J. C., Holland, N. D., and Hayami, I. 1985. Fine structure of the stalk of an isocrinid sea lily (*Metacrinus rotundus*) (Echinodermata, Crinoida). *Zoomorphology* **105**, 39–50.

Hamann, O. 1887. Beiträge zur Histologie der Echinodermen. *Jenaische Zeitschrift für Medizin und Naturwissenschaften* **21**, 87–266.

Harkness, M. L. R. and Harkness, R. D. 1959. Changes in the physical properties of the uterine cervix of the rat during pregnancy. *Journal of Physiology, London* **148**, 524–47.

Harkness, M. L. R. and Harkness, R. D. 1965. Functional alterations in the mechanical properties of collagenous frameworks. In *Structure and function of connective and skeletal tissue* (ed. S. F. Jackson, R. D. Harkness, S. M. Partridge, and G. R. Tristram), pp. 376–81. Butterworths, London.

Hayashi, Y. and Motokawa, T. 1986. Effects of ionic environment on viscosity of catch connective tissue in holothurian body wall. *Journal of Experimental Biology* **125**, 71–84.

Holland, N. D. and Grimmer, J. C. 1981. Fine structure of the cirri and a possible mechanism for their motility in stalkless crinoids (Echinodermata). *Cell and Tissue Research* **214**, 207–17.

Johnson, W. S. 1973. Respiration rates of some New Zealand echinoderms. *New Zealand Journal of Marine Biology and Freshwater Research* **7**, 165–9.

Kesling, R. V. 1967. Cystoids. In *Treatise on invertebrate paleontology, part S, Echinodermata 1, Vol. 1* (ed. R. C. Moore), pp. 85–267. Geological Society of America and University of Kansas Press, Lawrence, Kansas.

Koehl, M. A. R. 1977. Mechanical diversity of connective tissues of the body wall of sea anemones. *Journal of Experimental Biology* **69**, 107–25.

LaBarbera, M. 1982. Metabolic rates of suspension feeding crinoids and ophiuroids in a unidirectional laminar flow. *Comparative Biochemistry and Physiology* **71A**, 303–7.

Lane, N. G. and Burke, J. J. 1976. Arm movement and feeding mode of inadunate crinoids with biserial muscular arm articulations. *Paleobiology* **2**, 202–8.

Lane, N. G. and Macurda, D. B. 1975. New evidence for muscular articulations in Paleozoic crinoids. *Paleobiology* **1**, 59–62.

Macurda, D. B. 1973. The stereomic microstructure of the blastoid endoskeleton *Contributions from the Museum of Paleontology, University of Michigan* **24**, 69–83.

Macurda, D. B. 1976. Skeletal modifications related to food capture and feeding behavior of the basketstar *Astrophyton*. *Paleobiology* **2**, 1–7.

Macurda, D. B. and Meyer, D. L. 1975. The microstructure of the crinoid endoskeleton. *The University of Kansas, Paleontological Contributions, Paper* **74**, 1–22.

Macurda, D. B. and Meyer, D. L. 1976. The morphology and life habits of the abyssal crinoid *Bathycrinus aldrichianus* Wyville Thomson and its paleontological implications. *Journal of Paleontology* **50**, 647–67.

Macurda, D. B., Meyer, D. L., and Roux, M. 1978. The crinoid stereom. In *Treatise on invertebrate paleontology, part T, Echinodermata 2, Vol. 1* (ed. R. C. Moore and C. Teichert), pp. 217–28. Geological Society of America and University of Kansas Press, Lawrence, Kansas.

Minns, R. J., Soden, P. D., and Jackson, D. S. 1973. The role of the fibrous components and ground substance in the mechanical properties of biological tissues: a preliminary investigation. *Journal of Biomechanics* **6**, 153–65.

Minns, R. J. and Steven, F. S. 1977. The effect of calcium on the mechanical behaviour of aorta media elastin and collagen. *British Journal of Experimental Pathology* **58**, 572–9.

Motokawa, T. 1982. Factors regulating the mechanical properties of holothurian dermis. *Journal of Experimental Biology* **99**, 29–41.

Motokawa, T. 1984a. Connective tissue catch in echinoderms. *Biological Reviews* **59**, 255–70.

Motokawa, T. 1984b. The viscosity change of the bodywall dermis of the sea-cucumber *Stichopus japonicus* caused by mechanical and chemical stimulation. *Comparative Biochemistry and Physiology* **77A**, 419–23.

Motokawa, T. 1984c. Viscosity of holothurian body wall. *Journal of Experimental Biology* **109**, 63–75.

Motokawa, T. 1984d. Viscosity increase of holothurian body wall in response to photic stimulation. *Comparative Biochemistry and Physiology* **79A**, 501–3.

Motokawa, T. 1985. Catch connective tissue: the connective tissue with adjustable mechanical properties. In *Echinodermata. Proceedings of the fifth International Conference* (ed. B. F. Keegan and B. D. S. O'Connor), pp. 69–73. Balkema, Rotterdam.

Motokawa, T. 1986. Morphology of spines and spine joint in the crown-of-thorns starfish *Acanthaster planci* (Echinodermata, Asteroida). *Zoomorphology* **106**, 247–53.

Motokawa, T. 1987. Cholinergic control of the mechanical properties of the catch connective tissue in the holothurian body wall. *Comparative Biochemistry and Physiology* **86C**, 333–7.

Naftolin, F. and Stubblefield, P. G. 1980. *Dilatation of the uterine cervix*. Raven Press, New York.

Oji, T. 1986. Skeletal variation related to arm regeneration in *Metacrinus* and *Saracrinus*, Recent stalked crinoids. *Lethaia* **19**, 355–60.

Paul, C. R. C. 1967. The functional morphology and mode of life of the cystoid *Pleurocystites*, E. Billings, 1854. *Symposia of the Zoological Society of London* **20**, 105–23.

Paul, C. R. C. and Smith, A. B. 1984. The early radiation and phylogeny of echinoderms. *Biological Reviews* **59**, 443–81.

Perrier, E. 1889. Mémoire sur l'organisation et le développement de la comatule de la Méditerranée (*Antedon rosacea* Linck). Part 3. Organisation de l'*Antedon* adults. I–IX. *Nouvelles Archives du Muséum d'Histoire Naturelle, Paris*, Series 3 **1**, 169–286.

Peters, B. H. 1985. The innervation of spines in the sea-urchin *Echinus esculentus* L. An electron-microscope study. *Cell and Tissue Research* **239**, 219–28.

Rasmussen, H. W. 1978. Evolution of articulate crinoids. In *Treatise on invertebrate paleontology, part T, Echinodermata 2, Vol. 1* (ed. R. C. Moore and C. Teichert), pp. 302–16. Geological Society of America and University of Kansas Press, Lawrence, Kansas.

Reed, C. G. and Cloney, R. A. 1977. Brachiopod tentacles: ultrastructure and functional significance of the connective tissue and myoepithelial cells in *Terebratalia*. *Cell and Tissue Research* **185**, 17–42.

Roux, M. 1974. Observations au microscope électronique à balayage de quelques articulations entre les ossicules du squelette des crinoïdes pédonculés actuels (Bathycrinidae et Isocrinina). *Travaux du Laboratoire de Paléontologie, Orsay*.

Rudwick, M. J. S. 1970. *Living and fossil brachiopods*. Hutchinson, London.

Rüegg, J. C. 1971. Smooth muscle tone. *Physiological Reviews* **51**, 210–48.

Serra-von Buddenbrock, E. 1963. Etudes physiologiques et histologiques sur le tegument des holothuries (*Holothuria tubulosa*). *Vie et Milieu* **14**, 55–70.

Sevastopulo, G. D. and Keegan, J. B. 1980. A technique for revealing the stereom microstructure of fossil crinoids. *Palaeontology* **23**, 749–56.

Smith, A. B. 1980. Stereom microstructure of the echinoid test. *Special Papers in Paleontology* **25**, 1–81.

Smith, A. B. 1984. Classification of the Echinodermata. *Palaeontology* **27**, 431–59.

Smith, A. B. 1985. Cambrian eleutherozoan echinoderms and the early diversification of edrioasteroids. *Palaeontology* **28**, 715–56.

Sprinkle, J. 1983. Patterns and problems in echinoderm evolution. *Echinoderm Studies* **1**, 1–18.

Sprinkle, J. and Moore, R. C. 1978. Echmatocrinea. In *Treatise on invertebrate paleontology, part T, Echinodermata 2, Vol. 2* (ed. R. C. Moore and C. Teichert), pp. 405–7. Geological Society of America and University of Kansas Press, Lawrence, Kansas.

Takahashi, K. 1967a. The catch apparatus of the sea-urchin spine I. Gross histology. *Journal of the Faculty of Science, Tokyo University, Section IV* **11**, 109–20.

Takahashi, K. 1967b. The catch apparatus of the sea-urchin spine II. Responses to stimuli. *Journal of the Faculty of Science, Tokyo University, Section IV* **11**, 121–30.

Tasch, P. 1980. *Paleobiology of the invertebrates*, 2nd edn. Wiley, New York.

Ubaghs, G. 1967. Eocrinoidea. In *Treatise on invertebrate paleontology, part S, Echinodermata 1, Vol. 2* (ed. R. C. Moore), pp. 455–95. Geological Society of America and University of Kansas Press, Lawrence, Kansas.

Ubaghs, G. 1978a. Skeletal morphology of fossil crinoids. In *Treatise on invertebrate paleontology, part T, Echinodermata 2, Vol. 1* (ed. R. C. Moore and C. Teichert), pp. 58–216. Geological Society of America and University of Kansas Press, Lawrence, Kansas.

Ubaghs, G. 1978b. Camerata. In *Treatise on invertebrate paleontology, part T, Echinodermata 2, Vol. 2* (ed. R. C. Moore and C. Teichert), pp. 408–519. Geological Society of America and University of Kansas Press, Lawrence, Kansas.

Uexküll, J. von 1900. Die Physiologie des Seeigelstachels. *Zeitschrift für Biologie* **39**, 73–112.

Van Sant, J. F. and Lane, N. G. 1964. Crawfordsville (Indiana) crinoid studies. *The University of Kansas, Paleontological Contributions, Article* **35**, 1–136.

Viidik, A. 1980. Interdependence between structure and function in collagenous tissues. In *Biology of collagen* (ed. A. Viidik and J. Vuust), pp. 257–80. Academic Press, London.

Vogel, H. G. 1978. Influence of maturation and age on mechanical and biochemical parameters of connective tissue of various organs in the rat. *Connective Tissue Research* **6**, 161–6.

Wainwright, S. A. 1980. Adaptive materials: a view from the organism. *Symposium of the Society for Experimental Biology* **34**, 437–53.

Wilkie, I. C. 1976. A study of the process of ophiuroid arm autotomy. PhD thesis, University of Glasgow.

Wilkie, I. C. 1979. The juxtaligamental cells of *Ophiocomina nigra* (Abildgaard) (Echinodermata: Ophiuroidea) and their possible role in mechano-effector function of collagenous tissue. *Cell and Tissue Research* **197**, 515–30.

Wilkie, I. C. 1983. Nervously mediated change in the mechanical properties of the cirral ligaments of a crinoid. *Marine Behavioural Physiology* **9**, 229–48.

Wilkie, I. C. 1984. Variable tensility in echinoderm collagenous tissues: a review. *Marine Behavioural Physiology* **11**, 1–34.

Wilkie, I. C. and Emson, R. H. 1987. The tendons of *Ophiocomina nigra* and their role in autotomy (Echinodermata, Ophiuroida). *Zoomorphology* **107**, 33–44.

Wilkie, I. C., Emson, R. H., and Mladenov, P. V. 1984. Morphological and mechanical aspects of fission in *Ophiocomella ophiactoides* (H. L. Clark) (Echinodermata: Ophiuroida). *Zoomorphology* **104**, 310–22.

Winkelman, L. 1976. Comparative studies of paramyosins. *Comparative Biochemistry and Physiology* **55B**, 391–7.

25

Origins of the deep-sea holasteroid fauna

B. DAVID

U.A. CNRS 157, Centre des Sciences de la Terre, 6 bd. Gabriel
F-21100 Dijon, France and GRECO CNRS 'Ecoprophyce'

INTRODUCTION

7340 m! That's the greatest depth recorded for an irregular sea urchin (Mironov 1978b).

The echinoderms are amongst the best adapted and the most abundant organisms in the deep-sea: often constituting more than a third of the deep epibenthic macrofauna, as illustrated by recent quantitative investigations in several areas of the Atlantic (Sibuet 1985). Compared with the other classes of echinoderms, the deep-sea echinoids are generally neither very numerous, nor very diverse (Sibuet 1980; David and Sibuet 1985). However, they are often represented by forms that are very specialized and which show unusual features, arousing the enthusiasm of former echinologists: 'At first glance this appears one of the most remarkable of sea urchins' (Agassiz 1881, p. 195, about the aeropsid *Aceste bellidifera*); 'In the Pourtalesiidae the test becomes extremely elongated, reaching the limit in the almost tubular *Echinosigra*' (Clark 1917, p. 94).

Among the recent irregular echinoids, the holasteroids are the most successful deep-sea group and include species with strangely transformed architectures and shapes (Fig. 25.1). A histogram of their bathymetric distribution shows that they are mainly bathyal and abyssal, with an optimum between 1500 and 4500 m (Fig. 25.2). Thus, the recent holasteroids represent a well delimited ecological entity. Is this entity the result of a common historical pathway or the result of an opportunistic gathering? That is to say: is the origin of these sea-urchins unique or not? And what are the relationships between the Recent forms? Despite their importance and originality, almost nothing is known about the origin of

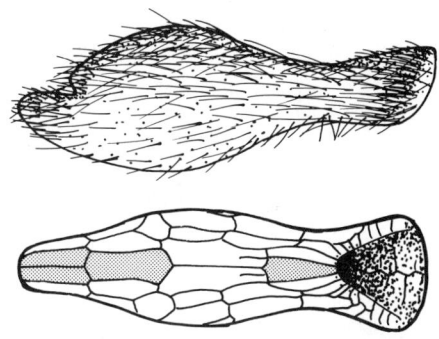

Fig. 25.1 Echinosigra phiale: lateral view (redrawn from Mortensen 1907), and architecture of the adoral side (interambulacrum 5 is stippled).

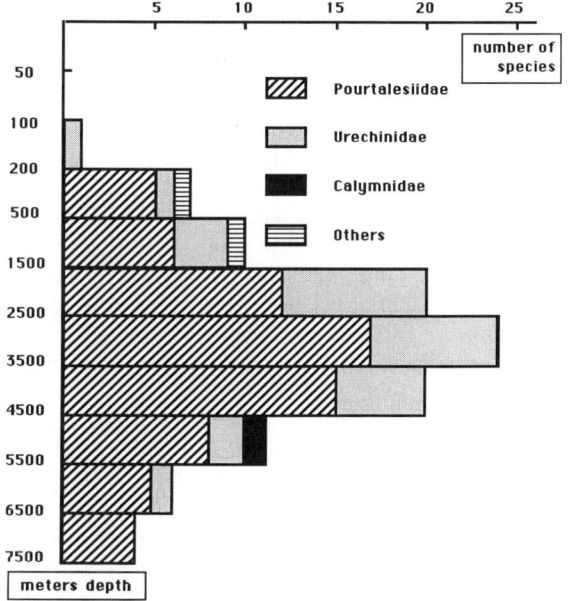

Fig. 25.2 Histogram of bathymetric distribution of the Recent species of Holasteroida (data from Mortensen 1950–51; Mironov 1976, 1978a).

Echinoderm phylogeny and evolutionary biology (ed. C. R. C. Paul and A. B. Smith). Clarendon Press, Oxford, 1988.

Recent holasteroids and these questions have previously remained unanswered.

The order Holasteroida appeared at the beginning of the Cretaceous (Fig. 25.3). The earliest forms were epicontinental, such as *Holaster*, which is the only genus of Holasteroida until the Aptian (except the closely related genus *Taphraster*). A great diversification of the group occurred during

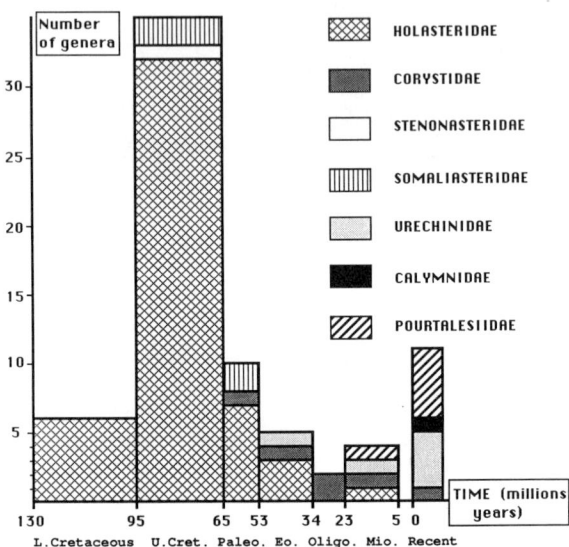

Fig. 25.3 Stratigraphic distribution of the genera of Holasteroida (data from various sources).

the Upper Cretaceous with a large expansion of the family Holasteridae. Subsequently, there was a progressive decline of the group in the Tertiary, particularly after the Eocene. Compared to the Tertiary forms, the Recent deep-sea holasteroids appear well diversified. However, the scarceness of the Tertiary fauna is partly due to the great difficulty in preserving deep-sea fossils, which needs great elevation of oceanic deposits without strong metamorphism. Moreover, the density of echinoid settlement from abyssal regions is generally low: 0 to 78 individuals per 10^4 m^2 in the Bay of Biscay (David and Sibuet 1985), up to 397 ind./10^4 m^2 in the Norwegian Sea which is a particularly rich area (Dahl *et al.* 1977). Considering these two restrictive conditions the discovery of a deep-sea fossil is extremely fortuitous. Indeed, *Chelonechinus* (Gregory 1889; Bather 1934), *Sanchezaster* (Sanchez-Roig 1924) and *Pourtalesia* (Kikuchi and Nikaïdo 1985) represent the only known occurrences of holasteroids in Tertiary deep-sea deposits.

Such a fossil record provided few constraints in the formulation of hypotheses about the origin of the abyssal holasteroids. Mortensen (1907) derived the Pourtalesiidae from the 'Ananchytidae' (viz. the Cretaceous Holasteridae). Some years later, he suggested in his monograph (1950–51) that they derived from the Disasteridae. Termier and Termier (1953) presented a phylogenetic chart where the Recent families (Pourtalesiidae, Urechinidae, and Calymnidae) were shown to arise from Lower Cretaceous forms. Mironov (1980) suggested that the deep-sea echinoid fauna has a Miocene origin. All these hypotheses were based on biogeographic or palaeonto-morphological arguments. The biogeographic argument is questionable because current knowledge of the biogeography of abyssal forms is linked to the opportunity of cruises and samplings which give a partial view of distributions. The palaeontological argument is very tenuous because of the sparse fossil record. There remains the morphology. However, morphology was never explicitly used to understand the history of these sea urchins. Previous workers were mainly discoverers and describers who made occasional comparisons. Their aims were not to ask the question: how did the colonization of deep-sea environments by Holasteroida happen?

MATERIALS AND METHODS

The order Holasteroida includes 61 genera belonging to seven families (Table 25.1, according to Wagner and Durham, 1966, with slight modifications inspired by Ernst 1972, and Foster and Philip 1978). There are 13 living genera distributed within four families.

Concerning the living forms, I have made direct observations from the collections of the *Challenger* (British Museum) and the material collected during several oceanographic cruises (BIOGAS program, INCAL, MD32, NORBI, WALDA organized by the IFREMER or by the TAAF and the Muséum National d'Histoire Naturelle). Complementary information is provided by precise descriptions and illustrations from Agassiz (1881, 1904), Lovén (1883), de Meijere (1904), Mortensen (1907, 1950–51). Concerning the fossil genera, my information is mainly bibliographic (Lambert

TABLE 25.1 List of the families and genera of Holasteroida, according to the usual classifications. Families and genera are listed alphabetically. The letter in front of the generic name in the group column corresponds to the groups used on the cladograms (Fig. 25.7–25.9). Recent forms are shown by **R**. The third column gives the nature of information on which the study is based.

Families	Group	Genera	Support of the study
Calymnidae	I	*Calymne* **R**	Museum collections: British Museum Bibliography: Agassiz 1881; Mortensen 1950
Corystidae	G	*Corystus*	Museum collections: British Museum
	G	*Cardabia*	Bibliography: Foster and Philip 1978
	G	*Huttonechinus*	
	G	*Stereopneustes* **R**	Museum collections: British Museum Bibliography: De Meijere 1904
Holasteridae	C	*Aurelianaster*	
	A	*Basseaster*	
	A	*Cardiaster*	
	C	*Cardiotaxis*	
	A	*Cibaster*	
	A	*Echinocorys*	Museum collections: miscellaneous
	A or E?	*Entomaster*	Bibliography: works of of Cotteau,
	H	*Galeaster*	Lambert, Orbigny ..., *Treatise*
	A	*Galeola*	*on Invertebrate Paleontology*
	A	*Ganbirretia*	
	H	*Guettaria*	
	D	*Hagenowia*	
	C or A	*Hemipneustes*	
	A	*Holaster*	
	D	*Infulaster*	
	A	*Ismidaster*	
	E	*Jeronia*	
	A	*Labrotaxis*	
	B	*Lampadaster*	
	B	*Lampadocorys*	
	A	*Messaoudia*	
	A	*Offaster*	
	A	*Opisopneustes*	
	A	*Paronaster*	
	A or C	*Pseudananchys*	
	A	*Pseudholaster*	
	B	*Pseudoffaster*	
	E	*Rispolia*	
	A	*Scagliaster*	
	H	*Stegaster*	
	C	*Sternotaxis*	
	A	*Taphraster*	
	E	*Tholaster*	
	?	*Titanaster*	no information
	A	*Toxopatagus*	
	A	*Zumoffenia*	
Pourtalesiidae	L	*Ceratophysa* **R**	Museum collections: British Museum Bibliography: Agassiz 1881; Lovén 1883
	L	*Cystocrepis* **R**	Bibliography: Agassiz 1904; Mortensen 1907
	L	*Echinocrepis* **R**	Museum collections: British Museum Bibliography: Agassiz 1881; Lovén 1883

334   B. David

TABLE 25.1   contd.

Families	Group-Genera		Support of the study
Pourtalesiidae	L	*Echinosigra* **R**	Oceano. cruises: Biogas prog. Incal
	L	*Helgocystis* **R**	Museum collections: British Museum Bibliography: Agassiz 1881; Mortensen 1907, 1950
	L	*Pourtalesia* **R**	Oceano. cruises: Biogas prog. Incal Norbi, Walda
	L	*Spatagocystis* **R**	Museum collections: British Museum Bibliography: Agassiz 1881, 1904
Somaliasteridae	F	*Brightonia*	Museum collections: British Museum
	F	*Iraniaster*	Bibliography: Kier 1957
	F	*Leviechinus*	Treatise on invertebrate paleontology
	F	*Somaliaster*	
Stenonasteridae	?	*Stenonaster*	Museum collections: British Museum Bibliography: Lambert 1928
Urechinidae	I	*Chelonechinus*	Museum collections: British Museum Bibliography: Bather 1934
	A? or J?	*Pilematechinus* **R**	Bibliography: Agassiz 1881, 1904; Mortensen 1950
	J	*Plexechinus* **R**	Oceano. cruises: Biogas prog. Bibliography: Agassiz 1904; Mortensen 1907
	K	*Sanchezaster*	Museum collections: British Museum Bibliography: Kier 1984
	K	*Sternopatagus* **R**	Oceano. cruise: Réunion MD32 Bibliography: De Meijere 1904
	J	*Urechinus* **R**	Museum collections: British Museum Bibliography: Agassiz 1904; Mortensen 1907; Mironov 1978

and Thiéry 1909–25; Mortensen 1950–51; Wagner and Durham 1966, among others).

The classification of Holasteroida is traditionally based on peristomial, architectural, apical, or ambulacral features. It is thus possible to undertake a cladistic analysis, including only some major features, which clarify the great subdivisions of the order. The results will be presented according to the following three steps.

(1) Analysis of the variation of some major features: plastronal architecture, apical structure, ambulacral pores.
(2) Construction of general cladograms restricted to the main branching points, with emphasis on the relationships of Recent deep-sea forms. Several cladograms, illustrating different hypotheses will be presented.
(3) Discussion about the implications of these hypotheses and their relationship to general geological events.

VARIATION OF MAIN FEATURES

Orientation of the peristome

The peristome of sea urchins is usually flush with the test. However, ambulacrum III can form a groove on the oral side. In some genera, this groove becomes very deep and sharp, and the peristome opens vertically at the end of an oral invagination; it can even become reversed in orientation. Almost all stages exist between a flush peristome and a reversed one.

Plastronal architecture

Atelostome echinoids (orders Disasteroida, Holasteroida, and Spatangoida) generally have an anterior mouth and the ventral plates of interambulacrum 5 form an area called the plastron. Holasteroids are characterized by the possession of a meridosternous plastron (Fig. 25.4A). The opening of the peristome is bordered by an

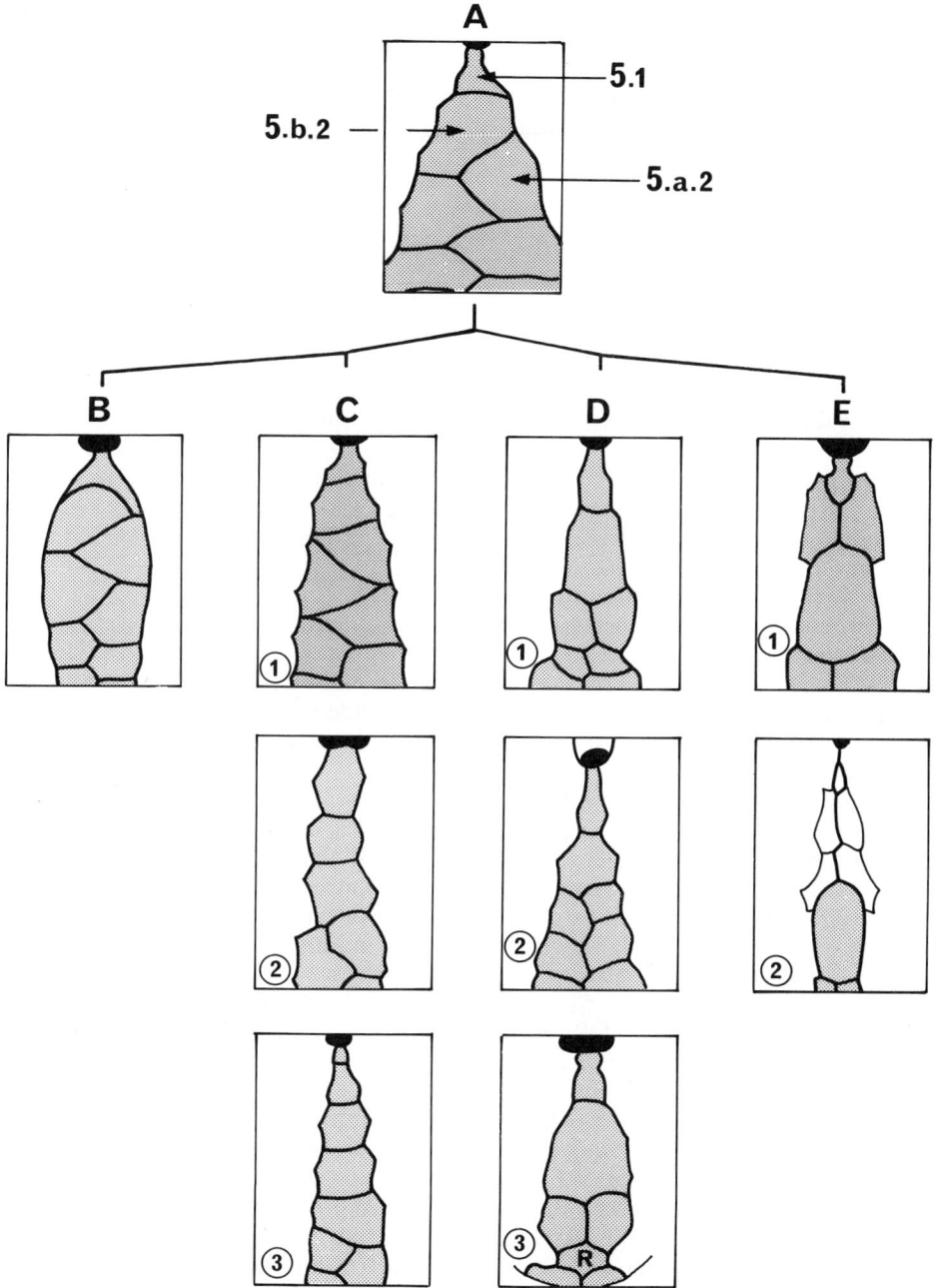

Fig. 25.4 Architecture of the plastron of Holasteroida: plesiomorphic (A) and apomorphic (B–E) conditions. (A) Basic meridosternous plastron of *Holaster intermedius* (from Lambert 1893, fig. 7). (B) Protamphisternous condition of *Labrotaxis cantianus* (from Casey 1960, fig. 1d). (C) Metasternous conditions of *Sternotaxis icaunensis* (C1) (from Raabe 1964, fig. 6), *Pseudananchys credneriana* (C2) (from Elbert 1901, pl. 4, fig. 2), and *Cardiotaxis heberti* (C3) (from Ernst 1972, fig. 20). (D) Orthosternous conditions of *Urechinus naresianus* (D1) (from Lovén 1883, fig. 239), *Stegaster heberti* (D2) (from Seunes 1889, pl. 25, fig. 1b), and *Corystus dysasteroides* (D3) (from Foster and Philip 1978, fig. 1a). (E) Interrupted interambulacrum 5 of *Plexechinus spectabilis* (E1) (from Mortensen 1950–51, fig. 107b), and *Pourtalesia miranda* (E2).

unpaired plate, the labrum (plate 5.1). The following plates are the sternal plates. They are paired, but only one (5.b.2) meets the labrum. Behind them, the episternal plates and pre-anal plates are alternate. This plate array is the basic *meridosternous* pattern of Holasteroida. However, it shows some variations.

Occasionally, the second sternal plate (5.a.2) can join the labrum and develop a small boundary (Fig. 25.4B). The plastron exhibits thus a *protamphisternous* condition, recalling the condition of the supposed protosternous ancestor (Mintz 1966). However, different species within the same genus vary in this character (Lovén 1883; Lambert 1893) and thus it can hardly be used as an apomorphy.

The structure of the plastron can be modified to reach a uniserial arrangement of plates called *metasternous* (Fig. 25.4C). The plates 5.b.2, 5.a.2, 5.b.3 ... extend and occupy the whole breadth of interambulacrum 5, the sutures becoming transverse. The transformation always begins adorally, and extends backwards, affecting more and more plates (Fig. 25.4C1–C3). These patterns occur usually as intergeneric variations.

One modification of the plastron corresponds to an architectural trend towards the individualization of a true sternum, the labrum being followed by a single large plate in unpaired position (Fig. 25.4D). Two symmetrical plates in episternal position follow. Behind, the pre-anals are arranged alternately. Such a pattern is called *orthosternous*. The sternum is classically viewed as an unpaired plate, composed by the coalescence of plates 5.a.2 and 5.b.2 (Mortensen 1950–51; Devriès 1960). Intermediate patterns, not fully symmetrical, exist with sutures obliquely disposed (Fig. 25.4D2). They show that the sternum corresponds more likely to the sole plate 5.b.2, the episternals being, respectively, 5.a.2 and 5.b.3. This interpretation corresponds to that of Lovén (1883), and recalls the numbering used by Bather (1934) for *Chelonechinus*. It places emphasis on homology of plates rather than analogy of position. An orthosternous plastron is a stable feature in several genera, but it also occurs as a variation in some Cretaceous forms.

Some genera shared a very strange plastron (Fig. 25.4D3). The two episternal plates are followed by a small central unpaired plate: the rostral plate. This plastron must be viewed as a derived orthosternal condition (and not metasternal as argued by Foster and Philip 1978). The rostral plate seems to be a very stable feature.

The most obvious transformation of the plastron of Holasteroida involves the separation of the labrum and sternum (Fig. 25.4E1). During growth, interambulacrum 5 is progressively interrupted by ambulacral plates of areas I and V which shift towards the plane of symmetry and meet (David 1987). The interruption of interambulacrum 5 can be more or less achieved, and the separation of labrum and sternum more or less wide (Fig. 25.4E2). These disrupted patterns are strongly correlated with the orthosternous configuration of several Recent and fossil genera. Slight breaks represent intrageneric or even intraspecific variations. Wide breaks—when the labrum is separated from the sternum by two or more rows of ambulacral plates—are architecturally more stable and appear as intergeneric variation.

Ambulacral pores

Each ambulacral plate of an echinoid is basically perforated by two pores. In Holasteroida, the paired ambulacra are flush with the test and form open-ended petals on the apical surface. The pores are frequently reduced on the oral side where the ambulacra become uniporous, whereas around the peristome they conserve the double pores of phyllodes. The major transformation which affects this ambulacral configuration is the reduction in size of pores and the change from a biporous to a uniporous condition. This transformation concerns mainly the petals and less often the phyllodes.

Structure of the apical system

The apical system of Holasteroida is typically elongated, ocular plates II and IV separating anterior and posterior genital plates (Fig. 25.5A). There are generally four genital pores, sometimes three or two, and exceptionally more than four (Fig. 25.5b).

The apical system of some genera is ethmophract (i.e. compact), recalling a spatangoid structure (Fig. 25.5C). In spite of that, these genera are commonly classified as Holasteroida, the authors arguing the importance of the gross morphology or of the plastronal architecture (Lambert and Thiéry 1909–25; Kier 1957; Wagner and Durham 1966). The strongest transformations of the apex occur in some Recent deep-sea genera and in the greatly modified *Hagenowia* from the

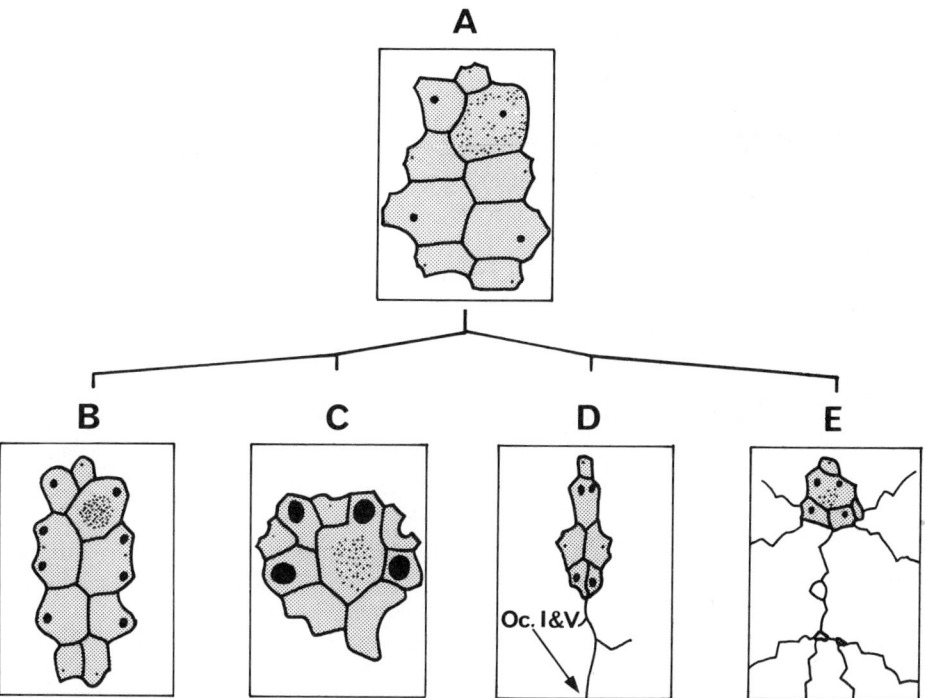

Fig. 25.5 Structure of the apical system: plesiomorphic (A) and apomorphic conditions (B–E). (A) Typical holasteroid structure of *Echinocorys sulcata* (from Lovén 1874, fig. 97). (B) Apex with numerous genital pores of *Guettaria rocardi* (from Lambert 1896, fig. 1). (C) Ethmophract apex of *Somaliaster cf. magniventer* (BM E 43775). (D) Disjunct apex of *Hagenowia rostrata* (from Gale and Smith 1982, fig. 3b). (E) Apex with coalesced genital plates and disjunct oculars of *Pourtalesia laguncula* (from Lovén 1883, fig. 52).

Upper Cretaceous. In *Hagenowia*, the apex is always disjunct (Fig. 25.5D). The disjunction is linked to the development of a long and slender rostrum. This trend leads to an architectural reorganization of the apical region, such that interambulacra 1 and 4 meet on the midline of the test separating the apical disc into two. This apical disjunction is more or less extensive in the different species of *Hagenowia*. In some Recent deep-sea genera, the apical system is twice transformed. First, the four genital plates join together and more or less completely coalesce; there are two to four genital pores. Secondly, the apex becomes disjunct with the posterior ambulacra (the oculars are not visible) situated far back. Interambulacra 1 and 4 meet on the midline (Fig. 25.5E). Intermediate structures between a typically holasterid apex and the latter type exist.

Peristomial architecture of paired interambulacra

In most atelostomes, each paired interambulacrum joins the peristome by one plate. This plate is bordered aborally by two other plates. But in some cases, only one large plate follows the first one, the plates of the interambulacrum becoming thus paired only from the third row. These two patterns are, respectively, called amphiplacous and meridoplacous by analogy with the plastronal structures.

Interambulacrum 1 is often meridoplacous in Holasteroida (Fig. 25.6A). Interambulacrum 4 is sometimes meridoplacous, but generally only when interambulacrum 1 is also meridoplacous (Fig. 25.6B). The architectural variation of these interambulacral regions is largely uncorrelated with generic subdivisions and it cannot be used to understand relationships without further complementary studies. The anterior interambulacra (2 and 3) are basically amphiplacous, and meridoplacous only in some well delimited genera. Moreover, in this case all the interambulacra are meridoplacous and the architecture is named holomeridoplacous (Fig. 25.6C).

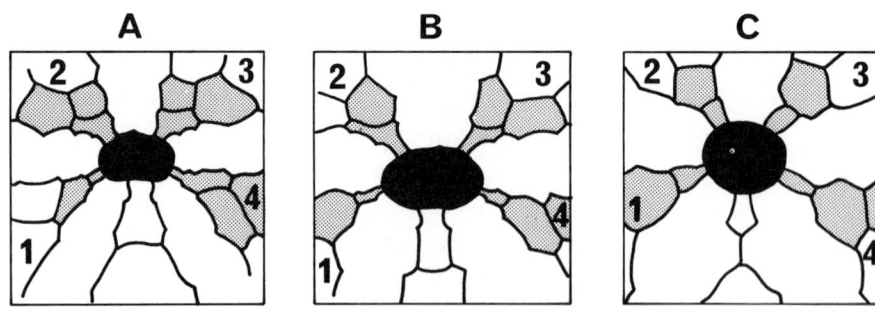

Fig. 25.6 Peristomial architecture of paired interambulacra: plesiomorphic (A and B) and apomorphic (C) conditions. (A) Meridoplacous interambulacrum 1 of *Holaster suborbicularis* (from Lovén 1874, fig. 54). (B) Meridoplacous interambulacra 1 and 4 of *Echinocorys sulcata* (from Lovén 1874, fig. 51). (C) Holomeridoplacous structure of *Plexechinus hirsutus* (from Mortensen 1907, pl. 7, fig. 19).

RELATIONSHIPS

The paraphyletic stem group

The genus *Holaster* has characters which are the most general in distribution among the Holasteroida:

(1) basic meridosternous plastron (Fig. 25.4A). I agree here with Raabe (1966) and include those species of *Holaster* with a uniserial plastron in the genus *Sternotaxis*;
(2) amphiplacous interambulacra 2 and 3 (Fig. 25.6A);
(3) peristome flush with the test;
(4) apical system elongated (Fig. 25.5A);
(5) double ambulacral pores.

Several other genera (*Echinocorys*, *Cardiaster*, etc.) share these general characters with *Holaster*. All together compose the primitive sister group of Holasteroida. However, they share only plesiomorphic features and constitute a paraphyletic assemblage which is not resolved and has still to be worked out in detail.

However, biostratigraphy provides a useful clue for understanding this stem group. Many species belong to the genus *Holaster* (sixty or so, according to Lambert and Thiéry 1909–25) and they are the only holasteroid echinoids known from the Valanginian to the Aptian. During this period (approximately 25 million years), marine deposits containing various atelostome echinoids are very abundant and the lack of diversity of holasteroids cannot be blamed on a poor fossil record. Hence, the ancestor of the whole group may be a species of the genus *Holaster* (probably among the species *H. cordatus*, *H. grasanus*, or relatives). In fact another genus, *Taphraster*, occurs during the Valanginian. However, although closely related with *Holaster*, *Taphraster* has aborally depressed paired ambulacra, a derived feature which eliminates it as a possible ancestral type.

The general framework

A list of characters used in the construction of cladograms is given (Table 25.2). It provides an interpretation of the primitive or derived state of each feature: the more general states have been determined as plesiomorphic and the less general as apomorphic. Several hypotheses of relationships have to be presented. They vary according to the weighting given to architecture, to the peristomial groove or to ambulacral pores, and according to the respective importance attributed to convergences and reversions. Figures 25.7–25.9 are general cladograms of Holasteroida. They present the major branchings and leave unresolved some paraphyletic sets of genera which are not directly involved in the origins of the deep-sea holasteroids. I shall describe each step of these cladograms successively from the left to the right. The groups formed are simply named by a letter in order to avoid any systematic discussion. (The aim of this paper is not a revision of the Holasteroida). The Stenonasteridae are not included in the cladograms because they have many strange features, and according to the characters used here, their position within the Holasteroida is unknown without further, more complete studies.

TABLE 25.2 Primitive and derived character states in Holasteroida. Lettered derived character states are independently derived from the primitive state

Primitive character state	Derived character state
1 Peristome flush with the test	Peristome in a more or less deep groove (a) oral groove not very sharp (b) deep and sharply delimited oral groove
2 Meridosternous plastron	Transformed meridosternal architecture (a) metasternous plastron (b) orthosternous plastron (c) large break in interambulacrum 5 (d) presence of a rostral plate
3 Ambulacra with double pores Petals open distally	Transformed ambulacra (a) ambulacra with single pores adapically (b) uniporous phyllodes (c) closely petaloid paired ambulacra
4 Apical system elongate with oculars II and IV juxtaposed	Transformed apex (a) apex compact (b) apex disjunct with interambulacra 1 and 4 meeting adapically; genital plates compact
5 Amphiplacous paired interambulacra	Meridoplacous paired interambulacra (a) holomeridoplacous condition (b) interambulacra 1 and 4 meridoplacous (5b is not a good apomorphy, because the feature is too variable)

Hypothesis 1 (Fig. 25.7)

The basic branching of the cladogram is a fan-shaped polytomy involving four branches: the paraphyletic group (A) including *Holaster* discussed above; and three sister groups, each characterized by their own apomophies.

The first (group B) is characterized by the presence of a peristomial groove with sharp edges. It includes genera such as *Lampadaster* and *Lampadocorys*. The fact that this apomorphy will appear several times independently in the cladogram is discussed below.

The second group (C+D) is based on the uniserial arrangement of plates of the plastronal area (metasternal condition). That transformation occurs for the first time during the Cenomanian in the genus *Sternotaxis* (including here the species *S. gregoryi* and *S. trecensis*) and concerns five other fossil genera.

Metasternous holasteroids are thus considered as a natural grouping and *Sternotaxis* may be their primitive sister group. Among them the genera *Infulaster* and *Hagenowia* (group D) share several advanced characters that distinguish them: test acuminate to form an apical rostrum, slight buccal groove, anterior apical system. Moreover, they have related ontogenetic pathways (their differences result from heterochronies, Dommergues *et al.* 1986) and they are biogeographically and stratigraphically linked (Chalk from the Upper Cretaceous of northwestern Europe).

The third set (comprising groups E–L) is the most complex. It includes all the holasteroids with an orthosternous plastron. It is noteworthy that this architecture is often linked with breaks of interambulacrum 5; and raises the probability that in many genera separation of labrum and sternum is likely to arise independently from forms with an orthosternous plastron and thus should not be considered a reliable apomorphy on its own.

Three orthosternous genera (*Jeronia*, *Rispolia*, and *Tholaster*) share no other derived character and constitute a primitive paraphyletic stem group (E). The Somaliasteridae (group F) have other apomorphies: petaloid paired ambulacra, compact apical system (Fig. 25.5C). They look very similar in gross morphology and have a good stratigraphic homogeneity. They probably form

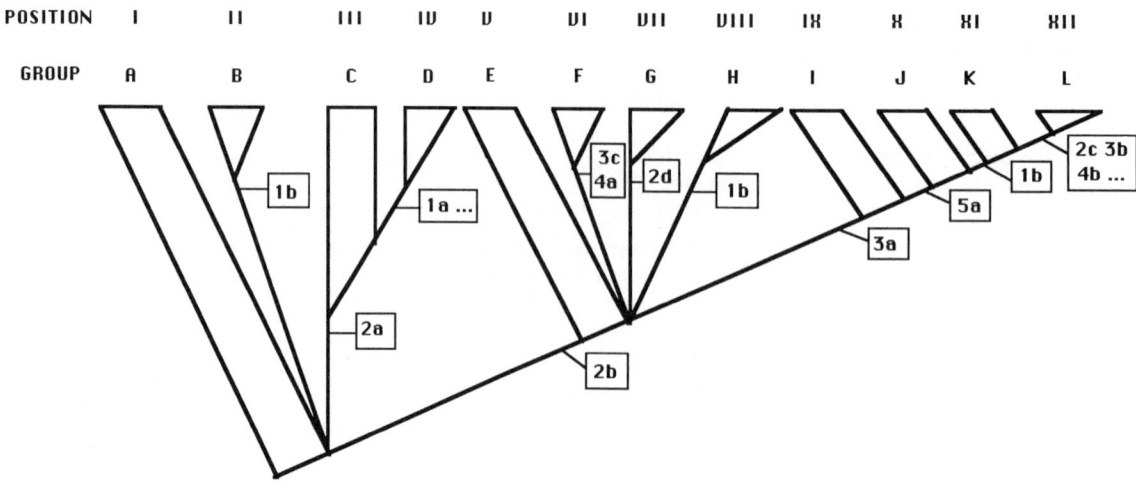

Fig. 25.7 Cladogram of major subdivisions of Holasteroida (hypothesis 1). In this and the next two figures, the apomorphic groups are represented by triangles (real or virtual), and the plesiomorphic sets (not still resolved) by trapezoidal areas. The groups are named by a capital letter, and their position is indicated by a roman numeral. The list of genera belonging to each group is given in Table 25.1. Numerals included in squares refer to the characters given in Table 25.2.

an homogeneous group. Nevertheless, among them, the genera *Brightonia* and *Leviechinus* have a special type of interrupted plastron which is not orthosternous. Even so the homogeneity of the group can be retained if one supposes that their interrupted plastron is derived from an orthosternous type, using the argument given above that breaks of interambulacrum 5 affect mainly orthosternous sea urchins. Otherwise, the affinities of the Somaliasteridae to the Holasteroida can be challenged. Either they should be included within Spatangoida (considering the structure of the apical area, the petals and the general morphology) or they belong to the orthosternous holasteroids (considering the plastron).

The Corystidae (group G) is a family established by Foster and Philip (1978) for some genera from the Tertiary of Australia and New Zealand. It is a well delimited natural grouping with a good apomorphy: the rostral plate (Fig. 25.4D3). The group H comprises orthosternous holasteroids with a sharply defined peristomial groove (*Stegaster*, *Galeaster*, and *Guettaria*). This feature appears for the second time as an apomorphy in this cladogram and will appear a third time in the most derived group. It is thus strongly convergent. Conversely, Mortensen (1950–51) used it to establish the subfamily Stegasterinae. In my opinion, the Stegasterinae include an heterogeneous assemblage of genera. Indeed, holasteroids with a peristomial groove are distributed among meridosternous s.s. (group B), metasternous (group D), and orthosternous (groups H to L) sea urchins. All these architectural transformations seem to me drastic and less likely to converge, than a mere deepening of the oral groove. Moreover, the hypothesis accepted here is also the more parsimonious.

The members of the next groups share the following successive apomorphies: simple ambulacral pores on the apical side of the test (groups I–L); holomeridoplacous interambulacra (groups J–L); sharp and deep peristomial invaginations (groups K and L); uniporous phyllodes and a compact apical system with a disjunct bivium (group L). Three convergences, concerning the peristomial groove, occur in this first hypothesis.

Hypothesis 2 (Fig. 25.8)

The groupings A to G remain unchanged. Within orthosternous holasteroids (above apomorphy 2b), once the architectural threshold had been crossed, the character 'peristomial groove' can become an apomorphy. If it does, then group H switches position with group I and shifts to position IX. Groups J and K can also be switched; L remains unchanged. Groups J and K are here monophyletic, not paraphyletic, the former

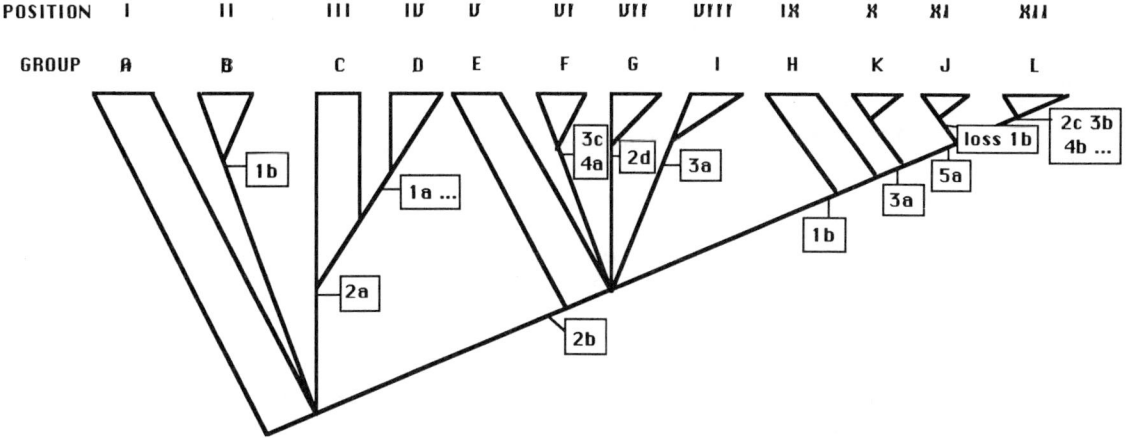

Fig. 25.8 Cladogram of major subdivisions of Holasteroida (hypothesis 2).

because the lack of a peristomial groove is interpreted as a reversion (and thus becomes an autapomorphy at this level), the latter because of a more accurate interpretation of the relationships between *Sanchezaster* and *Sternopatagus* (see the next section). Two convergences and one reversion occur in this second hypothesis. The convergences concern the peristomial groove (apomorphy 1b) and the ambulacral pores (apomorphy 3a); the reversion concerns the peristomial groove.

Hypothesis 3 (Fig. 25.9)

As previously, groups A to G remain unchanged. This hypothesis again retains the peristomial groove as an apomorphy, but reversion of this character is also found in Group I. In comparison with the previous hypothesis, a reversion replaces a convergence.

Other circular permutations between different taxa, leading to other hypotheses, could be tried. However, they do not affect the problem of the origins of deep-sea holasteroids and would need to be resolved by the integration of new characters and probably by using a numerical cladistic approach.

The Recent forms

The goal of this section is to examine the placement of the Recent forms on the three different hypotheses of relationships established above.

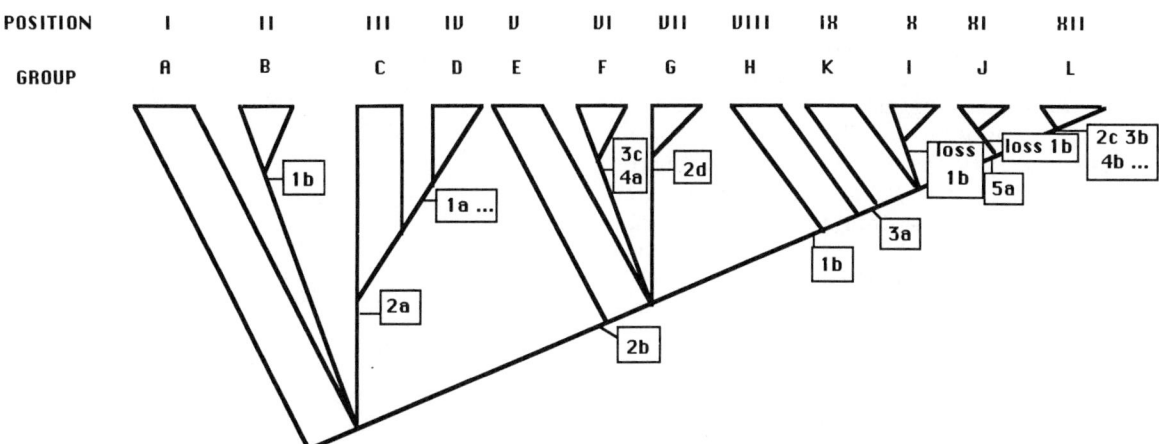

Fig. 25.9 Cladogram of major subdivisions of Holasteroida (hypothesis 3).

Pourtalesia and its relatives. The family Pourtalesiidae is well known. It includes seven genera which share numerous apomorphies: twice transformed apical system, having a compact anterior part and a disjunct bivium and trivium (Fig. 25.5E); uniporous phyllodes; large break in interambulacrum 5 in the adults (architecturally always more advanced than in any other orthosternous holasteroids, Fig. 25.4E2); vertical or reversed periproct with a periproctal invagination and two associated structures, the hood and the subanal rostrum (Fig. 25.1); characteristic ophicephalous pedicellariae. The Pourtalesiidae thus constitute a monophyletic group. They also have an orthosternous plastron, uniporous ambulacra, holomeridoplacous interambulacra and a deep oral groove with a vertical peristome, features which place them at the top of the three cladograms (group L). Although not apparent in the adults, the holomeridoplacous nature of interambulacra has been clearly recognized in juveniles (David 1987). The adult condition can thus be interpreted as a more derived pattern, on the argument that there is a peramorphic evolutionary trend within the family (David 1986).

Sternopatagus. *Sternopatagus* contains with certainty only a single species *S. sibogae*, described by de Meijere (1904). It is conical in shape with an orthosternous interrupted plastron, uniporous ambulacra, but biporous phyllodes, and a peristome that opens in a sharply demarcated, deep groove. The posterior paired interambulacra are meridoplacous, the anterior ones are hardly known, because their adoral ends are hidden in the peristomial groove. Nevertheless, their architecture could be holomeridoplacous. Indeed, *Sternopatagus* shows a plate array similar to that of an adult pourtalesiid, suggesting that juveniles are truly holomeridoplacous. In this hypothesis, *Sternopatagus* belongs to the group K in the eleventh position, just below the Pourtalesiidae (Fig. 25.7). It is interesting to note here that *Sternopatagus* was originally referred to the Pourtalesiidae by de Meijere (1904) and Mortensen (1907). Conversely, cladograms 2 and 3 reject this hypothesis and push group K towards a lower position (Figs 25.8, 25.9).

Sternopatagus seems related to *Sanchezaster* from the Eocene of Cuba. Although *Sanchezaster* is not completely known (Kier 1984), it possesses an orthosternous interrupted plastron, adorally uniporous ambulacra and a ventral fasciole close to that of *Sternopatagus*. If this last feature is viewed as apomorphic, the two genera can be classified together in group K (Fig. 25.8).

Urechinus. *Urechinus* is a diverse genus (eight species according to Mortensen 1950–51; Mironov 1978a). The geographical distribution of the genus is almost worldwide. *Urechinus* has the following apomorphies: orthosternous plastron, uniporous ambulacra (except peristomially), holomeridoplacous interambulacra, apical system sometimes with intercalated plates, and a slight subanal rostrum (mainly in juveniles). Its position on the cladograms is linked to that of the following genus and is discussed in the next section.

Plexechinus. *Plexechinus* is a small sea urchin which looks like a young *Urechinus* in gross morphology, but with hyper-adult features in architecture or apical structure. Indeed, *Plexechinus* has the same apomorphies as *Urechinus*, with in addition: an apical system more anterior and generally disjunct, a peristome sometimes opening at the end of an oral groove, an interambulacrum 5 often interrupted, and a subanal rostrum always well developed. Accordingly, these two genera appear very closely related and must be classified together in group J (tenth position on Fig. 25.7).

Nevertheless, other observations lead to alternative hypotheses of relationships. Namely, a knowledge of post-larval ontogeny adds the following observations.

(1) An adult of *Urechinus*, a small *Plexechinus* and a very young *Pourtalesia* (with a continuous interambulacrum 5) are similar: their architecture is completely identical, all have a supramarginal periproct and a subanal rostrum.
(2) The architectural pattern of an adult *Plexechinus* is the same as that of a small *Pourtalesia* (with a single break of interambulacrum 5).
(3) *Plexechinus*, and to a lesser extent *Urechinus*, have intercalated apical systems, a configuration recalling that of Pourtalesiidae.

These remarks suggest that *Urechinus*, *Plexechinus*, and the Pourtalesiidae probably share the same ontogenetic process followed to varying degrees. They could, thus, be very closely related

(Figs 25.8 and 25.9). However, *Urechinus* never has and *Plexechinus* only exceptionally has an oral groove. Thus a reversion, which could be attributed to neoteny, must have affected these two genera if this relationship is correct. It is also important to notice that their position on the cladograms depends on the architectural interpretation of *Sternopatagus*.

Pilematechinus. *Pilematechinus* is a very strange genus known from two species. It is a large sea urchin with a very high, almost conical test, a flush ventral surface and adapically uniporous ambulacra. The architecture of the oral side is a very difficult feature to interpret phylogenetically. It is more primitive in ambulacral plate arrangement than any other holasteroid, recalling the Collyritidae. From an ontogenetic point of view, the general architectural pattern of *Pilematechinus* is incompatible with that of *Urechinus*, *Plexechinus*, *Sternopatagus* and the Pourtalesiidae. On the other hand, the paired interambulacra are perfectly meridoplacous and interambulacrum 5 can be interpreted as orthosternous, although the sternum is very small. This leads to an alternative. Either the 'primitive' features of *Pilematechinus* architecture are autapomorphies secondarily developed and *Pilematechinus* becomes a completely derived urechinid (group J); or these features are really plesiomorphic and *Pilematechinus* becomes a basic holasteroid (group A). The former hypothesis is not in accordance with the ontogenetic sequences, and the latter implies that the holomeridoplacous structure is a convergence. Both eventualities can be integrated on the three cladograms.

Calymne. The genus *Calymne* is based on the single species *C. relicta*, known from some fragments dredged by the Challenger in the Atlantic at 4845 m depth. This sea urchin has an orthosternous plastron, uniporous ambulacra, and a flush peristome. *Calymne* seems closely related to *Chelonechinus*, a Miocene genus from Barbados, the Fiji Islands and Java, which comes from fine deposits of foraminiferal or radiolarian ooze, usually regarded as deep-sea deposits (Bather 1934). These two genera can be included in group I on the cladograms where they occupy a position varying between VIII and X, according to the hypothesis chosen.

Stereopneustes. *Stereopneustes* is based on a single bathyal species, *S. relictus*, widely distributed in the Malay region (Mortensen 1950–51). It has an orthosternous plastron with a rostral plate and must be included in the Corystidae (group G at the seventh position in all the hypotheses presented).

HISTORY AND CONCLUSIONS

The different hypotheses discussed above suggest that Recent deep-sea holasteroids belong to between two and four monophyletic groups. Thus, according to these choices, and namely to the position of *Pilematechinus*, the Recent holasteroids appear as a more or less polyphyletic grouping. Two extreme cases will be presented here.

The most polyphyletic grouping

This condition supposes the greatest disparity of Recent forms. It is illustrated in Fig. 25.10A. The genus *Pilematechinus* becomes a member of the most primitive sister group that also includes

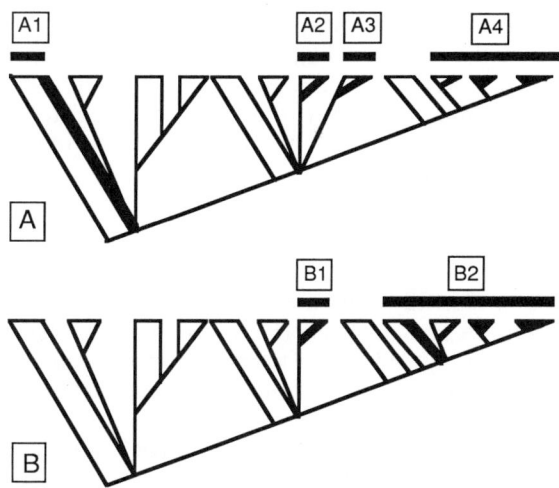

Fig. 25.10 Positions and relationships of Recent deep-sea holasteroids. (A) The most polyphyletic grouping with four sets of Recent forms: A1 = *Pilematechinus*, A2 = *Stereopneustes*, A3 = *Calymne*, A4 = *Sternopatagus* + *Urechinus* + *Plexechinus* + the Pourtalesiidae (the frame is that of hypothesis 2, illustrated in Fig. 25.8). (B) A least polyphyletic grouping with two sets of Recent forms: B1 = *Stereopneustes*, B2 = the other 12 Recent genera (the frame is that of hypothesis 3, illustrated in Fig. 25.9).

344   B. David

Holaster. The holomeridoplacous interambulacra and the uniporous ambulacra are here interpreted as convergent. The latest forms of this group, from which *Pilematechinus* could have been derived, are Eocene, but any Cretaceous form could also be ancestral (*Toxopatagus* from the Miocene is much too specialized to be the direct ancestor).

Stereopneustes probably derives from the Corystidae of Australia and New Zealand. This ancestor-descendant relationship is based on the belief that the fossil record gives sufficiently reliable information in epicontinental environments to be of use. It allows me to minimize the stratigraphic gap between *Stereopneustes* and its ancestors, and is in accordance with the geographic distribution of these sea urchins.

Calymne is most closely related to *Chelonechinus*, but the scarcity of deep-sea deposits in the fossil record makes it extremely dangerous to identify *Calymne* as the direct descendant of *Chelonechinus*. More probably, they have derived together from some generalized orthosternous holasteroid of the late Cretaceous or Palaeocene.

The other Recent genera (*Urechinus*, *Plexechinus*, *Sternopatagus*, and the seven Pourtalesiids) form a fourth set which could come from the group that includes *Stegaster* and its relatives (viz: the orthosternous holasteroids with an oral groove). Nevertheless, it is almost impossible to find evidence to support this in the biogeography of the *Stegaster* group, because its distribution is too wide.

The least polyphyletic grouping.

Figure 25.10B illustrates a less heterogeneous condition and provides the fewest monophyletic groupings of the Recent forms. The architecture of *Pilematechinus* is here viewed as derived from an orthosternous condition. All the Recent forms become thus orthosternous holasteroids and can be grouped into two sets:

(1) *Stereopneustes* remains a corystid;
(2) the others appear as more or less derived holasteroids with single ambulacral pores. Among them the two fossil genera *Sanchezaster* (Eocene) and *Chelonechinus* (Miocene) cannot be considered as direct ancestors of their relatives *Sternopatagus* and *Calymne*. They simply give the latest possible date of origin of the whole group: the ancestor of this second set of deep-sea holasteroids is Eocene or older.

This second hypothesis of relationships suggests two evolutionary migrations into the deep-sea for holasteroids.

(1) The move of *Stereopneustes* into the bathyal zone could have occurred around the middle of the Tertiary from an oceanic stock of the Corystidae.
(2) The downward migration of other Recent holasteroids may be pre-Eocene, since *Sanchezaster* is probably itself a deep-sea form (Kier 1984).

This agrees well with the supposed relationship between this group of holasteroids and the *Stegaster* group. In this context the migration down could have occurred during the Upper Cretaceous. This eventuality is supported by the fact that *Stegaster* is known in bathyal deposits (Marnes de Bidart, from the Basque country, SW France). Moreover, bioturbation traces attributed to sea urchins have been reported from Cretaceous and Eocene flysch deposits of Spain (Smith and Crimes 1983). As already suggested by Smith (1984), the migration may be linked with the great Upper Cretaceous transgressions. Indeed, these transgressions opened large bathyal areas on epicontinental shelves: several hundred meters in depth (Hancock and Kauffman 1979); more than 300 m in the Chalk Sea (Pomerol 1984). The benthic fauna, previously living in shallow waters, progressively came to be situated in deeper and deeper waters. Some holasteroids could have been well adapted to these great depths, initiating thus a bathyal adaptative radiation. Moreover, these forms could have been trapped under bathyal conditions by the strong Late Cretaceous regression which induced drastic reduction of the neritic zone, leading to an increase in competition between benthic organisms. A lot of them disappeared, but some may have migrated out of the epicontinental shelf, supplying the abyssal fauna with new forms. The stratigraphic distribution of the Holasteroida (Fig. 25.3) supports the hypothesis of an Upper Cretaceous migration into the deep-sea environment, since the low diversity of the fauna during the Tertiary could be due to a lack of deep-sea fossils.

This historical scenario can also be applied to the former hypothesis (Fig. 25.10A). Indeed different phyletic origins do not necessarily suppose the same number of migration events into the deep-sea. Global geological events could have induced the migration of several independent groups. It is thus possible that the migrations of *Pilematechinus*, *Calymne*, and of the fourth set of genera (made of *Urechinus*, *Plexechinus*, *Sternopatagus*, and the Pourtalesiidae) have been induced in the same way. But concerning *Pilematechinus* and *Calymne*, such a scenario is purely hypothetical and later (Tertiary) or earlier (Lower Cretaceous) migrations can also be supposed.

Although bathymetrically well delimited, the Recent deep-sea holasteroids are in any case polyphyletic. These last sections show that according to the degree of heterogeneity of the selected hypothesis, deep-sea holasteroids belong to two to four independent phyletic branches. Noticing the relative independence between phyletic diversity and the history of the deep-sea colonization, the Recent holasteroids have at least a two-fold origin: Upper Cretaceous and Middle Tertiary. There remains now the task of understanding more completely the relationships within the paraphyletic sets not resolved herein, and especially within the basic group of *Holaster* and the *Stegaster* group. This could produce more accurate phyletic trees and a more detailed history.

ACKNOWLEDGEMENTS

I am obliged to A. B. Smith for his help, and correction of the English manuscript. I thank A. Clark, A. Guille, D. Lewis, G. Paterson, and C. Vadon for making specimens available to me, and the IFREMER for the organization of oceanographic cruises.

REFERENCES

Agassiz, A. 1881. Report on the Echinoidea dredged by H.M.S. Challenger, during the years 1873–1876. In *Reports of Scientific Results of the Voyage of HMS Challenger, Zoology* **3**, 1–321.

Agassiz, A. 1904. The panamic deep sea echini. *Memoirs of the Museum of Comparative Zoology, Harvard* **31**, 243 pp. + atlas, 110 pls.

Bather, F. A. 1934. *Chelonechinus* n.g. a Neogene urechinid. *Bulletin of the Geological Society of America* **45**, 799–874, pls 108–10.

Casey, R. 1960. A new echinoid from the Lower Cretaceous (Albian) of Kent. *Palaeontology* **3**, 260–4, pl. 44.

Clark, H. L. 1917. Hawaiian and other Pacific echini. *Memoirs of the Museum of Comparative Zoology, Harvard* **46**, 85–283, pls 144–61.

Dahl, E., Laubier, L., Sibuet, M., and Strömberg, J. O. 1977. Some quantitative results on benthic communities of the deep Norwegian Sea. *Astarte* **9**, 61–79.

David, B. 1986. Jeu en mosaïque des hétérochronies: variation et diversité chez les Pourtalesiidae (Echinides abyssaux). In *Ontogenèse et Evolution* (ed. J. Chaline et al.), pp. 137–58. *Colloque International du CNRS*.

David, B. 1987. Dynamics of plate growth in the deep-sea echinoid *Pourtalesia miranda* Agassiz: a new architectural interpretation. *Bulletin of Marine Science* **40**, 29–47.

David, B. and Sibuet, M. 1985. Distribution et diversité des Echinides. In *Peuplements profonds du golfe de Gascogne* (ed. L. Laubier, and C. Monniot), pp. 509–34. Fromer, Paris.

Devries, A. 1960. Contribution à l'étude de quelques groupes d'échinides fossiles d'Algérie. *Publications du Service de la Carte Géologique de l'Algerie* (nlle sér), **3**, 278 p., 39 pl.

Dommergues, J. L., David, B., and Marchand, D. 1986. Les relations ontogenèse-phylogenèse: applications paléontologiques. *Géobios* **19**, 335–56.

Elbert, 1901. Das untere Angoumien in den Osninbergketten des Teutoburger Waldes. *Verhandlungen des Naturhistorischen Vereines der Preussischen Rheinlande und Westphalens* **8**, 77–167, pl. 2–5.

Ernst, G. 1972. Grundfragen der Stammesgeschichte bei irregulären Echiniden der nordwesteuropäischen Oberkreide. *Geologisches Jahrbuch* A, **4**, 63–175.

Foster, R. J. and Philip, G. J. 1978. Tertiary holasteroid echinoids from Australia and New Zealand. *Palaeontology* **21**, 791–822, pls 85–93.

Gale, A. S. and Smith, A. B. 1982. The palaeobiology of the Cretaceous irregular echinoids *Infulaster* and *Hagenowia*. *Palaeontology* **25**, 11–42, pls 3–6.

Gregory, J. W. 1889. *Cystechinus crassus*, a new species from the Radiolarian Marls of Barbados, and the evidence it affords as to the age and origin of the deposits. *Quarterly Journal of the Geological Society of London* **45**, 640–50.

Hancock, J. M. and Kauffman, E. G. 1979. The great transgressions of the late Cretaceous. *Journal of the Geological Society of London* **136**, 175–86.

Kier, P. M. 1957. Tertiary Echinoidea from British Somaliland. *Journal of Paleontology* **31**, 839–902, pl. 103–5.

Kier, P. M. 1984. Fossil spatangoid Echinoids of Cuba. *Smithsonian Contributions to Paleobiology* **55**, 336 p., 90 pls.

Kikuchi, Y. and Nikaido, A. 1985. The first occurrence of the abyssal echinoid *Pourtalesia* from the middle Miocene Tatsukuroiso Mudstone in Ibaraki Prefecture, northeastern Honshu, Japan. *Annual Reports in*

Geoscience, *University of Tsukuba* **11**, 32–4.

Lambert, J. 1893. Etudes morphologiques sur le plastron des Spantangidés. *Bulletin de la Société des Sciences Historiques et Naturelles de l'Yonne* **46** (1892), 55–98.

Lambert, J. 1896. Note sur quelques échinides Crétacés de Madagascar. *Bulletin de la Société Géologique de France* **24**, 313–32, pl. 10–13.

Lambert, J. and Thiery, P. 1909–25. *Essai de nomenclature raisonnée des Echinides.* L. Ferriere, Chaumont.

Lovén, S. 1874. Etudes sur les Echinoidées. *Bihang till Kongelige Svenska Vetenskapsakademiens Handlingar* **11**, 91 pp. + atlas (1875), 53 pls.

Lovén, S. 1883. On *Pourtalesia* a genus of Echinoidea. *Bihang till Kongelige Svenska Vetenskapsakademiens Handlingar* **19** (1881), 95 p., 21 pl.

Meijere, C., de 1904. Die Echinoidea der Siboga-Expedition. *Siboga Expeditie* **43**, 251 pp., 23 pls.

Mintz, L. W. 1966. The origins, phylogeny, descendants of the echinoid family Disasteridae, A. Gras, 1848. PhD Thesis, University of California, Berkeley.

Mironov, A. N. 1976. Deep-sea urchins of the northern Pacific. *Trudȳ Instituta Okeanologii im PP Chirchova* **99**, 140–64. [In Russian.]

Mironov, A. N. 1978a. Meridosternous Echinoids collected by the 16th Cruise of the R/V 'D. M. Mendeleyev'. *Trudȳ Instituta Okeanologii im PP Chirchova* **113**, 208–26. [In Russian.]

Mironov, A. N. 1978b. The most deep-sea species of sea urchins (Echinoidea, Pourtalesiidae). *Zoologicheskii Zhurnal* **57**, 721–6. [In Russian.]

Mironov, A. N. 1980. Two ways of formation of deep-sea echinoid fauna. *Oceanologia* **20**, 703–8. [In Russian.]

Mortensen, T. 1907. Echinoidea (part 2). *Danish Ingolf-Expedition* **4**, 200 pp., 19 pls.

Mortensen, T. 1950–51. *A monograph of the Echinoidea*—Spatangoida., **5** (1) (1950), 432 pp., 25 pls; **5** (2) (1951), 593 pp., 64 pls. C. A. Reitzel, Copenhagen.

Pomerol, B. 1984. Géochimie des craies du Bassin de Paris. Utilisation des éléments traces et des isotopes stables de Carbone et de l'oxygène en sédimentologie et en paléoocéanographie. *Memoires des Sciences de la Terre de l'Université P. et M. Curie* **84–21**, 531 p. + annexes (73 pp.).

Raabe, H. 1964. Untersuchungen am plastron von '*Holaster*'. *Neues Jahrbuch für Geologie und Paläontologie. Monatshefte* **5**, 306–11.

Raabe, H. 1966. Die irregulären Echiniden aus dem Cenoman und Turon der Baskischen Depression (Nordspanien). *Neues Jahrbuch für Geologie und Paläontologie. Abhandlungen* **127** 82–126.

Sanchez Roig, M. 1924. Revision de los Equinidos fossiles Cubanas. *Memorias dela Sociedad Cubana de Historia Natural 'Felipe Poey'* **5**, 6–50, pls 1–7.

Seunes, J. 1889. Echinides crétacés des Pyrénées occidentales (II). *Bulletin de la Société Géologique de France* **17**, 804–24, pl. 24–7.

Sibuet, M. 1980. Adaptations des échinodermes à la vie abyssale. In *Echinoderms: present and past* (ed. M. Jangoux), pp. 233–40. Balkema, Rotterdam.

Sibuet, M. 1985. Quantitative distribution of echinoderms (Holothuroidea, Asteroidea, Ophiuroidea, Echinoidea) in relation to organic matter in the sediment, in deep sea basins of the Atlantic Ocean. In *Echinodermata. Proceedings of the fifth International Conference* (ed. B. F. Keegan and B. D. S. O'Connor), pp. 99–108. Balkema, Rotterdam.

Smith, A. B. 1984. *Echinoid palaeobiology.* Allen and Unwin, London.

Smith, A. B. and Crimes, T. P. 1983. Trace fossils formed by heart urchins—a study of *Scolicia* and related traces. *Lethaia* **16**, 79–92.

Termier, H. and Termier, G. 1953. Classe des Echinides. In *Traité de Paléontologie* (ed. J. Piveteau) **3**, 857–947, 5 pls. Masson, Paris.

Wagner, C. D. and Durham, J. W. 1966. Holasteroids. In *Treatise on invertebrate paleontology, part U. Echinodermata 3*, (ed. R. C. Moore), pp. 523–43. Geological Society of America and University of Kansas Press, Lawrence, Kansas

26

Feeding and respiratory strategies in Stylophora

RONALD L. PARSLEY

Department of Geology, Tulane University, New Orleans, Louisiana 70118, USA

INTRODUCTION

In recent years echinoderms of the class Stylophora have been extensively reviewed, not only from the view point of their functional morphology and their affinities as echinoderms, but as early chordate-like organisms, the calcichordates (Jefferies 1967, and numerous subsequent papers), as well. Among workers who view stylophorans as echinoderms there seems to be little argument concerning the gross aspects of their morphology and classification, but considerable differences of opinion exist about their functional morphology. The fairly numerous examples of preserved cornutes with aulacophore cover plates open to expose the food groove makes them directly comparable with other recumbent, essentially bilaterally symmetrical, echinoderms with feeding appendages, e.g. rhipidocystids, pleurocystitids, and homoiosteleans. Far more difficulty is encountered when dealing with the Mitrata. Even though they are almost certainly derived from early cornutes, their puzzling, putative non-echinodermal morphology invites a rather wide range of speculation concerning their mode of life.

Two schools of thought regarding 'echinodermal' mitrates are currently in vogue (for an excellent succinct review see Kolata and Jollie 1982). The first holds that the stylophoran aulacophore is not a feeding structure and both mouth and anus are located at the opposite end of the theca from the aulacophore, and by direct inference, a 'U'- or loop-shape gut was present. For discussions supporting this viewpoint see Phillip (1979), and Kolata and Jollie (1982). The second school maintains that the aulacophore was a feeding structure and the anus was located at the opposite end of the theca from the aulacophore. This interpretation was proposed by Ubaghs (1961) and subsequently discussed by him (1968, 1969, 1979), and supported by Caster (1983), Parsley and Caster (1982), Parsley (in prep.), Carlson and Fisher (1981), and Chauvel (1981).

Many workers who support the 'echinodermness' of stylophorans are in agreement that they represent an early offshoot of the phylum and gave rise to no other higher taxa (e.g. Ubaghs 1968). The primary reason for this taxonomic isolation is that stylophorans are asymmetrical organisms that approach bilateral symmetry, but rarely, if ever, truly achieve it. This lack of familiar echinoderm symmetry patterns has been responsible for their assignment by divers authors to non-echinodermal taxa and functional morphological attributes which are totally different from those of any other echinoderm. Symmetry plasticity displayed by echinoderms is more varied than in any other phylum: that of the Stylophora and some of the other Homalozoa, e.g. Homoiostelea, simply represents one end of a spectrum that ranges from asymmetry through bilateral, triradial, and bipentaradial patterns. Symmetry aside, all other stylophoran morphological features fit into a broad syndrome of accepted echinodermal taxonomic characters. Accordingly, a number of features shared among primitive recumbent echinoderms have been independently (isochronously and heterochronously) evolved in Stylophora. These include the following.

(1) Differentiation of the theca into marginal framing plates and central plates (centralia) that form tesselated surfaces or scale-like shingled surfaces with the plates distally overlapping adjacent plates.
(2) Respiratory structures manifest as sutural pores, cothurnopores or lamellate organs.

Echinoderm phylogeny and evolutionary biology (ed. C. R. C. Paul and A. B. Smith). Clarendon Press, Oxford, 1988.

(Cothurnopores appear to be highly derived sutural pores, lamellate organs may well be derived from cothurnopores.)
(3) Anal pyramids or a complex anal valve apparatus (especially in the cornutes).
(4) Aulacophore with its food groove and cover plates is, in its gross aspects, a typical echinoderm feeding structure.

In addition, there is good circumstantial evidence for a water vascular system (especially in the cornutes) extending the length of the aulacophore, but if one supports the concept of Blastozoa, this may be an unusual feature in the composition of a recumbent echinoderm.

A point that I want to re-emphasize is that all of the morphological features in stylophorans are echinodermal and I do not see a single morphological feature that is shared in any way with protochordates or any other deuterostome group. For example, respiratory structures are just that, they had nothing to do with the alimentary system, just as respiratory and alimentary systems are not closely associated in other echinoderms (see Bonik et al. 1978). I wish to eliminate any special pleadings and base my arguments on, the most logical and most 'echinodermal', comparative functional morphology and functional continuity, (for a superb practitioner of these concepts, see Manton 1977).

The thrust of this paper is to illustrate the functional morphology of stylophorans, especially their feeding and respiration. And, to examine the probable evolutionary developments which led to the origin of mitrates from a primitive, symmetrical cornute stock.

SALIENT MORPHOLOGICAL FEATURES

Cornutes

The theca in most cornute genera is usually markedly asymmetrical, while contrastingly a few genera are quite bilateral in outline, but not in plate arrangement. For our purposes we shall refer to cornutes as symmetrical or asymmetrical. Cornute thecae are made up of narrow marginal plates and an asymmetrically placed strut or zygal on the inferior face. These enclose, in almost all genera, tesselated superior and inferior surfaces of very small plates (centralia) which were assumed to have been flexible. The few exceptions are some of the earliest known cornutes; *Ceratocystis* from the Middle Cambrian of Bohemia, which has large rigidly sutured centralia on both surfaces, and *Nevadaecystis* from the Upper Cambrian of North America, which has large infracentrals and large sprocket-shaped supracentrals.

Under surfaces of marginal framework plates usually have cuspate knobs that are directed aborally (away from the aulacophore) and ventrolaterally. Most asymmetrical cornutes also have three large spatulate, aborally directed, spines that are part of the marginal framework. An interesting exception is the genus *Chauvelicystis* which lacks spatulate spines and the cuspate knobs of which are greatly reduced. Here, the lateral and aboral marginals have numerous short, slightly curved spines that appear to have been directed into the substratum (Fig. 26.1).

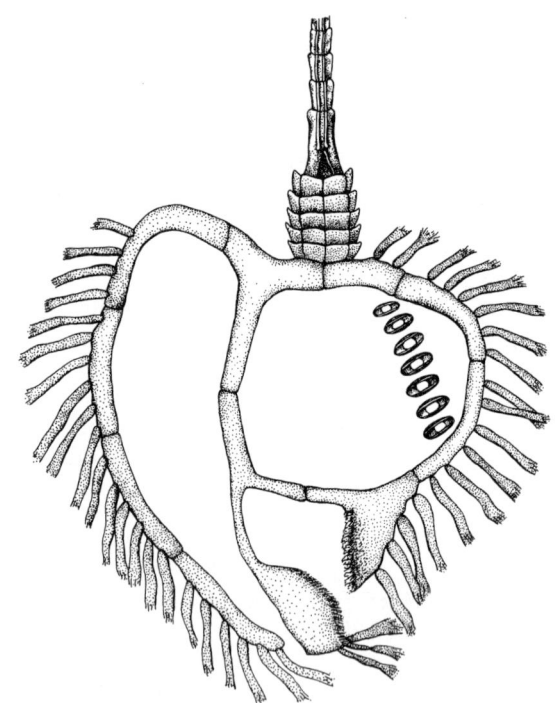

Fig. 26.1 Partial reconstruction of superior surface of the cornute *Chauvelicystis spinosa* Ubaghs. In this genus there is a total lack of sediment-piercing pegs on the marginals and minimal development of aborally directed marginal spines. Anchoring was done entirely by short buttressing spines which offset the motion of the aulacophore. Both surfaces were covered by flexible tesselated pavements that floored and partially roofed over the large cothurnopores.

Symmetrical cornutes, *Phyllocystis*, *Amygdalotheca*, and *Reticulocarpos* have neither long aboral spatulate spines nor sediment-piercing spines on the marginals. Instead, the marginals are bowed around the sagittal axis, so that most of the infracentralia were raised off the seafloor and the thin marginals made minimum bottom contact.

All cornutes have in the adoral right quarter of the supracentralia, arcuately arranged sutural pores, cothurnopores, or lamellate organs (essentially paralleling the marginals) which, as discussed below, are taken to be respiratory.

The basic composition of the aulacophore is the same in cornutes and mitrates, however functional modifications in each group have altered its appearance. In cornutes the proximal aulacophore is made up of tetramerous segments; in each segment the inferior pair of the tetrameres (inferolaterals) are about twice as long in total circumference as the superior (tectales) pair. Distal flaring on the lateral surfaces of the inferolaterals strongly indicates the principal movement of the proximal aulacophore was in the lateral plane (parallel to the substratum). The large lumen of the proximal aulacophore most likely contained large powerful muscles. The cavity extends from the aulacophore apophyses formed by marginals M1/M′1 distally to approximately the mid-length of the stylocone. Proximal aulacophores in cornutes usually have five or six segments. In both cornutes and mitrates the mouth is located at the junction of the stylocone/styloid and proximal aulacophore.

The stylocone, like the styloid of mitrates, is made up of fused uniserial ossicles which are doubtlessly derived from the distal aulacophore. Usually, three segments are incorporated into this structure. Cover plates are usually present but reduced in size relative to those of the distal aulacophore. Stylocones of cornutes, with the exception of *Reticulocarpos*, are smooth and conical. *Reticulocarpos* has a styloid similar to that of mitrates.

The distal aulacophore is made up of segments each comprising a uniserial basal ossicle and a pair of erectable cover plates that cover the superior (upper) surface of the ossicle. Imprinted on the upper surface of the ossicles is the continuous longitudinal food groove, flanked on each ossicle, by side branches that end in a lateral depression. These depressions are probably ampullar pits for feeding podia. The ampullar pits are deeper parts of a lateral groove system that parallels both sides of the food groove. On the outer margin of the lateral groove are the articulating bases for the cover plates. They tend to be located near the proximal end of each ossicle. Cornute cover plates are broad, somewhat fan-shaped, with a greatly narrowed and rounded articulating surface. They appear to be articulated so that when they were deployed they could open widely to expose the food groove and subvective podia. Cover plates in cornutes are often preserved in this fully opened position. All stylophoran cover plates have considerable distal overlap, so that if the entire series were open, closure would progress from the distal end to the proximal.

Mitrates

Mitrates have, or nearly have, a bilaterally symmetrical theca (in outline) while the arrangement of thecal plates, especially the marginals, is usually nearly bilaterally symmetrical (Fig. 26.2). Some mitrates have flexible polyplated centralia, in most cases they are distal to the adaulacophoral plates on the superior surface. When such polyplating is present it is usually slightly shingled, the overlap is always distal, as it is in some of the Mitrocystitidae or in the genus *Lagynocystis*. Rarely is the inferior surface polyplated; a notable exception is *Chinianocarpos*. The tesselated surfaces of cornutes are not found in mitrates. Most mitrates, however, have rigid thecae and, even in genera like *Anomalocystites* which have polyplated supracentralia, the plates are tightly sutured to form a rigid surface.

Unlike cornutes, mitrates are not flattened in cross-section or profile, but are inflated, convexly curved on the superior surface, concavely curved on the inferior (in cross-section) and have a curved hydrodynamic wedge-shaped profile. The aulacophore extends out from the thickest (proximal) end of the theca. In mitrates, the laterally positioned marginal plates are ventrally produced into fore and aft directed runners or skids.

Most mitrates have a single, or a pair, of articulating spines at the aboral end. These spines are variously articulated with the theca but always appear to have been deployed straight out from the theca, to have moved paralleled to the substratum, and to have been non-sediment piercing. Mitrates do not have sutural pores, cothurnopores or lamellate organs on the superior surface of the theca as cornutes do. The ctenoid organs of *Lagynocystis* bear superficial resemblance to

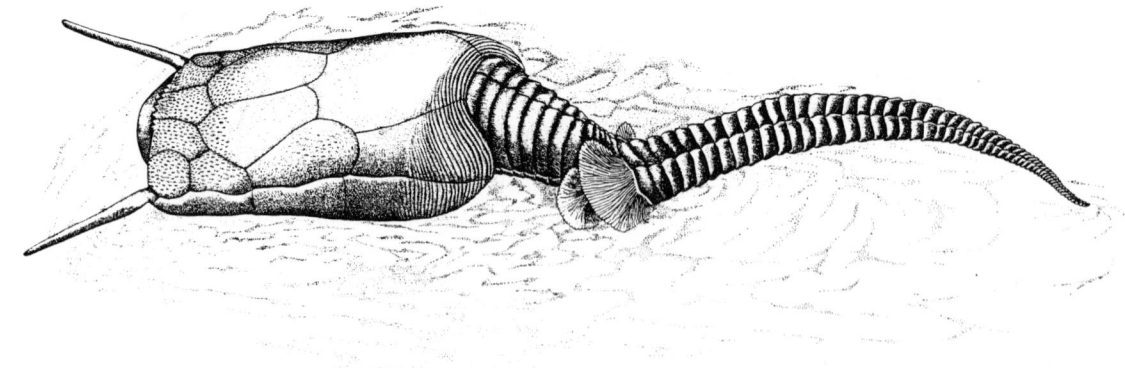

Fig. 26.2 Living orientation of the mitrate *Enoploura popei*. The aulacophore fed with the cover plate pairs closed and food entered, between the pairs, through chevron-shaped to arcuate openings caused by convexly arching the structure. The only sediment-piercing structures are on the styloid and proximal ossicles of the distal aulacophore. The theca narrowly rests on the substratum on its lateral margins while the distal articulating spines help buttress it. They, too, are non-sediment-piercing.

lamellipores, but they are better explained as feeding structures.

Mitrates show a definite tendency toward reduction of the anal pyramid/anal valvular apparatus. Anal valves are best developed in some of the Mitrocystitidae (Ordovician of Europe) in genera which have large, flexible, polyplated supracentralia. *Lagynocystis* (Ordovician of Bohemia) also has a prominent anal valve closely associated with its polyplated supracentralia. Anomalocystitidae (Ordovician–Devonian of North America and Europe) and Allanicytidiidae (Silurian–Devonian of Australia and New Zealand) usually show a distinct reduction in valvular structures. In Ordovician and Silurian anomalocystitids triangular platelets of the anal valvular apparatus are frequently fused to the plates forming the aboral opening and may indicate a reduced or vestigial valvular structure. In the genus *Anomalocystites*, from the Devonian of New York, Pennsylvania, and Maryland, the wedge-shaped anal platelets are completely missing; the area where the plates would be located is folded back on itself. In the Kirkocystidae (Ordovician of North America and Europe) and in the mitrocystitid *Chinianocarpos* the subanal plate apparently opened sufficiently to allow faecal elimination.

The aulacophore, while compositionally the same as that of cornutes, has modified into a rather uniform structure throughout the Mitrata. Proximal aulacophores are tetramerous as they are in cornutes, but the quarter segments are essentially equalized. In *Lagynocystis*, from the Middle Ordovician of Bohemia, the segments are polymerous. In this case it is probably a secondarily derived condition, and the aulacophore appears to have been unusually flexible. In most genera this structure seems to be capable of effective movement in almost any direction. Also, as in the cornute proximal aulacophore, the lumen is large and probably housed the major muscle apparatus for the entire aulacophore.

Between proximal and distal aulacophores is the styloid. It is homologous with the stylocone of cornutes but differs in having pronounced flanges or spines for anchoring in the substratum. The more proximal basal ossicles of the distal aulacophore are produced into spines, in some cases rivalling in size those of the styloid. These spine projections are undercut on their proximal face so that the entire length of the distal aulacophore could have been convexly arched without the spines binding against each other. The food groove on the superior surface of the ossicles is prominent, but the lateral groove and ampullar pits are much less distinct. The ossicles are taller and ventrally more attenuate than in cornutes. Cover plates in mitrates usually have a broad base, often extending the length of the basal ossicle, with the articulating surface on the cover plate being slightly concave and articulating against the corresponding convex base on the ossicle. The curvature of these articulations is greater on segments close to the styloid, indicating greater vertical movement. In contrast to those of cornutes, the paired cover plates in mitrates, are tightly sutured and form a triangular vault over the food groove. There are virtually no known examples of cover plates in mitrates being preserved in the open position, whereas it is a relatively common condition in cornutes. Also, it is very rare for the cover plates to be missing: their suturing to the basal ossicle and to each other seems to have been very tight. The tightness of suturing and preserved state of mitrate distal aulacophores strongly indicate that the cover plates were non-erectible.

The brief précis is by no means complete but touches on those aspects which I deem important for a grasp of feeding and respiratory functions in stylophorans. For a complete discussion of stylophoran morphology see Ubaghs (1968).

FUNCTIONAL MORPHOLOGY

From the brief précis above it is obvious that homologous and analogous elements of theca and aulacophore function differently in cornutes and mitrates. Many of these differences are directly related to their different feeding strategies. Also, related to feeding are differences in anchoring and maintaining stability in and on the substratum.

Cornutes

Cornutes deployed their relatively cumbersome aulacophore straight out along the sea floor. As feeders, judging by the wide subvective area exposed, they were generalists. The cover plates were apparently widely opened, as indicated by the mode and placement of their articulatory apparatus, to expose the food groove and the putative water vascular system. The podia, as they were envisaged by Ubaghs (1968), were capable of feeding on a wide range of detritus and microbiota. Food capture probably was somewhat similar to that employed by crinoids. Podia, coated with mucus, captured detritus and microbiota and transported it to the branches of the ambulacral groove. Food transport from the podia and along the ambulacral groove system was carried out by the ciliated ambulacral epidermis, (presumed to be present), which conducted food, in mucous strings, to the mouth.

Even if the water vascular system with its podia were absent, the wide catchment surface of the dorsal ossicles and widely opened cover plates would have been quite effective in supporting the nutritional needs of the organism. Internal markings on the cover plates may indicate that considerable tissue was present and an associated ciliary-mucoid apparatus may have played a role in feeding. The major disadvantages to this system were the problem of sediment fouling and the vulnerability to the widely exposed food gathering area. The latter point I am not prepared to comment on extensively. Predation in the Lower Palaeozoic is controversial. The exposed food catchment area was doubtlessly inviting and vulnerable. Cornutes do not survive the Upper Ordovician. On the former point, i.e. fouling by sediment, I believe there is considerable circumstantial evidence concerning how cornutes coped.

Proximal aulacophores in cornutes have slip planes, primarily developed in the larger inferolaterals, well suited for lateral back and forth movement, i.e. wagging. In many cases, the inferolaterals are flared laterally to allow greater lateral movement. Cornute proximal aulacophores could clearly move vertically and obliquely, but movement in these directions was probably less frequent. The smooth stylocone between the proximal and distal aulacophore allowed unimpeded lateral movement. While undergoing wagging the distal aulacophore with cover plates open would have been cumbersome, and clearly the advantages of such a weighty condition must have outweighed the disadvantages. To close the cover plates, which by their overlap would have had to commence at the aulacophore's distal end and proceed as a wave proximally to the stylocone, would have been a ponderous task, whether done by musculature or perhaps in conjunction with an extension of the water vascular system (cf. Paul and Smith 1984, pp. 472–5). To prevent fouling by 'normal' means may well have meant the cover plates would have been closed most of the time. Thus, by wagging, the feeding apparatus would have been kept clean of sediment. Also, feeding itself may have been augmented by the wagging process. In at least one species, *Scotiaecystis griffei* Ubaghs, from the Lower Ordovician of France, numerous proximal ossicles of the distal aulacophore appear to be narrowed and fused into a non-flexible structure (Fig. 26.3). The functional ramifications would appear to be similar to that of a circus whip, the fused shaft of ossicles imparted greater movement with less energy down the length of the distal aulacophore: hence, more efficient wagging.

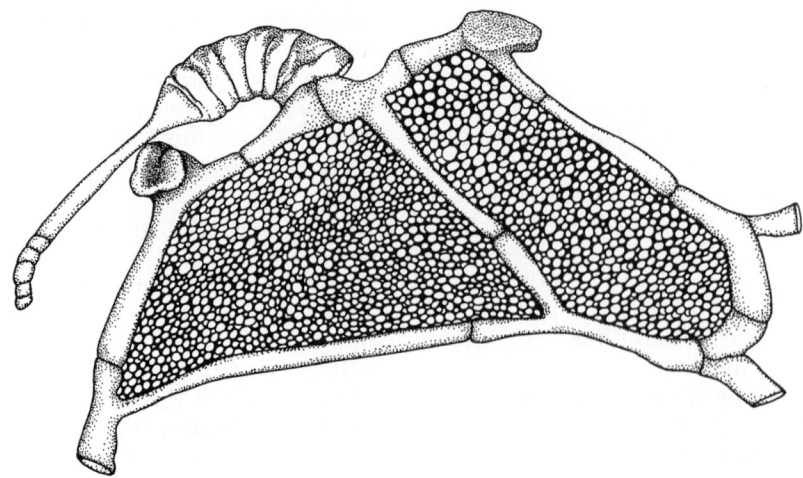

Fig. 26.3 Partial reconstruction of the inferior surface of *Scotiaecystis griffei* Ubaghs. The large sediment-piercing pegs close to the proximal aulacophore and three laterally directed spines, shown truncated, assured sessility of the animal by minimizing the effects of thrust and torque generated by wagging of the aulacophore. Note that the styloid and proximal ossicles of the distal aulacophore are fused so that less energy needed to be expended in wagging the structure.

In summary, the morphology of the cornute aulacophore indicates it opened its cover plates widely to expose the feeding system (food groove, water vascular system, podia). While the system, so deployed, was cumbersome, it must have been efficient and it is suggested it was easier to wag the structure to prevent fouling than ponderously to close the large distally overlapped cover plates.

The morphology of the theca is directly related to the proposed functional aspects of the aulacophore. The asymmetrical cornutes seem to have been sessile animals. Their asymmetrical to extremely asymmetrical thecal configuration suggests that they were hydrodynamically unstable organisms. In asymmetrical cornute genera the marginals tend to be rather flat on their inferior surfaces while in symmetrical cornutes the inferior surface of the marginals is more runner-like, as it is in mitrates, minimizing contact with the substratum. Marginals adjacent to the aulacophore bow upwards, so that the insertion area of the proximal aulacophore is elevated slightly above the surface of the seafloor and presumably served to facilitate the wagging movements of the proximal aulacophore.

Extending from the ring of marginals in asymmetrical cornutes are sediment-piercing, peg-like projections. Often, these pegs are cuspate (Fig. 26.3). The pegs are slanted aborally if located close to the insertion of the aulacophore and laterally outwards on the lateral faces of the theca. In addition, extensions of the marginals into a flat spatulate spine (spinal) or spatulate plates rigidly articulated with the marginals (glossal and digital) all of which are directed aborally, were most likely distally embedded in the substratum during life (Fig. 26.4). The function of the pegs and spatulate spines was to assure stability and to counteract thrust and torque forces generated by the aulacophore. An interesting exception that seems to verify the role of the spines and pegs is seen in the asymmetrical genus *Chauvelicystis* Ubaghs from the Lower Ordovician of France where the spines and pegs are missing. Instead, along the aboral and lateral sides of the theca short sediment-piercing spines extend outwards and function in exactly the same way as pegs and spatulate spines (Fig. 26.4).

Symmetrical cornutes *Phyllocystis* and *Amygdalotheca* are without spatulate spines or large cuspate pegs. While some species of *Phyllocystis* do have short marginal spines for stability at the oral end of the theca, the overall contact of marginal plates with the substratum was minimal. These genera are assumed to have had limited vagile capabilities. To move effectively, they probably closed the aulacophore cover plates and then, by wagging, they were able to move some-

Reticulocarpos is also a symmetrical cornute, but because of its numerous convergent morphologic features with mitrates will be discussed below.

Mitrates

Compositional elements of the mitrate aulacophore are the same as those of cornutes, but they functioned somewhat differently. Mitrates deployed the distal aulacophore arched convexly over the sea floor. Cover plates, in contrast to those of cornutes, did not open, but formed a tightly sutured vault over the food groove (Fig. 26.2).

Previous authors have variously dealt with the functional aspects of the mitrate aulacophore. Because the distal aulacophore is often preserved concavely arched, some have interpreted its feeding position as straight up, or slightly arched, over the theca, presumably with the cover plates opened (see for example Sprinkle 1976; Ubaghs 1979) (Fig. 26.5). Parsley and Caster (1975) suggested that, in this overhead, concave position, the cover plates rotated outwards, venetian blind fashion, on their curved bases to make ingestive openings. Two immediate problems confront us with the distal aulacophore aloft interpretation. First, the animal in such a position is quite unstable in any, but very slow currents. Secondly, as pointed out by Kolata (1982), concave arching of the distal aulacophore, jams the cover plates together and would have prevented their orderly opening.

Haude (1980) suggested that since the cover plates could not open in mitrates that perhaps the end of aulacophore was open and feeding could have taken place by branchial pumping resulting in pipetting of food off the muddy bottom. Numerous known examples of the distal-most ends of mitrate aulacophores clearly indicate they were not open tubes.

A more probable method of feeding was simply to flex the aulacophore convexly. The dorsally fused cover plate pairs would have separated from adjacent fused pairs, and resulted in arcuate- to chevron-shaped openings leading directly into the aulacophore vault (Fig. 26.6). Cilia, ?podia, or both, would conduct food to the longitudinal groove (Fig. 26.7). The water vascular system would have played a lesser role than in the cornutes, as evidenced by the less well defined pits and transverse grooves on the superior surface of the uniserial ossicles. The narrow openings

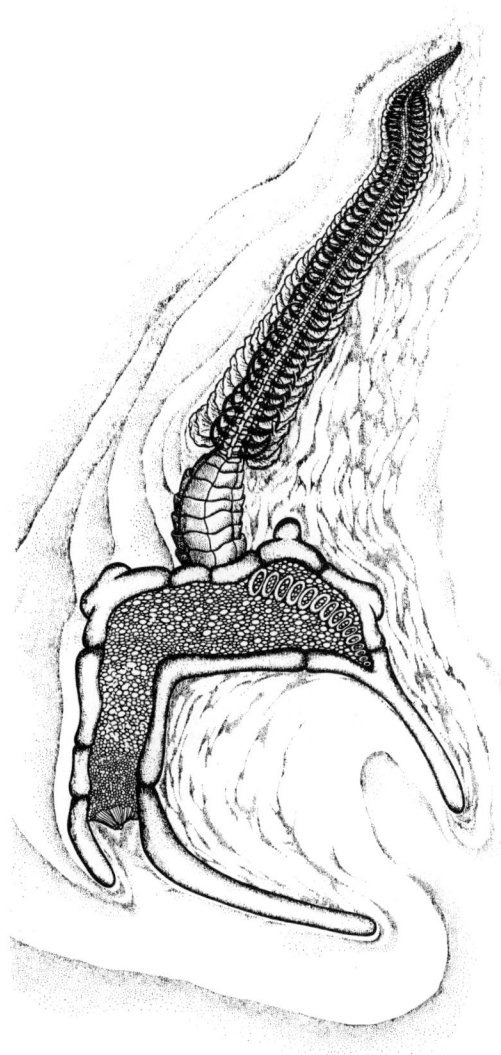

Fig. 26.4 Living orientation of the cornute *Cothurnocystis*. The aulacophore fed with the cover plates widely opened, aided by podia (in black). All sediment-piercing anchoring structures are on the theca. The anal opening and cothurnopores, which are respiratory structures, are closely associated with the tesselated supracentralia and are located in distinct lobes of the theca.

what clumsily to more advantageous feeding localities. They probably were not as efficient in feeding as asymmetrical forms for they would have lacked stability. However, the ability to wriggle off to more productive or safer areas may well have offset any problems related to feeding or stability.

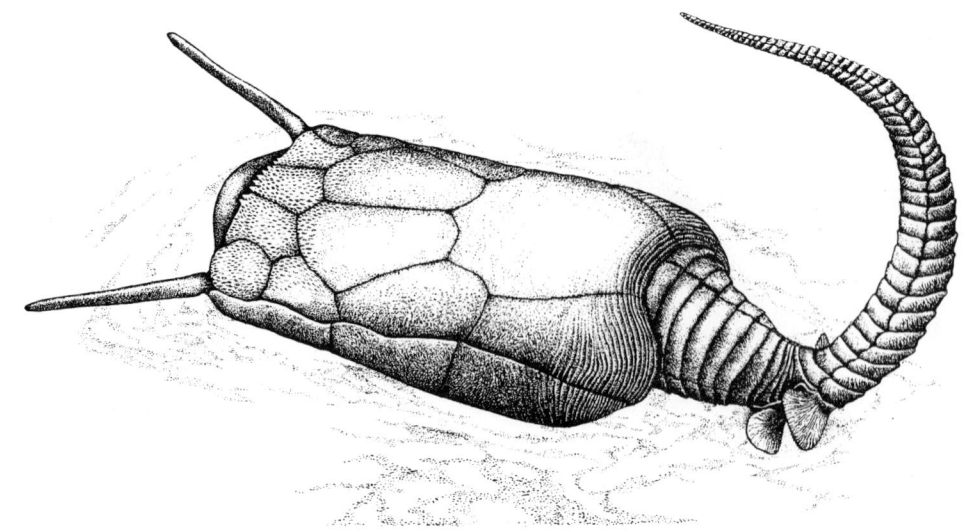

Fig. 26.5 Orientation of *Enoploura popei* Caster in a distress situation where the cover plate pairs are now jammed together and the openings into the food vault are closed. The frequent discovery of mitrate aulacophores concavely arched in this manner has led previous authors to argue that the aulacophore fed while being held aloft. In this orientation the animal was unstable in any but the slowest of currents.

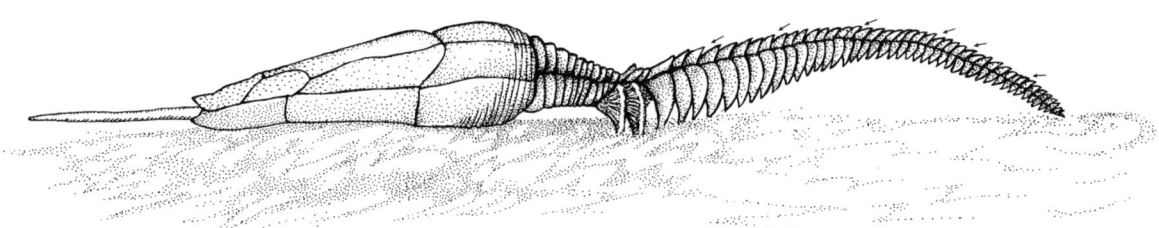

Fig. 26.6 Lateral view of *Enoploura popei* Caster in feeding orientation. Note that only the aulacophore flanges and spines anchor the animal. Currents moved across the animal from right to left.

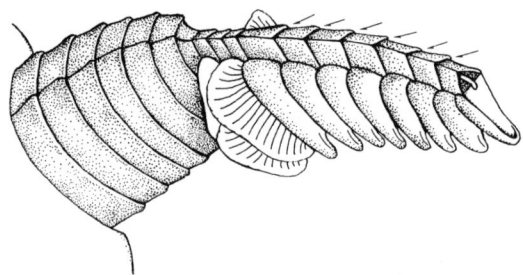

Fig. 26.7 An oblique view of the proximal aulacophore, styloid, and a portion of the distal aulacophore of *Enoploura popei*. Food entered into the food vault between the fused cover plate pairs and the moved to the mouth, at the junction of the proximal aulacophore and styloid, via the longitudinal food groove.

between cover plates, indicate that a much smaller size range and diversity of food was ingested than in cornutes; fine particulate food appears to have been their mainstay. Some mitrates appear to have incompletely fused cover plates resulting in a narrow triangular space at the distal end of the dorsal cover plate suture. This would have provided an additional opening into the vault, especially when the aulacophore was highly curved, and this space would have reduced binding of cover plates when the vault was closed by reflexing the aulacophore into a concave curve. Size of the openings leading into the vault, formed between cover plate pairs, was controlled by the degree of convexity assumed by the distal aulacophore: the greater the convexity, the farther the cover plate pairs would have been forced apart. A crude analogy to this feeding strategy is seen in rhynchonellid brachiopods where food passes through narrow zig-zag openings of the commissure (excluding larger fouling particles) induced by internally produced intake currents.

Movement of food inside the vault would have been, in some ways, similar to that of the cornutes; food particles would have moved along the food groove in mucus strings generated by the ciliated ambulacral epithelium (assumed to have been present). It would have differed in that the shortened podia would not have protruded between the cover plates, but were entirely contained in the vault. Cilia on the podia and along the food groove would have produced the incurrent. Water, entering along with the food particles, would have travelled inside the vault and through the gut, facilitated by gut pumping, and was voided at the anal opening.

This seemingly drastic modification of cover plate function and concomitant change in feeding habits, did not alter the functional continuity of the aulacophore. Probably, the driving adaptive force was that sieving through narrow conduits was a more efficient method of controlling sediment fouling than was wagging the open structure. With the aulacophore deployed straight out, the overlapping surfaces of the cover plates became slip planes and these would enhance the strength of the structure and make it more capable of sustained movement (skulling). Thus, supplementary adaptive forces were brought into play which were instrumental in modifying distal aulacophore morphology (see below).

Of critical importance in understanding the function of the mitrate aulacophore is the role of the styloid. Instead of being smooth on its undersurface as in cornutes, it is equipped with sediment-piercing flanges or spines for anchoring. These spines are the main, and usually the only, sediment-piercing structures on the entire organism. In many cases, the proximal ossicles of the distal aulacophore augmented the anchoring function of the styloid.

Thecae of mitrates and symmetrical cornutes are, in contrast to the normally flattened theca of asymmetrical cornutes, elevated and streamlined to form a hydrodynamic lifting body. The lateral marginals in mitrates are produced into oral-aboral runners. They are, as runners on a sled, to provide stability but with minimum frictional contact with the substratum. Runners, with few exceptions lack sediment-piercing spines and where spines are present (e.g. *Chinianocarpos*, *Mitrocystites*) they are close to the insertion of the aulacophore and are relatively small.

Mitrates vary considerably in the possession of aboral articulating spines. Anomalocystitids have paired spines which seem to be, in a linear sense, continuous with the runner system of the marginals. While they provide buttressing support for the theca the spines do not seem to be sediment-piercing structures. They are not flattened as are the large distal spines of asymmetrical cornutes. In contrast, *Enoploura popei* (Fig. 26.2), from the Upper Ordovician of North America, has spines that are curved, blade-like (flattened in the vertical plane), with the distal pointed ends directed away from the substratum. Peltocystids and kirkocystids have a single elongate, sinuate spine which is interpreted to be non-sediment piercing. *Lagynocystis* has a short broad spine. It has an omnidirectional ball and socket type of articulation and the shape of the spine seems better adapted for buttressing against the surface than piercing the sediment. Mitrocystitids are without articulating spines.

In summary, the significant sediment-piercing (anchoring) structures in mitrates are on the aulacophore, while in cornutes they are on the marginal plates of the theca (Fig. 26.8). The anchoring styloid spines in mitrates are closely adjacent to the most muscular part of the animal, the proximal aulacophore. It is assumed that by oblique and lateral wagging, the flanges/spines of the styloid could have been freed from the substratum and by lateral wagging the animal was propelled off and over the sea floor. Skulling would have propelled the animal, tadpole or sperm cell-like, until it settled down on the sea floor and implanted the flanges/spines of the styloid by a vigorous downward ventral movement by the proximal aulacophore.

Discussion

A logical question to ask is, why did cornutes become increasingly asymmetrical and sedentary and mitrates become increasingly bilaterized and vagile? The answer is speculative, but there is considerable consistency in my interpretation. Feeding with the wide open food groove arrange-

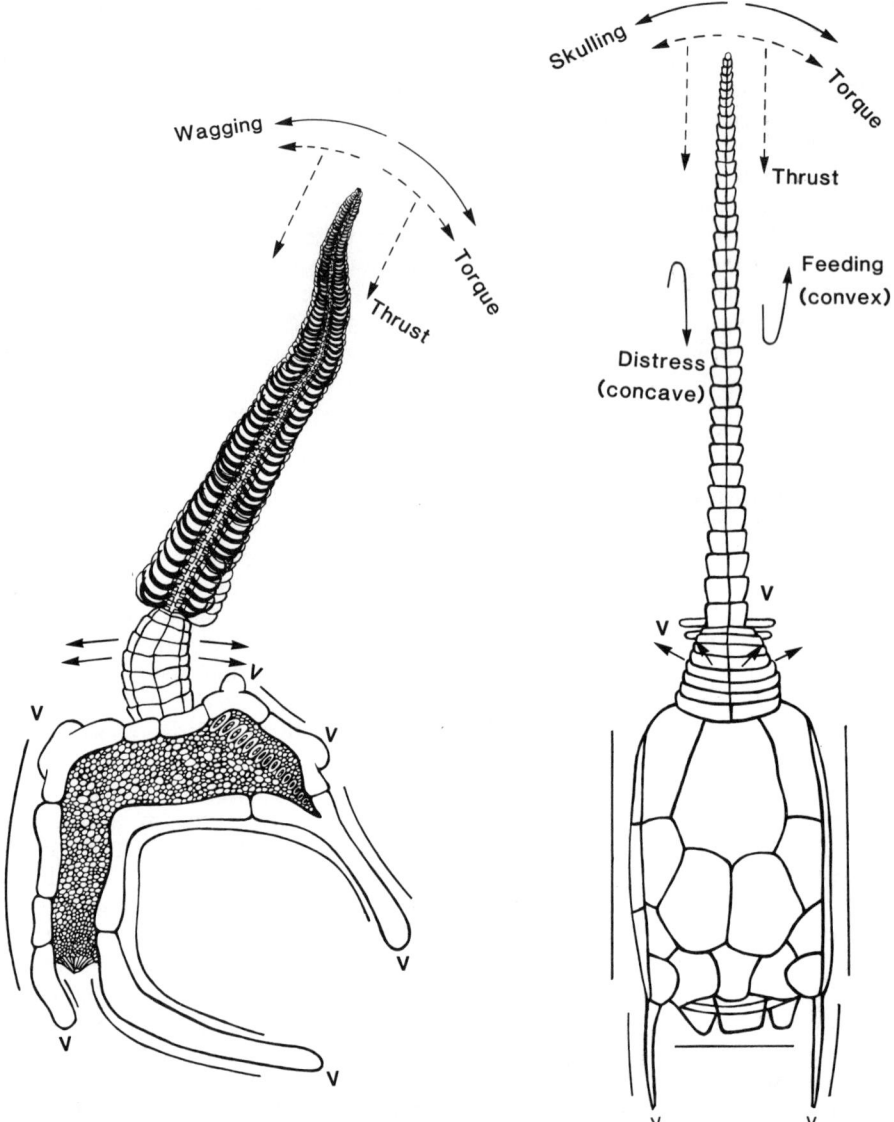

Fig. 26.8 Dynamics of cornutes (left, *Cothurnocystis*) and mitrates (right, *Enoploura*). Heavy spikes, V, indicate sediment-piercing structures; small spikes, v, are buttressing structures; lines parallel to the theca, resting surfaces; dashed lines, thrust and torque generated by the aulacophore; solid lines associated with the aulacophore, movements made by the structure.

The effect of wagging the aulacophore would produce a yawing motion, centred in the theca, if not checked by sediment-piercing structures. These counter-torque measures are especially evident in the asymmetrical cornutes. Torque generated by the distal aulacophore in mitrates would have helped the animal break away from the substratum. Once free of the bottom the mitrate theca would have yawed as a result of the skulling motion of the aulacophore.

ment of cornutes was efficient in that a wide size range of food could have been utilized and fouling by sediment particles could have been minimized by wagging the fully open structure. Sediment removal was probably efficient enough to allow feeding in rather turbid water. Secondary benefits would have included a greater volume of water swept by the feeding apparatus.

In order for the widely open system of cornutes to work efficiently, a stable platform was necessary. The asymmetrical cornute theca clearly bespeaks sessility and stability with its sediment-piercing knobs on the marginals and the large spatulate spines which would neutralize the thrust and torque generated by the wagging aulacophore.

Symmetrical cornutes (*Phyllocystis, Amygdalotheca*) appear to have had the capacity for a vagile mode of life, but had a typical cornute aulacophore. I would assume that symmetrical cornutes closed their distal aulacophore cover plates and wriggled along the bottom to find better feeding and environmental conditions. The higher diversity among asymmetrical than symmetrical cornutes would, however, seem to indicate the greater success of the former.

The feeding mechanism in mitrates may well represent a condition derived from symmetrical cornutes. The ability to feed in the 'closed and curved condition' may be a preadaptation from the cornute cover plate condition where the large overlapping cover plates, when closed, formed a leaky vault over the food groove allowing fine grained food particles to enter the vault. The permanent closing of the cover plates to form a vault (the mitrate condition) modified the entire distal aulacophore. The basal ossicles became relatively higher, narrower, and spinose. The cover plates became rectangular, rather than fan-shape; the articulating surfaces extended the length of the ossicle and bowed convexly upward, and the suture between the cover plate pairs essentially fused. The bilaterally symmetrical mitrate theca with its almost total lack of sediment-piercing spines and its narrow, parallel, resting surfaces or runners to minimize friction with the seafloor, and hydrodynamic shape, bespeaks mobility. The styloid, anchoring apparatus, adjacent to the main propulsive organ, the proximal aulacophore, also argues for a periodic vagile existence.

RESPIRATION AND RELATED FUNCTIONS

Respiratory gas exchange in Stylophora was apparently carried out in two ways. First, by direct gaseous exchange with the sea water across the integument (cornutes and mitrates). Secondly, through respiratory structures which, so far as is known, are restricted to the cornutes. These structures include aligned sutural pores (*Ceratocystis, Reticulocarpos* and some species of *Phyllocystis*); cothurnopores (*Cothurnocystis, Chauvelicystis, Nevadaecystis*, some species of *Phyllocystis*, and an unnamed form from the Middle Cambrian, Spence Shale; J. Sprinkle, personal communication), or lamellate organs (*Bohemiacystis, Scotiacystis*). All of these are located in, and open into, the centralia of the right adoral superior surface. They are always arranged in an arcuate pattern parallel to the marginal plates, but always separated from the marginals by a short distance, except in some cases, where the end members of an arcuate series abut the marginals. The opening of these structures to the exterior on the highest part of the superior surface was so that water intake would be devoid of as much sediment as possible.

Various interpretations have been made of cothurnopores, including genital bursae, multiple mouth openings, and branchial slits. Ubaghs (1968) has discussed them fully and I need not do it here. From a purely functional viewpoint, sutural pores, cothurnopores, and lamellate organs all seem best defined as purely respiratory structures.

Sutural pores and cothurnopores are elliptically shaped pits that extend to the inferior surface. Sutural pores in *Ceratocystis*, so far as known, are without covering apparatus, while those in *Phyllocystis crassimarginata* are partly constricted by supracentrals, which in pairs, form low chimney-like intake/exhaust openings. According to Ubaghs (1968), cothurnopores apparently had very small openings in the covering platelets in *Phyllocystis blayaci*, and, in my opinion, the same holds true in other genera with this respiratory structure. In the genera that have a lamellate organ, *Bohemiacystis* and *Scotiacystis*, it is difficult to determine whether the ambient seawater simply passed over the top of the structure or whether it passed between the narrowly spaced lamellipores as it did in some Rhombifera with pore rhombs, to effect gaseous exchange. I suspect that the latter was the case.

Current dynamics inside sutural pores and cothurnopores are conjectural and what follows is my interpretation. Water entered the sutural pore or cothurnopore at the admarginal end, travelled smoothly downwards toward the bottom of the structure where the current dissolved into a micromaelstrom of turbulent flow. Current was generated by ciliated tissue lining the inside of the pore structure and efficient gaseous exchange occurred across this tissue. Water left the pores at the abmarginal end. Currents moving in this direction would counter-flow the supposed body fluid circulation (see below) and would have resulted in high respiratory gaseous exchange efficiency.

The presence of respiratory structures in an organism that has a large flexible integument, a very high surface to volume ratio, and was probably gut pumping, is initially puzzling. Cornutes should have done well without them. However, these animals lived at the water/sediment boundary layer where oxygen content of the seawater is often diminished and currents (carrying oxygen) are greatly reduced. Respiratory requirements for muscles in the proximal aulacophore and in the theca where probably rather high and pore structures/lamellate organs would have helped to meet those requirements.

Cothurnopores appear to be highly modified sutural pores with specialized porous roofing plates. Lamellate organs are probably derived from sutural pores or cothurnopores (see Ubaghs 1968). Microscopic examination of their lamellipores on the holotype of *Bohemiacystis*, which are preserved in limonite, clearly show the same diffuse stereom pattern observed in pore rhombs of Rhombifera. The high ratio of stroma to stereom would allow efficient gaseous exchange across their lamellar surfaces.

Cothurnopores and lamellipores are always associated with tesselated surfaces and this pairing is presumably closely linked functionally. Tesselated surfaces that are bounded by well defined marginals are invariably seen on various echinoderm groups with recumbent habits.

Because tesselated surfaces in cornutes are made up of many thin, irregularly shaped platelets, that often do not appear to be tightly sutured, they are assumed by most workers to be flexible and associated with the pumping actions of the gut. (The term, tesselated surface, as used herein, is a surface made up of many small (about 1 mm in diameter), thin, often irregularly-shaped platelets, that are flexibly sutured, so that a surface composed of them would be quite pliable). Presumably, the purpose of such pumping actions, in addition to digestive functions, was to circulate body fluids, passing them past and through respiratory structures, e.g. pore rhombs, cothurnopores, lamellipores, and/or removal of oxygen with water associated with food, in the walls of the gut itself.

The functional nature of the tesselated surfaces is open to question and several functional modes were possible. First, the surfaces may have been passive and responsive to the presumed peristaltic waves moving down the gut. The polyplated surfaces simply allowed easier peristaltic wave movement to occur. Secondly, the tesselated surfaces may have been connected to muscles which generated waves that circulated body fluids and pumping, as such, was minimized in the gut. Thirdly, the tesselated surfaces and gut acted in concert to provide efficient circulation of body fluids and contents of the gut. While little direct evidence exists as to the true case, circumstantial evidence seems to support the third possibility in Stylophora (see below).

The anal opening is almost always located at an extremity of the polyplated field (Fig. 26.4) and in these cases the anus is associated with an elaborate valvular apparatus or anal pyramid. In cornutes and in many pleurocystitid rhombiferans the anus is in an attenuated part of the theca, or anal lobe, bounded by marginal plates. Frequently, the platelets adjacent to the anal apparatus are smaller and presumably were more flexible. Anal pyramids/valves seem to have been capable of withstanding rather high internal pressures, which would have been necessary if a peristaltic wave mechanism was efficiently employed. Most mitrates have rigid thecae and such forms have reduced (vestigial), or no anal valves or pyramids. Mitrates with parts of their superior surfaces retaining the shingled polyplated condition still have anal valves (e.g. *Mitrocystites* and *Mitrocystella*). *Lagynocystis* has numerous distally overlapping supracentrals and near the anal valve the polyplated field extends onto the lateral and inferior surfaces. The added flexibility adjacent to the anal valve would enhance the efficiency of constricting muscles and the circulation of body fluids. Here the morphology of the polyplated surface adjacent to a well developed anal valve supports the contention that in stylophorans the polyplated surface and gut acted in concert. In order for the progressively distal muscular action of the polyplates to work most efficiently, the gut contents would also have to be under an independent pressure system. A rough analogy is the muscular dynamics of milking a cow.

To come full circle, body fluids, pressurized by muscular tesselated surfaces, would also be more efficiently conveyed past the respiratory structures. The fact that respiratory structures too are in a lobe of the theca, but always slightly separated from the marginals probably indicates that body fluids were forced under pressure past the arcuately arranged respiratory structures and then collected on the opposite side in a lower pressure sinus adjacent to the marginals. It is also possible that turgor in the water vascular system, especially in the cornutes, was augmented by thecal pressurization. The vaulted aulacophore in mitrates probably had a reduced water vascular system and turgor was for them a minor problem.

The lack of respiratory structures in mitrates seems to be linked directly with the almost total lack of tesselated surfaces and elaborate anal structures. Their active mode of life selected for inflated, rigid hydrodynamic thecae. Water moving over the theca at much higher rates than in cornutes (Bernoulli effect) would provide adequate gaseous exchange. They were not, as were the asymmetrical cornutes, flattened and living essentially in the boundary layer at the bottom.

PHYLOGENETIC CONSIDERATIONS

As discussed above, cornutes appear to be of two distinct types, symmetrical and asymmetrical. The two earliest known cornutes, both of Middle Cambrian age, are divergent in their morphology. *Ceratocystis* from the Middle Cambrian, Skryje Beds, of Bohemia shows definite asymmetrical features in having distal horn-like marginals and a prominent embayment on its right side. Its respiratory structures are simple sutural pores and the centralia are large, non-flexible plates. Its proximal aulacophore is composed of polymerous segments. The second known type, has been illustrated (Sprinkle 1976), but remains undescribed; it is from the Spence Shale of Utah. The Utah specimen is probably more significant in terms of cornute evolution. This specimen is nearly symmetrical in outline except for an embayment on the right side. There are no spatulate spines and all of the thecal corners are rounded, much as in *Phyllocystis*. The centralia are organized into tesselated surfaces and the respiratory structures are cothurnopores. Its proximal aulacophore is tetramerous. Both of these Middle Cambrian genera have otherwise normal cornute aulacophores. Cornutes seem to be as conservative as the mitrates in maintaining their basic aulacophore pattern.

While *Ceratocystis* retains primitive sutural pores, the rest of the thecal morphology seems somewhat specialized and suggests that this genus is an early independent offshoot and has little to do with cornute evolutionary history. The Utah form seems to be much more typical of the lineage as a whole and seems easily relatable to both symmetrical and asymmetrical genera. Asymmetrical cornutes reflect a specialization towards sessility not only by their increasing asymmetry, but also by their tesselated surfaces and highly developed respiratory structures. Symmetrical cornutes persisted as somewhat more generalized forms. They displayed several different body plans as characterized by *Phyllocystis* and by *Amygdalotheca*. The former with a circlet of marginals enclosing the entire theca seems not to have given rise to any obvious succeeding morphological type(s); the latter body plan with a clear-cut distal gap in the marginals for the anal valve may well be ancestral to at least some of the mitrates.

Ubaghs (1968) has pointed out that mitrates and cornutes are clearly related, but are distinct and clearly defined when each order appears in the fossil record. The earliest mitrates *Peltocystis* and *Chinianocarpos*, from the Lower Ordovician of France, are indeed highly distinctive. They do not show close affinities with each other or with any other mitrate group. While *Chinianocarpos* has been included in the Mitrocystitidae by Ubaghs (1968), I would place the genus in its own family.

Despite all of the individual characteristics of early mitrates there is evidence of direct relationship between the two orders. The amygdalothecid genus *Reticulocarpos* is a mitrate-like cornute that clearly shows features common to cornutes and mitrates, especially to the Anomalocystitida. In fact, it is fairly safe to assume that thecal marginals $M1/M'1$ through $M4/M'4$ are homologous in the cornute Amygdalothecidae and mitrate Anomalocystitida. $M5/M'5$ marginals of the amygdalothecids are homologous with the articulating spines of anomalocystitids. The close relationship of *Reticulocarpos* to the mitrates has been recognized by Jefferies and Prokop (1972), Jefferies and Lewis (1973), and Philip (1979). *Reticulocarpos* is unique among cornutes in having a spinose styloid rather than a stylocone as well as having large ventrally spiked proximal ossicles on the distal aulacophore. Cover plates on the distal aulacophore seem to be non-erectile and formed a typical mitrate-like vault. From what we can see of the aulacophore it appears to have functioned in a mitrate-like manner. The theca in *Reticulocarpos* is quite symmetrical in outline, the zygal is reduced, as is the number of marginal plates. *Amygdalotheca*, its closest known relative, has seven marginals and a complete zygal. *Reticulocarpos* has five marginals with M4/M'4 connected by a transverse bar made up of extensions of those plates. Marginals M5/M'5 are protrusive in character and as indicated above would appear to be homologous with the articulating spines in anomalocystitids. It seems quite likely that this arrangement of marginals, as seen in *Reticulocarpos*, is a stadium for the aboral end in the anomalocystitid lineage; they all have, if we exclude *Barrandeocarpos*, four marginals, and an articulating spine attached to the distal end of each M4/M'4.

I suggest that this is probably a clear-cut neotenous process. Marginal plates forming near and adjacent to the aulacophore girdle would have formed earlier in embryology than those in a more distal position. The process seems to incur the loss of distal-most marginals and the 'retreat' of the anal apparatus, which is associated with somatic plates, to a more proximal position. It is interesting to note that along with the apparent 'shortening' of *Reticulocarpos*, by loss of distal marginals, the genus displays a mitrate-like characteristic in the lack of a well defined anal valve structure.

Amygdalothecids without shortening the theca by loss of distal marginals, may have given rise to the mitrocystitids. Mitrocystitids (excluding *Chinianocarpos*) have up to seven marginals, the number in *Amygdalotheca*, and are without articulating aboral spines. In these two groups anal valve structures are also somewhat similar in disposition. At least for the Anomalocystitidae and the Mitrocystitidae a common ancestor in the Amgydalothecidae seems probable.

The Peltocystidae, Lagynocystidae, and Kirkocystidae are far more difficult to place in terms of their origins. Their thecae are not only shortened relative to the number of marginal plates (when considering an amygdalothecid ancestral stock), but they are laterally compressed as well, i.e. they have lost numerous centralia and marginals. Marginal plate homologies in these groups are difficult, if not impossible, to work out at this time. These three families have a single spine. While the sigmoidal spine in peltocystids and kirkocystids may be homologous, the hollow lagynocystid spine appears to be of independent origin.

The aulacophores in all mitrates appear to be quite similar and broadly argue for a common origin in a cornute radical. Even the most differing aulacophore, that of *Lagynocystis*, appears to be a specialization of the basic structure. Mitrates are probably best considered as polyphyletic, generic-familial, groups derived from a common symmetrical cornute stock in which the development of the 'mitrate' aulacophore was well established.

It is interesting to note that evolution among mitrates has not always been directed towards a vagile life style. The mitrate *Diamphidiocystis* from the Upper Ordovician of Illinois displays convergent features of sessility with asymmetrical cornutes (Kolata and Guensburg 1979). The longitudinal axis of the theca is bowed into a broad U-shape so that the single spine base and the aulacophore are directed out from the top of the U. The animal seems devoid of hydrodynamic qualities that characterize a vagile organism. The aulacophore appears to have a greatly smoothed styloid, indeed it is a stylocone. The large flattened sickle-like spine is robustly articulated with the theca: its principal movement was lateral, parallel to the sea floor. I suggest it served as an anchor, minimizing thrust and torque generated by the aulacophore. If this interpretation of the genus is correct, it verifies the efficiency of the mitrate vaulted aulacophore in that it could support a sessile organism without opening the aulacophore cover plates.

ACKNOWLEDGEMENTS

Most of the inspiration for this paper occurred while the author was a National Academy of Sciences exchange scientist in Prague, Czechoslovakia, June–July, 1986. While in Prague, Dr Rudolf Prokop of the Národní Muzeum v Praze (National Museum in Prague) made important specimens available for study. Ann Marie Arnold ably made most of the illustrations and Dr Robert Horodyski, Department of Geology, Tulane University, aided in the photographic preparation of some of the illustrations. Dr Kenneth E. Caster, Department of Geology, University of Cincinnati and an anonymous reviewer greatly improved the quality of the manuscript. Their efforts are acknowledged.

REFERENCES

Bonik, K., Gutmann, W. F., and Haude, R. 1978. Stachelhauter mit Kiemen-Apparut: Der Beleg fur die Ableitung der Echinodermen von Chordatieren. *Natur und Museum, Frankfurt* **108**, 211–13.

Carlson, S. J. and Fisher, D. C. 1981. Microstructural and morphologic analysis of a carpoid aulacophore. *Geological Society of America, Abstracts with Programs* **13**, 422–3.

Caster, K. E. 1983. A new Silurian carpoid echinoderm from Tasmania and a revision of the Allanicytidiidae. *Alcheringa* **7**, 321–35.

Chauvel, J. 1981. Étude critique de quelques Echinodermes. Stylophores du Massif Amoricain. *Bulletin de la Société de Géologie et Minéralogie de Bretagne*, (C) **12**, 67–101.

Haude, R. 1980. A new Mitrate from the Lower/Middle Devonian, and evolutionary reconstruction of carpoid echinoderms. In *Echinoderms: past and present* (ed. M Jangoux), pp. 25–30. Balkema, Rotterdam.

Jefferies, R. P. S. 1967. Some fossil chordates with echinoderm affinities. *Symposia of the Zoological Society of London* **20**, 163–208.

Jefferies, R. P. S. and Lewis, D. N. 1973. The Ordovician fossil *Lagynocystis pyramidalis* (Barrande) and the ancestry of Amphioxus. *Philosophical Transactions of the Royal Society of London* **B265**, 409–69.

Jefferies, R. P. S. and Prokop, R. J. 1972. A new calcichordate from the Ordovician of Bohemia and its anatomy, adaptations and relationships. *Biological Journal of the Linnean Society* **4**, 69–115.

Kolata, D. R. 1982. The cornute and mitrate controversy—A review and commentary. In: *Echinoderms. Proceedings of the international conference, Tampa* (ed. J. M. Lawrence), p. 91. Balkema, Rotterdam.

Kolata, D. R. and Guensburg, T. E. 1979. *Diamphidiocystis* a new mitrate 'carpoid' from the Cincinnatian (Upper Ordovician) Maquoketa Group in southern Illinois. *Journal of Paleontology* **53**, 1121–35.

Kolata, D. R. and Jollie, M. 1982. Anomalocystitid mitrates (Stylophora–Echinodermata) from the Champlainian (Middle Ordovician) Guttenberg Formation of the Upper Mississippi Valley Region. *Journal of Paleontology* **56**, 631–53.

Manton, S. M. 1977. *The Arthropoda, habits, functional morphology and evolution.* Clarendon Press, Oxford.

Parsley, R. L. and Caster, K. E. 1982. Functional morphology of mitrate homalozoans (Echinodermata). *Geological Society of America, Abstracts with Programs* **14**, 583.

Paul, C. R. C. and Smith, A. B. 1984. The early radiation and phylogeny of echinoderms. *Biological Reviews* **59**, 443–81.

Philip, G. M. 1979. Carpoids–echinoderms or chordates? *Biological Reviews* **54**, 439–71.

Sprinkle, J. 1976. Biostratigraphy and paleoecology of Cambrian echinoderms from the Rocky Mountains. *Geological Studies of Brigham Young University* **23**, 61–73.

Ubaghs, G. 1961. Sur la nature d l'organe appelé tige ou pedoncule chez les Carpoïdes Cornuta et Mitrata. *Comptes Rendus Hebdomadaires de Seances de l'Academie des Sciences, Paris* **53**, 2738–40.

Ubaghs, G. 1968. Stylophora. In *Treatise on invertebrate paleontology, part S, Echinodermata 1* (ed. R. C. Moore), pp. 496–565. Geological Society of American and University of Kansas Press, Lawrence, Kansas.

Ubaghs, G. 1969. Les Echinodermes Carpoïdes de L'Ordovician Iférieur de la Montagne Noire (France). *Cahiers de Paléontologie* 112 pp.

Ubaghs, G. 1979. Trois Mitrata (Echinodermata: Stylophora) nouveaux de l'Ordovicien de Tchechoslovaquie. *Paläontologische Zeitschrift* **53**, 98–119.

Systematic Index

Abludoglyptocrinus 261
Abrachiocrinus 259
Abyssocrinus 251
Acanthaster planci 314
Acentrotremites 218, 232
  ellipticus 226
Aceste bellidifera 331
Achradocystites 209
Acolocrinidae 246, 252
Acolocrinus 258, 260–1, 263
acraniates 9, 39
Acrocrinidae 224
acrosaleniids 184
Actinocrinitidae 225, 260–1
Actinopyga echinites 315–16
Aethocrinidae 236, 238, 240
Aethocrinus moorei 236, 238–41, 258
Aethocrinus murchisoni 236, 238–42
Agostocrinus 260, 263
Ailsacrinus 278
Akadocrinus 200, 203–6
Allagecrinacea 246, 251, 260
Allagecrinidae 246–7, 252, 258
Allagecrinus 247
Allanicystidiidae 350
Allocatillocrinus 246, 247, 252
Ambolostoma 232
Ambolostomatidae 215
Amorphocrinus 225
amphioxus 8, 20
Amphiuridae 54
Amygdalocystites 209
Amygdalotheca 349, 352, 357, 359–60
Amygdalothecidae 359–60
Anamesocrinus 251
Anartiocrinus 258
Anemetocrinus 265
Angioblastus 219, 226–8, 233
  caddense 226, 227
  dotti 226
  ellesmerensis 226–7
Angulatoblastus 231
Annacrinus 275
Anomalocrinacea 252
Anomalocystites 349–50
Anomalocystitidae 350, 359–60
Antedon 255
Anthoblastus 233
Anthocidaris crassispina 46, 48–9, 51
Apatopygus 160
Apiocrinitidae 276–80
Apodasmocrinus 251
Aptilechinus 92–3
Arachnocystites 203
Arachnoides 158
Arachnoididae 115, 159

Arbacia punctulata 33, 37, 125–6, 141, 192–3
Arbacia stellata 302–3
Arbacioida 37
Arbucklecystis 208–9
Archaeocrinus 221, 261
Archaetaxocrinus burfordi 258
Archegonaster 89, 90, 91
Arcuoblastus 231
aristocystitids 202, 210
Articulata 46, 256–7, 269–83, 321–2, 327
Artuschisma 215, 233
  rossica 226
ascidians 39
Ascocystites 204–5
Aspidochirotida 46, 48, 76, 290, 293, 322
Aspidodiadema arcitum 301, 306–7
Aspidodiadema hawaiiense 301, 306–7
Aspidodiadema jacobyi 301, 307
Aspidodiadema meijerei 306
Aspidodiadema nicobaricum 306
Aspidodiadema tonsum 301, 307
Aspidodiadematidae 299, 301, 305, 307–10
Asterias 87
  amurensis 46, 47, 49, 50
  forbesi 32–3, 126–7
Asterina pectinifera 192
asteroblastids 210, 212
Asterozoa 29–38, 50–1, 69, 77, 294–6
Astriclypeidae 115, 159
Astrocrinus 218, 232
  tetragonus 226
Astrocystites 210
Astropecten 57
  scoparius 46, 47, 49
Astropectenidae 57, 60
Astropyga pulvinata 301, 305
Atelicrinidae 284
Atelostomata 334
Atubaria 8
Aulechinus 92–3
Auloblastus 231
Aulodonta 308
Aurelianaster 333
Austinocrinus 275–6
Australoblastus 233

Baerocrinus 259
Balanocrinus 275
Barrandeocarpos 360
Barycrinus 256, 325
Basseaster 333
Bathericrinus 256

Bathericystis 204
batocrinids 224
Baueria 183
Belemnocrinacea 246, 251
Belocrinus 231
Bilateria 6, 7
Billingsocystis 209
Bistomiacystis 209
blastoids 200, 202, 204, 206–7, 209–11, 324–5
Blastozoa 199, 348
Bockia 206–7, 210–11
  grava 207
  mirabilis 207
Bohemiacystis 357
bothriocidarids 91
Bothriocidaris 92, 95, 182
Bothryocrinidae 261
Bourgueticrinida/ina 269–73, 276, 279, 281, 284, 312–13
Bourgueticrinus 276
Brachiopoda 7
Brachyschisma 222, 231
Breynia 153–5, 159
  australasiae 159
  carinata 154
  aff. *carinata* 153–5
  desorii 153, 155, 159
  neanika 159
Brightonia 334, 340
Brockocystis tecumseth 202
Bromidocrinus nodosus 264
Bromidocystis 207–8
Bryozoa 7

Caenopedina cubensis 301, 306
Caenopedina mirabilis 301, 306
Cainocrinidae 270–1, 275, 281, 284
Cainocrinus 275, 284
Calathocrinus digitatus 277
Calceocrinacea/idae 245, 251, 258, 261, 267
calcichordates 38, 347
Calpiocrinus 262
Calycoblastus 233
Calymne 333, 343–5
  relicta 343
Calymnidae 331–3
Camarodonta 46, 48, 125
Cambraster 89, 90–1, 95
Cambrocrinus 204, 205
camerates 221–6, 235–9, 250, 256–67
Camptostroma 10, 23, 87–8, 321–3, 325
Canadocystis 209
Carabocrinus 261

364 Systematic Index

Cardabia 333
Cardiaster 333, 338
Cardiocystites 204, 206
Cardiotaxis 333
  heberti 335
Carpenteroblastus 231
Caryoblastus 221, 231
Caryocystites 203
cassiduloids 152–3, 157, 159, 160, 168–70
Catillocrinidae 252
Cenocrinus 275
Centrostephanus coronatus 301, 305, 310
Cephalochordata 20–1
Cephalodiscus 7, 8, 15–16, 21
Ceratoblastus 233
Ceratocystis 8, 10, 348, 357
Ceratophysa 333
Certonardoa semiregularis 46–7, 49
Chaetodiadema granulatum 301, 305
Chaetodiadema pallidum 301, 305
Chaetognatha 6, 7
Chauvelicystis 352, 357
  spinosa 348
Cheirocystella 204–5, 211
Chelocrinus 273
  schlotheimi 272
Chelonechinus 332, 334, 336, 343–4
Chinianocarpos 349–50, 355, 359–60
Chladocrinus 275
chordates 7, 20, 31–2, 38
Cibaster 333
Cidaridae/oida 34, 37, 57, 60, 63–4, 128, 182, 302
Cincinnaticrinacea/idae 249, 252, 267
Cincinnaticrinus 246, 252, 261
cinctans 21
Cladida 223–9, 245, 250, 257–67
Clathrocrinus 256
Clypeaster japonica 109, 192
Clypeaster rosaceus 185
Clypeasteridae/ina/oida 46, 107–8, 114–17, 151–3, 159–60, 166–7, 170, 184–5
Clypeasterina 185
Codaster 218, 232
  acutus 226
Codiacrinacea 249, 258
Collyritidae 343
Colpodecrinus 237–8
Columbocystis 207
Comanthus japonica 45–7, 49, 314
Comarocystites 209
Comatulida/idia/ina 265, 269–71, 274–6, 279–84, 312–13, 325
Comosteridae 276
Compagicrinus fenestratus 236, 238–42
Concentricycloidea 32
Conuloblastus 221, 231
Cornutes 8–10, 245, 347–60
coronates 201–2, 204, 206–11

Corystidae 332–3, 340, 344
Corystus 333
  dysasteroides 335
Coscinasterias acutispina 46–7, 49, 50
Costatoblastus 231
Cothurnocystis 353, 356–7
Cranocrinus 249
Cremacrinus 251
Cribroblastus 232
Crinoidea 211, 294–6
Crossaster 293
Crotalocrinites 260–1
Cryptoblastus 231
Cryptocrinites 208
cryptocrinoids 206–11
Cryptoschisma 221, 231
Cryptosyringida 88, 90–1
Ctenocrinus 256
ctenocystoids 21
Cupulocrinus 258
cyathocrinine cladids 258–67
Cyathocrinites 226
cyclocystoids 79, 80
Cyclohydrocoela 21–3
Cydonocrinus 226
Cydrocrinus 262
Cyrtocrinida/idia 269–71, 277, 280, 284
Cystidea nugatula 203
Cystocrepis 333
Cystoidea 79, 211, 221, 324–5
Cytidocrinus 260, 263, 265
Cyttarocrinus eriensis 249–50

Dadocrinidae 270–1, 277–8, 280, 284
Dadocrinus 273
  gracilis 277–8
  kunischi 277
Daedalocrinus 246, 251, 261
Decaschisma 221, 231
Decemoblastus 231
Deltoblastus 217, 219, 228, 233
Dendraster 167
  excentricus 108, 158
Dendrasteridae 114–15, 117, 159
Dendrochirotida 46, 48, 76, 290, 293, 312, 322
Dendrocrinidae 236, 238, 240
dendrocrinine cladids 258–67
Dentiblastus 231
Desmocrinus conifer 314
Deuterostomia 6, 7, 20, 29, 38
Devonoblastus 222, 231
Dexiothetica 6–9
Diabolocrinus arbucklensis 264
Diadema 291, 300, 302
  antillarum 300–2, 304
  mexicanum 300–4
  savignyi 301, 304
  setosum 46–7, 49, 302
Diadematacea/idae/oida 46, 299–302, 304–10

Dialutocrinus 225, 261
Diamphidiocystis 360
Dichocrinidae/inae 224, 264
Dichocrinus 265
Dichoporita 211
Diplobathrida 224, 236
Diploblastus 232
Diplocrinus 275
Diploporita 201–2, 210, 322, 324
Dipteroblastus 233
Disasteridae/oida 332, 334
Discoides 168
Disparida 235–9, 245–52, 257–67
Dolichoblastus 233
  shimanski 226
Doliocrinus 246
Doreckicrinus 275
Doryblastus 231
Drosophila 128, 131–2
  melanogaster 101

Echinarachnidae 115, 159
Echinarachnius 161
  laganolithinus 159
  parma 109, 159
Echinidae 126
Echinocardium cordatum 161
Echinocorys 333, 338
  sulcata 337–8
Echinocrepis 333
Echinocyamus pusillus 192
Echinoidea 29–38, 46, 48, 50–1, 294–6
Echinolampas 159, 168
  gambierensis 160
  morgani 160
  ovulum 160
  posterocrassa 160
Echinometra lucunter 181
Echinopluteus transversus 299, 302, 308
Echinosigra 153–4, 331, 334
  phiale 331
Echinosphaerites 203
Echinostrephus 183
Echinothrix calamaris 301, 305
Echinothrix diadema 301, 305
Echinothuriidae/oida 180, 299
Echinozoa 69, 70, 77, 91
Echinus 291
  esculentus 179, 318
Echmatocrinea/idae 236, 245
Echmatocrinus 250, 265, 323, 325
  brachiatus 235–6, 239–42, 257, 259
Ectenocrinus 246, 251
Ectinechinus 92
edrioasteroids 10, 79, 88, 323, 325
edrioblastoid 210
Edriocrinus 260
Eifelocrinus 261
Elasipoda 76
Eleutherocrinus 222, 231

Eleutherozoa 4, 21, 23, 30, 32, 69, 77, 78, 82, 321
Ellipticoblastus 215, 218, 225, 228
 ellipticus 225
Embryocrinus 249, 259
Encrinidae 269–74, 278–80, 283
Endoxocrinus 275
Enopleura 356
 popei 350, 354–5
enteropneusts 8, 54, 61, 70
Entomaster 333
Eocomatula 274, 284
Eocomatulidae 270–1, 274, 284
eocrinoids 324
Eophiura 90–2
Eopinnacrinus 263
Eoscutellidae 115, 117, 158
Eothuria 92, 182
Erisocrinidae 270–2, 278
eublastoids 201
Eucidaris 60
 metularia 64
 thouarsi 302–3
 tribuloides 33, 37
Eucladocrinus 263
eucoelomate phyla 38
euechinoids 60
Eumorphocystis 210, 212
Euonychocrinus 262, 264
 simplex 258
Eupatagus 159
Eupentacta quinquiesemita 314, 317–19
Eustenocrinidae 236, 237, 240, 252
Eustenocrinus 252, 259
Eustypocystis 200, 203, 206

Fibulariidae 115, 158
Fissiculata 215–33
fistuliporite rhombiferans 199–201
Flexibilia 226, 235, 250, 257–67, 322, 327
Florometra 55
Foerstecystis 207–8
Forcipulatida 46, 48, 55, 60, 64

Galeaster 333, 340
Galeola 333
Ganbirretia 333
Gillocystis 92–4
Globoblastus 231
Globocrinus 224
Globulocystis 208–9
Glyptocystitida 200, 202, 324
Glyptosphaerites 212
Gogia 199, 200, 202, 203, 205, 211, 323
 granulosa 202
 hobbsi 202
 kitchnerensis 203
 palmeri 203

prolifica 203
Gongyloblastus 232
 amoricanus 225
Granatocrinus 232
Grenprisia 256, 261
Guettaria 333, 340
 rocardi 337

Hadroblastus 231
 benniei 226
Hagenowia 153–4, 168, 333, 336–7, 339
 anterior 154
 blackmorei 154
 rostrata 337
Hallaster 93
Hallocrinus 261
Haplocrinites 246, 250–1
Haplosphaeronis 202
Hebrocidaris 183
Helgocystis 334
Helicoplacoidea 9, 10, 21, 23, 78–81, 87, 321, 324–5
Helicoplacus 88, 95
Heliocidaris 37–8
 erythrogramma 33, 38–9, 183
 tuberculata 33, 38
Heliocrinites 203
Heliophora 158
 orbiculus 154, 156
Hemiaster 157, 159–60
Hemicentrotus pulcherrimus 46, 48–9
Hemichordata 6, 7, 17–18, 20, 54, 70
hemicosmitids 202, 210–11
Hemipneustes 333
Henricia 191, 293
Hercocrinus 264
Heteroblastus 218, 232
 cumberlandi 226
Heterocrinacea/idae 249, 252
Heterocrinus 246
Heteroschisma 222, 231
Hexacrinites 256, 261
Holaster 156, 332–3, 338–9, 344–5
 cordatus 338
 grasanus 338
 intermedius 335
 suborbicularis 338
Holasteridae/oida 153, 157, 160–1, 168, 184, 331–45
holectypoids 168, 184
Holectypus 184
Holocrinidae 270–1, 273–4, 277, 279–83
Holopodinidia 284
Holothuria leucospilota 315, 317–19
Holothuria monacaria 45–7, 49
holothurians 29–38, 50–1, 166, 294–6
Holynocrinus 246
Homalocrinus 262
Homalozoa 21, 347
Homocrinacea 246, 250–2, 267

homoiosteleans 347
Homocrinus 245, 247–51
human 128, 131–2
Huttonechinus 333
Hybocrinida 220–1, 245, 250, 257, 259
Hybocrinus 220, 236, 238, 240
Hydreionocrinus 262–3, 265
Hyocrinidia 271, 277, 280, 284
Hypalocrinus 275
Hyperoblastidae 215
Hyperoblastus 221–2, 231

Ibexocrinus 246, 251
Inadunata 220, 226, 236, 250, 257–67, 322, 324–5
Indoblastus 233
Infulaster 153–4, 333, 339
Inyocrinus 258
 strimplei 236, 238–41
Iocrinidae 252
Iraniaster 334
irregular echinoids 153–62, 166, 184–5
Ismidaster 333
Isocrinida/idae/ina 269, 270–5, 278–81, 283–4, 312–3
Isocrinus 275
Isselicrinidae 270–1, 275–6, 281, 284
Isselicrinus 275

Jaekelicrinus 247
Jeronia 333, 339

Kallimorphocrinus 246–7
Katoblastus 231
 konincki 224
 puzos 224
Kazachstanoblastus 233
 carinatus 226
Kinzercystis 88, 199, 211
Kirkocystidae 350, 355, 360

Labrotaxis 333
 cantianus 335
Laeothetica 20
laganidae 115–17, 158
Lagynocystidae 360
Lagynocystis 349–50, 355, 358, 360
Lampadaster 333
Lampadocorys 333, 339
Lampadosocrinus 249
Lamprometra palmata 32–3
Lepidocystis 87, 199, 202, 211, 324
lepidocystoids 323, 325
Leptodiadema purpureum 305–6
Leptoschisma 231

Systematic Index

Leptosynapta 290–1
  inhaerens 32
Leviechinus 334, 340
Lichenoides 199, 200, 202–3, 210
Lingulocystis 204, 206
Linthia 157, 167
Lissodiadematidae 301, 305–6, 308–10
Litocrinus 246, 250
  angulatus 247
Lophoblastus 231
Lophophorate phyla 7
Loriolocrinus 276
Lovenia 159
Luidia 57
Luidiidae 57, 60
Lysechinus incongruens 150
Lysocystites 201, 206, 210
Lytechinus 291
  pictus 33, 37, 126, 141, 192

Macrocystella 204–5, 211
Macurdablastus 206, 210, 215, 220, 231
Malchiblastus 233
  australis 226
Malocystites 209
Manillocrinus 263
Marsupites 276
Marsupitidae 270–1, 273–4, 276, 278–9, 281, 284
Mastoblastus 233
  ornatus 226
Mediaster aequalis 190
Meekechinus 152
Mellita quinquiesperforata 109
Mellitidae 114–15, 117, 159
Melocrinites 256
Melocrinitidae 260, 263
Merriamaster 168
Mesoblastus 219, 225, 231
  angulatus 225
  crenularis 224
  elongatus 225
Mesothuria intestinalis 76
Messaoudia 333
Metablastus 231
Metabolocrinus 261
Metacrinus 275
  rotundus 314
Micraster 166
microcrinoids 226
Micropyga tuberculata 301, 306
Micropyga violacea 301, 306
Micropygidae/oida 299, 301, 305–6, 308–10
Millericrinida/idae/ina 269–74, 277–81, 284
mitrates 9, 10, 92, 347–60
Mitrocystella 358
Mitrocystites 355, 358
Mitrocystitidae 349, 355, 359–60

Molpadia intermedia 74
Molpadiida 57, 76
Monadoblastus 231
Monobathrida 219, 224–5, 236–7, 240, 260
Monobrachiocrinus 249, 259
Monophorasteridae 115
Monoschizoblastus 218, 228, 232
  rofei 225
mouse 128, 131–2
Myelodactylacea/idae 246, 252, 321
Myelodactylus 324
Myophiurida 46

Nannoblastus 233
Neilsenicrinus 275
Neoasteroidea 327
Neocrinus 275
Neolaganidae 158
Neolageniocrinus 226
neolampadids 161
Neoschisma 227, 233
Neozeacrinites 262
Nevadaecystis 348, 357
Nodoblastus 228, 233
  librovitchi 216, 226
Nolichuckia 199, 200
Notoblastus 233
Nowracrinus ornatus 272
Nucleocrinus 222, 231
Nymphaeoblastus 232

Offaster 333
Ohiocrinus 246, 252, 261
Oklahomacystis 209, 220
oligopygoids 149, 160, 171
Onychocrinus 256, 261–2
ophiocistioids 76, 78–9, 91–5
Ophiocoma pumila 55
Ophiocoma wendti 32–3
Ophiocomella ophiactoides 321
Ophiocomina 291–2
Ophioderma 55
Ophionereis 55
  annulata 54
Ophiophragmus filograneus 318
Ophioplocus japonicus 46–7, 49
Ophiothrix 288, 291–2
ophiuroids 29–38, 50–1, 318, 322
Opisopneustes 333
Orbiblastus 231
Orbitremites 215, 219, 228, 232
  derbiensis 226
Orbitremitidae 228
Orophocrinus 225, 231
  celticus 225
  orbignyanus 224–5
  pentangularis 225
  praelongus 225
Othneiocrinus 252, 261

Ova 155

Pachyblastus 231
Palaeocrinus hudsoni 261
Palaeocucumaria 92–5
Palaeocystites 208, 210
Palaeoholopodidae 260
Palaeura 91, 93
parablastoids 201, 210–11, 221
Paracentrotus lividus 129, 132, 135
Paracomatulacea/idae 270–1, 275, 280, 284
paracrinoids 200–2, 204, 206–8, 210–11, 220–1
Paradichocrinus 263
Parahexacrinidae 260
Parahexacrinus 260–1
Paraster 167
Paratalarocrinus 261
Pareocrinus 204–6
Paronaster 333
Paxillosida 46, 48
Pedinidae/oida 299, 301, 304–6, 308–9
Pellecrinus 256
Pelmatozoa 21, 23–4, 30, 78, 87, 199, 211, 321–7
peltocystids 355, 360
Peltocystis 359
Peniculocrinus 246, 252
Pentablastus 233
  supracarbonicus 226
Pentacrinites 270–2, 274, 280
  dargniesi 274
  dichotomus 274, 280
Pentacrinitidae 274, 278–81, 283
Pentacrinus interbrachiatus 274
Pentagonopterix insculptus 273
Pentremites 215, 219, 224, 228, 232
  rusticus 226
Pentremitidae 215
Pentremitidea 221, 231
Pentremoblastus 232
Pericosmus 160
Periechocrinus 261, 263
Perischoechinoidea 182
Perittocrinacea/idae 236–8, 240, 246, 252
Perittocrinus radiatus 236–7, 239–41
Peronella 158
  japonica 39, 46–7, 49, 51, 194
  orbicularis 158
  rictum 158
  tuberculata 158
Petaloblastus 231
Petalocystites 204, 206
Petraster 89–91
Phaenoblastus caryophyllatus 224
Phaenoschisma 225–6, 232
  acutum 225
Phanocrinus 265
Phoronida 7

Phyllocystis 349, 352, 357, 359
  blayaci 357
  crassimarginata 357
Pilematechinus 334, 343–5
Pisaster ochraceus 189–90
Pisocrinacea 246–51
Pisocrinus 246–7, 249, 324
Planacrocrinus 324
Platanaster 89, 90
platycrinids 224, 250
Platycrinites 225, 324
Platycystites 208–9, 220
Platyhelminthomorpha 7
Plesiodiadema antillarum 301–7
Plesiodiadema horridum 301–7
Plesiodiadema indicum 182, 301, 307
Pleurocystites 324
pleurocystitids 347, 358
Pleuroschisma 221–2, 231
Plexechinus 334, 342–5
  hirsutus 338
  spectabilis 335
Pogonipocrinus 257
  antiquus 236, 237, 239–41
Polycrinus 261
Polydeltoideus 221, 231
Poteriocrinina 26, 224, 226, 256, 258–67, 279, 324, 327
Pourtalesia 153–4, 332, 334–5, 342
  laguncula 337
  miranda 335
Pourtalesiidae 153, 331–4, 342–5
Pradesura 90–1
Procidaris edwardsi 183
Proexenocrinus inyoensis 236–7, 239–41, 258, 263–4
Proisocrinus 272
Promelocrinus 263
  anglicus 263
Pronechinus 152
Protenaster 157, 159
  australis 157
  preaustralis 157
Proterocidaris 152
Protocrinitida 210, 212
Protoscutellidae 115
protostomes 38
Psammechinus 291
  miliaris 126, 135, 192
Pseudanachys 333
  credneriana 335
Pseudholaster 333
Pseudocentrotus depressus 46, 48–9
Pseudoffaster 333
Pteraster tesselatus 191
Pterasteridae 57
Pteroblastus 233
pterobranchs 8, 13, 15–16, 61, 70, 74, 76
Ptychoblastus 232
Pycnopodia helianthoides 189–90
pyrgocystids 323
Pygorhytis 156–7

Pyramiblastus 232

Radialia 6, 7
Ramacrinus 251
Ramseyocrinidae 236–8, 240
Ramseyocrinus 236–7, 239–41, 246
  cambriensis 236–41, 258
  vizcainoi 236–7, 239–41
Regnellicystis 210
regular echinoids 151, 153
Reteocrinidae 263
Reticulocarpos 349, 353, 357, 359–60
Rhabdopleura 15
Rhipidocrinus 256, 263
rhipidocystids 204, 206, 347
Rhipidocystis 204, 206
Rhodocrinitidae 236
Rhombifera 202, 324, 357–8
Rhopaloblastus 233
Rhopalocystites 204–5
Rhopocrinus 262
Rhyncholampas 168
Ridersia 204–5
Rispolia 333, 339
Ristnacrinus 237
Rotasaccus 92–4
Rotulidae 115–17, 154–6
Rotuloidea 158
  fimbriata 154, 156
  fonti 154, 156
  vierai 154, 156
Roveacrinida/idae 269, 271, 276–9, 284

Saccocomidae 278
Sagenocrinitacea 262
Sanchezaster 332, 334, 314, 344
sand dollars 167–8, 184
Saracrinus 275
Scagliaster 333
Schizaster 155–7, 159–60, 166–7
Schizoblastus 232
Schizotremites 231
Schuchertocystidae 208
Schuchertocystis 208
Scotiacystis 357
  griffei 351–2
Scutasteridae 159
Scutellidae/ina 115, 185
Seirocrinus 270–1, 274, 280
Siluraster 92
Simplococrinus 256
Sinclairocystis 209
Solaster stimpsoni 190–1
Sollasina 92
solutes 21–2
Somaliaster 334
  cf. *magniventer* 337
Somaliasteridae 332, 334, 340
Somphorocrinidae 278
Spatangocystis 334

Spatangoida 149, 151, 153, 157, 159–61, 166–8, 170, 172, 184, 334, 340
Sphaeronites 202
Sphaeronitida 199–200, 202–3, 212
Sphaeroschisma 233
Spinulosida 55, 60, 64
Spiraculata 215–33
Spiralian eubilaterians 7
Springericystis 207–8
Steganocrinus 263
Stegaster 333, 340, 344–5
  heberti 335
Stegasterinae 340
Stellarozoa 69, 77
Stenonaster 334
Stenonasteridae 332, 334, 338
Stereopneustes 333, 343–4
  relictus 343
Sternopatagus 334, 341–5
  sibogae 342
Sternotaxis 333, 338–9
  gregoryi 339
  icaunensis 335
  trecensis 339
Stichocystis 203
Stichopus californicus 69, 74, 76
Stichopus japonicus 45–7, 49, 315
Stirodonta 125
Streblocrinus 249
Stromatocystites 87–8
Strongyloblastus 232
Strongylocentrotidae 126, 129, 302
Strongylocentrotus 64, 291
  droebachiensis 46, 48–9, 64, 102, 110, 116, 192–4
  franciscanus 111, 316
  intermedius 111
  pallidus 150, 152
  purpuratus 33, 37, 64, 101, 108–10, 114, 116, 123–32, 141, 192–4
Strotocrinus 261
Styela clava 128
Stylophora 21, 38, 347–61
Sycocrinitidae 259
Synbathocrinidae 251
Synbathocrinus 226, 247

Talarocrinus 224
tanaocrinids 237
Taphraster 332–3, 338
Tasmanocrinus mariensis 272
Teliocrinus 275
Temnopleuroida 37, 46, 48, 50
Temnopleurus toreumaticus 46, 48–9
Tetracionocrinus transistor 236–41
Tetragonocrinus pygmaeus 236–7, 239–42
Thaminocrinus 247
Thamnocrinus 262
Thaumatoblastus 233
Theeliidae 92

Theloreus 251
Thiolliericrinidae 275–6
Tholaster 333, 339
Thyone briareus 32
Thyone sp. 46–9
Timorechinus 261
Timoroblastus 233
Titanaster 333
Togocyamus 115, 152
Tornatilicrinus 261
Toxaster 157
Toxopatagus 333, 344
Toxopneustes pileolus 45–6, 48–9
Toxopneustidae 126
Trachelocrinus 206–7, 211
Trichinocrinus terranovicus 237, 264
Trichotocrinus 233, 256
Tripneustes gratilla 141, 179–81
Tripneustes ventricosus 150

Triturus cristatus 101
Troosticrinidae 215
Troosticrinus 221, 231
Trybliocrinus 256, 263
Tryssocrinus 261
Tunguskocrinus 251
tunicates 9, 20
Tympanoblastus 233

Uintacrinida/idae 269, 276, 284
Uintacrinus 276
Ulrichocystis 208
Unibothriocidaris 91, 94
Uranaster 89
Urechinidae 331–2, 334
Urechinus 334, 342–5
  naresianus 335
Urochordata 20–1, 128

Valvatida 46, 48
Vertebrata 9, 20–1, 31, 39
Villebrunaster 89–92
Volchovia 92
Vosekocrinus 258
  granulatus 237

Wellerocystis 209

Xenoblastus 225, 232
xenocrinid camerate 237
Xenopus 128, 131–2

Zumoffenia 333
Zygometridae 275–6

Subject Index

aberrant variation 182
aboral stalk 199, 201
accelerated development 150, 157, 170
actin genes 142, 144
adambulacrais 89–90
adhesion 287–97
adhesive disc, larval 55
adult shape 55–6
adultation 13
allometric growth 153, 156, 158–60, 167, 169–70
ambitus, position of 183
ambulacra 21, 23, 72, 78, 81, 87, 89–91, 181, 199, 201, 207, 210
   mode of enclosure 69, 74
   width 183
ambulacral articulations, asteroid 327
   plate growth 158
   plating 155, 308
   pores 182, 336, 338–9
amino acid composition of collagen 45, 47–8
   evolution of 43
   rate of substitution 132–4
   relatedness 48, 50
   sequences, mitochondrial genes 131–2
   similarity, mitochondrial genes 134
   substitution 44
   variation in 44
anal migration 184
   plate 270
   pyramid, stylophoran 348, 350, 358
   series, crinoid 246–7
ancestors 86
ancestral articulate crinoid 278
apical system 154, 161, 182
   of holasteroids 336–9, 342
apical tuft cells 290–6
archenteron 57
arrested variation 182
Aristotle's lantern 69, 77, 79, 92
arm facets 247–9
arm homology 256
arms 87, 199, 200, 202–3, 207, 272–3
   articulation 256
   branching pattern 256–8, 261, 274
   earliest appearance of 266
   evolution of 265–7
   number of 258–60, 265
   of crinoids 246, 255–67, 324
   of ophiuroids 91, 312, 318–20
   order of inception 248–50

uniserial arms 235
symmetry, of larval development 13, 14, 18–24, 86
atomy 260
atrial cavities 20
attachment organ, larva 58, 61, 72, 74
attachment, larval 23–4
aulacophore 347–60
auricularia 33, 51, 55–7, 59, 61, 75
autaporphies of the Articulata 269
   of the Chordata 11
   of the Cyclohydrocoelia 21
   of the Deuterostomata 11
   of the Dexiothetica 11
   of echinoderms 11, 21
   of the Laeothetica 21
   of the Radialia 11
   definition 5
autotomy 281, 312, 315, 318, 320–2
axial complex 72, 76, 78
axial nerves 270, 273
axial torsion 71
axocoel 3, 7, 9, 11, 13, 16–17, 61, 72, 74, 78, 81, 86–7
axohydrocoel 55, 60–1, 74–5, 81, 87
azygous tube feet 23

bathymetric distribution, holasteroids 331
batyl alcohol 77
bilateral symmetry 184
bipinnaria 51, 56–7, 59, 63
blastoid genera, ranges and distribution 231–3
blastomeres 191–4
blastopore 11, 54, 60
blastula 63, 142
blastulation 38
blot hybridization 125–6
body size 172
body wall, of holothurians 319
bourrelets, evolution of 158–60
bouyancy of larvae 54
brachia 324–5
brachial articulations 275–6
brachiolar arms 55, 57
brachioles 199, 200–1, 204–7, 210, 220, 324–5
branching pattern, arms 255–67
branching, phyletic 36
branchiolaria 60
brooding chambers 185
brooding, in ophiuroids 64
buccal podia 73–5, 81, 87, 93

buccal slits 89, 91
burden 63

calcareous ring 69, 77, 79, 93
calcification 178–9
cataspires 211
catch connective tissue 312, 315
cell fates, map of 142–3
cell lineages 139, 142–3
cephalic shield 16–17, 23–4
character loss 34
characters of diademataceans 308
   inconsistent choice of 34
   of Holasteroida 339
   of larvae, for phylogenetic analysis 56–9
   for phylogenetic analysis of classes 72–3
chordates, origin 19
ciliated bands 54, 58, 60
circum-oesophageal ring 11
circumoral ciliated band 54
circumoral plating 202–5
cirral articulation 273, 324
   ligaments 312, 322
cirri 273
cladids, ontogeny 249
class diversity 327
   relationships 29–40, 50, 77–80, 294–6
classification of Articulata 283
clavulae 160
cleavage pattern 38
codons, in mitochondrial genes 128–9
coelomic organisation 81
   pores 76
   spaces 324, 326
   symmetry 81
coeloms 11, 13, 16, 19, 21–23, 54, 60–1, 74, 77–8, 81
collagen, preparation of 45
   proteins 43
   structure 43–4
   significance for palaeobiology 321–7
   unique properties 320
column, columnals 202, 204–6, 274
communities, Lower Carboniferous 223–4
   Ordovician 220
complexity of mRNA 140–1
compound plating 182
   radials 245–9, 251–2
connective tissues 311–27
conserved domains 134

convergence, between classes 55
  in development 53, 56, 62
  problems in phylogeny 29, 70
coronal growth 149–52
  shape 151
cothurnopores 348–9, 357–8
cover plates, stylophora 349–55, 357
crinoid characters, phylogenetic
    analysis 239
  classification 245
  communities 224
  evolution 219–20
  larvae 54
crown group, definition 4, 5, 85
cryptorhombs 211
crystallographic axes 299–309
ctenoid organ 349–50
cup plating, Articulata 273
cuticle 290
cystidean stage 255
cytochrome c 122
cytochrome oxidase 125–6, 129

deformities, pollution-induced 181
dental apparatus 94
deposit feeding in evolution 18
dermis 325
  of holothurians 312, 315, 317, 321–2
deuterostomes, phylogenetic tree 11, 21
developmental complexity 64
  direct 33–8, 69–70
  gene regulation 139–40
  indirect 39, 69–70
  patterns and phylogeny 38–9
  rate 192–4
  regulation 63, 149
dexiothetism 8, 11
dilator muscles 57, 60
dipleurula 13–17, 81
diplopores 203
disc integument 315, 318, 322–4
disparid relationships 250
dispersal capability 190
distance, molecular 37
divergence time,
    asteroid/cryptosyringid 88
  calculation 125, 128–9, 132
  clypeasteroids 115–17
  crinoid/eleutherozoan 87
  echinoid/holothurian 92
  estimating 86
  from amino acid sequence
      similarity 132–5
  ophiuroid/echinozoan 91
diversity of camerate crinoids 219
  cladid crinoids 219
  crinoids 236–9
  blastoids 215–19
DNA coding sequences 101, 104
  duplexes 101, 103, 110

intraspecific variation 102, 109, 112
multiple substitutions in 114
mutation events in 111–12
sequence data 104
sequence divergence 103–4, 114
sequence homology 109
single copy 101–3, 108–14, 121, 123, 132
DNA–DNA hybridisation 107–17
doliolaria 51, 54–5, 57–9, 61
domed test 153
dorsal coelom 57
dorsal flexor muscles, larvae 60
dorsal, definition 14
  homology 10
DS value, definition 45

echinoid developmental pathways 39
echinopluteus 55, 58, 60, 62
echinozoan characteristics 69
egg organic content 192
  size 64, 189–94
embryological characters in
    phylogenetic analyses 35
embryonic traits, homology of 33
energy-sparing adaptations 326
enterocoel 74
enterocoelic pouch 72
enteropneust larvae 56
enzyme immunochemistry 50
epaulettes 58, 60
epineural sinus 74, 78, 81
epispires 200, 202–4
epithecal food grooves 202–3
eta-globin psuedogene 103–4
evisceration 319
evolutionary distance tree 31
evolutionary pathways, articulate
    crinoids 279
extraembryonic membranes 60, 63

fascioles, evolution of 160
fecundity 190, 193–4
feeding larvae 53, 55, 60
feeding, evolution of tube feet 157, 292
  mode in evolution 17–20, 22–4
fertilization 63
fertilization membranes 191
filtration fans 326
fistulipores 203
flattening of test 152, 183–4
foerstepores 209
food grooves 158–9, 167
fossil record of clypeasteroids 115–17

gastrulae 63
gene 'batteries' 141, 144
gene conversion 121–3

expression in the embryo 142
  order 121
  regulation 139, 144
  transposition, in evolution 125
generic diversity, blastoid 216–20
  duration, blastoids 207–8
  survivorship, blastoids 217–18
genes, actin 142, 144
  histone 103, 113, 142
  orthologous genes 121–2
  paralogous genes 30–1, 121–2
  regulatory 81–2, 144, 169
  specification of 139
  transcription of in development 140
genital cord 76
  plates 300
  number of plates 160–1
  pores 336
  rachis 72, 76, 87
genome evolution 113
  organization 101–5, 108, 121
  size 101, 108, 111, 140–1
gill slits 9, 11, 17–20, 308
glycocalyx 290, 295
goblet cells 290–6
gonads 72, 76, 78, 87
gonopores 80, 87, 202, 205, 207
growth gradients 69
  lines 178
  of echinoid plates 176
  rates 177
  rings 150

Haeckel's biogenetic law 62
haploid DNA content 101
heterochrony 149–62, 169–71
  in crinoid evolution 281
  in larval development 13, 38
  of crinoid arms 265
heterologous hybridization of DNA 110
histone genes 103, 113, 142
holdfast 203
homologous DNA sequences 103, 110–1
homology 85
  distinction from convergence 53
  identification of 36
  in larval development 53
  of mitochondrial DNA 123
  of protein sequences 132–4
  recognition 70
homoplasy, recognition of using
    DNA 117
horizontal transfer 113
humatirhombs 200, 203, 208
hyaline layer 63
hydrocoel 3, 11, 13, 16–17, 20–1, 23, 61, 69, 78, 81, 86–7
hydropore 13, 16–17, 57, 72, 76, 78, 80, 86, 201–2, 205, 207–8, 210

hydrospires 201, 215
hydrostatic pressure, internal 177–9
hypermorphosis 150, 152, 170
hyponeural coelom 72, 77
  nervous system 325

imbricate skeletons, significance of 324
imino acid content, of collagen 45–6
incremental growth 153
interambulacral plates, number of 160
internal buttressing 184–5
  tethering 179
interspecific hybridization 102, 109
intervertebral muscle 312
intraspecific variation 191
introns 113
introvert dermis 312, 314
ionic environment, effect on collagen 316–18, 20

jaw apparatus 88, 91–2
juxtaligamental cells 312, 315, 325

labrum 154, 336
lamellate organs 348–50, 357–8
lancet plate 210
Lange's nerves 10
lantern 92
lantern musculature 179–80
larvae 53–65, 140, 189–93
  arms 61
  characters in phylogenetic analysis 53–65
  coeloms 3, 7–10
  development 33, 142
  fedding mode 53, 55
  form 303, 306
  mode of development 38
  mortality 190, 193–4
  size 192–3
  skeleton 33–4, 54–5, 57, 60–2, 72, 75, 299–309
  success 189, 193
lateral arm plates 91
lateral gene transfer 31
lecithotrophy 54, 190, 194
left side dominance 81
life history 171–2
ligament attachment areas 325
ligaments, crinoid 311–14
locomotion of stylophora 356–7
lunules 154–5, 167

madreporite 72, 76, 78, 87, 89, 90, 160
marginal frame, cystoid 206

mechanical properties of collagen 315–18, 320
median DNA sequence divergence 103–4
mesenchyme, ingression of 56–7, 60
mesenterial threads 179–80, 185
mesocoel 7, 8, 10–11
mesodermal neuro-effector system 325
metacoel 8, 11
metamorphosis 55, 62, 79
microphagous feeding 292
mitochondrial DNA 114
  advantages in phylogenetic studies 122–3
mitochondrial RNA 113, 140–1
  complexity 121, 140–1
  gene order 125–8
  genome 123–7
mobile genetic elements 101
mode of life, articulate crinoids 279–81
molecular clock 37, 108, 112–13, 121–3, 125, 128–9, 132
  drive 121–2
  rate of evolution 37
  sequence data, imperfections of 36
  trees 36
monophyly, of echinoderms 32
morphogenesis 62–3, 175–86
  of rudiment 141–2
morphology of Disparida 245–6
morula cell 291
mouth angle plates 89, 91
mouth frame 87, 91
mouth, migration of 19
  position 21
  site of formation 54–6, 58–9, 61
mucous feeding 292
mucus secretion 287–8
multibrachiate arms 260
multigene families 121–2

neoteny 150, 159, 170
nervous system 318, 325
neutral evolution 102, 113
  mutation theory 43–4, 46, 48
niche partitioning 220
non-coding DNA 113
non-feeding larvae 53, 55, 60
normalized percent hybridization, definition 103
nucleotide substitutions 36
  rate of 128

odontophore 89
offspring fitness 189
ontogeny 139, 149
  of camerates 249
  of cladids 249
  of disparids 246

of crinoid arms 255
oogenesis 38, 140–1
ophiopluteus 33, 54–5, 57, 60, 62
oral area plating 203, 207–8
oral cavity, larval 60, 69
orphons 122
outgroup, choice of 34
  of echinoderms 70

paedomorphosis 150, 157, 159–62
  in clypeasteroids 152
  in spatangoids 154
palaeobiogeography, Carboniferous blastoids 223, 227
  Devonian blastoids 222
  Ordovician crinoids 239–40
  Permian blastoids 227
  Silurian blastoids 221
papillae 290, 295
parallel evolution 182
  of larval skeleton 33
  problems in phylogeny 29, 34
paraphyletic groups, recognition using DNA 116
parental investment 189–93
paroral bands 57, 61
parsimony 53, 58, 71, 73
  reliability 35
  sensitivity to molecular clock rate 37
paxillae 89
pectinirhombs 201, 205
pedicellariae 34, 60, 72, 76, 78, 342
peduncle 20
pentacrinoid stage 255
pentacula 14
pentameral symmetry 10, 11, 86
peramorphosis 156–62
  definition 150
  in echinoids 153–4
perianal coelom 72, 77
perihaemal vessels 10
peripharyngeal coelom 72, 77
perioral plating 201
periproct, evolution 160–1
  invagination 342
  position 184, 201, 205, 209
peristome, architecture in holasteroid 337–8
  evolution 160
  groove 339–40, 342–3
  orientation 334
perivisceral coelom 77
  coelomic pores 72, 78–81
petals 153, 155–7, 168, 336
  formation of 185
pharngeal gill slits 17
pharyngula stage 63
phyllodes 167, 336, 342
  evolution of 157
phylogenetic implications of tube feet 295–6

phylogenetic (cont.)
  relationships of holasteroids 340–5
  tree of articulate crinoids 270
  trees for crinoids 235, 239, 242
  trees for echinoderm classes 30, 32, 50, 58–9, 71, 79, 95, 322–3
pillars, clypeasteroid 184–5
pinnate ambulacra 207–8, 210
pinnulate arms 270
pinnulation 237, 258, 263–6
pinnules 246, 266, 325
  homology 256
planktotrophic larvae 38, 189–90, 192, 194
plastron 153–4
  architecture 334–6, 338–9, 342–3
plate growth 149–50, 153, 161, 176, 179–80
  migration 184
  occlusion 155–6
  size 150
  sutures 177–8
  translocation 150, 153–5, 159–61
plated tube feet 92
plesion 4–6
pluteus 30, 33, 39, 54–6, 59, 63–4, 75, 142
pneus 175–80, 182
podia 55
  number of 264
  primary 72–3, 75
  stylophoran 349, 351, 355
podial basins 89–91
point mutation 111
polarity, determination 29, 34
  problems of defining 61
  using fossils 86
polarization of egg 144
Polian vesicles 72, 75
pores, ambulacral 208–9, 336, 338–9
  numbers of 155
post-displacement 150, 159, 170
posterodorsal larval arms 299–300, 302, 308
postoral larval arms 299–300, 302
pre-displacement 150, 170
preoral lobe 61, 80
primary podia 78, 255
proboscis of enteropneusts 74
progenesis 150, 152, 161, 170
progressive variation 182
protein sequence data 44
  homology 132–4
protein-coding genes, mitochondrial 123
proteins, collagen 43
protocoel 11
protostomy 6
pseudogenes 104, 122
pseudopetals 185
pseudopinnules 208–10

radial canals 23, 255
  cleavage 6, 12
  plates 255
  symmetry 4, 10
  water vessel 86, 88, 92
ramules, homology 256
rank, categorial 6
rapid phyletic splitting, effect of 36
rate of amino acid divergence 134
  of amino acid substitution 132
  of change of DNA 104, 112
  of cleavage 193
  of DNA evolution 102, 116–17
  of molecular sequence evolution 37, 107, 112
  of plate growth 152–3
  of plate production 155
  of silent nucleotide substitution 128–9, 132
  of single-copy DNA evolution 113–14
ray trunk 256, 258, 263
regressive variation 182
regulation in development 63
  of genes 139
regulative processes 63
regulatory genes 81–2, 144, 169
relationships of clypeasteroid families 115–17
  of primitive crinoids 238–9
relative rate test 102, 112
repetitive DNA 108
resorption of skeleton 151
  of larval body 62
respiratory coelom 81
  rates 326
  structures 347–9
restriction map of mitochondrial genome 124, 127
retrovirus, DNA transfer 114
reversions, problems in phylogeny 29, 34
ribosomal RNA 29, 123–5, 128
  5S rRNA 31
  12S rRNA 125–6
  16S rRNA 125–6
  18S rRNA 30–32, 36–37, 39
RNA complexity, of mature eggs 102
rock boring adaptations 183
rostrum, evolution of 153–4, 337
rudiment 61, 71–3, 78, 140–1, 300

sequence complexity of mitochondrial RNA 121, 140–1
sequence data, DNA 104
shape of test 167
  adult 72, 75
  in evolution 165, 169, 171, 175–86
silent substitutions 103–4, 113
similarity of protein sequences 133–4
  divergence 104, 112

size at metamorphosis 192, 194
  increase during development 65
  of egg 189–94
  of genome 108, 140
  of juveniles 192
  selection for 171
skeletogenic mesenchyme 144
skeleton 9, 11–12, 20, 72, 76, 87, 325
  complexity 65
  deformities 181
  growth 80, 178
  of holothurian larvae 62
  of larvae 33–4, 54–6, 60–2, 72, 75, 299–309
  morphogenesis 176
  muscles 327
  reduction of 86
  rods, larva 59
  spiculation 92
somatocoel 3, 8, 10, 13, 16, 23, 72, 78
speciation event, effect on DNA variation 102–3
species per genus, blastoids 218
spherical tests 152
spicule secretion 33
spine ligament 312, 314, 316–18, 321–2
spine thickness 181
spines 72, 78, 91, 159, 166, 308
  of mitrate 349–50, 355
stalk 200, 202–3, 312–15, 318, 322, 324
stele 21
stem 87, 200–3, 205, 208, 326
  articulation 274–6, 324
stem group, definition 4, 5, 85
  of holasteroids 338
stem lineage, definition 4, 5
stereom 3, 9, 11, 20, 65, 87, 178
  microstructure 324–5
sternal plates 336, 343
stomodaeum 69
stone canal 7, 11, 13, 16, 76, 78, 80, 86–7, 89–90, 255
stratigraphic distribution of Articulata 271
  blastoids 231–3
  holasteroids 332
  primitive crinoids 240–1
streamlining of shape 166
stress fields in growth 177
structural genes 81–2, 113, 142
stylocone 349–51
styloid 349–50, 354–5
subanal rostrum 342
subdermal muscle layer 326
suckered tube feet 288
suspension feeding in evolution 17, 326
sutural compression 176–7
  ligaments 321

pores 357–8
suture lines 180
symmetry 10, 72, 347–9, 355, 357
   deviation from 181
   of coeloms 81
   of larva 13–14, 16
   patterns 69
symplectial junctions 315, 324
synapomorphies, definition of 5
synarthrial articulations 274
synostosial junctions 315, 324
syzygial ligaments 312

taxonomic significance of arm patterns 265
taxonomic survivorship, of blastoids 215–17
teeth 92
   microstructure 308
tegmen 270
   crinoid 246

tendons, of ophiuroids 312, 315
tensile properties of collagen 311–27
tension-induced growth 177–8
tentacles 16–19, 288–91, 294–5
teratological phenomena 181–2
tesselate plating, functional significance 358
test elongation 153
   flattening 167
   shape 151–3, 165–6, 179, 181–3
   thickness 181
Tiedemann's bodies 72, 75
thecal plating 200–7, 210, 347
thermal stability of DNA duplexes 109, 111–12
tornaria 54, 56–7, 61, 70
torsion 10–11, 23, 69–71, 78–9, 86
total group, definition 4
trace fossils, deep sea 344
transfer RNA 123–5
translocation of plates 150, 153–5, 159–61

transverse ciliated bands 54
trimery 6, 11
tube feet 72, 74, 77, 92 235, 287–97
   effect on growth 179
   evolution of 157
   funnel building 166
   plated 93
tubercules 308
   evolution 159
ventral, definition 14
   homology 10
vertebrae 91
vertical tiering 326
vestibule 20, 34, 60, 69, 78
virgalia 89
visceral torsion 71

water vascular system 3, 7, 10, 255–6, 348, 351–3
wheel spicules 92